Alkaline Anion Exchange Membranes for Fuel Cells

Alkaline Anion Exchange Membranes for Fuel Cells

From Tailored Materials to Novel Applications

Edited by Jince Thomas, Alex Schechter, Flavio Grynszpan, Bejoy Francis, and Sabu Thomas

WILEY-VCH

Editors

Dr. Jince Thomas
International and Inter University Center for Nanoscience and Nanotechnology, Mahatma Gandhi University, Kottayam 686560, Kerala, India

Prof. Dr. Alex Schechter
Ariel University
Ariel 40700
Israel

Prof. Dr. Flavio Grynszpan
Ariel University
Ariel 40700
Israel

Dr. Bejoy Francis
Mahatma Gandhi University
St. Berchmans College
Changanassery
Kottayam 686101, Kerala
India

Prof. Dr. Sabu Thomas
International and Inter University Center for Nanoscience and Nanotechnology,
Mahatma Gandhi University
Priyadarshini Hills P.O.
Kottayam 686560, Kerala
India

Cover Image: © Forance/Shutterstock; Olivier Le Moal/Adobe Stock Photos

Library of Congress Card No.: applied for

British Library Cataloguing-in-Publication Data
A catalogue record for this book is available from the British Library.

Bibliographic information published by the Deutsche Nationalbibliothek
The Deutsche Nationalbibliothek lists this publication in the Deutsche Nationalbibliografie; detailed bibliographic data are available on the Internet at <http://dnb.d-nb.de>.

Print ISBN: 978-3-527-35039-1
ePDF ISBN: 978-3-527-83760-1
ePub ISBN: 978-3-527-83759-5
oBook ISBN: 978-3-527-83758-8

Typesetting: Straive, Chennai, India
Printing and Binding: CPI Group (UK) Ltd, Croydon, CR0 4YY

C9783527350391_290124

Contents

Preface

The urgent need to tackle the global challenges of climate change and resource depletion has made the development of sustainable and efficient energy technologies imperative. Fuel cells are one such clean energy solution that holds great promise for revolutionizing the way we generate electricity and power our vehicles. The advancement of fuel cell technology largely depends on the pivotal role played by membranes, particularly anion exchange membranes (AEMs), which have recently gained significant attention. *Alkaline Anion Exchange Membranes for Fuel Cells: From Tailored Materials to Novel Applications* is an all-inclusive exploration of the science and engineering of AEMs and their crucial role in the fuel cell industry. The book delves into the intricate world of AEMs, examining their design, synthesis, properties, and applications in detail. It covers fundamental principles as well as the most cutting-edge research, providing a roadmap for researchers, engineers, and innovators in the field of renewable energy. Notable contributions from esteemed academics, industry experts, and government–private research labs worldwide make this book an essential tool for university and college faculty members, as well as post-doctoral research fellows. Within the upcoming pages, readers will embark on an exhilarating journey through the dynamic realm of AEMs. We will explore the historical landscape of fuel cells and membrane technologies, observing their evolution and transformation into viable, sustainable energy solutions. This book unites esteemed experts and visionaries who have devoted their careers to unraveling the intricacies of AEMs, unlocking their potential for a more environmentally friendly future.

Chapter 1 provides both a concise summary and a comprehensive analysis of the current state of AEMs. Additionally, the chapter discusses the latest challenges and prospects in the field. It offers an overview of the fundamental principles of AEMs, including their unique properties, synthesis techniques, performance features, and essential design factors. Chapter 2 delves into the rich history and latest developments in alkaline fuel cell applications utilizing AEMs. This book chapter has been inspired by the remarkable progress and research undertaken in the membrane field for alkaline fuel cells in recent decades. The focal point of the chapter is primarily centered on the fundamental requirements for AEMs in alkaline fuel cell assembly, along with their modifications and methodologies. Chapter 3 covers the challenges and procedures for creating AEMs. It assesses and compares different fabrication processes based on factors like cost, durability, scalability, and performance targets. Chapter 4 analyzes the various types of polymeric electrolytes that can be utilized in alkaline exchange membrane fuel cell (AEMFC), water electrolyzers, and CO_2 electrolyzers. The chapter covers the benefits and drawbacks of each type of polymeric electrolyte, as well as their chemical composition and properties. Furthermore, the chapter delves into the suitability of each type of electrolyte for specific applications and how they can affect the performance of the respective electrochemical devices. Chapter 5 explores proton and AEMs, their characteristics, investigative techniques, measurements, transport mechanisms, and water management strategies for proton exchange membrane fuel cell (PEMFC) and AEMFC. Chapter 6 covers transportation and conductive mechanisms in AEMs, as well as the carbonation process and methods to regulate its effects. Additionally, the chapter evaluates the use of electrochemical impedance spectroscopy for monitoring ionic conductivity and stability during fuel cell operations. Chapter 7 discusses recent approaches to polymer membrane modification using quaternary ammonium (QA) cations for improved stability and electrochemical performance. Techniques, including tethering alkyl chains and fabricating multi-cation side chains using different quaternization strategies, are also discussed at length. Chapter 8 highlights recent applications of guanidinium-based alkaline exchange membranes in energy conversion. It enriches the systematic study of AEMs and provides a valuable reference for the design of new cationic groups. Chapter 9 provides a comprehensive and in-depth critical analysis of AEMs that utilize imidazolium and thiazolium cations. The chapter discusses various aspects of these membranes, including their properties, performance, and potential applications. Chapter 10 covers the grafting method and synthesis parameters for optimized AEMs. It discusses radiation-induced graft copolymerization for anion-exchange membranes and crosslinked structures for fuel cells and other applications. The insight toward the pathways leading to the degradation of organic and inorganic cationic functionalities, including β-hydride elimination, nucleophilic substitution reactions, and ring-opening reactions in heterocyclic systems, is discussed in Chapter 11. Chapter 12 confidently asserts the critical role of computational simulations in comprehending the underlying mechanisms, optimizing the design, and forecasting the performance of AEM systems. Chapter 13 delves into the characteristics and challenges of commercial membranes. Additionally, a comparison between commercial and non-commercial AEMs is explored in detail, highlighting the obstacles faced by commercial AEMs. Furthermore, the current research on non-commercial AEMs emphasizes their distinct properties and traits. Chapter 14 discusses the recent advancements in preparing membrane electrode assemblies (MEA) for AEMFCs. The chapter covers different methods, such as catalyst-coated membrane (CCM) and gas diffusion electrode (GDE), to prepare MEAs. Chapter 15 expertly examines the multiple applications of AEMs in various cutting-edge technologies, including alkaline water electrolyzers, CO_2 electrolysis, redox flow batteries, alkali metal–air batteries, reverse electrodialysis, electrodialysis, diffusion dialysis, and microbial fuel cells. Chapter 16 of this book discusses cutting-edge innovations in crafting high-performance AEMs using polymer-based substances. It also explores potential avenues for further research in this area, as well as the endurance of AEMFCs and the challenges of designing durable AEMs.

We extend our gratitude to all the contributors who have shared their expertise, as well as to the readers who embark on this journey with us. We sincerely thank Wiley-VCH GmbH, Germany, for their invaluable support in bringing this book to fruition. Our collaborative efforts will explore the exciting realm of alkaline AEMs and their potential to transform the energy industry with eco-friendly and renewable power generation. Step into the future of energy with us.

Thanks to All for the successful completion of this book!

February 2024

Jince Thomas
Alex Schechter
Flavio Grynszpan
Bejoy Francis
Sabu Thomas

1

An Introduction to Polymeric Electrolyte Alkaline Anion Exchange Membranes

Jince Thomas[1], Minu Elizabeth Thomas[2], Bejoy Francis[3], and Sabu Thomas[1]

[1] *Mahatma Gandhi University, International and Inter University Center for Nanoscience and Nanotechnology, Kottayam 686560, Kerala, India*
[2] *Mahatma Gandhi University, School of Polymer Science and Technology, Kottayam 686560, Kerala, India*
[3] *Mahatma Gandhi University, St. Berchmans College, Department of Chemistry, Kottayam 686101, Kerala, India*

1.1 Introduction

Currently, the magnitude of energy usage cannot be denied. It is indispensable in every aspect of life, and a booming population results in increasing energy demand. Recognizing that nonrenewable sources will eventually run out, the value of renewable sources cannot be underestimated because they are sourced from unlimited sources. The most crucial consideration of renewable sources is their environmental impact while using them. The proper use of energy appears to be a major topic these days, and one must decide which type of energy should be used, and why it is vital?

The urgent necessity of researching, developing, and commercializing renewable energy sources and the technologies accompanying them is universally acknowledged as a prime focus. Time and place are essential components of most renewable energy systems. Therefore, it is crucial to build relevant energy conversion and storage devices to capture these unreliable energy sources effectively. The most prominent electrochemical energy storage and energy conversion devices are batteries, electrochemical super capacitors, and fuel cells.

Electrolytes are vital components of electrochemical energy storage and conversion devices, and their properties and performance can significantly impact the overall efficacy, safety, and longevity of these systems. Although electrolytes have been recognized and researched for centuries, their physiological role was not fully understood until the late nineteenth and early twentieth centuries [1, 2]. Starting from that point, researchers delved deeper into the behavior of electrolytes, focusing

Alkaline Anion Exchange Membranes for Fuel Cells: From Tailored Materials to Novel Applications, First Edition. Edited by Jince Thomas, Alex Schechter, Flavio Grynszpan, Bejoy Francis, and Sabu Thomas.

on their conductive properties. This led to the creation of innovative electrochemical techniques and tools, like the pH meter and the potentiometer. Later on, advancements in materials science and engineering resulted in the development of solid-state electrolytes with high ion conductivity, which are widely used in batteries, fuel cells, and electrochromic devices. Nowadays, there is a growing interest in crafting new electrolytes for emerging electrochemical applications, such as energy storage and conversion, electrochemical water treatment, sensing, and biosensing.

1.2 Different Types of Electrolytes

Understanding the classification of electrolytes can provide valuable information about their behavior and properties in different applications, depending on the specific context and purpose. There are various ways to classify electrolytes, such as the type of ions they contain, their physical form, and their conductivity [3]. Below are some of the most frequently used methods for categorizing electrolytes,

- Based on the origin
- The type of ions present
- Its physical state
- Conductivity measure
- The contrast between acidic and alkaline properties

The most popular is the classification based on the physical nature of electrolytes (Figure 1.1). During the early 1970s, researchers began exploring the potential of solid-state materials such as ceramic, glass, crystalline, and polymer electrolytes. This led to various types of polymer electrolytes with different compositions and structures, including solid-state polymer electrolytes (polymer–salt complex), gel

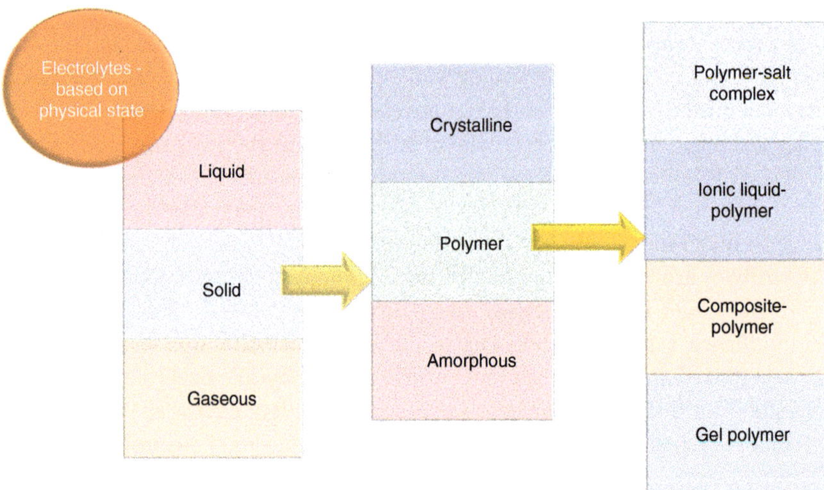

Figure 1.1 Classification of electrolytes.

Table 1.1 Advantages and disadvantages of liquid-state and solid-state electrolytes.

Electrolyte	Advantages	Disadvantages
Liquid	Effortless processing	Gas solubility is low
	Low cost	Change on concentration
	Ionic conductivity is high	Potential window is short
	Using different ions	Parallel reactions
		Interactions with other gases
Solid	A wider potential window	Complex processing
	No parallel reactions	Low ionic conductivity
	Gas solubilization is not required	Expensive
	Low external interferences	

polymer electrolytes, composite polymer electrolytes, and ionic liquid polymer electrolytes.

The solid-state electrolytes have more advantages than conventional liquid electrolytes. The advantages and disadvantages of solid-state and conventional liquid-state electrolytes are presented in Table 1.1.

Our focus is on solid-state electrolytes, specifically polymer electrolytes. A polymer electrolyte is a membrane that has alkali–metal–ion conductivity. It is composed of a polymer matrix as a solvent and solutions of salts that are dissociated within the polymer matrix. Polymer electrolyte is a remarkable solid-state system that showcases impressive ionic conduction abilities, making it an ideal choice for a wide range of electrochemical devices like rechargeable batteries [4], solid-state batteries [5], fuel cells [6], supercapacitors [7], electrochemical sensors [8], electrochromic windows, and analog memory devices [3, 9, 10].

1.3 Why Polymer Electrolytes Are Important?

Polymer electrolytes have distinct advantages over conventional electrolytes, making them essential in many fields and applications. They offer increased safety, especially in high-energy applications like lithium-ion batteries, as they are solid or gel-like, reducing the risk of fire or explosion. Polymer electrolytes also exhibit improved chemical and electrochemical stability, which minimizes electrode degradation and corrosion, leading to better device performance and longer lifespan. They have a broad electrochemical stability window, which enables operation at higher voltages without decomposition, which is critical for high-voltage batteries and supercapacitors. Although their ionic conductivity is lower than that of liquid electrolytes, advancements in polymer chemistry have improved conductivity, expanding their suitability for diverse applications.

Another benefit of polymer electrolytes is their ability to be processed into various shapes and forms, such as thin films or membranes, making them versatile for different device configurations and facilitating complex system integration and device miniaturization. Additionally, many polymer electrolytes are eco-friendly and are made from sustainable and recyclable materials, providing a greener alternative to liquid electrolytes containing hazardous components. Their compatibility with different materials and ability to function in various conditions make them ideal for numerous technologies. Some important benefits to using polymer electrolytes in comparison to traditional liquid electrolytes are

- High ionic conductivity
- Solvent-free
- Reduced leakage
- Safety
- Easy processability
- Thin-film forming ability and transparency
- Light-weight and flexibility

Polymer electrolyte membranes are mainly categorized into two types: anion exchange membranes (AEMs) and cation exchange membranes (CEMs). They are designed in a way that enables them to selectively transport either anions or cations, depending on their characteristics. The two membranes have significant functions in different electrochemical devices. The choice of the membrane is determined by the system's unique needs and ion transport goals. In Table 1.2, a comparison between AEM and CEM is presented.

The choice between AEMs and CEMs is often driven by the electrochemical device's specific requirements and operating conditions. Although AEMs have advantages, like lower cost, the ability to utilize various renewable fuels directly, and higher pH operation, CEMs are also beneficial due to their higher proton conductivity and compatibility with acidic environments. Choosing between AEMs and CEMs depends on

Table 1.2 Comparison of anion exchange membranes and cation exchange membranes.

Properties	AEM	CEM
Ion transport selectivity	Selective transport of anions	Selective transport of cations
pH operating range	Suitable for alkaline environments	Suitable for acidic environments
Requirement of the catalyst	Non-precious metal catalysts	Precious metal catalysts
Hydration tolerance	Excessive hydration can lead to swelling and instability, while insufficient hydration can hinder ion transport	More tolerant to varying water content compared to AEMs
Fuel flexibility	High	Limited

factors like the required pH range, cost, fuel accessibility, and performance objectives. These two types of membranes are essential in the progress of sustainable energy technologies, and their unique features aid in the overall enhancement and refinement of electrochemical devices. To reap the full benefits of ion exchange membranes in electrochemical applications, it is imperative to have a comprehensive comprehension of the pros and cons of AEMs versus CEMs. Thoughtful contemplation is critical.

1.4 Anion Exchange Membrane (AEM)

AEMs are made up of ion-conducting polymers that possess functional groups that attract and transport anions. Their primary purpose is to facilitate the movement of anions, such as hydroxide (OH^-) and bicarbonate (HCO_3^-), while hindering the transfer of cations. The selectively permeable AEMs contain positively charged functional groups that facilitate anion transport while preventing cation crossover. Various materials are utilized in the production of AEMs, including quaternary ammonium ions [11], imidazolium ions [12, 13], guanidinium ions [14–16], phosphonium ions [17], spirocyclic cations [18], carbocations [19], ionic liquids [20], multi-cations [21, 22], and metal cations [23], which are all functionalized with polymers [24].These remarkable functional groups exhibit the fundamental capability of ion exchange by attracting and transporting anions across the membrane while simultaneously maintaining the balance of the overall charge.

1.4.1 Fundamental Concepts of Anion Exchange Membranes as Polymer Electrolytes

Understanding the fundamental concepts underlying AEMs is crucial in grasping their significance in polymer electrolyte applications. These specialized membranes possess unique characteristics that enable them to function effectively and are critical in ensuring effective ion conduction. By recognizing their value, we can unlock the full potential of polymer electrolyte systems. It is crucial to carefully evaluate the membrane's performance under operational conditions, durability, and cost targets to ensure the successful commercial development of the AEM as an electrolyte. These key factors significantly influence the desired properties and demand careful consideration. The membrane is designed to facilitate the transport of anions while blocking the flow of cations, gases, and electrons. Choosing the cationic group is essential to achieve a high concentration of charges in the membrane and ensure sufficient ionic mobility. It is even more critical to select the right cationic group as it directly impacts the chemical stability of the membrane in optimal operating conditions. Also, a viable approach is to augment the ionic groups within the membrane to improve ionic conductivity.

However, the high concentration of the cationic group can adversely affect the mechanical properties due to excessive water absorption. Therefore, it becomes imperative to implement stringent control measures over the membrane's morphology to enhance its mechanical attributes. However, when the membrane is too hydrated or brittle due to intensive dryness, it will deteriorate the membrane's mechanical properties and have a profoundly negative impact on cell performance. Therefore, maintaining the appropriate level of water uptake is crucial for optimal membrane performance.

Thus, in a nutshell, for optimal performance of the AEM, the following requirements are the desired prerequisites:

- High anion conductivity is essential for efficient ion transportation in materials.
- Maintain chemical stability to withstand the effects of the electrolyte environment and avoid degradation.
- Strong and stable materials are essential for device longevity and proper functioning, maintaining structural integrity and minimizing deterioration.
- Need high selectivity for anions to avoid cation crossing.
- Function as an absolute barrier to prevent the passage of undesirable particles like electrons and gases.
- A balance of water uptake and retention is necessary.
- Manufacturable using scalable and cost-effective techniques to ensure their commercial viability.

Thus, to fully maximize the effectiveness of polymer electrolytes in various electrochemical applications, it is crucial to have a solid grasp on the core concepts of AEMs. This includes understanding their selective ion transport, ion exchange capacity, and water uptake. Through the use of precise and meticulous characterization techniques, the researchers were able to confirm these essential parameters.

Several techniques and measurements are employed to evaluate the structural, morphological, chemical, and electrochemical properties of AEMs. These characterizations offer valuable insights into the behavior and performance of AEMs, facilitating their optimization and comprehension. To identify the AEM's chemical structure, the polymer backbone and functional groups are analyzed through spectroscopic techniques like Fourier transform infrared (FTIR) and nuclear magnetic resonance (NMR). Furthermore, membrane morphology can be analyzed using microscopic techniques such as scanning electron microscopy (SEM) and transmission electron microscopy (TEM). To evaluate a membrane's chemical stability over time, subjecting them to harsh chemical environments for desired time is necessary. Meanwhile, their mechanical strength can be analyzed by measuring parameters like tensile strength, Young's modulus, and elongation at break. To measure the thickness of AEMs, techniques such as micrometers or profilometers can be used. The swelling behavior of the membrane is related to its dimensional changes upon water absorption, which can also be examined by measuring its thickness.

The ion exchange capacity is an important factor that measures the amount of ion exchange locations in the AEM material. It is calculated using titration methods, wherein the membrane is subjected to a known concentration of an ion, and the number of exchanged ions is measured. The determination of membrane conductivity can be achieved through either electrochemical impedance spectroscopy (EIS) or conductivity measurements conducted under controlled conditions. Water uptake can be evaluated by comparing the weight or volume of the dry membrane to that of the fully hydrated membrane.

Currently, computational simulations and machine learning methods are utilized to explore the properties and efficiencies of AEM. These techniques aid in proposing novel membrane structures, transport mechanisms, and stability, as well as identifying discrepancies and factors that may be challenging to determine through experimental results.

1.4.2 Classification of AEM

Researchers are currently experimenting with new techniques to create membranes that meet the basic requirements necessary for their function. As a result, there are now numerous types of membrane structures, each classified according to their unique morphology [24, 25]. They are

- Heterogeneous
  - Ion solvating polymers
  - Hybrid membranes
- Interpenetrated polymer networks
- Homogeneous
  - Copolymerization of monomers
  - Radiation grafting
  - Chemical modification

A heterogeneous membrane consists of an anion-exchange material embedded within an inert compound. These membranes can be categorized into two types based on their composition. The first type comprises ion-solvating polymers composed of a water soluble polymeric matrix containing electronegative heteroatoms, hydroxide ions, and plasticizers. This combination results in a material that possesses the polymeric matrix's mechanical properties and the hydroxide ions' electrochemical properties. Materials fabricating ion-solvating polymers include polyethylene oxide (PEO), polyvinyl alcohol (PVA), and chitosan. These polymers exhibit good mechanical properties, and their ionic conductivity at contact electrodes tends to be low due to their thickness and high electrical resistance. On the other hand, hybrid membranes are composed of both organic and inorganic segments. The organic components contribute to the electrochemical properties, while the inorganic elements, typically silica or siloxane, enhance the mechanical properties. Examples of hybrid membranes include combinations of PEO, PVA, and polyphenylene oxide (PPO) with silica (SiO_2) or titanium dioxide (TiO_2). Despite the good mechanical properties of incorporating inorganic components, membranes in this category still suffer from the nonuniformity issue observed in ion-solvating polymers. Consequently, their ionic conductivity remains similar or even lower.

Interpenetrated polymer network membranes, which belong to another class of AEMs, demonstrate higher ionic conductivities compared to heterogeneous membranes. These membranes are created by blending two polymeric materials through cross-linking without promoting the formation of covalent bonds between them.

One of the polymers is hydrophobic and possesses excellent chemical, mechanical, and thermal properties, while the other polymer acts as an ionic conductor. Fabrication of interpenetrated polymer network membranes is simple, and various polymers may be employed, making them cost-effective. The primary focus of research on membranes in this category includes materials such as polyethylene, PVA, polysulfone, and PPO. These membranes exhibit low electrical resistance, high mechanical strength, chemical stability, and durability. However, due to the absence of covalent bonds between the constituent materials, the conductive polymer slowly diffuses out of the membrane over time. This leads to gradual decreases in conductivity and ion exchange capacity. Homogeneous membranes represent another class of AEMs. These membranes are formed using polymers composed of a single material modified to possess ion exchange capacity. The modification involves covalently attaching cationic functional groups to the polymer backbone, creating ionic sites within the membrane along with associated mobile counterions. The classification of homogeneous membranes is based on the specific methods used for functionalization, including copolymerization, radiation grafting, and chemical modification. Detailed discussions on these classifications will be presented in the forthcoming chapters dedicated to this subject.

1.4.3 Pros and Cons of AEM

Understanding the advantages and disadvantages of AEMs is crucial for enhancing their utilization in various electrochemical applications and advancing the development of efficient and long-lasting energy conversion devices. Table 1.3 highlights the pros and cons of using AEMs as an electrolyte.

Table 1.3 Pros and cons of anion exchange membranes.

Pros	Cons
• High ionic conductivity	• Excessive water uptake can lead to swelling and reduced mechanical stability
• Precise control over selective ion transport	• Limited stability at high temperatures
• Operate over a broad pH range	• The diffusion of mobile counterions out of the membrane decreased conductivity and ion exchange capacity over time
• Good resistance to chemical degradation	• Require periodic cleaning or replacement due to the susceptibility to fouling by organic and inorganic species
• Long-term stability in aggressive environments	• Manufacturing complexity and limited material options compared to other membranes
• Exhibit high mechanical strength and resilience	
• Compatible with aqueous electrolytes	
• Manufactured at a reasonable cost	

Researchers and engineers can make educated judgments about selecting, optimizing, and building AEMs for specific applications by knowing their strengths and limits. Addressing the issues with these membranes will result in improved performance, longer lifespan, and wider use of AEM-based systems in the field of energy conversion and storage.

1.4.4 Application of AEM

AEMs are highly adaptable membranes with the unique ability to hinder cations while permitting a selective flow of anions. Their versatility makes them indispensable in numerous industries, ranging from energy to water treatment and beyond. Their ability to control the movement of ions makes them a crucial tool for many vital processes, including separation, purification, and desalination, in addition to their role as electrolytes. The following are some of the significant applications of AEMs.

1. **Electrochemical Energy Conversion:** Alkaline fuel cells (AFCs) and alkaline water electrolyzers frequently use AEMs, which are electrochemical devices. In AFCs, AEMs facilitate the electrochemical reaction by transporting hydroxide ions (OH^-) from the cathode to the anode. In alkaline water electrolyzers, AEMs play a crucial role in separating gases by selectively conducting hydroxide ions.
2. **Water Treatment:** AEMs are useful in different water treatment methods like electrodialysis and electro-deionization. They serve to eliminate undesired anions, specifically nitrates, sulfates, and chlorides, from water sources, which is beneficial for both water purification and desalination.
3. **Electrodialysis:** AEMs have the vital function of letting only anions pass through the membrane while preventing the migration of cations in electrodialysis. This crucial process facilitates separating and eliminating undesired ions from a solution.
4. **Electrochemical Sensors:** AEMs can be integrated with ion-selective electrodes to identify specific anions present in solutions. These AEM-based sensors are extensively used for environmental monitoring, water quality analysis, and industrial process control.
5. **Chlor-alkali Industry:** The chlor-alkali industry is responsible for producing chlorine gas (Cl_2), sodium hydroxide (NaOH), and hydrogen gas (H_2) via the electrolysis of saltwater. AEMs are utilized as ion exchange membranes in the electrolytic cells to facilitate this process. These membranes permit the migration of chloride ions (Cl^-) and hydroxide ions (OH^-) while preventing the passage of sodium ions (Na^+), which guarantees the desired separation of products.

The most significant application of AEMs is prominently displayed in the diagrammatical representation of Figure 1.2. AEMs find extensive usage in fuel cell applications, specifically in AFCs. The primary area of research for AEMs is centered on fuel cells, aiming to improve their performance, stability, selectivity, and cost-effectiveness.

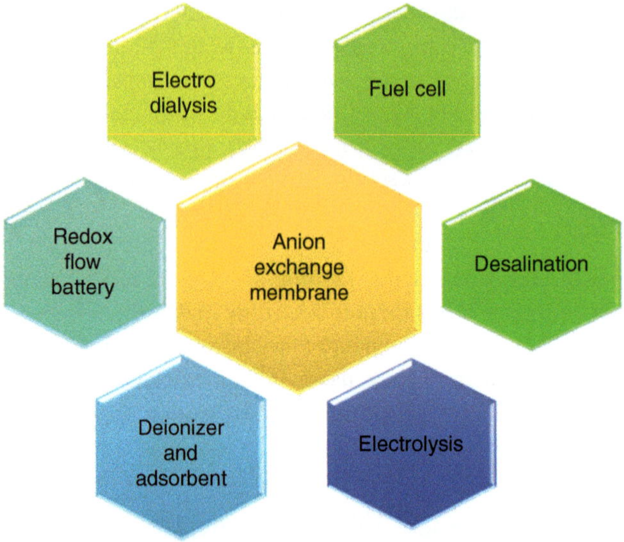

Figure 1.2 Important application of anion exchange membranes.

1.5 AEMs in Fuel Cells

A fuel cell is an energy conversion device that directly converts the fuel's chemical energy into electrical energy by chemically reacting a fuel with an oxidant, usually oxygen from the air. Although the origins of the fuel cell invention are uncertain, two notable individuals are associated with its discovery. Christian Friedrich Schönbein's work on the concept was published in the January 1839 issue of Philosophical Magazine, according to the United States Department of Energy. Meanwhile, Sir William Grove developed the fuel cell and published his findings in the February 1839 issue of the same magazine, as noted by Grimes. Fuel cell technology has grown substantially since its discovery in the nineteenth century. Through the years, researchers and engineers have made significant strides in enhancing fuel cells' efficiency, durability, and practicality. This progress results in the emergence of diverse fuel cell types, each with its distinct features and uses.

The fundamental design of a fuel cell involves an electrolyte layer that separates two electrodes – the anode and the cathode. The electrolyte facilitates the movement of ions while simultaneously preventing any mixing of the fuel and oxidant gases. In the early stages, AFCs utilize a liquid electrolyte solution containing potassium hydroxide (KOH), as it is the most conductive among alkaline hydroxides. Compared to other fuel cells, these AFCs have certain benefits. They are easier to manage as they operate at a relatively lower temperature, have electrodes made of inexpensive metals, and exhibit higher reaction kinetics at the electrodes compared to acidic conditions. However, the sensitivity of the KOH solution to CO_2 limits the AFC's use of liquid electrolytes. Optimal operation requires low CO_2 concentrations in the oxidant stream. If oxygen is replaced with air, the hydroxyl ions may react with CO_2 present in the air, leading to the formation of K_2CO_3. This leads to the precipitation of K_2CO_3 crystals and reduces the availability of hydroxyl ions, leading to decreased efficiency. Incorporating solid electrolytes in proton fuel cells (PFCs) has paved the way for anion-conducting polymer electrolyte membranes to be integrated into AFCs. This move effectively resolves most issues caused by liquid electrolytes, giving rise to a new subfield of AFCs – the anion exchange membrane fuel cell (AEMFC) – and leading to unprecedented growth in the industry. Using a membrane instead of a liquid electrolyte has several advantages. One significant benefit is eliminating the adverse effects of CO_2, which reduces electrode weeping and corrosion. Other membrane benefits include leak-proof properties, volumetric stability, solvent-free conditions, and easy handling. Additionally, the size and weight of the fuel cell are reduced, which expands its potential uses. Thus, AEM is used in an AFC to enhance efficiency and lifespan by slowing down performance degradation over time. It is essential to continue researching and developing AEMs to advance fuel cell technology and facilitate its widespread use as a reliable and eco-friendly energy conversion solution. Although AEM fuel cells have potential advantages and commercial significance, they are still in their early stages of commercialization and have not yet been widely deployed compared to other fuel cell types, such as proton exchange membrane fuel cells (PEMFCs). Nevertheless, there has been growing interest and research in AEM fuel cells.

The demand for clean and sustainable energy solutions is rising, and AEMFCs are gaining commercial significance due to their advantages over other types of fuel cells. AEMFCs use a polymer membrane that is more resistant to degradation, reducing maintenance and replacement costs. They also use low-cost materials, such as non-precious metal catalysts, making them more cost-effective than other fuel cell types. AEMFCs offer fuel flexibility, allowing for the use of various fuels, making them adaptable to different energy sources, and enabling the utilization of existing infrastructure and fuel distribution networks. They have shown promising efficiency levels, potentially converting more fuel energy into usable electrical energy, leading to improved overall energy conversion. Additionally, the polymer membrane used in AEMFCs allows for efficient control of water transport, preventing flooding and facilitating better performance under varying operating conditions. They produce clean electricity without the combustion of hydrocarbon fuels, resulting in lower greenhouse gas emissions and improved air quality.

1.6 Conclusion and Outlook

AEMs as the polymer electrolyte have garnered much attention in electrochemical devices such as fuel cells, batteries, and electrolyzers. AEMs possess unique properties that make them well-suited for various energy conversion and storage applications. Their primary function is to conduct negatively charged ions or anions while impeding the transport of positively charged ions or cations. This selective ion transport is achieved through positive charges or functional groups embedded in the polymer matrix of the membrane. These polymers have excellent mechanical strength, chemical stability, and suitable conductivity for hydroxide (OH^-). The OH^- conductivity is a crucial characteristic of AEMs, enabling efficient anion transport in alkaline environments. Due to their cost-effectiveness, durability, and fuel flexibility, they are highly valued and widely used in commercial settings.

AEM has a broad range of applications. For instance, AEM-based AFCs have shown improved tolerance to carbon monoxide poisoning, enhanced catalyst kinetics, and reduced reliance on expensive platinum catalysts. Compared to other fuel cell types, AEMFCs have these benefits, making them a commercially significant option for clean energy generation in various industries. Additionally, AEMs are widely used in alkaline water electrolyzers, which split water into hydrogen and oxygen using electricity.

One of the significant advantages of AEMs as polymer electrolytes is their ability to operate at low temperatures. Unlike proton exchange membranes (PEMs), which require high operating temperatures, AEMs can function effectively at room temperature or even lower. This feature offers opportunities for developing energy conversion devices that are more cost-effective, efficient, and durable. Moreover, AEMs have environmental benefits, making them a desirable and viable option for the future.

Despite holding great promise in various electrochemical applications, AEMs face challenges such as alkaline stability, ion transport, water management, mechanical stability, and catalyst compatibility. These challenges need to be addressed for widespread adoption. However, with continued research and development, AEM-based devices have the potential to contribute significantly to a cleaner and more sustainable energy future.

References

1 Kreysa, G., Ota, K.-i., and Savinell, R.F. (2014). *Encyclopedia of Applied Electrochemistry*. New York: Springer https://doi.org/10.1007/978-1-4419-6996-5.

2 Chen, S., Zhang, M., Zou, P. et al. (2022). Historical development and novel concepts on electrolytes for aqueous rechargeable batteries. *Energy Environ Sci* 15: 1805.

3 Ngai, K.S., Ramesh, S., Ramesh, K., and Juan, J.C. (2016). Review of polymer electrolytes: fundamental, approaches and applications. *Ionics* 22: 1259.

4 Lu, T., Guan, L., Zhan, Q. et al. (2023). Designing polymer electrolytes for advanced solid lithium-ion batteries: recent advances and future perspectives. *Mater Chem Front* 7: 3937–3957.

5 Zhou, D., Shanmukaraj, D., Tkacheva, A. et al. (2019). Polymer electrolytes for lithium-based batteries: advances and prospects. *Chem* 5 (9): 2326–2352.

6 Grandi, M., Rohde, S., Liu, D.J. et al. (2023). Recent advancements in high performance polymer electrolyte fuel cell electrode fabrication – novel materials and manufacturing processes. *J Power Sources* 562: 232734.

7 Salleh, N.A., Kheawhom, S., Ashrina, A. et al. (2023). Electrode polymer binders for supercapacitor applications: a review. *J Mater Res Technol* 23: 3470.

8 Wang, C., Yang, J., Li, J. et al. (2023). Solid-state electrochemical hydrogen sensors: a review. *Int J Hydrog Energy* 48: 31377.

9 Zhao, X., Wang, C., Liu, H. et al. (2023). A review of polymer-based solid-state electrolytes for lithium-metal batteries: structure, kinetic, interface stability, and application. *Batter Supercaps* 6: e202200502.

10 Zhao, Y., Wang, L., Zhou, Y. et al. (2021). Solid polymer electrolytes with high conductivity and transference number of Li ions for Li-based rechargeable batteries. *Adv Sci* 8: 2003675.

11 Peltier, C.R., You, W., Fackovic Volcanjk, D. et al. (2023). Quaternary ammonium-functionalized polyethylene-based anion exchange membranes: balancing performance and stability. *ACS Energy Lett* 8: 2365.

12 Li, X., Wang, Z., Chen, Y. et al. (2023). Imidazolium-based AEMs with high dimensional and alkaline-resistance stabilities for extended temperature range of alkaline fuel cells. *J Membr Sci* 670: 121352.

13 Salma, U. and Shalahin, N. (2023). A mini-review on alkaline stability of imidazolium cations and imidazolium-based anion exchange membranes. *Results Mater.* 17: 100366.

14 Sajjad, S.D., Hong, Y., and Liu, F. (2014). Synthesis of guanidinium-based anion exchange membranes and their stability assessment. *Polym Adv Technol* 25: 108.

15 Hibbs, M.R. (2013). Alkaline stability of poly(phenylene)-based anion exchange membranes with various cations. *J Polym Sci B Polym Phys* 51: 1736.

16 Chen, Y., Tao, Y., Wang, J. et al. (2017). Comb-shaped guanidinium functionalized poly(ether sulfone)s for anion exchange membranes: effects of the spacer types and lengths. *J Polym Sci A Polym Chem* 55: 1313.

17 Barnes, A.M., Du, Y., Zhang, W. et al. (2019). Phosphonium-containing block copolymer anion exchange membranes: effect of quaternization level on bulk and surface morphologies at hydrated and dehydrated states. *Macromolecules* 52: 6097.

18 Pham, T.H., Olsson, J.S., and Jannasch, P. (2017). N-spirocyclic quaternary ammonium ionenes for anion-exchange membranes. *J Am Chem Soc* 139: 2888.

19 Thomas, J., Francis, B., Thomas, S. et al. (2021). Dependable polysulfone based anion exchange membranes incorporating triazatriangulenium cations. *Solid State Ion* 370: 115731.

20 Cao, M., Chu, J., Fan, X. et al. (2023). Poly (ionic liquid) filled and cross-linked bacterial cellulose-based organic-inorganic composite anion exchange membrane with significantly improved ionic conductivity and mechanical strength. *J Membr Sci* 675: 121558.

21 Tsehaye, M.T., Choi, N.H., Fischer, P. et al. (2022). Anion exchange membranes incorporating multi N-spirocyclic quaternary ammonium cations via ultraviolet-initiated polymerization for zinc slurry-air flow batteries. *ACS Appl Energy Mater* 5: 7069.

22 Li, Z., Li, C., Long, C. et al. (2020). Elastic and durable multi-cation-crosslinked anion exchange membrane based on poly(styrene-b-(ethylene-co-butylene)-b-styrene). *J Polym Sci* 58: 2181.

23 Zha, Y., Disabb-Miller, M.L., Johnson, Z.D. et al. (2012). Metal-cation-based anion exchange membranes. *J Am Chem Soc* 134: 4493.

24 Merle, G., Wessling, M., and Nijmeijer, K. (2011). Anion exchange membranes for alkaline fuel cells: a review. *J Membr Sci* 377: 1.

25 Ramírez, S. and Paz, R.R. (2018). Hydroxide Transport in Anion-Exchange Membranes for Alkaline Fuel Cells. *InTech.* https://doi.org/10.5772/intechopen.77148.

2

Historical and Recent Developments in Anion Exchange Membranes (AEM)

Priya Goel[1], Priyabrata Mandal[1], Sujay Chattopadhyay[1], and Vinod K. Shahi[2]

[1] *Indian Institute of Technology Roorkee, Department of Polymer and Process Engineering, Roorkee 247667, Uttarakhand, India*
[2] *CSIR-Central Salt and Marine Chemicals Research Institute, Membrane Science and Separation Technology Division, Bhavnagar 364002, Gujarat, India*

2.1 Introduction

The current growing concern of greenhouse emissions and limited availability of fossil fuels draws the attention to develop alternative sources of energy for the sustainable growth of our society. Among the various ways to solve energy issues, fuel cells have been considered as one of the most promising environment-friendly solutions of the twenty-first century. It is a clean and efficient power generation technology that enables the direct transformation of chemical energy into electrical energy [1]. Therefore, numerous efforts have been made to mature this technology from the laboratory to real applications. Nowadays, five major types of fuel cells are distinguished by the type of electrolyte: alkaline fuel cells (AFCs), proton-exchange membrane fuel cells (PEMFCs), phosphoric acid fuel cells (PACFs), solid oxide fuel cells (SOFCs), and molten carbonate fuel cells (MCFCs). The AFC (using aqueous potassium hydroxide [KOH] as electrolyte) was the first type to be put into practical service, at the start of the twentieth century, but the development of solid anion-exchange membrane (AEM) as an alternative to liquid KOH solution as electrolyte paved the way for the renewed interest in AFC applications. These solid AEMs are also called alkaline-exchange membrane when they are particularly used in AFCs.

This chapter presents an outline of the history and recent advancements in the area of AEMs used for AFC applications. The inspiration for this book chapter came from the copious research and countless growths in the membrane field for AFCs during the last few decades. Primary attention is paid to the requirements for AEM

Alkaline Anion Exchange Membranes for Fuel Cells: From Tailored Materials to Novel Applications, First Edition. Edited by Jince Thomas, Alex Schechter, Flavio Grynszpan, Bejoy Francis, and Sabu Thomas.

in the AFC assembly, their fundamentals, modifications, and methodologies. It is necessary to discuss the various types of AEMs, their mechanisms of anion transport; and its effect on electrochemical as well as physicochemical properties, and the methods that were used to fulfill the requirement of efficient AEMs for fuel cell application. Lastly, a summary is given where technical difficulties are defined, and the remaining technical challenges and future directions are highlighted.

2.2 Fuel Cell: Conventional Versus Modern Approach

The AFC is an electrochemical device in which the fuel (usually hydrogen) reacts at the anode side and oxygen from air reacts at the cathode side to produce electricity and heat. The universal components of fuel cells are cathode, anode, and electrolyte (liquid or solid electrolyte). The fuel cell is one of the oldest energy conversion technologies which originated alongside with the concept of electrolysis in the early 1800s (Table 2.1). A German scientist, named Christian Friedrich Schoenbein (January 1839), first described the fuel cell concept, but the exact inventor is not known clearly [4, 5]. Just after the Schoenbein, Sir William Grove (February 1839) published a similar report about fuel cell in the Philosophical Magazine. The idea of fuel cell originated from Sir William Grove during his research on electrolysis. Hydrogen and oxygen can be generated from water molecules if electricity is supplied to electrodes, then the electricity certainly can be created by the invert response. Afterward, the first practical fuel cell device was developed (1811–1896) which was adopted by the US Navy and the British Royal Navy [4]. Thereafter, a mixture of hydrogen and oxygen was the primary fuel, utilized within the fuel cell framework created by William Grove to produce electricity. Then Francis Thomas Bacon (1904–1992) introduced a type of fuel cell that was built with anode (sintered nickel)/cathodes (lithiated nickel oxide) and KOH electrolyte. The system of 5 kW showed significant performance. Bacon first introduced the AFC design using KOH solution as an electrolyte. KOH is used as an electrolyte for AFCs due to its higher solubility and ionic conductivity than rest of the electrolytes such as sodium hydroxide (NaOH) and potassium carbonate (K_2CO_3). After the discovery, many designs and modifications have been developed for various applications.

Conventional AFCs offer several advantages, e.g. easy handling, low operating temperature range, higher electrodes reaction, and high electrical efficiency, than other types of fuel cells (Figure 2.1). However, the use of liquid electrolytes became the main drawback. The liquid electrolyte here, KOH, is very sensitive toward CO_2. When air is used at the cathode side, the generated OH^- ions may react with CO_2 present in the air to form K_2CO_3. The K_2CO_3 precipitation affects the composition of the electrolytes and decreases the number of OH^- ions for anodic reaction. The precipitation also resulted in a reduction in ionic conductivity and overall the cell performance. Additionally, the lack or excess of liquid electrolytes also leads to electrode drying or flooding. Therefore, the low CO_2 concentration in the feed stream or the presence of pure oxygen is essential for the best cell performance; however, this may increase the overall cost.

Table 2.1 AFCs outline of chronicled advancement and accomplishment of AEMs and fuel cell areas including AFCs.

Year	Progress
1800	W. Nicolas et al. explained the electrolysis of water
1839	C. F. Schönbein et al. reported the concept of fuel cell
1842	William Grove invented first fuel cell consisting of a gas voltaic battery prototype
1889	Charles Langer and Ludwig mond introduced fuel cell; advanced Grove's invention
1890	Ostwald discovered first IEM (cation- or anion-exchange membranes)
1893	F. W. Ostwald defined theoretical fuel cells performance
1896	W. Jacquse developed first fuel cell
1911	Donnan described the existence of boundary and "Donnan exclusion potential"
1925	Synthesis of ion-exchange membrane for the first time
1932	F. T. Bacon developed the AFC
1940	First industrial implementation of electro-membrane
1947	Oganes Davtyan developed fuel cell prototype which oxidizes in the presence of carbon monoxide with oxygen to form air at 700 °C and can run for several days
1959	A farm tractor was the first electric vehicle powered by an AFC
1960	The NASA Gemini and Apollo space programs started using KOH-based AFC systems
1970	Oil crises, and advanced additional potent AFC for NASA's space shuttle Orbiter, IFC, Windsor Connecticut
1970	Fuel cell developed by Siemens for submarines and German submarine industry
2000	The world's first alkaline-fuel-cell based ship, the Hydra, used an AFC system
>2000	Improvement of new AEMs with enhanced selectivity, lower electrical resistance, improved properties
2018	GenCell, commercial system

Source: Xu [2] and Cheng et al. [3].

Due to the aforementioned disadvantages of a liquid electrolyte, some efforts have been made by using solid electrolyte as AEM. Cheng et al. [6] who worked with PEMFCs depicted the possibility of utilizing a strong polymer electrolyte over a fluid electrolyte. Since then, the investigation has focused on the promising advancement of AEM-based AFCs over PEMFCs such as enhanced electrochemical kinetics of oxygen reduction reaction and the use of inexpensive non-platinum electrocatalysts. The most important advantage of using a membrane instead of a liquid electrolyte is, as mentioned, elimination of the negative effects of CO_2. The conducting species is now in a fixed solid polymer; therefore, there will be some carbonates due to the reaction of the $-OH$ with CO_2. Furthermore, no liquid caustic is present; hence, electrode weeping and corrosion are minimized. Additional benefits include leakproofness, volumetric stability, solvent-free conditions, and easy handling.

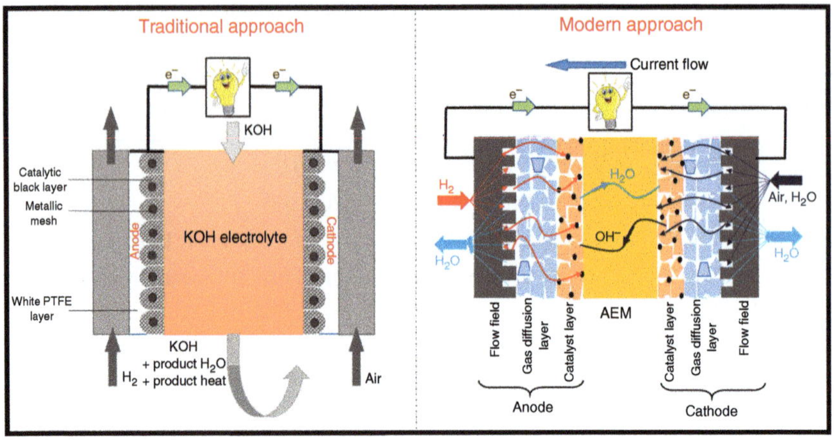

Figure 2.1 Schematic representation of liquid-electrolyte-based AFC versus anion-exchange-membrane-based AFC.

The size and weight of the fuel cell are reduced which widens the domain of application. The main idea behind employing an AEM in an AFC is to improve the AFC's efficiency and life (slow down performance degradation with time).

2.3 Role of AEM in Fuel Cell Technology

Despite AFCs' early success, interest in this technology was dropped initially due to economic factors, material problems, and certain inadequacies in the operation of electrochemical devices. But the use of AEMs has drawn some potential in this direction. The research on AEMs for AFCs is encouraged by the fact that anion systems are capable to improve the efficiency of the current outcomes. AEMs have the potential to solve the problem of the upcoming energy crisis. AEMs are typically thin polymeric films containing cationic fixed charged groups which are ionizable in water [2]. Hence, covalently attached cationic fixed charge sites allow the oppositely charged ions to pass through them in the presence of an electric field [7]. The concept of AEM in AFCs came with drastic modifications in the field of fuel cell as well as in the membrane field when the basic structure of fuel cell is proposed with the integration of membrane to replace liquid electrolyte. This turn also made an impressive growth in the number of publications related to AEMs or AFCs over the past few decades (Figure 2.2). The increased number of publications in the past decades indicates the growing interest in the area of fuel cell technology. In an upgraded strategy of a fuel cell, a polymeric membrane is placed between the cathode and anode electrodes, and also provides a conductive path between two electrodes for oxidation and reduction reaction of hydrogen and oxygen, respectively. A broad range of liquid/solid fuels may be used in AFC setup (e.g. hydrogen, hydrazine, methanol, ethanol, ethylene glycol, glycerol, borohydride, ammonia, dimethyl ether, potassium formate, 2-propanol), but hydrogen, methanol, ethanol are being

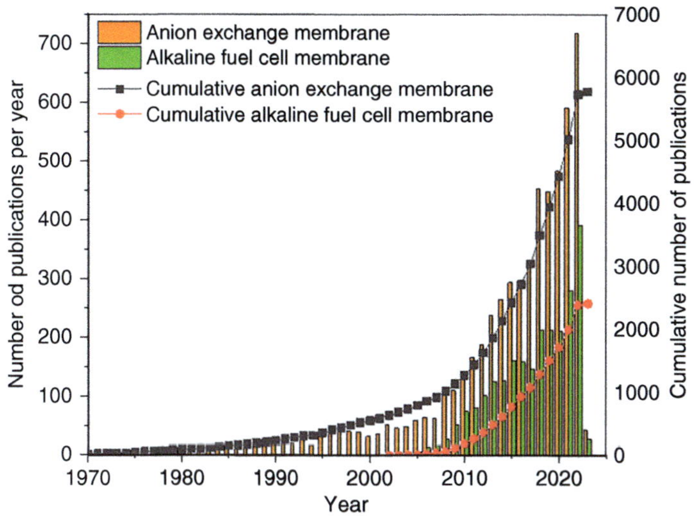

Figure 2.2 The statistics of the number of published articles indexed in the "Scopus" based on studies related to (a) anion-exchange membrane, and (b) those focused on alkaline fuel cells during the past 50 years (1971–2021) extracted on 11 December 2022.

used as primary fuels in recent years. This combined system is popularly known as membrane electrode assembly (MEA). The main advantage of using MEA is the replacement of liquid electrolyte and the elimination of CO_2 poisoning because the conducting species is a fixed solid polymer. This solid polymeric membrane is known as an AEM.

AEMs are polymer electrolytes in which cationic functional sites are incorporated into the polymeric chain by chemical means. In AEMs, the fixed functional sites allow the hydroxide ion (counter-ion) move across it and block cations (co-ion) in the presence of an electric field [2]. Conventionally, different cationic functional groups have been integrated into the polymer matrix which acts as an anion-exchange counterpart. Irrespective of AEMs, AEMs (a similar concept with anions different than hydroxide ions) are widely used in electrochemical devices including AFC technology [2, 3], redox flow batteries [8, 9], water treatment such as water electrolysis [10, 11], electrodialysis (ED) [12], reverse ED [13, 14], diffusion dialysis [15, 16], and many other applications [17–20].

The chemical and thermal stability of AEM are very important criteria in the design and synthesis of membranes for fuel cell application. The characterization of an AEM is intimately related to two parameters, namely the polymer matrix and the nature and concentration of the cationic charges. Parameters of the polymer matrix determine mechanical and thermal, stability, while parameters related to the ionic groups determine anion-exchange capacity, transport number, and ionic conductivity. The main concern for anion AEM is that it relies on OH^- ion, a relatively strong base and nucleophile, for conductivity. With the use of a stable polymer backbone, the stability and degradation of AEMs then rest chiefly on the cation stability. The chemical stability is determined by both the polymer matrix and especially the cationic groups.

2.4 Preparation of AEMs

Several methods have been applied to fabricate AEMs including heterogeneous membranes (ion-solvating polymers, and organic–inorganic membranes), interpenetrating polymer networks (IPNs), and homogeneous membranes [5]. Heterogeneous membranes are good in terms of mechanical strength, but such kinds of membranes are usually thick and uneven with different concentrations of exchangeable ionic groups, which leads to non-uniform transport of anions and poor electrochemical performance. In the direction of hybrid membranes (organic–inorganic), improvements are further needed in terms of a better hold over the reaction parameters and the precursors that are required to affix [21]. Furthermore, IPNs are comparatively easy to fabricate using a variety of polymers. Additionally, these membranes show excellent chemical stability, high mechanical strength, high electrical conductivity (due to the presence of polyelectrolyte), long life (due to the stable polymer backbone), and are inexpensive [22]. But polyelectrolyte slowly diffuses out from the membrane which gradually decreases the ion-exchange sites as well as the conductivity of the membrane. Therefore, IPNs are better than heterogeneous membranes, but still remain unsatisfactory to fulfill the requirements of AFC applications [23].

To overcome the aforementioned difficulties, a new class of membranes was introduced which probably encountered the previously introduced membrane. Homogeneous membranes are made of polymer materials and fixed charge groups attached to their polymeric chain and homogeneously distributed over it. They can be synthesized by polymerization or polycondensation, radiation-induced grafting, or chemical grafting methods [24–26]. But homogeneous membranes are the most widely adopted technique for further development of membranes for AFCs. The most successful AEM was synthesized by copolymerization of chloromethylstyrene and divinyl benzene, but, regrettably, limited availability of chloromethyl styrene and divinyl benzene use increases the cost extensively [5, 27]. However, the success of such membranes at a larger stage is restricted. Here, quaternary ammonium groups were familiarized by chloromethylation of aryl carbon or bromination to benzyl carbon after amination (Figure 2.3). Typically, AEM synthesis with this method is a two-step process for the introduction of the cationic functional sites (i) chloromethylation or bromination followed by (ii) quaternization of chloro- or bromo-substituent [28]. Generally, the route of chloromethylation is achieved by chloromethyl methyl ether (CMME), because chloromethylation reaction kinetics using CMME is very fast but CMME has been banned due to its toxicity, carcinogenicity, and harm to human beings [29]. To avoid CMME, other chemicals such as chloromethyloctylether are used for chloromethylation [30]. Additionally bromination is a good alternative to chloromethylation. Except chloromethylation, bromination is another method where liquid bromine (Br_2) or *N*-bromosuccinamide is generally used as a brominating agent. Bromination is a fast method as compared

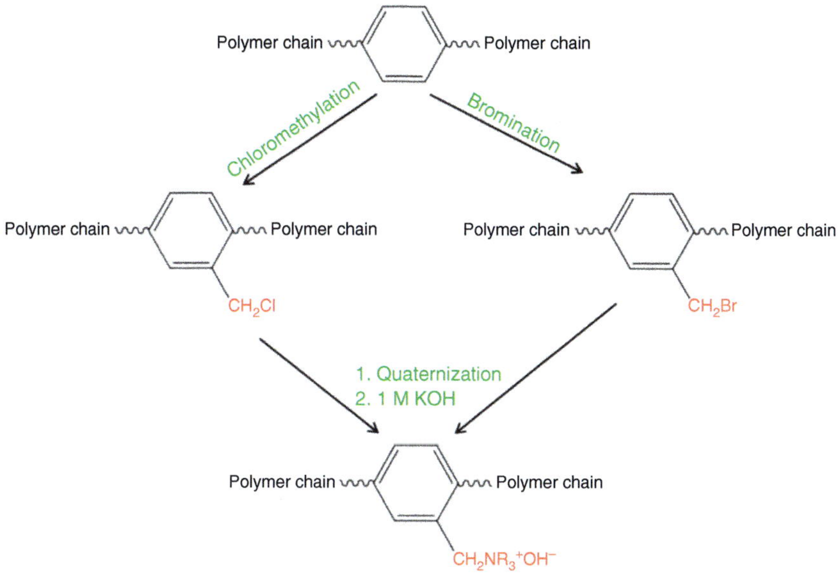

Figure 2.3 The general reaction steps for the preparation of homogeneous AEM where a polymer chain is transformed into a polymeric AEM with quaternary functional site after chloromethylation or bromination.

to chloromethylation because it follows a radical mechanism. Regardless of the rapid method, radical bromination is restricted to certain polymers, e.g. PPO and PEEK, because radical reaction required the presence of benzyl carbon in the polymer chain to react [31].

2.5 Challenges in Existing AEMs

An ideal AEM requires high hydroxide conductivity, high ion-exchange capacity (IEC), good anion stability, optimum water uptake, swelling ratio, thermo-mechanical stability, and good performance at long-term operation under optimum temperature range. Compatibility between all the properties for an ideal AEM is difficult to maintain practically. If the IEC is high, conductivity will be high because conductivity depends on the cationic functional sites affixed in the polymer matrix, but this may lead to dimensional or thermo-mechanical instability. Additionally, water uptake and swelling ratio also rises after a certain limit of quaternization. Beyond a certain limit of functionalization, polymer suffers degradation which leads to inadequate performance. Many factors are responsible to influence the physicochemical and electrochemical properties of AEMs. Besides this, type of polymer and associated cationic functional groups attached in the polymer backbone [32, 33] also influence the overall properties of the AEMs (Table 2.2). It is required to engineer a well-designed polymer architecture that fulfills most of the desired requirements and a balance between physico-chemical and electrochemical properties.

Table 2.2 A range of polymer and cationic functional groups used in preparation of AEMs

Polymer	Cationic functional groups
Poly (ether ketone), poly(phenylene oxide), poly(ether sulphone), polysulfone, poly(ether ether ketone), polyphenyene, polyvinyl difluoride, poly(aryl piperidinium), polybenzimidazole, poly(tetrafluoroethylene), polyethylene, polynorbornene, polyimide, polyphosphazene, poly vinyl alcohol, etc.	Quaternary ammonium, guanidinium, imidazolium, phosphonium, piperidinium, 1,4-diazabicyclo[2.2.2]octane, *N,N,N,N*-tetramethylethylenediamine (TMEDA), 1-azabicyclo[2.2.2]octane, ruthenium, cobaltocenium, benzimidazole, sulfonium, spirocyclic groups, etc.

2.6 Recent Advancement

As mentioned in Section 2.5, an ideal AEM requires a good compatibility between the physico-chemical and electro-chemical properties. Additionally, a good sustainable AEM for AFC operation mainly demands high hydroxide conductivity with good anion stability for the long-term operation of AFC assembly. As the AEM started to use in AFCs, a lot of research and development in terms of membrane performance has taken place over the last 30 years. Noteworthy progress has been made in developing AEMs to improve AFC performance, but it should be noted that the AEMs are still facing a variety of challenges, including hydroxide-ion conductivity, membrane stability (mechanical, chemical, and thermal stability), constant high performance, and low cost. Overall, streamlined design approaches are required to fabricate AEMs to avoid incompatibility of physicochemical and electrochemical properties along with long-term performance. To avoid such problems, nowadays a variety of schemes have been conceived to conquer these issues, e.g. crosslinking [34–37], microphase separation [38], and organic–inorganic composites [4]. Cross-linking is a very useful route to develop mechanically stable membranes. Usually, high cross-linking leads to a decrease in IEC but increases mechanical stability. A moderate crosslinking was utilized in creating a breakthrough AEM without the major impact on other properties.

Moreover, cross-linking can restrict the swelling of the membrane up to a limit and provides good chemical stability. The cross-linked polymer creates a network that reduces the free space between the polymer chains and therefore makes additional access of OH^- ions to the labile cations. Ma et al. [39] prepared a series of novel AEMs modified with bulky rigid β-cyclo-dextrin and long flexible multiple quaternary ammonium which showed higher tensile strength (29.3 MPa), without sacrificing its hydroxide conductivity, than the unmodified membrane (16.7 MPa).

The idea of microphase separation has contributed to remarkable development in AEM overall performance in recent time. In microphase separation, hydrophilic (cationic polymer) and hydrophobic (polymer backbone) segments are generated. This type of architecture generates an area where water molecules can be captivated locally to enhance the OH^- ion conductivity. Even though the microphase separation enhances hydroxide conductivity of particular membrane and adjusts their dimensional stability, IEC remains almost the same [40, 41]. Mainly four types of architecture can be introduced to develop microphase separation in AEMs including, block polymer, graft polymer, clustered polymer, and comb-shaped polymer. Despite the numerous useful AEMs performances along with microphase separation, there is no proof of causality between this approach and AFC durability or peak power density. Creating a composite (organic–inorganic) membrane by incorporating nanoparticles or layered double hydroxide as filler in AEMs signifies another approach for improving its mechanical properties. Due to insufficient mobile molecules, the overall performance of fuel cells related to this technique, it has now no longer garnered noteworthy attention.

Besides the above methods, many other changes have also been explored to make an efficient AEM for AFC performance. A variety of cross-linkers, different or multiple cationic functional groups and their influence, blending of polymers, tuning backbone and side-chain, steric hindrance effect, surface modification, effect of degree of cross-linking, integration of pristine, or functionalized nanoparticles are seen [42–46]. Table 2.3 shows the properties of various AEMs synthesized recently and their performance in AFCs.

2.7 Major Challenges

Even though AEMs offer many advantages in the AFC system over conventional liquid electrolyte system a few challenges still need to be addressed (Table 2.4). As AEMs become the most crucial component of MEA in AFCs, it is necessary to develop high-performance, stable, and inexpensive production methods. Most of the techniques suffer few limitations such as relatively low anion-exchange capacity and poor conductivity besides compromising the mechanical and thermal properties of the synthesized membranes. The poor stability of the ammonium employed leads to a decrease in the ionic conductivity and a loss of efficiency in the system. A lot of research are being done to develop an efficient AEM for AFCs. A membrane with high conductivity and long-term anion stability is highly required for the AFC applications. The OH^- conductivity depends on the membrane thickness and IEC value of functionalized cationic groups. However, a trade-off with mechanical stability owing to the high swelling in presence of more functional groups is a challenge. Additionally, AEMs are vulnerable to hydroxide ion attack and undergo rapid polymer or functional group degradation via Hofman elimination, nucleophilic substitution, etc. The hydroxide attack in membrane architecture drops the conductivity as well as overall performance. High temperature and harsh alkali pH conditions promote the faster degradation of quaternary ammonium through Hofmann elimination, E1 elimination reaction, or via nucleophilic substitution [5]. The AEM stability against alkali attack is a crucial property to promising long run and consistent AFCs process. The stability is frequently checked by measuring the change in different physico-chemical properties including IEC treating AEM in a harsh alkali

Table 2.3 Recent studies on synthesis of AEMs for AFC application.

Membrane ID	Modification	Method	Properties	OH^- conductivity	Power density	References
QPAF-C3-Pip	Different ammonium head groups	Phase-separation	IEC = 1.26 mmol g^{-1} mechanical strength, and anion stability in 4 mol l^{-1} KOH at 80 °C for 736 h	55.4 mS cm^{-1}	232 mW cm^{-2} at 60 °C	[32]
5% qPPO-C_1/qPPO-C_6	Complementary side chain grafting	Blending	IEC = 1.45 mmol g^{-1}	37.5 mS cm^{-1} at 30 °C	192 mW cm^{-2} at 60 °C	[38]
PQP-100	Poly(aryl piperidinium) with extended aromatic segments	Superacid-catalyzed polymerization	IEC = 2.3 mmol g^{-1} excellent anion stability in 1 mol l^{-1} KOH solution over 1344 h	118.72 mS cm^{-1} at 80 °C	496 mW cm^{-2} at 60 °C	[47]
QPAEK/*f*-TiO_2-5 wt%	Functionalized TiO_2 as inorganic segment	Solution casting	IEC = 1.36 mmol g^{-1} improved mechanical (Young modulus: 200.7 ± 1.8 MPa) and thermal anion stability in a 2 mol l^{-1} KOH solution at 70 °C for 14 d, excellent oxidation stability	74.6 mS cm^{-1} at 90 °C	137.3 mW cm^{-2} at 60 °C	[21]
NPBI-QA55-B45	−NH−effect in polybenzimidazole	Grafting	IEC = 1.56 mmol g^{-1} excellent stability in 5 mol l^{-1} NaOH at 80 °C	31.7 mS cm^{-1} at 25 °C 69.2 mS cm^{-1} at 80 °C	260.1 mW cm^{-2} at 60 °C	[48]
70-EB-PIM 25-γ-SM	High degree of crosslinking	Radiation-induced grafting	IEC = 2.55 mmol g^{-1} improved mechanical properties	265 mS cm^{-1} at RH100% 240 mS cm^{-1}	1.43 W cm^{-2} at 80 °C	[49]

OBImPPO26-PEG0.5	Cation dipole interactions between imidazolium and electronegative alkoxy	Grafting	IEC = 1.69 mmol g^{-1}, with highly stable in 2 mol l^{-1} aqueous NaOH for 216 h at 60 °C	23.92 mS cm^{-1} at 30 °C (90% RH)	407 mW cm^{-2} at 60 °C	[50]
PITP-C10Q85	Comb shaped copolymer	Ternary copolymerization	IEC = 2.26 mmol g^{-1}, high stable in 1 mol l^{-1} NaOH at 80 °C	65 mS cm^{-1} at 25 °C in water 134.5 mS cm^{-1} at 80 °C	445 mW cm^{-2} at 870 mA cm^{-2} at 60 °C	[51]
QPPO-NGO$_2$	Functionalized graphene oxide (cross-linker)	Solution casting	IEC = 2.39 mmol g^{-1}, less water uptake and swelling ratio but higher ion conductivity and anion stability	123 mS cm^{-1} at 80 °C in water 83 mS cm^{-1} at 2 mol l^{-1} KOH at 60 °C for 384 h-QPPO-NGO$_1$	397.8 mW cm^{-2} at 60 °C	[52]
PFAP-x AEIs	Copolymerization	Solution casting	IEC PFTP-13 (2.81 mmol g^{-1}) PFBP-0 (3.52 mmol g^{-1}) PFPN-100 (3.09 mmol g^{-1}) mechanical stability of the membrane. PX75-T50 displays only 8% water uptake and 3% swelling ratio at 80 °C	208 mS cm^{-1} at 80 °C	2.34 W cm^{-2} (H_2–O_2 fuel) and 1.25 W cm^{-2} (H_2–air, CO_2-free) at 80 °C	[53]
PHFB-VBC-DQ-80%	C=C bond end-group crosslinked structure	Polymerization, grafting reaction	IEC = 2.24 mmol g^{-1}, dimensional stability, chemical stable at 2 mol l^{-1} NaOH solution at 80 °C for 500 h	135.80 mS cm^{-1} at 80 °C	141.7 mW cm^{-2} at 80 °C (current density of 260 mA cm^{-2})	[54]

(Continued)

Table 2.3 (Continued)

Membrane ID	Modification	Method	Properties	OH^- conductivity	Power density	References
PDTP-25	Poly(aryl-*co*-aryl piperidinium) (c-PAP) copolymers	*In situ* styrene-crosslinked	IEC = 2.80 mmol g^{-1}, mechanical strength maintaining in 1 mol l^{-1} NaOH at 80 °C for 2160 h	75 mS cm^{-1} at 30 °C	2.26 W cm^{-2} (H_2–O_2 fuel) and 1.32 W cm^{-2} (H_2–air) at 80 °C	[55]
c-91	Acetylation	Macromolecular crosslinking	IEC = 1.54 mmol g^{-1} Stable in 1 M NaOH mol l^{-1} at 80 °C for 1000 h	78.8 mS cm^{-1} at 80 °C	105 mW cm^{-2} at 80 °C	[56]
Etherfree polyfluorene-based AEMs (PFPE-Pi-based AEM)	Block copolymer design	Suzuki cross-coupling reaction	IEC = 2.49 mmol g^{-1} High anion stability 2 mol l^{-1} KOH solution at 80 °C for long time, low swelling ratio of 9.4% at 80 °C	80.44 mS cm^{-1} at 80 °C with tremendous anion stability in 2 mol l^{-1} KOH solution	661 mW cm^{-2} at 80 °C	[57]
HWU-AEM	Anisotropic changes	*In situ* polymerization	IEC = 1.86 mmol g^{-1} high water uptake, adequate mechanical stability (Young modulus: 289 ± 21 MPa) and chemically stable in 2 mol l^- NaOH solution at 80 °C for 1000 h	49.8 mS cm^{-1} (30 °C) and 99.4 mS cm^{-1} (80 °C) in CO_2-free NaOH solutions (1.0 mol l^{-1})	1.1 W cm^{-2} (H_2–O_2 fuel) at 80 °C	[58]

Table 2.4 Advantages and disadvantages of AEM based AFCs.

Advantages	Disadvantages
1. Solid-state electrolyte	1. Average hydroxide conductivities
2. Better CO_2 tolerance	2. Thin membrane fabrication (~10 mm)
3. Decreased gas crossover	3. Low conductivity at low humidity
4. Increased efficiency	4. Long-term durability
5. No flooding and weeping of electrolyte	5. Embrittlement if stored in dry conditions after activation in 1.0 $mol\,l^{-1}$ KOH
6. Easier to handle	6. Mechanical/chemical stability of AEM
7. Relatively low operating temperature	7. Membrane degradation
8. Very less carbonate precipitation	

solution (usually 2–4 $mol\,l^{-1}$ aqueous KOH) at high temperatures (65–85 °C). After drop in conductivity, IEC represents the degradation of cationic groups and mechanical property degradation shows the possibility of polymer chain scission.

Inhibition of SN_2 reactions by introducing a long spacer chain between the nitrogen atom of the cationic site and the benzene ring of the main polymer chain is reported. Besides, alkylene spacers longer than propylene were also found to make the membrane less susceptible to OH^- attack. Hofmann elimination can be avoided by the use of diazabicyclo-octane (DABCO) as the quaternization regent [51, 52].

Different polymer backbone structure can also have a significant influence on the anion stability and plays a crucial role in determining the membrane's mechanical strength and chemical stability. It is interesting to note that different polymer backbones with the same cationic functional group show different behavior in alkali. Similarly, different cationic groups in the same polymer backbone display different outcomes in anion stability, even though noteworthy progress in developing AEMs with improved performances has been achieved.

Researchers in this field are still facing challenges including trade-off between membrane stability and conductivity as well as cost. Great efforts have been made in recent years to overcome these issues. Crosslinking, microphase separation (to increase ionic conductivity and adjust their dimensional stability), and organic–inorganic composites (alter the mechanical stability) are a few of them. Additionally, to improve the anion stability, different approaches such as novel membrane architecture, stable cationic sites, polymer backbone modification, morphological changes, steric hindrance effect, electronic effect and fixing non-organic cations in polymer matrix, and so on have been implemented in the laboratory scale. Despite this, a variety of novel cationic groups were explored to accomplish comprehensive avoidance of decomposition [28]. Although numerous polymer backbones have been explored for AEMs, majority of them possessed limited overall performance especially aryl-ether-polymer-based AEMs because they are susceptible to chain cleavage as well as subsequent mechanical degradation. A remarkable improvement in chemical stability has been made recently with the development of any ether-free polymers and highly durable cationic groups [59].

The real-life application of fuel cells turns around their consistency coupled with their longevity, and it is very important for industrial/customer acceptance. The fuel cell industry requires a scheme to improve the operational tolerance for customer acceptance. There is still a break between existing fuel cell development and the customer requirements.

2.8 Commercially Available AEMs

Globally, there are many companies which are providing commercially available ion-exchange membranes including AEMs (Table 2.5). Low-cost polymer backbones such as PEEK, PPO, and polysulfone have been explored for AEMs development in the laboratory and industrial scale. Many of the commercially available membranes present in the market are dedicated to the chloralkali industry, desalination plants, or electrolyzer applications, and very few of them are designed especially for AFC application. In addition, many other research groups are also developing various high-quality advanced AEM structures. Currently, the commercially available AEMs that are being tested for AFCs are having good physico-chemical and electrochemical properties, but most of them are expensive. Most of the available commercial AEMs are supplied on a reinforcing material that not only provide additional mechanical strength, but also help in transportation. FuMA-Tech, Xergy, Orion (Orion™), DiOxide Materials (Sustainion®), Asahi Chemical Industry Co, etc. are few companies that are successfully working in the field of membrane production.

Table 2.5 Commercially available AEMs for fuel cell applications.

Country	Company	AEMs
Canada	Ionomr	AFN-HNN8-25-X, AFN-HNN8-50-X, etc.
Czech Republic	MEGA	Ralex AMH-5E, Ralex MH-PES
USA	Xergy	Xion Durion™, Pention™
	DiOxide Materials	Sustainion®
	Orion	Orion™
China	Tianwei	TWEDG, TWDDG, TWAEDI, TWANS, etc.
Japan	Asahi Glass Industry Co.	Selemion AMV, Selemion DSV, Selemion ASV, etc.
	Asahi Chemical Industry Co.	Aciplex A-501-SB, Aciplex A221, Aciplex A-192, etc.
Germany	FuMA-Tech	FAS, FAA, FAN, FAB, FAD, etc.

Source: Adapted from Ferriday and Middleton [59].

The Pention AEM (Xergy Inc.), depending on the thickness and crosslinking degree, has demonstrated up to 3.37 $W\,cm^{-2}$ power density in AFCs with excellent durability and lifetime. For AFCs, a UK-based company AFC Energy is also developing AFC technology based on AEMs.

2.9 Current Scenario and Future Market

The development pathway over 220 years ago started with Sir William Grove through Francis Bacon's power plant to the fuel-cell-powered tractor to the Space Race to electric vehicles. The AFC was the first fuel cell technology that was used in practical application and make the generation of electricity from hydrogen feasible. Starting with applications in space missions to submarines, the anion cell offered the most effective solution for high-energy conversion efficiency with no moving parts and high reliability [60]. AFCs were used as the basis for the first experiments with vehicular applications of fuel cells [61], starting with a farm tractor in the late 1950s equipped with an Allis Chalmers AFC. In 1959, a farm tractor, the first electric vehicle powered by an AFC, was invented. However, despite early success of AFC technology in space missions, submarines, and automotive vehicles, AFCs have fallen out of favor with researchers and were obscured by the quick growth of the PEMFC as the latter technology opts for vehicular applications. However, the limitation related to PEMFCs directed researchers' interest to the development of AFCs. AEM-integrated AFC is one of the most established and cheaply available fuel cell technologies that have more than 60% operational efficiencies. The growth of the AFC market is primarily driven by its advantages such as quick start even at extremely low temperatures, increased conductivity, better fuel efficiency, and cheaply available electrodes as compared to other fuel cells. Some of the major companies in the AFC market are Fuel Cell Energy, AFC Energy Plc Bloom Energy, Ballard Power Systems Inc., and Toshiba. To date, few companies are supplying AFC solutions including GenCell Energy from Israel and the UK-based company AFC Energy. Four years ago in 2018, GenCell Energy demonstrated a commercial AFC system of 4 kW in which cracked ammonia (~99.5%) was used as a hydrogen source as an immobile off-grid power supply. The project was named Project Alkammonia because cracked ammonia is a promising fuel option.

In 2020, the total market size of the fuel cell business amounted to USD 263 million on a global scale and is expected to reach up to USD 848 million by 2025 [62]. Asia Pacific is assessed to be one of the largest consumers from 2020 to the upcoming five years for use of fuel cell vehicles in the region. The growing demand for clean and green technology is endorsed by the growing demand for environmental friendly clean energy generation in developed regions increased by public transportation systems, booming power sector, fuel-cell-powered vehicles, and enlarged power generation capacities globally. Based on application, the market has been segmented into stationary, portable, and transport categories where the fuel cell is potentially used for fuel cell-power vehicles, utilities, defense, etc.

2.10 Summary and Concluding Remarks

The increasing interest in AFCs in past 50 years is associated with the development, improvement, and production of new AEMs. The progress of AEMs has added momentous attention over the last five years. AEM is one of the important and major key components of AFC systems. The long-term chemical stability, low cost, and constant performance will determine whether the AEM-integrated AFC technology can reach up to final commercialization and daily life application. Performance of AEM has increased remarkably in AFCs, though it has yet to be satisfied for ongoing demands. However, several challenges have to be pointed out before AEMs become a conventional product, including chemical stability, high performance, adequate mechanical strength, long-life durability, and low-cost production methods. The primary restraint of the AEMs in fuel cell operation is the life of membrane and the high ionic conductivity of the membrane under anion conditions. AEMs could play a key role in the search for substitute power sources like AFCs that can be competitive with other existing power sources. Multi-kW anion alkaline cells could be used in our daily lives to power our homes, electric vehicles, offices, apartments, small malls, and hospitals. Moreover, partial capacity loaded fuel cells could be used for powering cell phones, computers, and wireless connections in the initial phase. Still, commercial AEMs are considered to be their early stage of enlargement for fuel cell applications. Numerous studies demand situations that need to be triumphed over earlier than AEMs may be efficiently implemented in pre-commercial fuel cells. If the aforementioned limitations of currently existing systems can be overcome, AFCs will become a clean and green alternative to fossil fuels in upcoming years.

References

1 Wang, Z., Zhang, X., and Rezazadeh, A. (2021). Hydrogen fuel and electricity generation from a new hybrid energy system based on wind and solar energies and alkaline fuel cell. *Energy Rep* [Internet] 7: 2594–2604. https://doi.org/10.1016/j.egyr.2021.04.060.

2 Xu, T. (2005). Ion exchange membranes: state of their development and perspective. *J Membr Sci* 263 (1–2): 1–29.

3 Cheng, J., He, G., and Zhang, F. (2015). A mini-review on anion exchange membranes for fuel cell applications: stability issue and addressing strategies. *Int J Hydrog Energy* [Internet] 40 (23): 7348–7360. https://doi.org/10.1016/j.ijhydene.2015.04.040.

4 Zakaria, Z., Shaari, N., Kamarudin, S.K. et al. (2020). A review of progressive advanced polymer nanohybrid membrane in fuel cell application. *Int J Energy Res* 44 (11): 8255–8295.

5 Merle, G., Wessling, M., and Nijmeijer, K. (2011). Anion exchange membranes for alkaline fuel cells: a review. *J Membr Sci* 377 (1–2): 1–35.

6 Cheng, X., Shi, Z., Glass, N. et al. (2007). A review of PEM hydrogen fuel cell contamination: impacts, mechanisms, and mitigation. *J Power Sources* 165 (2): 739–756.

7 Sata, T. (2004. 314 p). *Ion Exchange Membranes: Preparation, Characterization, Modification and Application* [Internet]. Royal Society of Chemistry http://books.google.com/books?id=QIzRdn0lGFUC&pgis=1.

8 Yuan, Z., Li, X., Zhao, Y., and Zhang, H. (2015). Mechanism of polysulfone-based anion exchange membranes degradation in vanadium flow battery. *ACS Appl Mater Interfaces* 7 (34): 19446–19454.

9 Ren, J., Dong, Y., Dai, J. et al. (2017). A novel chloromethylated/quaternized poly(sulfone)/poly(vinylidene fluoride) anion exchange membrane with ultra-low vanadium permeability for all vanadium redox flow battery. *J Membr Sci* [Internet] 544 (August): 186–194. https://doi.org/10.1016/j.memsci.2017.09.015.

10 Mandal, M. (2021). Recent advancement on anion exchange membranes for fuel cell and water electrolysis. *ChemElectroChem* 8 (1): 36–45.

11 Miller, H.A., Bouzek, K., Hnat, J. et al. (2020). Green hydrogen from anion exchange membrane water electrolysis: a review of recent developments in critical materials and operating conditions. *Sustain Energy Fuels* 4 (5): 2114–2133.

12 Goel, P., Bhuvanesh, E., Mandal, P. et al. (2021). Di-quaternized graphene oxide based multi-cationic cross-linked monovalent selective anion exchange membrane for electrodialysis. *Sep Purif Technol* [Internet] 276 (June): 119361. https://doi.org/10.1016/j.seppur.2021.119361.

13 Guler, E., Zhang, Y., Saakes, M., and Nijmeijer, K. (2012). Tailor-made anion-exchange membranes for salinity gradient power generation using reverse electrodialysis. *ChemSusChem* 5 (11): 2262–2270.

14 Besha, A.T., Tsehaye, M.T., Aili, D. et al. (2020). Design of monovalent ion selective membranes for reducing the impacts of multivalent ions in reverse electrodialysis. *Membranes (Basel)* 10 (1): 7.

15 Zhang, C., Zhang, W., and Wang, Y. (2020). Diffusion dialysis for acid recovery from acidic waste solutions: anion exchange membranes and technology integration. *Membranes (Basel)* 10 (8): 1–23.

16 Güler Akgemci, E., Ersöz, M., and Atalay, T. (2005). Transport of formic acid through anion exchange membranes by diffusion dialysis and electro-electro dialysis. *Sep Sci Technol* [Internet] 39 (1): 165–184. http://www.tandfonline.com/doi/abs/10.1081/SS-120027407.

17 Shen, Y.K., Li, X.Q., Wang, N. et al. (2021). Introducing and investigation of a pumped hydro-compressed air storage based on wind turbine and alkaline fuel cell and electrolyzer. *Sustain Energy Technol Assess* 47 (June): https://doi.org/10.1016/j.seta.2021.101378.

18 Chen, N. and Lee, Y.M. (2021). Anion exchange polyelectrolytes for membranes and ionomers. *Prog Polym Sci* 113: https://doi.org/10.1016/j.progpolymsci.2020.101345.

19 Li, F.R., Jia, Y.X., Guo, R.Q., and Wang, M. (2021). Preparation of composite anion-exchange membrane with acid-blocking performance for brine reclamation by bipolar membrane electrodialysis. *Sep Purif Technol* [Internet] 254 (July 2020): 117587. https://doi.org/10.1016/j.seppur.2020.117587.

20 Jin, H., Yu, Y., and Chen, X. (2020). Membrane-based electrochemical precipitation for water softening. *J Membr Sci* [Internet] 597 (August 2019): 117639. https://doi.org/10.1016/j.memsci.2019.117639.

21 Lee, K.H., Chu, J.Y., Kim, A.R. et al. (2021). Functionalized TiO_2 mediated organic-inorganic composite membranes based on quaternized poly(arylene ether ketone) with enhanced ionic conductivity and alkaline stability for alkaline fuel cells. *J Membr Sci* [Internet] 634 (May): 119435. https://doi.org/10.1016/j.memsci.2021.119435.

22 Malik, R.S., Soni, U., Chauhan, S.S. et al. (2021). Semi-interpenetrating polymer networks of poly (vinyl alcohol)-functionalized nanocrystals/sulfonated poly (ether ether ketone) (PVA-FNCs/SPEEK) as fuel cell membrane. *Mater Today Commun* 29 (October): 102897.

23 Chikh, L., Delhorbe, V., and Fichet, O. (2011). (Semi-)Interpenetrating polymer networks as fuel cell membranes. *J Membr Sci* 368 (1–2): 1–17.

24 Wang, C., Xu, C., Shen, B. et al. (2016). Stable poly(arylene ether sulfone)s anion exchange membranes containing imidazolium cations on pendant phenyl rings. *Electrochim Acta* [Internet] 190: 1057–1065. https://www.sciencedirect.com/science/article/pii/S0013468615311099 (accessed 29 May 2019).

25 Lin, C.X., Zhuo, Y.Z., Hu, E.N. et al. (2017). Crosslinked side-chain-type anion exchange membranes with enhanced conductivity and dimensional stability. *J Membr Sci* [Internet] 539 (April): 24–33. https://doi.org/10.1016/j.memsci.2017.05.063.

26 Sharma, S., Dinda, M., Sharma, C.R., and Ghosh, P.K. (2014). A safer route for preparation of anion exchange membrane from inter-polymer film and performance evaluation in electrodialytic application. *J Membr Sci* [Internet] 459: 122–131. http://linkinghub.elsevier.com/retrieve/pii/S0376738814001173 (accessed 5 May 2018).

27 Sata, T. (2000). Studies on anion exchange membranes having permselectivity for specific anions in electrodialysis — effect of hydrophilicity of anion exchange membranes on permselectivity of anions. *J Membr Sci* [Internet] 167 (1): 1–31. https://www.sciencedirect.com/science/article/pii/S037673889900277X (accessed 5 June 2019).

28 You, W., Noonan, K.J.T., and Coates, G.W. (2020). Alkaline-stable anion exchange membranes: a review of synthetic approaches. *Prog Polym Sci* [Internet] 100: 101177. https://doi.org/10.1016/j.progpolymsci.2019.101177.

29 Laskin, S., Kuschner, M., Drew, R.T. et al. (1971). Tumors of the respiratory tract induced by inhalation of bis(chloromethyl)ether. *Arch Environ Health* 23 (2): 135–136.

30 Dai, J., He, G., Ruan, X. et al. (2016). Constructing a rigid crosslinked structure for enhanced conductivity of imidazolium functionalized polysulfone hydroxide exchange membrane. *Int J Hydrog Energy* [Internet] 41 (25): 10923–10934. https://www.sciencedirect.com/science/article/pii/S0360319916304657 (accessed 24 July 2019).

31 Goel, P., Mandal, P., Bhuvanesh, E. et al. (2021). Temperature resistant cross-linked brominated poly phenylene oxide-functionalized graphene oxide nanocomposite anion exchange membrane for desalination. *Sep Purif Technol* [Internet] 255 (September 2020): 117730. https://doi.org/10.1016/j.seppur.2020.117730.

32 Ahmed Mahmoud, A.M. and Miyatake, K. (2022). Highly conductive and alkaline stable partially fluorinated anion exchange membranes for alkaline fuel cells: effect of ammonium head groups. *J Membr Sci* [Internet] 643 (November 2021): 120072. https://doi.org/10.1016/j.memsci.2021.120072.

33 Liu, F.H., Lin, C.X., Hu, E.N. et al. (2018). Anion exchange membranes with well-developed conductive channels: effect of the functional groups. *J Membr Sci* 564 (May): 298–307.

34 Fang, S.C., Zhang, H., Wang, F. et al. (2012). Synthesis and properties of self-crosslinking anion exchange membranes based on quaternary poly(arylene ether sulfone)s. *Adv Mater Res* 608–609: 857–860.

35 Zhang, S., Zhu, X., Jin, C., and Hu, H. (2019). Pyridinium-functionalized crosslinked anion exchange membrane based on multication side chain tethered elastomeric triblock poly(styrene-*b*-(ethylene-*co*-butylene)-*b*-styrene). *React Funct Polym* [Internet] 138: 62–69. https://www.sciencedirect.com/science/article/pii/S1381514818310733 (accessed 6 June 2019).

36 Lin, B., Qiu, L., Lu, J., and Yan, F. (2010). Cross-linked alkaline ionic liquid-based polymer electrolytes for alkaline fuel cell applications. *Chem Mater* 22 (24): 6718–6725.

37 Wang, L., Liu, Y., and Wang, J. (2019). Crosslinked anion exchange membrane with improved membrane stability and conductivity for alkaline fuel cells. *J Appl Polym Sci* 136 (44): 1–10.

38 Zhou, R., Ren, Y., Gong, S. et al. (2022). Complementary side chain promotes microphase separation in the membranes for alkali fuel cells. *Polymer (Guildf)* 238 (November 2021): 124403.

39 Ma, L., Qaisrani, N.A., Hussain, M. et al. (2020). Cyclodextrin modified, multication cross-linked high performance anion exchange membranes for fuel cell application. *J Membr Sci* 607 (April): https://doi.org/10.1016/j.memsci.2020.118190.

40 Yang, C., Liu, L., Huang, Y. et al. (2019). Anion-conductive poly(2,6-dimethyl-1, 4-phenylene oxide) grafted with tailored polystyrene chains for alkaline fuel cells. *J Membr Sci* [Internet] 573 (December 2018): 247–256. https://doi.org/10.1016/j.memsci.2018.12.013.

41 Yang, Q., Lin, C.X., Liu, F.H. et al. (2018). Poly (2,6-dimethyl-1,4-phenylene oxide)/ionic liquid functionalized graphene oxide anion exchange membranes for fuel cells. *J Membr Sci* [Internet] 552 (November 2017): 367–376. https://doi.org/10.1016/j.memsci.2018.02.036.

42 Goel, P., Mandal, P., Bhuvanesh, E. et al. (2021). Sulfonated poly (ether ether ketone) composite cation exchange membrane for NaOH production by electro-electrodialysis using agro-based paper mill green liquor. *J Environ Chem Eng* 9 ([Internet], 6): 106409. https://doi.org/10.1016/j.jece.2021.106409.

43 Mandal, P., Bhuvanesh, E., Goel, P. et al. (2021). Caustic recovery from green liquor of agro-based paper mills using electrolysis. *Sep Purif Technol* [Internet] 262 (November 2020): 118347. https://doi.org/10.1016/j.seppur.2021.118347.

44 Mandal, P., Goel, P., Bhuvanesh, E. et al. (2021). Caustic production from industrial green liquor using alkali resistant composite cation exchange membrane. *J Environ Chem Eng* [Internet] 10 (1): 107016. https://doi.org/10.1016/j.jece.2021.107016.

45 Eswaraswamy, B., Mandal, P., Goel, P., and Chattopadhyay, S. (2021). Potential of dipicolinic acid as a water-dissociating catalyst in a bipolar membrane. *ACS Appl Polym Mater* 3: 6218.

46 Eswaraswamy, B., Goel, P., Mandal, P. et al. (2021). Nanocomposite interface coupled with thickness optimization promoting water dissociation in heterogeneous bipolar membrane. *Polym Adv Technol* 33 (August): 1–15.

47 Liu, M., Hu, X., Hu, B. et al. (2022). Soluble poly(aryl piperidinium) with extended aromatic segments as anion exchange membranes for alkaline fuel cells and water electrolysis. *J Membr Sci* 642 (October 2021): 119966.

48 Wang, Y., Qiao, X., Liu, M. et al. (2021). The effect of –NH– on quaternized polybenzimidazole anion exchange membranes for alkaline fuel cells. *J Membr Sci* [Internet] 626 (December 2020): 119178. https://doi.org/10.1016/j.memsci.2021.119178.

49 Biancolli, A.L.G., Barbosa, A.S., Kodama, Y. et al. (2021). Unveiling the influence of radiation-induced grafting methods on the properties of polyethylene-based anion-exchange membranes for alkaline fuel cells. *J Power Sources* 512 (September): https://doi.org/10.1016/j.jpowsour.2021.230484.

50 Yu, W., Zhang, J., Liang, X. et al. (2021). Anion exchange membranes with fast ion transport channels driven by cation-dipole interactions for alkaline fuel cells. *J Membr Sci* [Internet] 634 (May): 119404. https://doi.org/10.1016/j.memsci.2021.119404.

51 Zhou, X., Wu, L., Zhang, G. et al. (2021). Rational design of comb-shaped poly(arylene indole piperidinium) to enhance hydroxide ion transport for H_2/O_2 fuel cell. *J Membr Sci* 631 (January): 119335.

52 Zhang, D., Xu, S., Wan, R. et al. (2022). Functionalized graphene oxide cross-linked poly(2,6-dimethyl-1,4-phenylene oxide)-based anion exchange membranes with superior ionic conductivity. *J Power Sources* [Internet] 517 (November 2021): 230720. https://doi.org/10.1016/j.jpowsour.2021.230720.

53 Chen, N., Wang, H.H., Kim, S.P. et al. (2021). Poly(fluorenyl aryl piperidinium) membranes and ionomers for anion exchange membrane fuel cells. *Nat Commun* [Internet] 12 (1): https://doi.org/10.1038/s41467-021-22612-3.

54 Sun, L.X., Gou, W.W., Gao, X.L. et al. (2021). End-group crosslinked hexafluorobenzene contained anion exchange membranes. *Int J Hydrog Energy* [Internet] 46 (80): 39921–39931. https://doi.org/10.1016/j.ijhydene.2021.09.223.

55 Chen, N., Hu, C., Wang, H.H. et al. (2021). Chemically & physically stable crosslinked poly(aryl-*co*-aryl piperidinium)s for anion exchange membrane fuel cells. *J Membr Sci* [Internet] 638 (June): 119685. https://doi.org/10.1016/j.memsci.2021.119685.

56 Yang, W., Yan, J., Liu, S. et al. (2021). Macromolecular crosslink of imidazole functionalized poly(vinyl alcohol) and brominated poly(phenylene oxide) for anion exchange membrane with enhanced alkaline stability and ionic conductivity. *Int J*

Hydrog Energy [Internet] 46 (74): 37007–37016. https://doi.org/10.1016/j.ijhydene.2021.08.184.

57 Xu, F., Chen, Y., Lin, B. et al. (2021). Highly durable ether-free polyfluorene-based anion exchange membranes for fuel cell applications. *ACS Macro Lett* 10 (10): 1180–1185.

58 Zhu, Z.Y., Gou, W.W., Chen, J.H. et al. (2021). Crosslinked naphthalene-based triblock polymer anion exchange membranes for fuel cells. *J Membr Sci* [Internet] 636 (May): 119569. https://doi.org/10.1016/j.memsci.2021.119569.

59 Ferriday, T.B. and Middleton, P.H. (2021). Alkaline fuel cell technology - a review. *Int J Hydrog Energy* [Internet] 46 (35): 18489–18510. https://doi.org/10.1016/j.ijhydene.2021.02.203.

60 Gülzow, E. (1996). Alkaline fuel cells: a critical view. *J Power Sources* 61 (1–2): 99–104.

61 Kordesch, K., Gsellmann, J., Cifrain, M. et al. (1999). Intermittent use of a low-cost alkaline fuel cell-hybrid system for electric vehicles. *J Power Sources* 80 (1): 190–197.

62 Fuel cells market size, share, industry report to 2025. (2022) [Internet]. https://www.marketsandmarkets.com/Market-Reports/fuel-cell-market-348.html?gclid=Cj0KCQiA8vSOBhCkARIsAGdp6RSGTORX-yxbKpXJAnvC0NLSyWuVrLXg-_jCmIAQOBteNEwFlDOPEHMaAuafEALw_wcB (accessed 11 January 2022).

3

Fabrication Processes and Characterization Procedures of Anion Exchange Membranes

Graciela C. Abuin[1] *and Roxana E. Coppola*[2]

[1] *Instituto Nacional de Tecnología Industrial (INTI), Departamento de Almacenamiento y Conversión de la Energía, Av. General Paz 5445, San Martín, B1650KNA, Buenos Aires, Argentina*
[2] *Instituto Nacional de Tecnología Industrial (INTI), Departamento de Nanomateriales Funcionales, Micro y Nanotecnologías, Av. General Paz 5445, San Martín, B1650KNA, Buenos Aires, Argentina*

3.1 Introduction

Climate change increasingly exacerbates the impact of other promoters on nature and human well-being. The frequency and intensity of extreme weather events, and the fires, floods, and droughts that they can bring, have increased in the last 50 years, while the global average sea level has risen by between 16 cm and 21 cm since 1900, and at a rate of more than 3 mm per year over the past two decades. These changes have contributed to widespread impacts on many aspects of biodiversity, including species distribution, phenology, population dynamics, community structure, and ecosystem function. According to observational evidence, the effects are accelerating in marine, terrestrial, and freshwater ecosystems and are already impacting agriculture, aquaculture, fisheries, and nature's contributions to people [1].

A consensus has formed that an energy transition grounded in renewable sources and technologies that increases efficiency and conservation is the only way to give us an opportunity to fight and limit global warming to 1.5 °C by 2050. Only a few years ago, the renewables-centered approach was considered idealistic. Today, even some of the most conservative energy players have realized it as the only realistic option for a climate-safe world. We do not have time. The window is closing and the path to a net-zero carbon future is narrowing. The science is clear: 45% of global greenhouse gas emissions from 2010 levels must be reduced by 2030. Unfortunately, the recent trends show that the gap between where we are and where we should be is widening. We need to change the course now [2].

Alkaline Anion Exchange Membranes for Fuel Cells: From Tailored Materials to Novel Applications, First Edition. Edited by Jince Thomas, Alex Schechter, Flavio Grynszpan, Bejoy Francis, and Sabu Thomas.

Hydrogen energy devices like fuel cells (FCs) and electrolyzers, and e-fuels applications are key pieces in the movement toward a net-zero carbon future, playing a major role in the energy transition strategy, allowing energy storage for a long time without significant losses [3]. Furthermore, green hydrogen and e-fuels can decarbonize several industrial sectors and general uses, either directly as part of industrial processes or by replacing fossil fuels for heating demand or transportation [4].

The search of efficient and cost-effective materials and components for energy devices such as FCs and electrolyzers, particularly alkaline-membrane-based devices as anion-exchange membrane (AEM) FCs and water electrolyzers have gained interest, demonstrated by the continuous increase in the number of verified publications in the last decade.

Apart from meeting cost, durability, and performance targets, AEMs should meet other important requirements related to the fabrication process, as shown in Figure 3.1. It must be ensured that the desired structure is reached by the chosen synthetic route, selecting the appropriate ion-exchange groups and their positioning on the polymer matrix. The control of membrane morphology ensures good mechanical properties and scalability. For industrial fabrication, low-cost materials, mild reaction conditions, and one-pot processing methods should be employed as far as possible.

The main challenge for successful fabrication of AEMs is the combination of high OH^- ion conductivity and good mechanical properties [5]. Anion-exchange groups do not dissociate strongly as do SO_3H groups in proton-exchange membranes (PEMs) like Nafion®, and the inherent electrochemical mobility of hydroxide ions in water is much lower than that of protons. Thus, it is very difficult for the

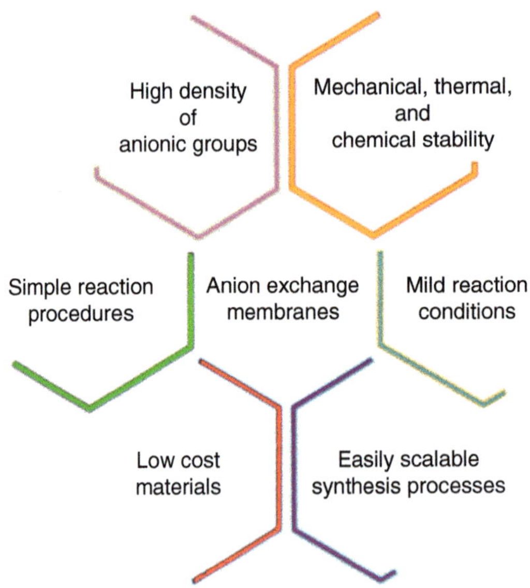

Figure 3.1 Main requirements for fabrication of good alkaline fuel cell membranes.

ionic (hydroxide) conductivity of AEMs to approach that of typical commercial PEMs. While the ionic conductivity of AEMs can be enhanced by increasing the number of cationic sites, i.e. by increasing the polymer's ion-exchange capacity (IEC), this approach is limited due to the concomitant deterioration of its mechanical properties with increasing fixed charge concentration; causing excessive swelling with loss of resistance in fully hydrated state, and brittleness in partially hydrated or dry membranes [5].

3.2 Fabrication Processes of Anion Exchange Membranes

3.2.1 AEM of Cationic Charged Polymers

In this kind of AEMs, the polymers have ionic sites bonded to the backbone of the polymer chain, with mobile counter anions. The cationic groups covalently bound to the polymer backbone can be obtained through different fabrication processes and starting materials [6, 7]:

- Polymerization of monomers resulting in *polymers with pendant anion-exchange groups* in their backbones or cationic polymers that can be obtained through a secondary reaction step.
- *Incorporation of cationic groups on a preformed film by grafting*, a monomer that has cationic groups or groups capable of subsequently reacting to give rise to anionic-exchange groups.
- *Chemical modification of a polymer or polymer blend to form cationic moieties*, followed by dissolution and casting.

Quaternary ammonium salts have been extensively used as anion-conducting groups (AEGs). In the first design and fabrication process of an AEM, patented by Tokuyama Soda Company, polychloropropene was crosslinked with divinylbenzene (DVD), functionalized with quaternary ammonium groups obtained from triethylamine [8].

Quaternary ammonium salts have been widely used as AEGs. In the first design and manufacturing process of an AEM, patented by Tokuyama Soda Company, polychloropropene was crosslinked with DVD, functionalized with quaternary ammonium groups obtained from triethylamine [8].

DVD can react to form a benzyl halide moiety, which in turn reacts with a tertiary amine to form a quaternary ammonium salt as anionic-exchange group. This route applies the Menshutkin reaction that converts a tertiary amine into a quaternary ammonium salt by reaction with an alkyl halide. Varcoe and Slade, pioneers in AEM for direct methanol fuel cells (DMFCs), obtained membranes by radiation grafting the vinylbenzylchloride (VBC) into poly(vinylidenefluoride) (PVDF) and poly(tetrafluoroethene-*co*-hexafluoro propylene) (FEP) and then the $-CH_2Cl$ moiety is quaternized with trimethylamine [9]. As they did not obtain sufficiently stable materials by this route, they employed electron-beam grafting of VBC onto poly(ethylene-*co*-tetrafluoroethylene) (ETFE), obtaining better results, with ionic conductivity $23\,mS\,cm^{-1}$ at 50 °C [10, 11].

A detailed synthesis procedure is found in Ref. 12. A known mass of FEP was immersed in excess VBC (or a 50% v/v solution of VBC in toluene) and purged with nitrogen. The grafting reaction was done at 40–70 °C during known periods of time (days). The grafted membranes were heated at 70 °C in toluene overnight, to remove any unbound poly(vinylbenzyl chloride) (PVBC) homopolymer, and toluene was then removed under reduced pressure. These films, denoted FEP-g-PVBC, were then immersed in a trimethylamine solution. The salt form of the membrane is reached by treatment with potassium hydroxide (KOH) aqueous solution to yield an OH^- ion conducting AEM.

Many known organic reactions can be applied to modify a polymer backbone, functionalizing it and/or joining it to other molecules in order to attain the desired properties in the resulting AEM. For example, Zhou et al. [13] synthesized a polyketone-based AEM with N-substituted pyrroles. These groups were directly introduced into polyketone backbones via Paal–Knorr reaction. Friedel–Crafts reaction can be used for adding an alkyl or acyl group to an aromatic ring by an electrophilic aromatic substitution [14, 15]. Many commercial AEMs are based on this scheme, which is advantageous in terms of low cost and facile synthesis [16–21].

Yang et al. [22] reported poly(phenylene oxide) (PPO) membranes with flexible spacers, using the easy and safe Suzuki–Miyaura coupling reaction. The membranes obtained showed good dimensional stability, with water uptake below 55 wt.% and conductivities above 60 mS cm^{-1} at 70 °C.

Biaryls groups can be introduced into AEM via the Ullmann coupling reaction. For example, Li et al. [23] reported an AEM with tetra-quaternary ammonium carbazole groups introduced into a poly(arylene ether ketone) with iodobenzene by Ullmann grafting reaction.

The Leuckart reaction, where aldehydes or ketones are reductively alkylated either with ammonium formate or with a mixture of dimethylformamide (DMF) and formic acid, was employed by Zhang et al. [24] for the modification of poly(aryl ether ketone)s to fabricate polyelectrolyte poly(aryl ether amine)s.

Copper-catalyzed azide-alkyne cycloaddition (CuAAC) is a simple pathway that allows for the covalent attachment of two molecular building blocks with different functional groups. The work by Yang et al. [25] reviews AEMs preparation methods using CuAAC.

A route based on the Vilsmeier reagent, a mixture of β-phosphoryliminium chloride and β-chloroiminium phosphate, was used by several authors such as Wang et al. [26] for the preparation of an AEM. Using this reagent, a guanidinium salt was incorporated into the poly(arylpiperidinium) (PAP) backbone through a simple and efficient synthetic route.

AEGs, such as 1,4-diazabicyclo[2.2.2]octane (DABCO), hexamethylenetetramine, *N*,*N*,*N′*,*N′*-tetramethyl-1,6-hexanediamine (TMHDA), and tetramethylethylenediamine (TMEDA), were proposed in order to obtain AEM with high ionic conductivity and chemically stability in high pH media (Figure 3.2) [27–31].

Some functional agents like TMHDA and DABCO have two tertiary amine groups and then can be used as crosslinkers to improve dimensional stability and mechanical properties [32–35]. In addition, other AEGs such as imidazolium and benzimidazolium

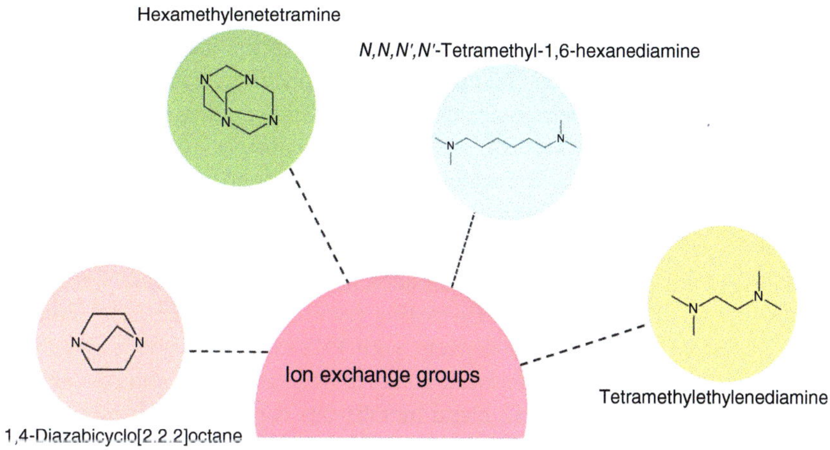

Figure 3.2 Ion-exchange groups usually employed in AEMs.

were reported. Benzimidazolium-based AEMs were prepared by methylation of poly[2-2′-(*m*-phenylene)-5-5′-bibenzimidazole] (PBI) [36–38]. In this membrane, the blocking of C2 position of PBI improved to some extent the alkali resistance by hindering the nucleophile attack by OH^- anions. Pyridine, phosphonium, tertiary sulfonium, and metal-cation-based AEGs were also reported.

3.2.2 AEMs of Ion-Solvating Polymers

This kind of AEMs is composed of a polymer matrix and an aqueous solution of a metal hydroxide, usually KOH, which combines the mechanical, chemical, and thermic properties of the polymer with the conductive properties of the alkaline solution. The polymer's electronegative heteroatoms such as nitrogen, oxygen, or sulfur interact with the cations in solution through a donor–acceptor mechanism, enabling ionic conduction through these heteroatom–cation interactions, and also through the mobility of the amorphous polymer chains [39].

Anion-solvating AEMs based on alkali-doped polybenzimidazoles are good options, due to their high chemical and mechanical properties, along with good conductivities. These linear heterocyclic polymers contain benzimidazole groups as part of the structurally repeating unit.

Alkali-doped PBI membranes were first proposed by Xing and Savadogo [40], followed by other authors for alkaline FCs [41,42].

Poly(2,5-benzimidazole) (ABPBI), the structurally simplest member of the polymer family, has a very simple synthesis route. It is prepared by condensation of the single 3,4-diaminobenzoic acid (DABA) monomer in polyphosphoric acid (PPA), according to the procedure reported by Gómez-Romero and coworkers [43].

In order to improve the mechanical properties, polybenzimidazole can be crosslinked. A scheme of the procedure to crosslink ABPBI with benzoxazine (BZ) [44], consisting in mixing ethanol solutions of BZ monomer and ethanol/NaOH solutions of ABPBI. After membrane formation, the BZ polymerization and crosslinking process is induced by thermal treatment at 180 °C for 10 h. The high temperature promotes the oxazine ring opening and further reaction between the methylene group and benzene ring of BZ monomer (oligomerization or polymerization) or ABPBI polymer (crosslinking). These membranes attained an anionic conductivity of 25 mS cm^{-1} at room temperature.

Blending allows a synergetic combination of the properties of base materials. Polyvinyl alcohol (PVA) is an attractive polymer owing to its hydrophilic character, good film-forming ability, and low cost, being as well biodegradable, biocompatible, and with a high solubility in water. Besides, it can react with different crosslinkers. ABPBI/PVA AEMs, prepared by casting from ethanol/NaOH and dimethyl sulfoxide (DMSO) solutions, respectively, with PVA crosslinked with glutaraldehyde (GA) [45, 46], allowed improved conductivity, mechanical properties, film formability, and reduced costs in blend membranes compared with pure ABPBI ones. Membranes attained anionic conductivity 90 mS cm^{-1} at 90 °C.

The characteristics of cationic charged and ion-solvating polymers are present in the ABPBI-crosslinked-PVBC membrane, quaternized with DABCO and prepared by casting [47]. The fabrication process begins mixing *N*-methyl-2-pyrrolidone (NMP) solutions of ABPBI and PVBC, along with $LiNO_3$ (1 wt.%), then placed in a mold and dried at 100 °C for 48 h. The membranes, denoted as ABPBI-*c*-PVBC, are quaternized by immersion in a DABCO ethanol solution. These films, denoted as ABPBI-*c*-PVBC/Cl, are converted to the OH^- form by immersion in a 1 M KOH aqueous solution, to obtain the ABPBI-*c*-PVBC/OH membranes.

For the PBI version (denoted as PBI-*c*-PVBC/OH) similar procedure was used to obtain PBI crosslinked with PVBC, quaternized with DABCO [48, 49], with a hydroxide conductivity greater than 25 mS cm^{-2} at room temperature and low swelling ratio. Hao et al. employed 1-butyl-4-aza-1-azaniabicyclo[2.2.2]octane bromide (BDABCO), synthesized using DABCO and 1-bromobutane, to prepare BDABCO substituted PBI/PVBC crosslinked membranes [50], that attained an anionic conductivity of 91 mS cm^{-2} at 80 °C and 340 mW cm^{-2} peak power density (PPD) in H_2/O_2 single FC.

3.2.3 AEMs with Nanofibers

Membranes with high porosity, good mechanical properties, and large size can be obtained by the electrospinning technique. By varying the size and the morphology of the fibers, the desired characteristics can be obtained with high reproducibility and easy upscaling.

In 2020, Mann-Lahav et al. [51] prepared ionomeric anion conducting nanofibers with different morphologies, varying the solvent used on its fabrication. The conductivity in the mats was related with the solid films cast, caused by variations in water absorption, as a consequence of morphological characteristics. The same group studied the influence of the relative humidity (RH) during the electrospinning process in these morphological characteristics [52]. They prepared fibers with the anion-conducting ionomeric commercial material (FAA-3 in its Br– form) purchased from

Fumatech BWT GmbH, a quaternary ammonium functionalized aromatic PPO-based polymer, with a branching effect in the electrospun fibers found to occur mostly at RH less than 30%.

Park et al. [53] reported diamine-crosslinked PPO polyelectrolyte nanofibers functionalized with either benzyl trimethylammonium or 1,2-dimethylimidazolium groups, embedded in a reinforcing matrix of polyphenylsulfone (PPSU), tested in H_2/O_2 FC.

The same group [54] reported a dual-fiber electrospinning procedure to prepare a composite AEM. They prepared chloromethylated polysulfone (CMPSf), converting a portion of chloromethyl groups to iodomethyl moieties via a modified version of the Finkelstein reaction, in order to increase the crosslinking reactivity of the CMPSF with an aliphatic diol (crosslinking agent). They electrospun simultaneously PPSU, as reinforcing polymer, and chloro/iodomethylated polysulfone into a dual-fiber mat, from separate spinnerets, with (1,6-hexanediol), as crosslinking agent added to the imidazolium functionalized polysulfone (IMPSF) spinning solution. The resultant membrane was quaternized in aqueous solution of trimethylamine or solution of either 1-methylimidazole or 1, 2-dimethylimidazole in methanol. Membranes exhibited anion conductivity of 49 mS cm^{-1} at room temperature.

Liao et al. [55] quaternized PVA (Q-PVA) with glycidyltrimethyl ammonium chloride (GTMAC), preparing composite membranes by immersing a mat of Q-PVA nanofibers in Q-PVA solution, doped with a 6 M KOH solution, that attained an anionic conductivity of 35 mS cm^{-1} at 60 °C in alkaline DMFC.

Yang et al. [56] compared PVA/chitosan (PVA/CS) blended and electrospun nanofiber composite membranes, prepared by soaking electrospun nanofiber PVA/CS crosslinked in GA vapor in PVA/CS solution, doped in 9 M KOH, for alkaline DMFC. The conductivity of the composite membranes was 19 mS cm^{-1} at 25 °C, with higher selectivity and lower methanol permeability compared with blended ones.

In 2021, we reported a facile preparation method of an AEM composed of PVA nanofibers crosslinked with GA, soaked in an ABPBI ethanol/NaOH solution, that showed good chemical stability and swelling resistance, with ionic conductivity of 41 mS cm^{-1} at 80 °C [57].

3.2.4 Hybrid AEMs

Composite organic inorganic AEMs, containing inorganic components taken up in a polymeric matrix, combine the advantages of both materials. Organic and inorganic components can be bonded covalently or through ionic interactions. Different inorganic components have been considered, like silica, metallic oxides, graphene oxide, or functionalized carbon nanotubes (CNTs), dispersed in a polymeric matrix through particle doping or *in situ* particle forming.

Fabrication processes include sol gel methods, blending, intercalation methods, *in situ* polymerization, and molecular auto assembling [27, 58].

Yang [59] reported the fabrication of hybrid PVA membranes with TiO_2, with good performance in DMFC. Huang et al. [60] reported a PVA-based membrane, with CNTs modified with PVA chains (m-CNTs) in order to improve the chemical compatibility. The m-CNTs improved KOH sorption and conductivity, while lowering swelling and permeability. The conductivity improvement was attributed to the influence of m-CNTs, hindering polymer chain alignment, thus promoting more amorphous regions.

Other authors studied the combination of PVA and CS with CNTs modified with double hydroxide (layered double hydroxides [LDHs]) hierarchical layers [61, 62], structured multi metallic cation and anion layers [63], positive charged crosslinking [64], or modification with alkyl spacers [65].

Zhang et al. [66], reported cross-linking of ammonium quaternized poly(2, 6-dimethyl-1, 4-phenylene oxide) (QPPO) with tertiary amine groups functionalized graphene oxide (graphene oxide nanosheets [NGOs]). NGO were obtained by mixing a DMF solution of GO with *N,N*-dimethyl-*p*-phenylenediamine (DMPD). Membrane was fabricated by casting of a NMP solution of synthetized bromomethylated PPO (BPPO) and NGO in DMF.

3.2.5 Recent Developments in AEMs

Many research groups have made efforts to design polymer architectures with high anionic conductivity and good alkaline resistance. As an example, the work of Gong et al. [67] about pendant imidazolium side chain functionalized polysulfone (PIMPSf) can be mentioned. Preparation of 1-(2-hydroxyethyl)-3-methylimidazolium chloride (HMIM-Cl) was performed by reaction of 1-methylimidazole and chloroethanol heated to 135 °C under microwave (200 W) for about 10 min with the protected nitrogen. After cooling to room temperature, the product was precipitated in acetone. PIMPSf was synthesized via Williamson etherification between CMPSf and HMIM-Cl. The membrane is prepared by casting from a solution of PIMPSf-Cl in NMP, with immersion in KOH aqueous solution for converting the material to the OH^- form. The resulting membrane had an anion conductivity of 72 mS cm^{-1} at 60 °C and good tensile strength in hydrated state. The presence of a long spacer chain also improved the alkaline stability.

Son et al. [68] reported a pore-filled membrane using PPO with long side chain and quaternary ammonium group through Friedel–Crafts acylation reaction, with polyethylene (PE) support. Acylated PPO (Ac-PPO), synthesized from PPO in 1, 2-dichloroethane, is dissolved in tetrahydrofuran (THF), with TMHDA, as crosslinking agent. The porous PE support is extended in a cylindrical centrifugal machine, and then the polymer solution is poured, penetrating into the pores of the PE support by centrifugal force. The solvent of polymer solution evaporates for 12 h with infrared radiant (IR) ramp simultaneously. As the solvent is volatilized, the polymer fills the pores to a uniform thickness. After being fully dried, the pore-filled membrane is soaked in trimethylamine solution and finally alkalinized by immersion in 1 M KOH. The ionic conductivity was 87 mS cm^{-1} at 80 °C, with good mechanical properties and 239 mW cm^{-2} PPD in H_2/O_2 single FC.

Kim et al. [69] reported the conversion of poly(methoxytriptycene ether sulfone) (OMe-PTES) to pyrazolium crosslinked poly(triptycene ether sulfone) (PXm-Tn), whose fabrication process consists in the demethylation of OMe-PTES to form poly(hydroxy-triptycene ether sulfone (OH-PTES), then crosslinked with 4-methyl-3,5-dichloro-1,2-diphenyl-pyrazolium tetrafluoroborate and 4-methyl-3-chloro-5-phenoxy-1,2-diphenyl-pyrazolium tetrafluoroborate. High conductivity along with mechanical and chemical stability is attained by means of a structure that promotes ionic highways formation. A cross-linked polymer network with cations as crosslinkers allows reaching conductivities as high as $0.11\,\mathrm{S\,cm^{-1}}$ at 80 °C with reduced hydration and IEC as low as $0.91\,\mathrm{mmol\,g^{-1}}$ and $730\,\mathrm{mW\,cm^{-2}}$ PPD in H_2/O_2 single FC with stable performance during 400 h tests.

Lin et al. [70] reported an AEM of poly(aryl piperidinium) (PAP) with clustered piperidinium cations (3QPAP-*x*, $x = 0.3$, 0.4, and 0.5), fabricated by mixing PAP, *N, N*-diisopropylethylamine (DIPEA), and 1-(6-bromohexyl)-1-methylpiperidinium bromide (BrPD); precipitated in ethyl ether from 1-methylpiperidine added to a solution of 1,6-dibromohexane (DBH) and acetonitrile (AR), stirred at 60 °C for 12 h; in DMSO and reacted at 80 °C for 12 h.

Yuan et al. [71] reported poly(isatin biphenylene) (PIB) membranes with [Cp-Fe-toluene] + organometallic cation fabricated by synthesis of PIB, preparation of PIB-alkyl alcohol in an alkaline environment provided by anhydrous potassium carbonate, and synthesis of PIB-alkyl-[Cp-Fe-toluene] by reacting PIB-bromoethanol (PIB-BE) with Br-[Cp-Fe-toluene] PF6.

Liu et al. [72] proposed soluble *p*-quaterphenyl-containing poly(aryl piperidinium) (PQP-100). PQPip-100 was prepared by superacid-catalyzed polymerization of *N*-methyl-4-piperidone and *p*-quaterphenyl, then quaternized with excess methyl iodide.

A facile and scalable method to make PE-reinforced poly(fluorenyl-*co*-terphenyl piperidinium (PFTP@PE) AEM was reported [73]. To penetrate through the hydrophobic PE substrate, ethanol was used as a co-solvent, since ethanol has low contact angle and interface tension on the PE substrate. EtOH-assisted PFTP@PE membranes achieved a tensile strength of 120 MPa, while membranes without ethanol assistance had a tensile strength of 80 MPa.

Following the scheme of Varcoe and Slade [9–11], who introduced by radiation grafting VBC onto PVDF obtaining materials not stable enough, Prakash et al. [74] worked with dehydrohalogenated PVDF matrix, henceforth termed as D-PVDF, by dissolving PVDF in *N,N*-dimethylacetamide (DMAc), with NaOH (0.1 M) in 2-methyl-1-propanol added to the solution. D-PVDF was then dissolved in DMAc, adding vinylbenzyltrimethylammonium chloride salt solution and initiator 2,20-azobis(2-methylpropionitrile) (AIBN) for chemical grafting. They obtained acceptable mechanical properties, IEC, and conductivity, but results are not shown in FC devices.

AEMs based on polynorbornene were explored because of their chemically and electrochemically stable backbone and facile functionalization by means of the Grubbs. Cao et al. [75] reported AEMs prepared via ring-opening metathesis polymerization (ROMP) by using Grubbs third-generation catalyst. The resultant AEMs exhibited high glass transition temperature (T_g) and tensile strength, and PPD $228\,\mathrm{mW\,cm^{-2}}$ without optimized membrane electrode assembly (MEA) fabrication process.

In an attempt to combine high conductivity and strength, interpenetrating polymer network (IPN) AEMs, with their mechanical support phase and ion conduction phase interlaced, were reported by Zeng et al. [76]. Both polymer networks are ionized to fabricate a kind of dual-cation IPN, and membrane preparation was done by casting. The resultant membrane showed excellent conductivity (160.5 mS cm^{-1} at 80 °C) and good mechanical strength (>20 MPa).

CMPSf-based AEM was crosslinked and quaternized by functionalizing DABCO with ether-containing flexible hydrophilic spacers [77]. The authors concluded that the long spacer influences the transport of hydroxide ions, the swelling and water uptake of the AEM.

Ma et al. [78] studied hydrophilic–hydrophobic phase separation in AEMs. They claim that in microphase-separated AEMs, the resistance against OH^- in the hydrophobic microphase restricts connection of ionic clusters to form a continuous pathway, and they proposed to lower packing density of the polymer chains by introducing a rigid bulky structure into AEMs. In this case, they worked with a cation-modified β-cyclodextrin. The resulting microphase separation and the enlarged free volume increase hydroxide conductivity.

Interestingly, an AEM can improve its performance-changing parameters of the FC. This was studied by Omasta et al. [79], who employed a MEA based in radiation-grafted ETFE-AEMs functionalized with benchmark benzyl trimethyl ammonium groups, and varying parameters like balancing membrane hydration while preventing electrode flooding reached PPDs higher than 1 W cm^{-2}.

The relationship between different kinds of reported AEMs, their fabrication process, properties, and performance is detailed in Table 3.1. The characteristics of main commercial AEMs are detailed in Table 3.2 [82–84].

3.3 Characterization Procedures of AEMs

The characterization of the AEM is important since it allows to know the relationship between the chemistry of the membrane, its structure, and transport properties, which determines its subsequent performance and stability.

The membrane and interface electrical properties, such as ionic conductivity, can be studied using electrochemical impedance spectroscopy (EIS). Ionic conductivity measurement provides a quick indication of the suitability of the membrane. However, it is important to complement this information by evaluating IEC, swelling ratio, water content, mechanical and thermal properties. On the other hand, the chemical stability under operating conditions is related to its performance and durability. Techniques such as ultraviolet–visible (UV–Vis) spectroscopy, attenuated total reflectance-Fourier transform infrared (ATR-FTIR), X-ray photoelectron spectroscopy (XPS), nuclear magnetic resonance (NMR), and elemental analysis allow to characterize AEM composition. Using microscopic techniques it is possible to evaluate its structures.

The different characterization techniques will allow a deeper understanding of the structure, conduction mechanism, morphology, and stability, which can help to design membranes with the desired characteristics. In Sections 3.3.1–3.3.6, some common characterizations techniques are described.

Table 3.1 Anion-exchange membranes reported in literature.

AEM name	Type of matrix used/ionic group name	Fabrication methods	IEC (meq g^{-1})	Cond. (mS cm^{-1})	Performance, PPD (mW cm^{-2})	Advantages	Disadvantages	References
PVDF-*g*-PVBTMAOH FEP-*g*-PVBTMAOH	PVDF/FEP-quaternary ammonium	Radiation grafting				Simple and robust synthetic route	Not stable materials	[9]
ETFE-*g*-PVBTMAOH	ETFE-quaternary ammonium	Radiation grafting	1.03	34/80 °C	130/H_2FC, O_2, 80 °C 8/DMFC,O_2, 80 °C	Simple synthetic route	Alkaline attack	[10, 11, 79]
QPSf	Polysulfone-quaternary ammonium	PSf chloromethylation, quaternization by TMA	0.8	50/70 °C	2/DMFC, air, 25 °C	Simple synthetic route	Limited conductivity	[14, 15]
QDPEEKOH	PEEK-DABCO	chloromethylated PEEK + DABCO crosslinking/quaternization	1.9	33/25 °C		Good conductivity, swelling, thermal stability	FC performance not informed	[28]
ABPBI:PVA/OH	Blend ABPBI, PVA crosslinked with GA-doped in KOH solution	Casting of solutions of ABPBI and PVA, crosslinking in GA solution		90/90 °C	76/DEFC, O_2, 90 °C	Simple synthetic route, low cost, good performance	Conduction owed to OH^- solvating in ABPBI	[45, 46]
ABPBI-*c*-PVBC/OH	ABPBI crosslinked with PVBC-DABCO-doped in KOH solution	Casting of solutions of ABPBI and PVBC, quaternization by DABCO	1.7	48/50 °C	70/DEFC, O_2, 90 °C	Good performance	Some conduction owed to OH^- solvating in ABPBI	[47, 48]

(Continued)

Table 3.1 (Continued)

AEM name	Type of matrix used/ionic group name	Fabrication methods	IEC ($meq\,g^{-1}$)	Cond. ($mS\,cm^{-1}$)	Performance, PPD ($mW\,cm^{-2}$)	Advantages	Disadvantages	References
PBI-*c*-PVBC/OH	PBI crosslinked with PVBC-BDABCO-doped in KOH solution	Casting of solutions of PBI and PVBC, quaternization by BDABCO	2.87	91/80 °C	340/H_2FC, O_2	Good performance	Some conduction owed to OH^- solvating in PBI	[50]
Crosslinked PPSU-Based NFs	Polyphenylsulfone (PPSU) NFs-benzyl trimethylammonium	Br-PPO and PPSU electrospun, imm. diamine crosslinker sol., quaternized	2.0	66/26 °C	320/H_2FC, O_2	Reproducibility, easy up-scalability	Special equipment	[53]
Q-PVA composite	Q-PVA NFs mat with casted Q-PVA-GTMAC	Quaternization of PVA with GTMAC, Q-PVA sol. poured onto Q-PVA mat	0.52	35/60 °C	54/DMFC, O_2, 60 °C	Reproducibility, easy up-scalability, low-cost polymer	Special equipment	[45]
NF CE-PVA/CS alkali doped	PVA/chitosan (PVA/CS) composite electrospun NFs	Soaking electrospun NFs PVA/CS crosslinked in GA in PVA/CS solution		19/25 °C	14/DMFC, O_2, 50 °C	Reproducibility, easy up-scalability, low-cost polymer	Low performance	[56]
PVA/TiO_2 composite	PVA crosslinked with GA and TiO_2-doped in KOH solution	Casting of an aqueous solution of PVA and TiO_2, crosslinking in GA solution		31/70 °C	7/DMFC, air, 60 °C	Simple synthetic route, low cost	Conduction owed to doping in KOH solution	[59]
PVA/m-CNT composite	PVA and CNTs modified with PVA chains doped in KOH solution	Casting, CNTs functionalized with PVA chains by an ozone-mediated method		160/60 °C	65/DEFC, O_2, 60 °C	Good performance, chemic compatibility	Conduction owed to doping in KOH solution	[60]

AEM name	Type of matrix used/ionic group name	Fabrication methods	IEC ($meq\,g^{-1}$)	Cond. ($mS\,cm^{-1}$)	Performance, PPD ($mW\,cm^{-2}$)	Advantages	Disadvantages	References
QPPO-NGO2	QPPO-crosslinked-NGO2-quaternary ammonium	Casting of a NMP solution of synthetized BPPO and NGO in DMF	2.14 ± 0.07	123/80 °C	398/H_2FC, O_2, 60 °C	Good performance, swelling	Need of degradation studies	[66]
PE/Ac-PPO pore-filled membrane	PE support, Ac-PPO crosslinked by TMHDA-quaternary ammonium	Porous PE support in a centrifugal machine, polymer solution poured	2.05 ± 0.06	87/80 °C	239/H_2FC, O_2	Good conductivity and mechanical properties	No simple fabrication, mild performance	[68]
PX*m*-T*n*	OH-PTES-Pirazolium groups	*In situ* functionalization	0.91	111/80 °C	730/H_2FC, O_2	Very high performance, conductivity, stability	No simple fabrication	[69]
3QPAP-0.5	poly(aryl piperidinium)-piperidinium groups	PAP functionalization by DIPEA, precipitation, react in DMSO/DBH/AR	2.26	117/80 °C	291/H_2FC, O_2	Fabricated by simple steps, conductivity, stability	Mild performance	[70]
PIB/Cp-Fe-toluene/ organometallic cation	poly (isatin biphenylene)-organometallic cation	Reacting PIB-bromoethanol (PIB-BE) with Br-[Cp-Fe-toluene] PF6	1.77	63/80 °C	64/H_2FC, air, 60 °C	1000 h alkali stability	Mild performance	[71]
PQP-100		Polymerization of *N*-methyl-4-piperidone and *p*-quaterphenyl, quaternized in excess methyl iodide	2.30	119/80 °C	496/H_2FC, O_2, 60 °C	Good performance	Chemical degradation of piperidinium groups	[72]

(Continued)

Table 3.1 (Continued)

AEM name	Type of matrix used/ionic group name	Fabrication methods	IEC (meq g^{-1})	Cond. (mS cm^{-1})	Performance, PPD (mW cm^{-2})	Advantages	Disadvantages	References
PFTP@PE	Poly (fluorenyl-*co*-terphenyl piperidinium	PE reinforced, EtOH helps solution to penetrate in the PE substrates	2.36	32/30 °C	1700/H_2FC, O_2, 80 °C	Facile and scalable	Non-conductive substrate inside	[73]
PBH-x-QA	Polynorbornene with arylene substituent-QA	Ring-opening metathesis polymerization (ROMP)	1.26–2.08	102/80 °C	228/H_2FC, O_2, 60 °C	Stable backbone and facile functionalization	Precise control of the crosslinking degree usually is difficult	[75]
cPVBMP–cQPPO	Interpenetrating polymer network VBMP and BPPO	VBMP, DVB, BPO, addition to BPPO sol., TMHDA added. Casting.	1.58–2.23	160/80 °C	576/H_2FC, air (CO_2 free), 80 °C	Good mechanical properties and cond.	Synthesis with many steps	[76]
PAES/DABCO	Multi cation cross linked PAES	CMPSf crosslinked and quaternized by chains obtained by reaction of DABCO with 1,6-dibromohexane	2.26	79/80 °C	76/H_2FC, O_2, 80 °C	Combination of crosslinking with multi-cation sites	Mild performance	[77]
PBN-CD/bpIL	Hydrophilic–hydrophobic bulky units modified AEM	PBN-CD reacts with excess 1-bromoadamantane in NMP, and quaternized with bpIL in NMP	1.81	122/80 °C	603/H_2FC, O_2, 60 °C	Good performance and durability	Synthesis with many steps	[78]

Table 3.2 Commercial anion-exchange membranes.

AEM name	Manufacturer	IEC (meq g^{-1})	Conductivity (mS cm^{-1})	Performance, PPD (mW cm^{-2})	Advantages	Disadvantages	References
Tokuyama A201	Tokuyama Co.	1.7	42	148/H_2FC, air, 50 °C	High conductivity, low-cost fabrication, thin film	Requires high concentration of alkaline condition, incorporation of carbonate inside it	[8, 60, 79, 80]
Fumasep® FAA-3	Fumatech	1.7–2.1	40–45	223/H_2FC, O_2, 60 °C	Good electrochemical properties, low resistance, compact structure	Low mechanical stability	[68, 79, 81]
Sustainion® 37–50	Dioxide materials		70–80		Good performance, 2000 h operation		[82, 83]
Ralex® AMH-PES	MEGA	1.8–2.0	>8.3		High resistance, long life cycle	Low conductivity	Technical data sheet
Aemion™	Ionomr	2.1–2.5	80			Mild performance in operation	[83]
Orion™	Orion polymer	2.19	60		Degradation <0.5% in conductivity and IEC loss after 1000 h operation at 80 °C in 1 M KOH		[83], technical data sheet

3.3.1 Ionic Conductivity

Transport phenomena in AEMs can be studied using the EIS technique. From each conductive and capacitive property in the equivalent circuit it is possible to quantitatively analyze the different structures. This non-invasive and non-destructive technique allows to obtain useful information on the interfacial phenomena of these systems, such as the properties of multilayer systems, the rate of electrochemical reactions, dielectric and transport properties of materials, and electrochemical reaction mechanisms [85, 86].

When an AEM is in contact with an aqueous electrolyte solution, the main contributions to its resistance come from the sum of three sublayers resistances: the diffusion boundary layer (DBL), the electrical double layer (EDL), and the AEM [87]. Besides the membrane physical properties, resistance is also affected by the AEM operating conditions, such as flow rate, electrolyte concentration, pH, and temperature [88].

Typical EIS experiments are performed in a two- or four-electrode cell, applying an alternating sinusoidal current or voltage and recording the response of the system for its analysis. Two-electrode cell allows in-plane or through-plane measurements, and only in-plane measurements are performed in a four-electrode cell. In the second configuration, the current or voltage is imposed to the external electrode pair, while the response is measured between the internal one, avoiding errors due to polarization phenomena and contact resistances [89]. Complex nonlinear regression analysis is used to interpret the electrochemical spectra [90]. From the high-frequency intercept of the impedance with the real axis, the value of the ionic resistance of the membrane (R_m) is obtained. This parameter is related to its ionic conductivity (σ) according to the following equation [91]:

$$\sigma = \frac{L}{R_m * A} \tag{3.1}$$

where L is the thickness of the membrane and A represents the active surface area of the sample.

To determine the hydroxide conductivity it is important to consider the effect of CO_2 presence on the measurement, due to the rapid formation of carbonates and bicarbonates when exposed to air. This causes a reduction in the diffusion coefficient and the conductivity of the anion [92–94]. Ziv and Dekel [95] found that the hydroxide conductivity value for a commercial membrane measured under bicarbonate-free conditions was twice the value obtained after the common practice of aqueous soaking in KOH followed by washing with water in a CO_2-free chamber (103 and 50 $mS\,cm^{-1}$, respectively). Likewise, Grew et al. [96] studied how the ionic conductivity of AEM is modified as a function of temperature and CO_2 concentration. However, when using AEMs for FCs, the atmospheric CO_2 effect can be reduced by operating at current densities above 1 $A\,cm^{-2}$, as reported by Siroma et al. [97].

3.3.2 IEC, Swelling Ratio, and Water Content

The IEC is a parameter that relates the number of moles of exchangeable ions per unit of dry membrane (m_{dry}), usually expressed in milliequivalent per gram or millimole per gram [91, 98].

Experimentally, the IEC can be determined by different methods such as spectroscopic analysis, ion-selective electrode potentiometry, elemental analysis, and titrimetry. When comparing different IEC values, it is necessary to consider the method by which it was obtained as well as the units, to obtain a correct interpretation. Karas et al. [99] compared the suitability of potentiometry with an ion-selective electrode (ISE) as a sensitive method, titration and spectrophotometry for the determination of IEC in both homogeneous and heterogeneous membranes. Of the three methods tested, spectrophotometry was the most accurate, while ISE showed the highest systematic error. However, the latter has some advantages, such as the possible evaluation of ion-exchange kinetics and membrane transport properties.

Currently the most used method is titration. There are different procedures to determine the IEC by this method. One of them is the OH^- back titration method [50]: AEM in the OH^- form is immersed in an aqueous solution of HCl so that the OH^- inside the membrane is transferred into the solution. Then, without removing the membrane, the resulting solution is titrated with a NaOH solution. The value of IEC is given by:

$$\mathrm{IEC} = \frac{\mathrm{moles}_{\mathrm{acid}} - \mathrm{moles}_{\mathrm{base}}}{m_{\mathrm{dry}}} \tag{3.2}$$

Direct titration of the exchanged OH^- is also possible by immersing the membrane in a NaCl solution. IEC can also be determined from a membrane in the Cl^- form, which is the way that AEM are usually synthetized. To do this, the membrane in the Cl^- form is immersed in a KOH or $NaNO_3$ aqueous solution, allowing the chloride ions being transferred into the solution, where they are titrated by argentometry [100].

It's important to note that IEC measurement works under the assumption that a complete exchange of ions takes place during the titration and that the ions are homogeneously distributed over the membrane. If these assumptions are not met, the uncertainty of the measurement will be affected [98].

Another type of characterization of the AEM is the determination of its water swelling ratio, measured from the dimensional difference between the membrane before and after immersing in water. Membrane swelling will change its pore size, affecting the diffusion coefficients and the concentration of electrolyte that the AEM may absorb. This means that variation in the expansion of a membrane influences the membrane transport, electrical conductivity, diffusion permeability, and the ion transport number [100]. The type of polymer matrix and ion-exchange groups as well as their counter ions, charge density, and crosslinking degree are parameters that influence the membrane's swelling ratio [91].

When exposed to water, membrane mass changes due to water sorption. Water uptake is therefore defined as the change in membrane mass after this exposure. It is usually determined gravimetrically by difference of the mass between wet and dry membranes [92].

IEC, swelling, and water uptake are deeply related parameters. Despite the water content enhancing mobility of ions, which has a positive effect in conductivity, a moderated value is desirable, since an excessive uptake will produce a greater swelling, affecting the mechanical stability and permeselectivity of the membrane. In addition, its ionic conductivity will also be affected as the concentration of fixed ions in the membrane is reduced [101,102]. A similar effect is seen with IEC. Although higher IEC values indicate a greater number of charged groups, this could lead to increased membrane swelling and thus a decrease in charge density, resulting in a reduction in ionic conductivity [91].

3.3.3 Mechanical and Thermal Properties

Assessment of mechanical properties is one of the common characterization procedures for an AEM. Since the response of a polymeric material can be very different depending on the type of stress applied, it is convenient to study its behavior according to its application. In the case of AEMs used in a FC, it is required that it can support the preparation of the MEA and its compression in the assembly of the stack. Taking this into account, Young's modulus is usually determined, which indicates the resistance of the material to be stretched or compressed [91, 103, 104].

Usually, Young's modulus, tensile strength, stress–strain curves, and elongation at break are obtained by stretching membrane samples, using a standard testing instrument for the tensility of plastics, controlling the temperature, and humidity of the chamber. Likewise, the nano-indentation technique using a nano-indenter equipment allows to determine the nano-hardness and Young's modulus of the membranes at the interface [105]. Franceschini and Corti [104] used the atomic force microscopy [AFM] nano-indentation technique to determine Young's modulus in Nafion membranes and different polybenzimidazoles.

The thermal properties of a membrane can be determined from a set of techniques in which a physical property is measured as a function of temperature (or time), while the membrane is subjected to a controlled temperature program in a given atmosphere, usually pure N_2. The most used to characterize AEMs are differential scanning calorimetry (DSC) and thermogravimetric analysis (TGA).

DSC is a thermo-analytical technique, in which the heat required to increase the temperature of the sample and a reference are measured as a function of temperature and at controlled heating rate. This technique allows to evaluate mainly the characteristic transition temperatures (such as glass transition temperature, T_g), cross-linking degree, the effects of thermal cycling, and the crystallinity of the sample. In addition, the stability of a membrane can be studied by TGA. In a typical TGA experiment, the mass of the sample is recorded continuously as a function of temperature in a controlled atmosphere, so changes in the membrane such as water loss, volatilization, and other chemical reactions in which weight changes occur are detected [103].

Despite DSC and TGA characterizations giving useful information about the membrane thermal stability, its analysis should be considered with care, since the conditions in which this experiments are performed do not exactly replicate the typical conditions of a FC.

3.3.4 Chemical Stability

Chemical stability of AEMs is a critical parameter and must be thoroughly evaluated taking into account their working conditions.

Frequently, the membrane's molecular structure contains cationic functional groups with low stability in strongly alkaline media. In presence of high OH^- concentration, mainly two types of degradation mechanisms take place. If there are no hydrogen atoms in the β position, nucleophilic substitution reactions occur, forming the corresponding tertiary alcohols and amines. On the other hand, the presence of β-hydrogens enables the Hofmann elimination reaction that produces vinyl groups (Figure 3.3). In both cases, the performance of the membrane will be affected as the number of cationic groups capable of exchanging anions is reduced and its mechanical properties will also change [106].

In order to study the effect of the alkaline medium, the properties of the membrane are regularly evaluated before and after immersing it in KOH aqueous solutions of different concentrations, at a certain temperature, for a given time. Changes in any of its properties such as morphology, IEC, conductivity, permeability, chemical structure, or mechanical stability will indicate that the membrane is not stable under the test conditions.

However, this test does not simulate a FC environment, which can overestimate the membrane stability. During the operation of the cell, a water gradient is generated, because the water is consumed at the cathode and produced at the anode. Thus, the solvation of hydroxyl ions is an important factor to consider [107]. Dekel et al. [108] studied this effect both experimentally and by density functional theory (DFT) simulations. When the hydration level is low, the OH^- are less solvated, increasing their nucleophilic character. Hence, a faster degradation will occur on the cathode, whereas the water content is lower.

An accelerated stress test frequently used is the Fenton test, which allows studying the oxidative stability of AEMs. This test consists of immersing the membranes in a solution of H_2O_2 3 wt.% and $FeSO_4$. The generated hydroxide radicals attack the polymer structure, making it possible to study polymer degradation under more unfavorable conditions than those found in a FC [109, 110].

3.3.5 Chemical Composition and Morphological Characterization

To determine the chemical composition of an AEM, different classical analytical methods are often used such as spectroscopies (UV–Vis, FTIR, XPS, Auger electron spectroscopy, Raman, energy dispersive X-ray spectroscopy, NMR) and elemental analysis, among others.

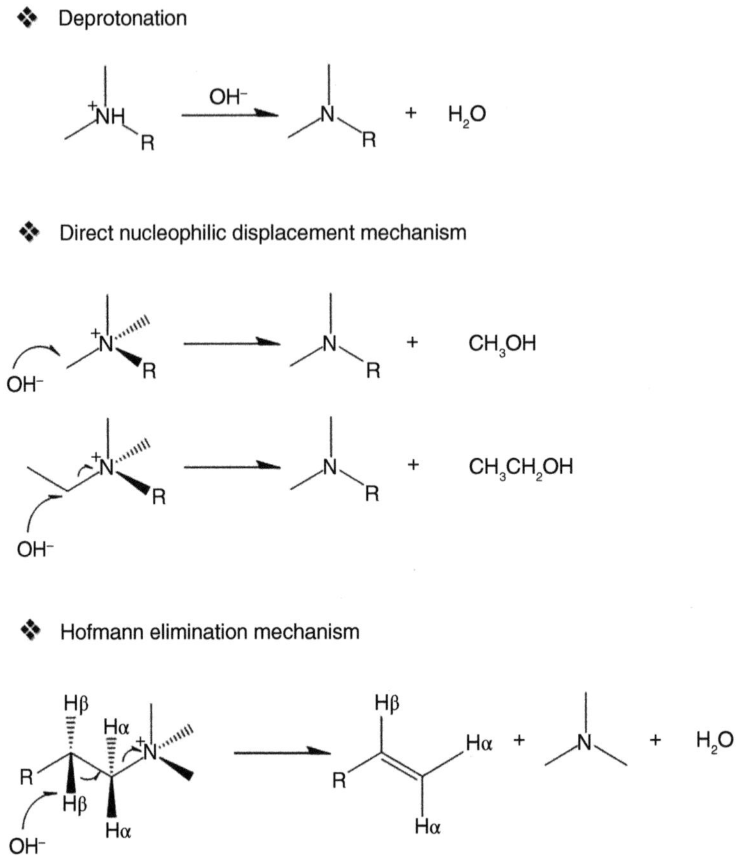

Figure 3.3 Degradation reactions of the quaternary ammonium group in the presence of a strong base.

Membrane structures can be examined by means of different microscopies such as scanning electron microscopy (SEM), field emission microscopy (FEM), transmission electron microscopy (TEM), and AFM.

In general, to study the homogeneity degree of AEMs, SEM images are used in the frontal and cross-sectional view of the membranes. This technique can achieve a resolution of up to 5 nm. However, as these materials are not electronic conductors, generally the films are sputtered with a thin conductor layer, such as Au or Ag, and are measured in high vacuum. On the other hand, low-vacuum SEM can also be used, without the need for sputtering the sample. When higher resolution is required, TEM and FEM techniques are used, which can reach resolutions of 0.4–0.5 nm and 0.6–0.7 nm, respectively [111].

To study the morphology and surface nanostructure of membranes, the AFM technique can also be used. In addition to topographic images, image phases are also available, which are sensitive to the material's surface properties, such as stiffness, viscoelasticity, and chemical composition [112–114].

3.3.6 Other Characterizations

In addition to the properties listed above, a membrane should ideally have a low fuel crossover. In the case of direct alcohol FC, alcohol crossover through the membrane will affect its efficiency due to the oxidation reaction of methanol at the cathode. However, in alkaline direct alcohol FCs, the alcohol crossover is reduced, as the OH^- ions within the membrane are transported in the opposite direction to the flow of alcohol [115].

Usually, alcohol permeability of the membranes is measured by using a two-compartment diffusion cell. One of them is charged with deionized water (compartment A), while the other compartment (B) has variable concentrations of alcohol or alcohol:KOH mixtures. To minimize the effects of concentration polarization, both compartments are kept under agitation during the test. Subsequently, the concentration is determined using gas chromatography [116] or a methanol concentration sensor [117]. The diffusion permeability of the membrane (P) is given by:

$$P = \frac{1}{C_a}\left(\frac{\Delta C_b(t)}{\Delta t}\right)\left(\frac{LV_b}{A}\right) \tag{3.3}$$

where C_a is the alcohol concentration in compartment A, $\Delta C_b(t)/\Delta t$ is the variation in concentration as a function of time of compartment B, V_b the volume of each compartment, A is the membrane area, and L is the membrane thickness [118].

Park et al. [119] measured the permeability of different membranes to hydrogen and oxygen using the constant volume or time delay method. However, since crossover gas affects open circuit cell power, most authors study cell performance directly.

The hydrophilicity/hydrophobicity of the surface of an AEM can be determined from the measurement of the contact angle with water, using sessile-drop technique. This test makes it possible to estimate the ability of liquids to flow or form drops on a solid surface depending on the balance between adhesive and cohesive forces. Adhesive forces cause the liquid to spread out on the solid surface, while cohesive forces cause the drop to contract [102, 120, 121].

Table 3.3 shows a summary of the most commonly measured parameters in AEMs.

3.4 Conclusions

Current developments in AEM are aimed at improving their chemical and mechanical properties, without compromising its ionic transport properties. The need to develop a scalable material, capable of operating in high pH environments without changes on its cell performance for long working periods, remains a challenge. In this chapter we reviewed several strategies currently employed for the AEM preparation.

The use of a polymer backbone with covalently bonded charged groups, mostly quaternary ammonium salts, has been extensively reported. The high density of

Table 3.3 Summary of the main characterization procedures of AEMs.

Parameter	Common technique
Ionic conductivity	EIS
IEC	Spectroscopic analysis, ion-selective electrode potentiometry, elemental analysis, titrimetry
Swelling ratio	Dimensional difference between the membrane before and after immersing in water
Water uptake	Gravimetrically: difference of the mass between wet and dry membrane
Mechanical properties	Tensile strength measurement, nano-indentation and AFM nano-indentation
Thermal properties	DSC, TGA
Alkali stability	Assay the membrane before and after Immersion in KOH solution
Oxidative stability	Fenton test
Chemical composition	UV–Vis spectroscopy, FTIR, XPS, Auger electron spectroscopy, Raman, energy dispersive X-ray spectroscopy, RMN, elemental analysis
Morphology	SEM, TEM, FEM, AFM

charged groups greatly improves IEC and conductivity. Nonetheless, a rise in the AEM swelling ratio and its water uptake could be observed. Moreover, the quaternary ammonium groups' stability needs to be considered.

Ion solvating polymers rely on the employment of electronegative heteroatoms capable of interacting with ions in the solution through an acid–base mechanism. Polybenzimidazole-based AEMs remain as the main alternative, mostly due to their good mechanical and chemical properties.

The morphology of the polymer plays a great role in several of its properties. In this sense, nanofiber-based AEMs show excellent mechanical properties. Electrospinning technique, which allows to varying the length and size of the fiber, and therefore modify the AEM properties, offers a scalable alternative for its production. Subsequent chemical modification of the polymer nanofibers allows to incorporate functional groups.

AEMs made of inorganic compounds in a polymeric matrix are known as composite AEMs. The use of CNTs and grapheme oxide as inorganic compound of composite AEMs was developed in this chapter.

We have taken into account the properties that have to be evaluated in an AEM and reviewed the main characterization techniques employed. A deep characterization of an AEM allows us to compare it to other previously reported AEM, helping us to understand its cell performance.

References

1 IPBES (2019). *Summary for Policymakers of the Global Assessment Report on Biodiversity and Ecosystem Services of the Intergovernmental*. Science-Policy Platform on Biodiversity and Ecosystem Services.

2 IRENA (2021). *World Energy Transitions Outlook: 1.5°C Pathway*. Abu Dhabi: International Renewable Energy, Agency.

3 Van Gerwen, R., Eijgelaar, M., and Bosma, T. (2020). The promise of seasonal storage. DNV GL Netherlands B.V, Position Paper.

4 Van Gerwen, R., Eijgelaar, M., and Bosma, T. (2019). Hydrogen in the electricity value chain. DNV GL Netherlands B.V, Position Paper.

5 Arges, C.B.G., Ramani, V., and Pintauro, P.N. (2010). The chalkboard: anion exchange membrane fuel cells. The electrochemical society interface, summer 2010. *Electrochem Soc Interface* 19: 31–35. https://doi.org/10.1149/2.f03102if.

6 Xu, T. (2005). Ion exchange membranes: state of their development and perspective. *J Membr Sci* 263: 1–29.

7 Strathmann, H., Richard, D.N., and Stern, S.A. (1999). Electrodialysis and related processes. In: *Membrane Science and Technology*, Chapter 6 (ed. D.N. Richard and S.A. Stern), 213–281. Elsevier.

8 Yamamoto, M. and Toi, K. (1986). Electrodialytic method. JP Patent 61192311 (assigned to Tokuyama Soda, K.K.).

9 Varcoe, J.R. and Slade, R.C.T. (2005). Prospects for alkaline anion-exchange membranes in low temperature fuel cells. *Fuel Cells* 5: 189–200.

10 Varcoe, J.R. and Slade, R.C.T. (2006). An electron-beam-grafted ETFE alkaline anion-exchange membrane in metal-cation-free solid-state alkaline fuel cells. *Electrochem Commun* 8: 839–843.

11 Varcoe, J.R., Slade, R.C.T., Yee, E.L.H. et al. (2007). Poly(ethylene-*co*-tetrafluoroethylene)-derived radiation-grafted anion-exchange membrane with properties specifically tailored for application in metal-cation-free alkaline polymer electrolyte fuel cells. *Chem Mater* 19: 2686–2693.

12 Herman, H., Slade, R.C.T., and Varcoe, J.R. (2003). The radiation-grafting of vinylbenzyl chloride onto poly(hexafluoropropylene-*co*-tetrafluoroethylene) films with subsequent conversion to alkaline anion-exchange membranes: optimisation of the experimental conditions and characterisation. *J Membr Sci* 218 (1–2): 147–163.

13 Zhou, Y.-C., Bao, R.-Y., Liu, Z. et al. (2021). Electrospun modified polyketone-based anion exchange membranes with high ionic conductivity and robust mechanical properties. *ACS Appl Energy Mater* 4: 5187–5200.

14 Abuin, G.C., Nonjola, P., Franceschini, E.A. et al. (2010). Characterization of an anionic-exchange membrane for direct methanol alkaline fuel cells. *Int J Hydrog Energy* 35 (11): 5849–5854.

15 Abuin, G.C., Franceschini, E.A., Nonjola, P. et al. (2015). A high selectivity quaternized polysulfone membrane for alkaline direct methanol fuel cells. *J Power Sources* 279: 450–459.

16 Couture, G., Alaaeddine, A., Boschet, F., and Ameduri, B. (2011). Polymeric materials as anion-exchange membranes for alkaline fuel cells. *Prog Polym Sci* 36: 1521–1557.

17 Hickner, M.A., Herring, A.M., and Coughlin, E.B. (2013). Anion exchange membranes: current status and moving forward. *J Polym Sci B Polym Phys* 51: 1727–1735.

18 Luo, Y., Guo, J., Wang, C., and Chu, D. (1992). Quaternized poly(methyl methacrylate-*co*-butylacrylate-*co*-vinylbenzylchloride) membrane for alkaline fuel cells. *J Power Sources* 195: 3763–3771.

19 Ko, B., Sohn, J., and Shin, J. (2012). Radiation-induced synthesis of solid alkaline exchange membranes with quaternized 1,4-diazabicyclo[2,2,2]octane pendant groups for fuel cell applications. *Polymer* 53: 4652–4661.

20 Kubota, H., Yano, K., Sawada, S. et al. (1996). Novel anion exchange resins with thermal stability: synthesis and characteristics. *Spec Publ R Soc Chem* 182: 182–192.

21 Watanabe, J., Kubota, H., Yano, K. et al. (1997). Anion exchange resins with heat and chemical stability. *Ultrapure Water* 14 (6): 39–43.

22 Yang, Z., Zhou, J., Wang, S. et al. (2015). Strategy to construct alkali stable anion exchange membranes bearing ammonium groups via flexible spacers. *J Mater Chem A* 2015 (3): 15015–15019.

23 Li, X., Wang, K., Liu, D. et al. (2020). Poly (arylene ether ketone) with tetra quaternary ammonium carbazole derivative pendant for anion exchange membrane. *Polymer* 195: 122456.

24 Zhang, F., Li, T., Chen, W. et al. (2021). Highly stable electron-withdrawing C––O link-free backbone with branched cationic side chain as anion exchange membrane. *J Membr Sci* 624: 119052–119061.

25 Yang, W., Chen, J., Yan, J. et al. (2022). Advance of click chemistry in anion exchange membranes for energy application. *J Polym Sci* 60: 627–649.

26 Wang, Q., Huang, L., Zheng, J. et al. (2022). Design, synthesis and characterization of anion exchange membranes containing guanidinium salts with ultrahigh dimensional stability. *J Membr Sci* 643: 120008–120018.

27 Ran, j., Wu, L., He, Y. et al. (2017). Ion exchange membranes: new developments and applications. *J Membr Sci* 522: 267–291.

28 Wang, J., He, G., Wu, X. et al. (2014). Crosslinked poly (ether etherketone) hydroxide exchange membranes with improved conductivity. *J Membr Sci* 459: 86–95.

29 Vengatesan, S., Santhi, S., Sozhan, G. et al. (2015). Novel cross-linked anion exchange membrane based on hexaminium functionalized poly (vinylbenzylchloride). *RSC Adv* 5: 27365–27371.

30 Lai, A.N., Wang, L.S., Lin, C.X. et al. (2015). Benzylmethyl-containing poly (arylene ethernitrile) as anion exchange membranes for alkaline fuel cells. *J Membr Sci* 481: 9–18.

31 Maurya, S., Shin, S.-H., Kim, M.-K. et al. (2013). Stability of composite anion exchange membranes with various functional groups and their performance for energy conversion. *J Membr Sci* 443: 28–35.

32 Lu, W., Shao, Z.-G., Zhang, G. et al. (2013). Preparation of anion exchange membranes by an efficient chloromethylation method and homogeneous quaternization/crosslinking strategy. *Solid State Ionics* 245–246: 8–18.

33 Zeng, L. and Zhao, T.S. (2013). High-performance alkaline ionomer for alkaline exchange membrane fuel cells. *Electrochem Commun* 34: 278–281.

34 Zhang, S., Li, C., Xie, X., and Zhang, F. (2014). Novel cross-linked anion exchange membranes with diamines as ionic exchange functional groups and crosslinking groups. *Int J Hydrog Energy* 39: 13718–13724.

35 Zhao, Y., Yu, H., Yang, D. et al. (2013). High-performance alkaline fuel cells using crosslinked composite anion exchange membrane. *J Power Sources* 221: 247–251.

36 Thomas, O.D., Soo, K.J., Peckham, T.J. et al. (2011). Anion conducting poly (dialkylbenzimidazolium) salts. *Polym Chem* 2 (8): 1641–1643.

37 Thomas, O.D., Soo, K.J., Peckham, T.J. et al. (2012). A stable hydroxide-conducting polymer. *J Am Chem Soc* 134: 10753–10756.

38 Wright, A.G. and Holdcroft, S. (2014). Hydroxide-stable ionenes. *ACS Macro Lett* 3: 444–447.

39 Armand, M.B., Chabagno, J.M., and Duclot, N.J. (1979). Polyether as solid electrolytes. In: *Fast Ion Transport in Solids* (ed. P. Vashishta, J.N. Mundy, and G.K. Shenoy), 52. Amsterdam: North Holland.

40 Xing, B. and Savadogo, O. (2000). Hydrogen/oxygen polymer electrolyte membrane fuel cells (PEMFCs) based on alkaline-doped polybenzimidazol (PBI). *Electrochem Commun* 2: 697–702.

41 Hou, H., Sun, G., He, R. et al. (2008). Alkali doped polybenzimidazole membrane for high performance alkaline direct ethanol fuel cell. *J Power Sources* 182 (2008): 95–99.

42 Modestov, A.D., Tarasevich, M.R., Leykin, A., and Filimonov, V. (2009). MEA for alkaline direct ethanol fuel cell with alkali doped PBI membrane and non-platinum electrodes. *J Power Sources* 188: 502–506.

43 Asensio, J.A., Borrós, S., and Gómez-Romero, P. (2002). Proton-conducting polymers based on benzimidazoles and sulfonated benzimidazoles. *J Polym Sci A Polym Chem* 40 (21): 3703–3710.

44 Diaz, L.A., Hnat, J., Heredia, N. et al. (2016). Alkali doped poly (2,5-benzimidazole) membrane for alkaline water electrolysis: characterization and performance. *J Power Sources* 312: 128–136.

45 Diaz, L.A., Coppola, R.E., Abuin, G.C. et al. (2017). Alkali-doped polyvinyl alcohol – polybenzimidazole membranes for alkaline water electrolysis. *J Membr Sci* 535: 45–55.

46 Herranz, D., Escudero-Cid, R., Montiel, M. et al. (2018). Poly (vinyl alcohol) and poly (benzimidazole) blend membranes for high performance alkaline direct ethanol fuel cells. *Renew Energy* 127: 883–895.

47 Coppola, R.E., Herranz, D., Escudero-Cid, R. et al. (2020). Polybenzimidazole-crosslinked poly(vinylbenzyl chloride) as anion exchange membrane for alkaline electrolyzers. *Renew Energy* 157: 71–82. https://doi.org/10.1016/j.renene.2020.04.140.

48 Herranz, D., Coppola, R.C., Escudero-Cid, R. et al. (2020). Application of crosslinked polybenzimidazole-poly(vinyl benzyl chloride) anion exchange membranes in direct ethanol fuel cells. *Membranes* 10 (11): 349.

49 Lu, W., Zhang, G., Li, J. et al. (2015). Polybenzimidazole-crosslinked poly (vinylbenzyl chloride) with quaternary 1,4-diazabicyclo (2.2.2) octane groups as high-performance anion exchange membrane for fuel cells. *J Power Sources* 296: 204–214.

50 Hao, J., Jiang, Y., Gao, X. et al. (2018). Functionalization of polybenzimidazole-crosslinked poly (vinylbenzyl chloride) with two cyclic quaternary ammonium cations for anion exchange membranes. *J Membr Sci* 548: 1–10.

51 Mann-lahav, M., Halabi, M., Shter, G.E. et al. (2020). Electrospun ionomeric fibers with anion conducting properties. *Adv Funct Mater* 30 (18): 1901733. https://doi.org/10.1002/adfm.201901733.

52 Halabi, M., Mann-Lahav, M., Beilin, V. et al. (2020). Electrospun anion-conducting ionomer fibers—effect of humidity on final properties. *Polymers* 12 (5): 1020. https://doi.org/10.3390/polym12051020.

53 Park, A.M., Wycisk, R.J., Ren, X. et al. (2016). Crosslinked poly (phenylene oxide)-based nanofiber composite membranes for alkaline fuel cells. *J Mater Chem A* 4 (1): 132–141.

54 Park, A., Turley, F., Wycisk, R., and Pintauro, P. (2015). Diol-crosslinked electrospun composite anion exchange membranes. *J Electrochem Soc* 162 (6): F560–F566.

55 Liao, G.M., Li, P.C., Lin, J.S. et al. (2016). Highly conductive quasi-coaxial electrospun quaternized polyvinyl alcohol nanofibers and composite as high-performance solid electrolytes. *J Power Sources* 304: 136–145.

56 Yang, J.M., Fan, C.S., Wang, N.C., and Chang, Y.H. (2018). Evaluation of membrane preparation method on the performance of alkaline polymer electrolyte: comparison between poly(vinyl alcohol)/chitosan blended membrane and poly(vinyl alcohol)/chitosan electrospun nanofiber composite membranes. *Electrochim Acta* 266: 332–340.

57 Coppola, R.E., Molinari, F., D'Accorso, N., and Abuin, G.C. (2021). Polyvinyl alcohol nanofibers reinforced with polybenzimidazole: facile preparation and properties of an anion exchange membrane. *Polym Adv Technol* 32: 3505–3514. https://doi.org/10.1002/pat.5361.

58 Ataollahi, N., Cappelletto, E., Vezzu, K. et al. (2018). Properties of anion exchange membrane based on polyamine: effect of functionalized silica particles prepared by sol-gel method. *Solid State Ion* 322: 85–92.

59 Yang, C.-C. (2007). Synthesis and characterization of the cross-linked PVA/TiO_2 composite polymer membrane for alkaline DMFC. *J Membr Sci* 288: 51–60.

60 Huang, C.-Y., Lin, J.-S., Pan, W.-H. et al. (2016). Alkaline direct ethanol fuel cell performance using alkali-impregnated polyvinyl alcohol/functionalized carbon nano-tube solid electrolytes. *J Power Sources* 303: 267–277.

61 Zhao, J., Chen, J., Xu, S. et al. (2014). Hierarchical NiMn layered double hydroxide/carbon nanotubes architecture with superb energy density for flexible supercapacitors. *Adv Funct Mater* 24: 2938–2946. https://doi.org/10.1002/adfm.201303638.

62 Pasquini, L., Becerra-Arciniegas, R.-A., Narducci, R. et al. (2020). Properties and alkaline stability of composite anion conducting ionomers based on poly(phenylene oxide) grafted with DABCO and Mg/Al lamellar double hydroxide. *ChemElectroChem* 7: 2917–2924. https://doi.org/10.1002/celc.202000523.

63 Gong, C., Zhao, S., Tsen, W.-C. et al. (2019). Hierarchical layered double hydroxide coated carbon nanotube modified quaternized chitosan/polyvinyl alcohol for alkaline direct methanol fuel cells. *J Power Sources* 441: 227176. https://doi.org/10.1016/j.jpowsour.2019.227176.

64 Han, X., Wang, J., and Wang, L. (2018). Preparation of anion exchange membranes based on pyridine functionalized poly (vinyl alcohol) crosslinked by 1, 4-dichlorobutane. *J Appl Polym Sci* 136 (16): 47395. https://doi.org/10.1002/APP.47395.

65 Hari Gopi, K. and Bhat, S.D. (2017). Anion exchange membrane from polyvinyl alcohol functionalized with quaternary ammonium groups via alkyl spacers. *Ionics* 24 (4): 1097–1109.

66 Zhang, D., Xu, S., Wan, R. et al. (2022). Functionalized graphene oxide cross-linked poly(2,6-dimethyl-1,4-phenylene oxide)-based anion exchange membranes with superior ionic conductivity. *J Power Sources* 517: 230720.

67 Gong, X., Yan, X., Li, T. et al. (2017). Design of pendent imidazolium side chain with flexible ether-containing spacer for alkaline anion exchange membrane. *J Membr Sci* 523: 216–224.

68 Son, T.Y., Kim, T.-H., and Nam, S.Y. (2020). Crosslinked pore-filling anion exchange membrane using the cylindrical centrifugal force for anion exchange membrane fuel cell system. *Polymers* 12: 2758. https://doi.org/10.3390/polym12112758.

69 Kim, Y., Wang, Y., Arthur France-Lanord, A. et al. (2019). Ionic highways from covalent assembly in highly conducting and stable anion exchange membrane fuel cells. *J Am Chem Soc* 141: 18152–18159. https://doi.org/10.1021/jacs.9b08749.

70 Lin, C., Cheng, W., Miao, X. et al. (2022). Clustered piperidinium-functionalized poly(terphenylene) anion exchange membranes with well-developed conductive nanochannels. *J Colloid Interface Sci* 608: 1247–1256.

71 Yuan, Y., Du, X., Zhang, H. et al. (2022). Poly (isatin biphenylene) polymer containing ferrocenium derivatives for anion exchange membrane fuel cell. *J Membr Sci* 642: 119986.

72 Liu, M., Hu, X., Hu, B. et al. (2022). Soluble poly(aryl piperidinium) with extended aromatic segments as anion exchange membranes for alkaline fuel cells and water electrolysis. *J Membr Sci* 642: 119966.

73 Wang, H.H., Hu, C., Park, J.H. et al. (2022). Reinforced poly (fluorenyl-*co*-terphenyl piperidinium) anion exchange membranes for fuel cells. *J Membr Sci* 644: 120160.

74 Prakash, O., Bihari, S., Keshav, T.S. et al. (2022). Dehydrohalogenated poly (vinylidene fluoride)-based anion exchange membranes for fuel cell applications. *Mater Today Chem* 23: 100640.

75 Cao, D., Yang, F., Sheng, W. et al. (2022). Polynorbornene-based anion exchange membranes with hydrophobic large steric hindrance arylene substituent. *J Membr Sci* 641: 119938.

76 Zeng, L., Yuan, W., Ma, X. et al. (2022). Dual-cation interpenetrating polymer network anion exchange membrane for fuel cells and water electrolyzers. *Macromolecules* 55: 4647–4655.

77 Lu, Y., Liu, L., Li, N. et al. (2022). Homologous flexible multi-cationic cross-linkers modified poly (aryl ether sulfone) anion exchange membranes for fuel cell applications. *Int J Hydrog Energy* 47: 17329–17340.

78 Ma, L., Li, L., Yuan, M. et al. (2022). Hydrophilic–hydrophobic bulky units modified anion exchange membranes for fuel cell application. *ACS Sustain Chem Eng* 10: 5748–5757.

79 Omasta, T.J., Wang, L., Peng, X. et al. (2018). Importance of balancing membrane and electrode water in anion exchange membrane fuel cells. *J Power Sources* 375: 205–213.

80 Ren, X., Price, S.C., Jackson, A.C. et al. (2014). Highly conductive anion exchange membrane for high power density fuel-cell performance. *ACS Appl Mater Interfaces* 6 (16): 13330–13333.

81 Carmo, M., Doubek, G., Sekol, R.C. et al. (2013). Development and electrochemical studies of membrane electrode assemblies for polymer electrolyte alkaline fuel cells using FAA membrane and ionomer. *J Power Sources* 230: 169–175.

82 Zakaria, Z. and Kamarudin, S.K. (2021). A review of alkaline solid polymer membrane in the application of AEM electrolyzer: materials and characterization. *Int J Energy Res* 45: 18337–18354.

83 Henkensmeier, D., Najibah, M., Harms, C. et al. (2020). Overview: state-of-the art commercial membranes for anion exchange membrane water electrolysis. *J Electrochem Energy Convers Storage* 18: 024001-1–024001-18.

84 López-Fernández, E., Gómez Sacedón, C., Gil-Rostra, J. et al. (2021). Recent advances in alkaline exchange membrane water electrolysis and electrode manufacturing. *Molecules* 26: 6326.

85 Park, J.S., Choi, J.H., Woo, J.J. et al. (2006). An electrical impedance spectroscopic (EIS) study on transport characteristics of ion-exchange membrane systems. *J Colloid Interface Sci* 300 (2): 655–662.

86 Macdonald, D.D. (2006). Reflections on the history of electrochemical impedance spectroscopy. *Electrochim Acta* 51 (8–9): 1376–1388.

87 Fontananova, E., Zhang, W., Nicotera, I. et al. (2014). Probing membrane and interface properties in concentrated electrolyte solutions. *J Membr Sci* 459: 177–189.

88 Zhang, W., Cheng, W., Tufa, R.A. et al. (2021). Studies on anion exchange membrane and interface properties by electrochemical impedance spectroscopy: the role of pH. *Membranes* 11 (10): 771.

89 Lee, C.H., Park, H.B., Lee, Y.M.N., and Lee, R.D. (2005). Importance of proton conductivity measurement in polymer electrolyte membrane for fuel cell application. *Ind Eng Chem Res* 44 (20): 7617–7626.

90 Zhang, W., Ma, J., Wang, P. et al. (2016). Investigations on the interfacial capacitance and the diffusion boundary layer thickness of ion exchange membrane using electrochemical impedance spectroscopy. *J Membr Sci* 502: 37–47.

91 Merle, G., Wessling, M., and Nijmeijer, K. (2011). Anion exchange membranes for alkaline fuel cells: a review. *J Membr Sci* 377 (1–2): 1–35.

92 Hagesteijn, K.F., Jiang, S., and Ladewig, B.P. (2018). A review of the synthesis and characterization of anion exchange membranes. *J Mater Sci* 53 (16): 11131–11150.

93 Dekel, D.R. (2018). Review of cell performance in anion exchange membrane fuel cells. *J Power Sources* 375: 158–169.

94 Ziv, N., Mondal, A.N., Weissbach, T. et al. (2019). Effect of CO_2 on the properties of anion exchange membranes for fuel cell applications. *J Membr Sci* 586: 140–150.

95 Ziv, N. and Dekel, D.R. (2018). A practical method for measuring the true hydroxide conductivity of anion exchange membranes. *Electrochem Commun* 88: 109–113.

96 Grew, K.N., Ren, X., and Chu, D. (2011). Effects of temperature and carbon dioxide on anion exchange membrane conductivity. *Electrochem Solid-State Lett* 14 (12): B127.

97 Siroma, Z., Watanabe, S., Yasuda, K. et al. (2011). Mathematical modeling of the concentration profile of carbonate ions in an anion exchange membrane fuel cell. *J Electrochem Soc* 158 (6): B682.

98 Giorno, L., Drioli, E., and Strathmann, H. (2016). Ion-exchange membrane characterization. In: *Encyclopedia of Membranes* (ed. E. Drioli and L. Giorno), 1052–1056. Berlin: Springer.

99 Karas, F., Hnát, J., Paidar, M. et al. (2014). Determination of the ion-exchange capacity of anion-selective membranes. *Int J Hydrog Energy* 39 (10): 5054–5062.

100 Wang, L., Magliocca, E., Cunningham, E.L. et al. (2017). An optimised synthesis of high performance radiation-grafted anion-exchange membranes. *Green Chem* 19 (3): 831–843.

101 Kozmai, A.E., Nikonenko, V.V., Zyryanova, S. et al. (2019). Modelling of anion-exchange membrane transport properties with taking into account the change in exchange capacity and swelling when varying bathing solution concentration and pH. *J Membr Sci* 590: 117291.

102 Jiang, S., Sun, H., Wang, H. et al. (2021). A comprehensive review on the synthesis and applications of ion exchange membranes. *Chemosphere* 282: 130817.

103 Sperling, L.H. (1992). *Introduction to Physical Polymer Science*. Wiley.

104 Franceschini, E.A. and Corti, H.R. (2009). Elastic properties of Nafion, polybenzimidazole and poly [2, 5-benzimidazole] membranes determined by AFM tip nano-indentation. *J Power Sources* 188 (2): 379–386.

105 Nayak, R., Dey, T., Ghosh, P.C., and Bhattacharyya, A.R. (2016). Phosphoric acid doped poly (2, 5-benzimidazole)-based proton exchange membrane for high temperature fuel cell application. *Polym Eng Sci* 56 (12): 1366–1374.

106 Varcoe, J.R., Poynton, S.D., and Slade, R.C.T. (2010). Alkaline anion-exchange membranes for low-temperature fuel cell application. In: *Handbook of Fuel Cells: Fundamentals, Technology and Applications*, 2695–2706. Wiley.

107 Diesendruck, C.E. and Dekel, D.R. (2018). Water–a key parameter in the stability of anion exchange membrane fuel cells. *Curr Opin Electrochem* 9: 173–178.

108 Dekel, D.R., Amar, M., Willdorf, S. et al. (2017). Effect of water on the stability of quaternary ammonium groups for anion exchange membrane fuel cell applications. *Chem Mater* 29 (10): 4425–4431.

109 Kim, Y.S. and Lee, K.S. (2015). Fuel cell membrane characterizations. *Polym Rev* 55 (2): 330–370.

110 Ayers, K.E., Anderson, E.B., Capuano, C.B. et al. (2013). Characterization of anion exchange membrane technology for low cost electrolysis. *ECS Trans* 45 (23): 121.

111 Giorno, L., Drioli, E., and Strathmann, H. (2016). Characterization of porous and dense membranes. In: *Encyclopedia of Membranes* (ed. E. Drioli and L. Giorno), 362–371. Berlin: Springer.

112 Binnig, G., Quate, C.F., and Gerber, C. (1986). Atomic force microscope. *Phys Rev Lett* 56 (9): 930.

113 Raghavan, D., Gu, X., Nguyen, T. et al. (2000). Mapping polymer heterogeneity using atomic force microscopy phase imaging and nanoscale indentation. *Macromolecules* 33 (7): 2573–2583.

114 Viswanathan, R., Tian, J. and Marr, D.W. (1996). Morphology characterization in multicomponent polymer systems using scanning probe microscopy. *MRS Online Proceedings Library (OPL)*, 461.

115 Antolini, E. and Gonzalez, E.R. (2010). Alkaline direct alcohol fuel cells. *J Power Sources* 195 (11): 3431–3450.

116 Xiong, Y., Fang, J., Zeng, Q.H., and Liu, Q.L. (2008). Preparation and characterization of cross-linked quaternized poly (vinyl alcohol) membranes for anion exchange membrane fuel cells. *J Membr Sci* 311 (1–2): 319–325.

117 Msomi, P.F., Nonjola, P., Ndungu, P.G., and Ramontja, J. (2018). Quaternized poly (2.6 dimethyl-1.4 phenylene oxide)/polysulfone anion exchange membrane reinforced with graphene oxide for methanol alkaline fuel cell application. *J Polym Res* 25 (6): 1–12.

118 Zakaria, Z., Kamarudin, S.K., Timmiati, S.N., and Masdar, M.S. (2019). New composite membrane poly (vinyl alcohol)/graphene oxide for direct ethanol–proton exchange membrane fuel cell. *J Appl Polym Sci* 136 (2): 46928.

119 Park, H.J., Chu, X., Kim, S.P. et al. (2020). Effect of N-cyclic cationic groups in poly (phenylene oxide)-based catalyst ionomer membranes for anion exchange membrane fuel cells. *J Membr Sci* 608: 118183.

120 Peter Atkins, P. and De Paula, J. (2011). *Physical Chemistry for the Life Sciences.* Oxford University Press.

121 Ismail, A.F., Khulbe, K.C., and Matsuura, T. RO membrane characterization. In: *Reverse Osmosis* (ed. N. Hilal, A.F. Ismail, T. Matsuura, and D. Oatley-Radcliffe), 57–90. Amsterdam: Elsevier.

4

Types of Polymeric Electrolyte Anion Exchange Membranes: Heterogeneous and Grafted Membranes, Interpenetrating Polymer Networks and Homogeneous Membranes

Liang Zhu

The Sherwin-Williams Company, 601 Canal Rd, Cleveland, OH 44113, USA

4.1 Heterogenous Anion Exchange Membranes

4.1.1 Ion-Solvating Polymers

In general, there are two types of heterogeneous membranes: ion-solvating polymers (ISPs) and hybrid membranes [1]. The ISP are polymer salt complexes, which were discovered more than 40 years ago by Fenton et al. and Armand et al. [1–3] The most extensively studied ISPs are based on polyethylene oxide (PEO) polymers, which have been developed for applications in solid-state Ni–Zn, Ni–MH, or Zn–air batteries [1]. ISPs will not be covered or discussed in this chapter since few ISPs for fuel cell application were reported during the last decade [1]. Readers with great interest on ISPs can look up the review written by Merle et al. [1]

4.1.2 Hybrid Membranes

Hybrid membranes consist of an organic and an inorganic segment. Typically, the organic part enables the ion conducting properties, while the inorganic part provides mechanical strength. The organic–inorganic hybrid anion exchange membranes (AEMs) can be prepared by a sol–gel process, intercalation, blending, *in situ* polymerization, and molecular self-assembly [1, 4–18].

Wu et al. [7] prepared a series of anion exchange hybrid membranes based on poly(2,6-dimethyl-1,4-phenylene oxide) (PPO) via bromination, hydroxylation, quaternization, and sol–gel reaction with monophenyl triethoxysilane (EPh) or/and

Alkaline Anion Exchange Membranes for Fuel Cells: From Tailored Materials to Novel Applications, First Edition. Edited by Jince Thomas, Alex Schechter, Flavio Grynszpan, Bejoy Francis, and Sabu Thomas.

tetraethoxysilane (TEOS), followed by heat treatment at 120–140 °C for different times. The chemical and physical properties of the resulting hybrid AEM, including the ion exchange capacity (IEC), hydrophilicity, hydroxide ion conductivity, and mechanical strength, can be tuned by adjusting the heating temperature and reaction time [7]. Although the resulting hybrid AEM demonstrated proper tensile strength (20 MPa), it only exhibited an ion conductivity of 8.5 mS cm^{-1} [7]. In order to improve the swelling resistance and enhance the ion conductivity without sacrificing the thermal and chemical stability, Wu et al. [9] optimized the content of silica and cross-linking degree between the PPO polymer and silica. The optimized PPO-based hybrid AEMs demonstrated a hydroxide conductivity of 12–35 mS cm^{-1} in the temperature range of 30–90 °C [9]. When a thickness of 140 μm hybrid AEM with 5.4% silica was incorporated into an alkaline membrane electrode assembly (MEA), a H_2/O_2 peak power density of 32 mW cm^{-2} was obtained [9].

It is well known that the PEO–SiO_2 composites displayed good flexibility and relatively high thermal stability in proton conducting membranes [6]. Inspired by this result, anion-exchange PEO–SiO_2 hybrid membranes were synthesized by Wu et al. [6] through the sol–gel reaction of different precursors: charged alkoxysilane-functionalized PEO-1000 (PEO-[Si$(OCH_3)_3]_2$(+)), EPh, *N*-triethoxysilylpropyl-*N,N,N*-trimethylammonium iodine (A-1100(+)), and in some cases also TEOS. The PEO–(PEO)–SiO_2 hybrid AEM showed good flexibility and tensile strength up to 20.55 MPa [6]. However, the ion conductivity of the hybrid membranes is extremely low (3 mS cm^{-1}) [6].

Wu et al. [8] prepared hybrid AEMs from poly(vinyl alcohol) (PVA), *N*-triethoxysilylpropyl-*N,N,N*-trimethylammonium iodine, and different types of cross-linker such as TEOS or glycidoxypropyltrimethoxysilane through the sol–gel process. By tuning the proportion of the cross-linker, the IEC ranged from 0.52 to 1.10 mmol g^{-1}, the tensile strength was in the range of 55–82 MPa, and the elongation at break can be as high as 16% [8].

Quaternized polysulfone (QPSU)/functionalized montmorillonite hybrid AEMs were prepared by Liao et al. [11] to evaluate the potential application for anion exchange membrane fuel cells (AEMFCs). Compared to the homogeneous polysulfone-based AEM, the hybrid AEM containing 5% montmorillonite modified by cetyl trimethyl ammonium chloride displayed lower water uptake, higher ultimate stress, and ion conductivity. It exhibited a hydroxide ion conductivity of 47.3 mS cm^{-1} at 95 °C [11]. A type of novel hybrid AEM was fabricated by Liu et al. [12] through incorporating quaternized graphenes (QGs) into the chloromethylated polysulfone (CMPSU) and followed by quaternization and alkalization. The hybrid AEM with 0.5 wt% QCs exhibited a bicarbonate ion conductivity of 18.7 mS cm^{-1} at 80 °C, which is nearly four times higher than that of the pristine AEM [12]. Moreover, the mechanical strength was also enhanced due to the incorporation of QCs. The hybrid AEM with 0.25 wt% QCs demonstrated a tensile strength as high as 205 MPa [12]. The enhancement in electrochemical and mechanical properties could be ascribed to the interconnected ionic transfer channels in the hybrid AEMs and the good compatibility between the polymer matrix and fillers [12].

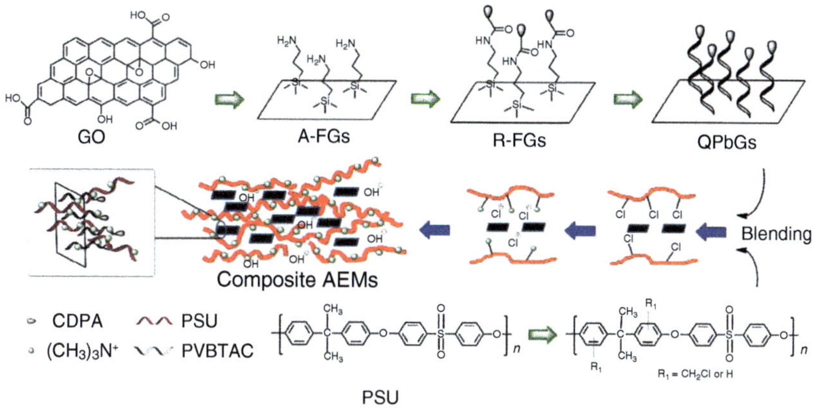

Figure 4.1 Synthetic route of QPSU/QPbGs hybrid anion exchange membranes. Source: © The Royal Society of Chemistry, 2016 [15]. Reproduced from Hu et al. [15] with permission from the Royal Society of Chemistry.

Hu et al. [15] designed a series of hybrid AEMs (Figure 4.1) for alkaline fuel cell application by incorporating different amounts of quaternized polymer brush functionalized graphenes (QPbGs) into QPSU. Compared to the pristine QPSU membranes, the hybrid AEMs containing an appropriate amount of QPbGs could maintain higher IEC, water uptake, and ion conductivity but lower methanol permeability because of the hydrophilic polymer brushes on QPbGs and the interaction between QPbGs and QPSU [15]. The bicarbonate ion conductivity of hybrid AEMs increased with the increasing content of QPbGs and exhibited a bicarbonate ion conductivity up to 56 mS cm^{-1} at 80 °C [15]. By doping the functionalized titanium dioxide (TiO_2) with QPSU, Yi et al. [13] prepared a hybrid AEM. The hybrid with 5 wt% content of functionalized TiO_2 displayed an ion conductivity of 50.7 mS cm^{-1} at 95 °C [13]. Miao et al. [18] synthesized a series of hybrid AEMs by incorporating quaternized polyhedral oligomeric silsesquioxanes (QPOSS) into the QPSU. Compared to the QPSU AEM, a 60% improvement in the hydroxide ion conductivity was observed for the QPSU3%–QPOSS [18]. It exhibited a hydroxide ion conductivity of 53.6 mS cm^{-1} at 80 °C [18]. The enhancement of the ion conductivity is attributed to formed ion clusters and connected ion transport channels of the hybrid AEMs [18].

The organic–inorganic hybrid AEMs can be fabricated by *in situ* copolymerization of functionalized monomers. Imidazolium-based organic–inorganic hybrid AEMs were synthesized by Feng et al. [14] through *in situ* copolymerization of 1-vinyl-3-butylimidazolium bromide ([VBIm][Br]), styrene, acrylonitrile, and 1-vinyl-propyl-triethoxysilaneimidazolium chloride ([VPSIm][Cl]), followed by anion exchange with hydroxide ions (Figure 4.2) [14]. The resulting organic–inorganic hybrid AEMs demonstrated a hydroxide ion conductivity as high as 41 mS cm^{-1} at 30 °C and 90.5 mS cm^{-1} at 90 °C [14]. Compared to the AEMs without incorporating [VPSIm][Cl], the hybrid AEMs exhibited much higher alkaline stability, and the hybrid AEMs can survive up to 480 h in 2 M KOH solution at 60 °C [14].

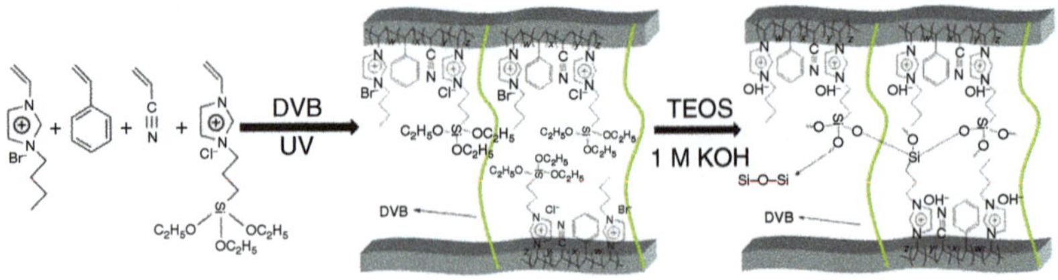

Figure 4.2 Synthetic route for the organic–inorganic hybrid anion exchange membranes. Source: © Elsevier, 2016 [14].

Organic–inorganic hybrid materials have advantages such as the unique combination of the remarkable functionality of organic materials with the stability of inorganic materials. Although the introduction of an inorganic component into the polymer matrix enhanced the mechanical strength of the membranes, the ion conductivity of the hybrid membranes is not enough for application in AEMFCs. Enhancement of hybrid membrane properties and better controlling of the morphology have to be done by choosing a suitable organic material and inorganic component.

4.2 Grafted Anion Exchange Membranes

Typically, a grafted membrane is composed of a polymer backbone and ion-containing side chains. Generally, the grafted membranes are prepared in two steps. The first step is to generate functional groups on the polymer backbones employing chemical or physical methods. Subsequently, the ion-containing side chains are grafted to the polymer backbones via the functional sites. The optimized combination of the polymer backbones and grafts can produce a graft membrane which exhibits high ion conductivity, excellent chemical stability, and long-term durability in devices.

4.2.1 Radiation-Grafted Membranes

The preparation of radiation-grafted AEMs starts from polymer films, which are exposed to irradiation, and is followed by grafting a functionalized monomer or a monomer that can be later functionalized [19–26]. A variety of polymer films, such as partially or fully fluorinated films, and polyolefin films, have been employed by the researcher to prepare the radiation-grafted AEMs (Figure 4.3) [19–26].

Vinylbenzyl chloride (VBC) is the most frequently used monomer to prepare the radiation-grafted AEMs based on partially or fully fluorinated films [19–21, 26]. Generally, the preparation of radiation-grafted AEMs consists of three steps. First, the film is irradiated with an electron beam, and after that the irradiated film is immersed in vinyl benzyl chloride solution. Subsequently, the resulting grafted poly(vinyl benzyl chloride) copolymer film is converted to quaternary ammonium (QA)-containing anion-conductive copolymers by immersing in trimethylamine solution. The last step

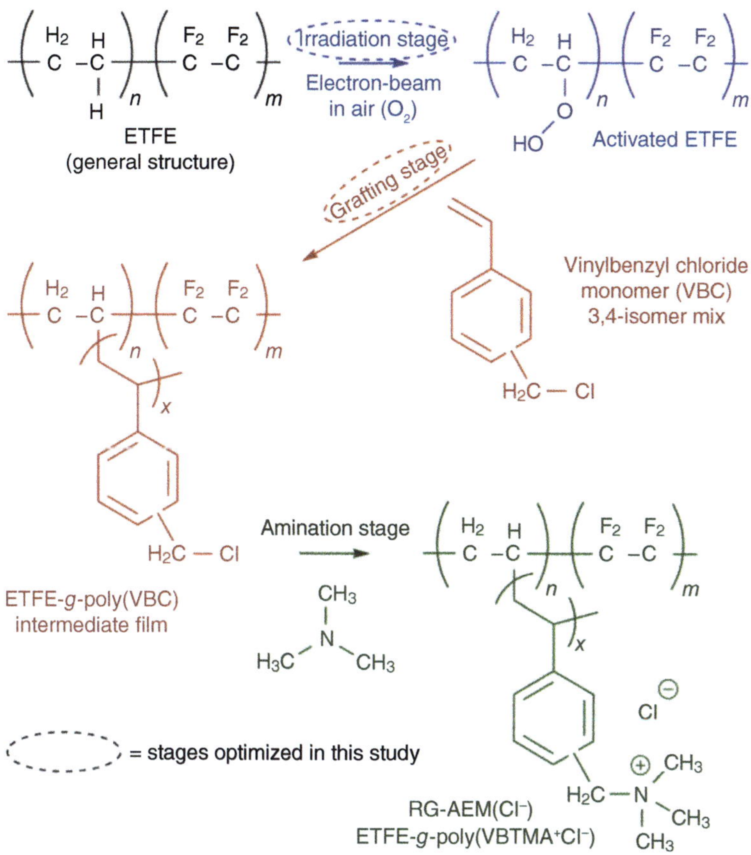

Figure 4.3 Outline of the synthetic route of the RG-AEM. Source: © The Royal Society of Chemistry, 2017 [24]. Reproduced from Wang et al. [24] with permission from the Royal Society of Chemistry.

is an alkaline exchange process carried out by soaking the membrane in NaOH/KOH solution. Danks et al. [19] grafted VBC to poly(vinylidenefluoride) (PVDF) and poly (tetrafluoroethene-*co*-hexafluoro propylene) (FEP) polymer films using radiation-grafting methodology to fabricate grafted AEMs [19]. The resulting grafted AEMs based on PVDF underwent undesirable degradation and were found to be unsuitable for preparation of AEMs. The grafted AEM based on FEP-based materials showed superior structural stability and ion conductivity up to 20 mS cm^{-1} at room temperature [19]. However, radiation-induced grafting requires large radiation absorbed doses to obtain high levels of monomer (e.g. styrene) grafting throughout the whole thicknesses of the polymer films, leading to a detrimental reduction of the mechanical strength. In addition, the vinyl benzyl chloride monomer is highly toxic and expensive. Therefore, it is desirable to reduce the concentration of VBCs and minimize the absorbed doses (for maximum mechanical stabilities). Wang et al. [24] synthesized grafted AEMs by the radiation grafting of VBC onto poly(ethylene-*co*-tetrafluoroethylene) (ETFE) films, using the organic-solvent-free method with reduced amounts of both monomer (VBC) concentration and electron-beam radiation absorbed dose (Figure 4.3) [24]. By using the grafted AEM, a remarkable peak power density of 1.16 W cm^{-2} at 60 °C was achieved in the H_2/O_2 AEMFCs test [24]. Inspired by the small-molecule study suggesting that AEMs containing aliphatic heterocyclic QA cationic head groups will exhibit enhanced alkali stabilities compared to the benzyltrimethylammonium head group, Ponce-González et al. [22] synthesized a series of benzyl-linked saturated-heterocyclic QA radiation-grafted ETFE-based AEMs in order to enhance the chemical stability. Compared to the prior benzyltrimethylammonium benchmark, the resulting benzyl-*N*-methylpyrrolidinium AEM exhibited higher relative alkali stability, ion conductivity, and *in situ* fuel cell performance [22].

Fluorinated polymer films are generally less suitable for radiation-induced grafting modification because of detrimental reduction of mechanical strength caused by radiation, leading to difficulty in maintaining long-term fuel cell application. Wang et al. [23] designed and synthesized radiation-grafted anion-exchange membranes (RG-AEMs) using the peroxidation of (non-fluorinated) low-density polyethylene (LDPE). Compared to the ETFE-based RG-AEM, the LDPE-based RG-AEM demonstrated enhanced mechanical robustness when exposed to alkali at 80 °C, yielding higher *in situ* performances when applied in single-cell AEMFCs [23]. Employing an LDPE-based RG-AEM alongside a Pt-based anode and cathode, a peak power density of 1.45 W cm^{-2} at 80 °C was achieved in the H_2/O_2 AEMFC test [23]. Moreover, with the use of a Ag/C cathode at 80 °C, a peak power density of 0.93 W cm^{-2} at 80 °C was obtained in the AEMFC test [23]. The moderate mechanical properties of LDPE-based RG-AEM limit their commercial application in AEMFCs [25]. In order to further enhance the mechanical properties, high-density polyethylene (HDPE) was selected as the precursor polymer to fabricate the HDPE-based RG-AEMs [25]. The switch from employing LDPE to HDPE as a precursor film resulted in enhanced cell performance when the AEM was tested in a single-cell AEMFC [25]. A peak power density of 2.55 W cm^{-2} at 80 °C was obtained in the AEMFC test using HDPE-based RG-AEMs [25]. Although the radiation-grafted AEMs exhibit remarkable performance in AEMFCs, a major challenge in the development of high-performance radiation-grafted membranes has generally been overcoming their poor mechanical properties as the high radiation doses required lead to damage of the polymer main chains in the radiation-treated precursor films. Moreover, no long-term device performance data were reported for the radiation-grafted AEMs.

4.2.2 Side Chain Grafted Membranes

Generally, the preparation of side chain grafted membranes is composed of two steps. The first step is to create functional groups on the polymer backbones employing chemical methods. Subsequently, the ion-containing side chains are grafted to the polymer backbones via the Menshutkin reaction [27–39].

Li et al. [27] produced a series of quaternized PPOs containing the terminal pendent *n*-alkyl group using a Menshutkin reaction (Figure 4.4). Compared to AEMs

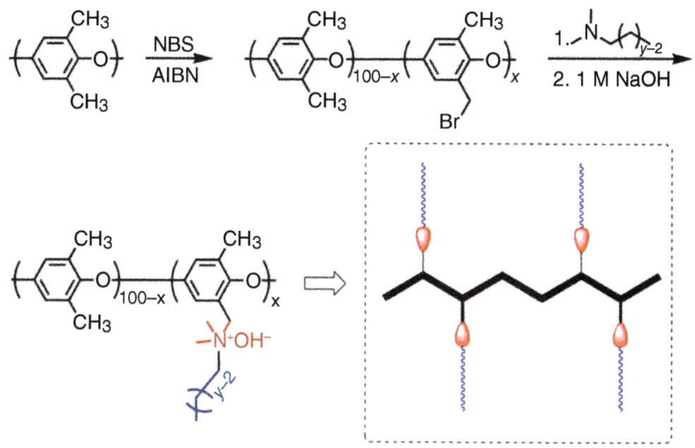

Figure 4.4 Chemical structure of comb-shaped CyDx copolymers. Source: © American Chemical Society, 2013 [27].

based on benzyltrimethyl ammonium, the comb-shaped CxDy AEMs exhibited higher hydroxide conductivity due to the formation of highly conductive ionic domains and phase separation [27]. More importantly, the comb-shaped AEMs demonstrated excellent alkaline stability, and more than 80% of their bicarbonate conductivities of the CxDy comb-shaped samples were retained after immersing in 1 M NaOH solution at 80 °C for 2000 h [27]. Reducing the degree of functionalization (DF) while increasing the charge number of the cationic groups was reported as a method to balance the ionic conductivity and the other required properties of AEMs [28]. Pan et al. [28] investigated a polysulfone (PS)-based AEM with a double quaternary ammonium groups (DQAPS) side chain. Compared to common AEMs, i.e. quaternary ammonium polysulfone (QAPS), DQAPS enabled higher IECs without increasing the DF and exhibited higher ionic conductivity as well as a lower swelling degree and higher chemical stability [28]. Although the DQAPS showed promising properties, the previous report did not detail systematic studies to demonstrate the effect of the arrangement of the charges on the performance of the AEMs. Zhu et al. [32] synthesized a series of PPO-based AEMs with different kinds of cationic side chains to investigate the relationship between the DF, the structure of the side chain (charge number and charge arrangement), and the AEM performance in terms of ionic conductivity, swelling degree, and chemical stability of the material (Figure 4.5) [32]. The highest conductivity, up to 99 mS cm^{-1} at room temperature, was achieved for a triple-cation side chain AEM with five or six methylene groups between cations [32]. In addition to high conductivity, the multication side chain AEMs showed good alkaline and dimensional stabilities. Based on the high-performance triple-cation side chain AEM, a Pt-catalyzed fuel cell with a peak power density of 364 mW cm^{-2} was obtained at 60 °C under 100% relative humidity (RH) [32]. In addition, it was found that the multication AEMs exhibited the highest ion conductivity with pentyl or hexyl spacers between cationic groups since too short or long alky spacers between cations decreases the water uptake, which is of great importance to ion transport [36].

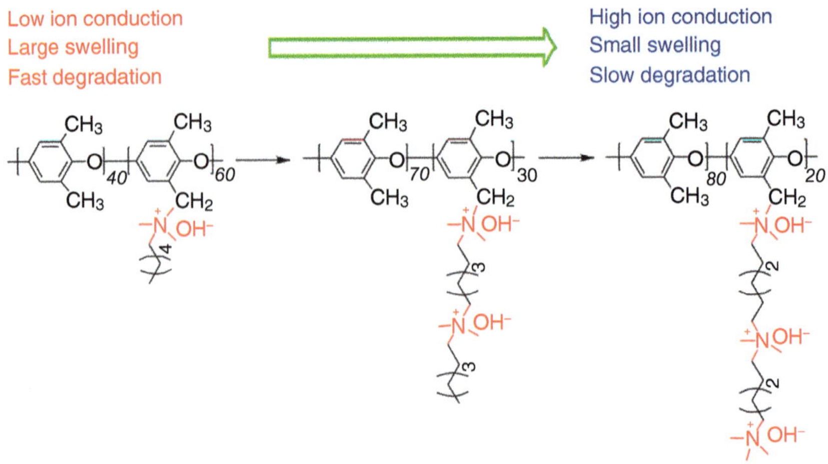

Figure 4.5 Multication side chain AEMs based on PPO. Source: © American Chemical Society, 2016 [32].

4.2.3 Long-side-chain Grafted Membranes

Typically, the long-side chain grafted AMEs are prepared by grafting long side chains to the main polymer backbones via polymerization, followed by functionalization of the long side chains [40–45].

Lin et al. [41] designed and fabricated long-side-chain grafted AEMs composed of poly(arylene ether sulfone)s backbone and pendent imidazolium-functionalized PPO via chemical graft reaction, bromination, functionalization, and alkalization (Figure 4.6). Distinct microphase separation was formed in the long-side-chain grafted AEMs because of the hydrophilic/hydrophobic differences between the backbone and side chain [41]. Based on the long-side-chain AEM, a Pt-catalyzed H_2/O_2 fuel cell with a peak power density of 56.8 mW cm^{-2} was achieved at 60 °C under 100% RH.

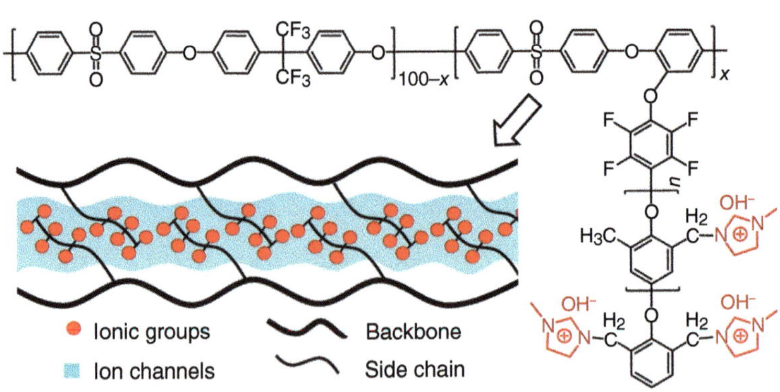

Figure 4.6 Structures of the long-side-chain grafted AEMs. Source: © Elsevier, 2016 [41].

With regard to the side chain grafted membranes, high ion conductivity, good dimensional stability, and excellent chemical stability can be achieved by properly tuning the chemical structures of side chains. However, the single-cell performance of the side chain grafted AEMs remains unsatisfactory.

4.3 Interpenetrating Anion Exchange Membranes

Compared to the acidic counterpart proton exchange membranes (PEMs), the AEMs typically exhibit inferior properties [46]. First of all, the lower mobility of hydroxide ion and the less-developed morphology of the aromatic polymers compared to perfluorinated structures (such as Nafion®) lead to the lower ion conductivity of AEMs [46]. In order to increase the ion conductivity of AEMs while maintaining the acceptable swelling ratio, the researchers have employed the cross-linking and controlled microphase separated strategies [46]. Second, the rigid aromatic polymer backbones of AEMs provide good mechanical strength (higher than 20 MPa), but the flexibility (elongation at break <30%) is usually unacceptable [46]. It is well known that the mechanical flexibility and toughness of AEMs is of great importance to the long-term fuel cell performance where the device is operated under wet–dry cycling conditions. The interpenetrating (semi-interpenetrating) polymer networks demonstrates outstanding mechanical strength and flexibility by introducing flexible polymer constituents. Thus, the interpenetrating polymer network (IPN) materials have drawn great interest because of the excellent combination of good mechanical strength, flexibility, and high ion conductivities.

The interpenetrating AEM is based on IPNs (Figure 4.7). An IPNs is a polymer comprising two or more networks which are at least partially interlaced on a polymer scale but not covalently bonded to each other [47–50]. The networks cannot be separated unless chemical bonds are broken. The semi-interpenetrating AEM is based on semi-interpenetrating polymer networks (semi-IPNs) (Figure 4.4). A semi-IPNs is a polymer comprising one or more networks and one or more linear or

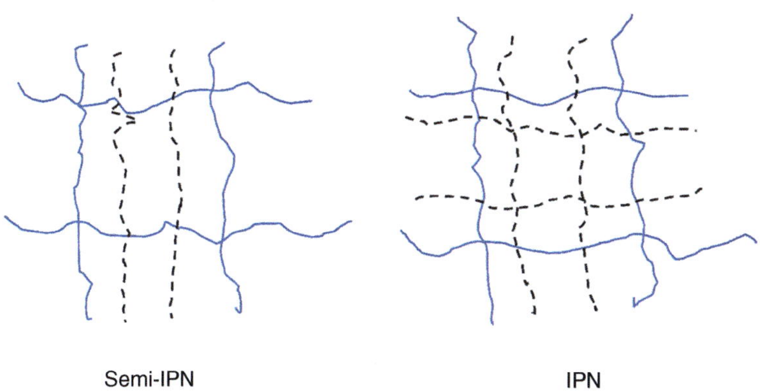

Figure 4.7 Structural representation of semi-interpenetrating polymer networks (semi-IPN) and interpenetrating polymer networks (IPN).

branched polymer(s) characterized by the penetration on a molecular scale of at least one of the networks by at least some of the linear or branched macromolecules [46, 51–56]. The constituent linear or branched polymers can be separated from the constituent polymer networks without breaking chemical bonds.

4.3.1 Anion Exchange Membranes Based on Interpenetrating Polymer Networks (IPN)

Gong et al. [47] prepared a type of anion-conducting IPN based on cross-linked QPSU and cross-linked PVA for AEMFCs. Compared to the QPSU membrane, the IPN demonstrated larger elongation at break and better dimensional and alkaline stability. Compared to the QPSU membrane, the swelling ratio of IPN55 at 95 °C decreased by nearly 50%, indicating better dimensional stability of the IPN [47]. In addition, the IPN55 exhibited an elongation at break as high as 56%, which is 4.5 times higher than that of the QPSU membrane [47]. Anion-conducting IPN based on cross-linked quaternized poly(vinylbenzyl chloride) (PVBC) and cross-linked PVA was designed and synthesized by Zeng et al. [50] The IPN with the IEC of 1.61 meq g^{-1} exhibited a hydroxide conductivity as high as 147.7 mS cm^{-1} at 80 °C because of the formation of a highly efficient ion channel within IPN AEMs [50]. By using of non-precious oxygen reduction reaction (ORR) catalyst (FeNx-CNTs), the IPN AEM demonstrated a peak power density of 0.64 W cm^{-2} in the H_2/Air (CO_2-free) AEMFC test with 0.2 MPa backpressure [50]. Zeng et al. [49] prepared IPN AEM-based cross-linked PVA and cross-linked poly (vinyl benzyl-*N*-methyl piperidinium) (Figure 4.8). The IPN with the IEC of 1.75 meq g^{-1} demonstrated an outstanding hydroxide conductivity of 257.8 mS cm^{-1} at 80 °C due to the formation

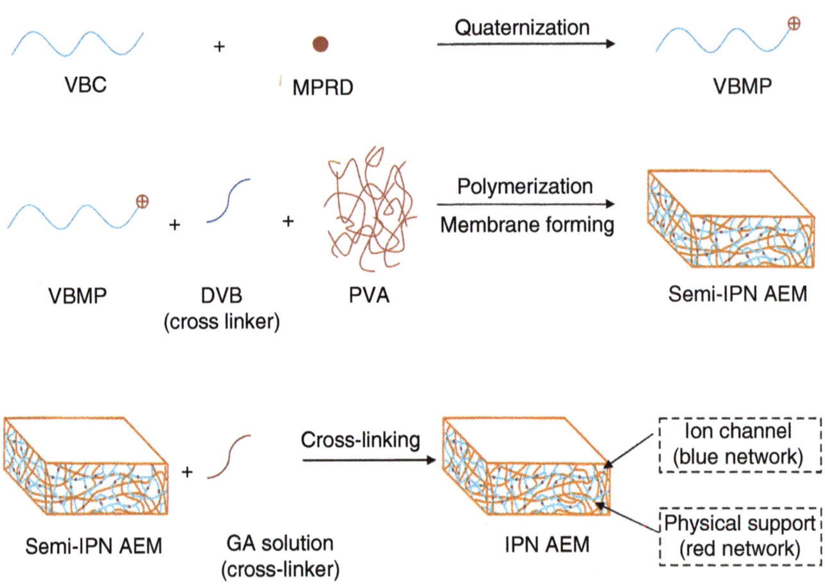

Figure 4.8 Fabrication process of IPN AEMs. Source: © Science Partner Journal, 2020 [49].

of highly efficient ion channels [49]. By using non-precious ORR catalysts (FeNx-CNTs), the IPN AEM demonstrated a remarkable peak power density of 1.20 W cm^{-2} at 60 °C in the H_2/Air (CO_2-free) AEMFC test with 0.1 MPa backpressure [49].

4.3.2 Anion Exchange Membranes Based on Semi-Interpenetrating Polymer Networks (Semi-IPN)

AEMs with semi-IPN based on quaternized chitosan (QCS) and polystyrene (PS) were synthesized by Wang et al. [51] by polymerization of styrene monomers in the emulsion of the QCS in an acetic acid aqueous solution, followed by post-cross-linking of the QCS. The swelling ratio of the membranes significantly decreased with the increasing content of PS [51]. The QCS-based AEM shows a swelling ratio of 555%, while the semi-IPN AEM with 38% PS exhibited a swelling ratio of 63% under the same condition [51]. However, the maximum hydroxide ion conductivity of the semi-IPN AEMs is only 52 mS cm^{-1} at 80 °C, which is unsatisfactory for AEMFC application [51]. In addition, the incorporation of the PS into the semi-IPN reduced the flexibility of the AEMs due to the rigid property of PS [51]. He et al. [52] prepared a type of semi-IPN AEM by reinforcing the highly charged benzyltrimethylammonium polysulfone with a cross-linked poly(styrene-*co*-divinylbenzene)network in order to maintain a desirable balance between IECs, ion conductivity, and dimensional stability (Figure 4.9) [52]. Compared to the benzyltrimethylammonium polysulfone-based AEMs, the semi-IPN AEM demonstrated significantly enhanced dimensional stability (water uptake 75% versus 301% at 25 °C) while maintaining good hydroxide ion conductivity (up to mS cm^{-1} at 25 °C) [52]. The highly charged semi-IPN membranes were stable up to 80 °C while operating in an MEA. A peak power density of 236 mW cm^{-2} under H_2/O_2 at 80 °C was obtained for the QA sIPN 70/30 MEA [52].

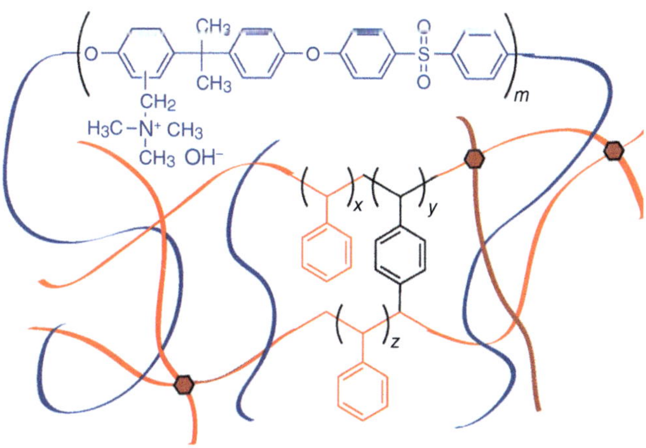

Figure 4.9 Illustration of semi-IPN AEM. Blue lines represent quaternary ammonium polysulfone, red lines mean poly(styrene-*co*-divinylbenzene), and hexagons denote covalent divinylbenzene crosslinks. Source: © Wiley-VCH Verlag GmbH & Co. KGaA, Weinheim, 2015 [52].

Pan et al. [46] designed a semi-interpenetrating network anion exchange membranes (SIPN AEMs) by introducing a hydrophilic, thiol-ene cross-linked, flexible poly(ethylene glycol)-*co*-poly(ally glycidyl ether) (xPEG-PAGE) network into the rigid quaternized poly-(2,6-dimethylphenylene oxide) (BTMA PPO) matrix [46]. Compared to the conventional BTMA PPO membranes, the SIPN exhibited excellent mechanical strength and flexibility due to the presence of both rigid and flexible polymer constituents [46]. As conventional rigid AEMs, QAPPO-40 demonstrated high strength (higher than 20 MPa), but low elongation at break (less than 40%) [46]. In contrast, the SIPN-60-2 membrane displayed much higher elongation at break (above 90%), while its tensile strength was maintained above 17 MPa [46]. However, no fuel cell performance was reported for the SIPN AEMs. Semi-IPN AEMs based on quaternized PS copolymer and cross-linked PPO networks were designed and fabricated by Xue et al. [54] In this semi-IPN, the quaternized PS ensured the ion conductivity, while the cross-linked PPO networks formed by azide-alkyne cycloaddition under organic base provided the mechanical properties and dimensional stability [54]. The sIPN-62/30 AEM with the IEC of 1.44 $meq\,g^{-1}$ exhibited a hydroxide conductivity of 37.2 $mS\,cm^{-1}$ at 20 °C, which was much higher than that of a control sample fabricated by physical blending (14.8 $mS\,cm^{-1}$ at 20 °C) [54]. In addition, the semi-IPN AEMs displayed much better mechanical properties than those of the control AEMs prepared by physical blending [54]. A peak power density of 110.6 $mW\,cm^{-2}$ under H_2/O_2 at 60 °C was achieved for the sIPN-62/30 MEA [54]. Yang et al. [57] designed and synthesized SIPN AEMs based on quaternized polysulfone/polyquaternium-10 (QPSU/PQ-10), which showed moderate swelling ratio, hydroxide conductivity, thermal oxidation stability, and mechanical properties under optimal condition. A peak power density of 75.6 $mW\,cm^{-2}$ under H_2/O_2 at 60 °C was achieved for the QP0.5–1 MEA [57].

Research on IPN has generated some promising results. IPNs exhibit a good mechanical strength, moderate ion conductivity, and good chemical stability at reasonable cost. Nevertheless, a slow extraction taken place over time was observed since the polyelectrolyte is not chemically bound to the other polymer, resulting in a loss of ion conductivity [1]. Therefore, although they are generally better than hybrid membranes, IPNs are still unsatisfactory in terms of ion conductivity and durability for AEMFC applications.

4.4 Homogenous Membranes

With the aim of enhancing the stability and mechanical properties of the membranes, extensive research has been undertaken on the design and synthesis of homogenous AEMs. In a homogeneous membrane, the cation groups are covalently bonded to the polymer matrix. Generally, the synthetic routes of homogenous membranes can be categorized into two pathways: direct polymerization of cationic monomers and/or comonomers and post polymerization functionalization, in which the cationic moieties are introduced into the polymer matrix. A variety of polymer backbones, such as poly(arylene ether)s [58–61], poly(phenylene)s [62], poly(styrene)s [63–65], poly(olefin)s [66–69], and poly(phenylene oxide)s [70–73], have been employed to prepare homogenous AEMs.

4.4.1 Homogenous Membranes Based on Poly(arylene ether)s

In the reported polymeric materials for homogenous AEMs, poly(arylene ether)s have been the most frequently used polymer backbones. Among them, polysulfones (PS) have garnered a lot of attention due to their toughness, chemical stability, and easy modification. Typically, the preparation of PS-based AEMs consists of three steps, as shown in Figure 4.10 [60]. First, chloromethyl groups were introduced to the aromatic position of the PS backbone via electrophilic substitution reaction in the presence of Lewis acid. Second, chloromethylated PSs reacted with trimethylamine to give the quaternized PSs, which were AEMs in the Cl^- form. Finally, the quaternized PSs were converted to PS AEMs by immersing in 1 M KOH solution. The PS-based AEMs exhibited hydroxide conductivity which was in the order of 10^{-2} S cm^{-1} at room temperature [60].

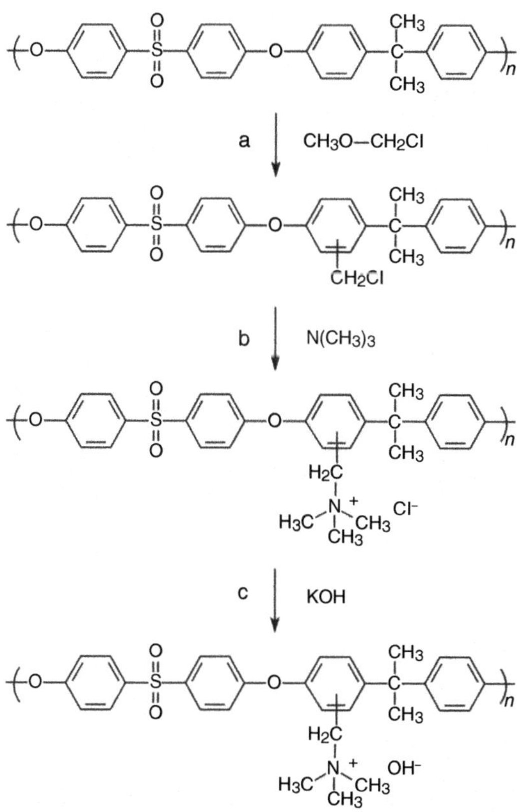

Figure 4.10 Synthetic route of quaternary ammonium polysulfone in hydroxide form. Source: © Wiley-VCH Verlag GmbH & Co. KGaA, Weinheim, 2010 [60].

In many reports, the PS-based AEMs were prepared via chloromethylation and quaternization of the benzylchloromethyl groups. However, producing new PS-based AEMs via the chloromethylation reaction has been hindered because of the use of toxic and carcinogenic chloromethyl methyl ether. Yan and Hickner [59] introduced the benzylmethyl moieties into the polymer during the polymer synthesis, thereby circumventing post-modification of poly(sulfone)s by chloromethylation (Figure 4.11). The ionic conductivity and water uptake of the PS-based AEMs could be tuned over a wide range by controlling the distribution of the cationic groups [59].

Mohanty et al. [58] reported the preparation of PS-based AEMs by methods of iridium-catalyzed C–H borylation followed by palladium-catalyzed Suzuki coupling (Figure 4.12). The side reactions could be minimized, and the DF of the polysulfones

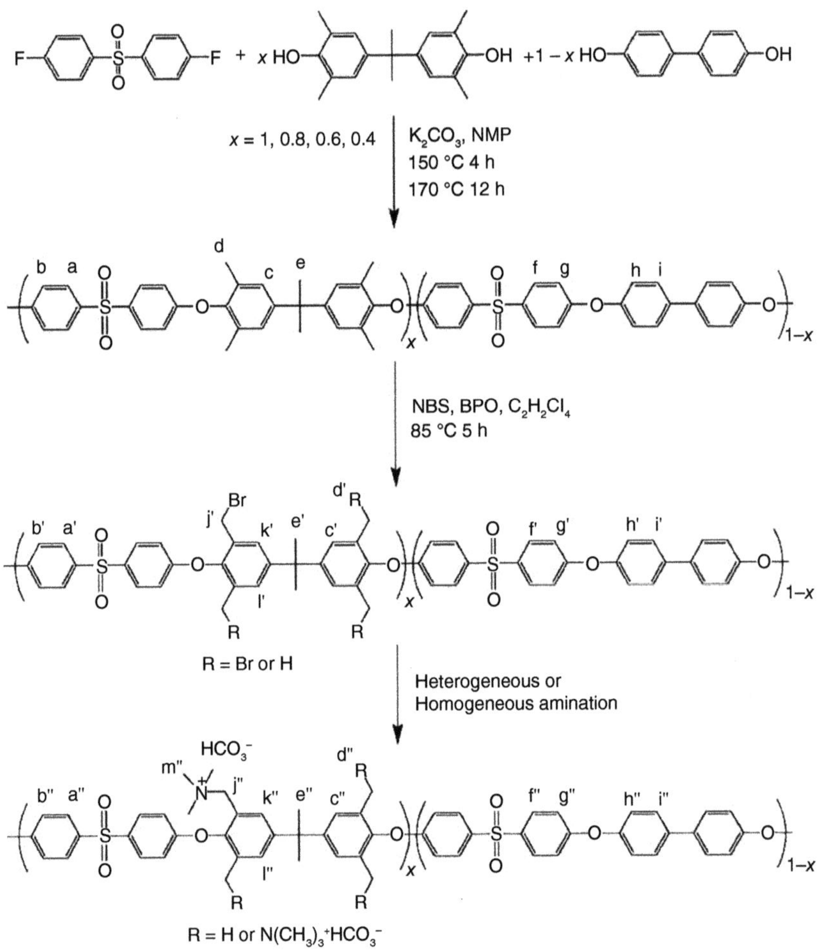

Figure 4.11 Synthetic route of quaternary ammonium-containing polymers from bromination of tetramethyl bisphenol A-based PS. Source: © American Chemical Society, 2010 [59].

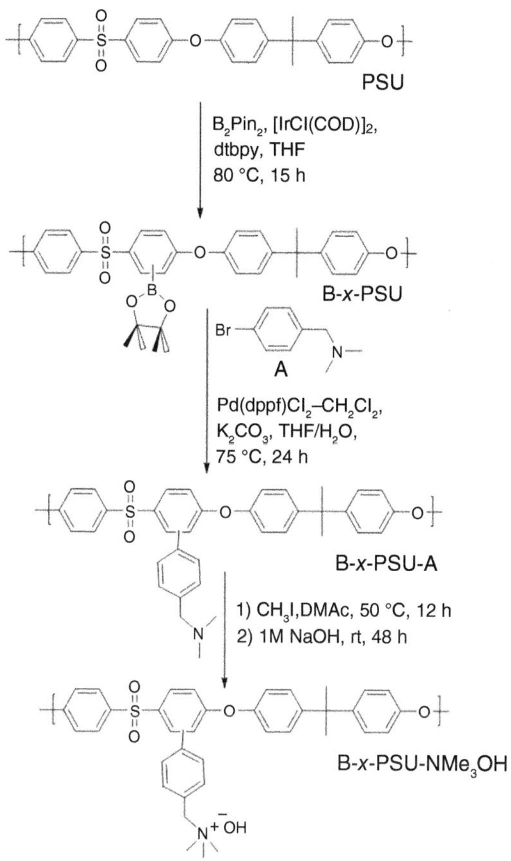

Figure 4.12 Synthetic route of trimethylbenzylammonium functionalized PS via C–H borylation and Suzuki coupling reactions. Source: © American Chemical Society, 2014 [58].

could be easily controlled due to the use of mild reaction conditions [58]. Compared to the corresponding AEM materials prepared via chloromethylation, these kinds of AEMs exhibited lower water uptake, while maintaining similar ionic conductivity [58]. These kind of AEMs exhibited hydroxide conductivity as high as 56 mS cm^{-1} in deionized water at 30 °C with IEC = 2.64 mmol g^{-1} [58].

Gu et al. [61] designed and synthesized a quaternary phosphonium-based polysulfone-AEMs: tri(2,4,6-trimethoxyphenyl) polysulfone-methylene quaternary phosphonium hydroxide (TPQPOH) (Figure 4.13), which was soluble in low-boiling-point water-soluble solvents. For this kind of AEMs, hydroxide conductivity as high as 45 mS cm^{-1} could be achieved in deionized water at 30 °C with IEC = 1.17 mmol g^{-1} [61]. The H_2–O_2 fuel cell based on TPQPOH AMEs using Pt/C as catalysts exhibited a peak power density of 258 mW cm^{-2} at 70 °C [61]. Although TPQPOH AEMs demonstrated high ionic conductivity, the large swelling ratio at elevated temperature could not ensure the mechanical strength of the AEM materials during the fuel cell operating conditions.

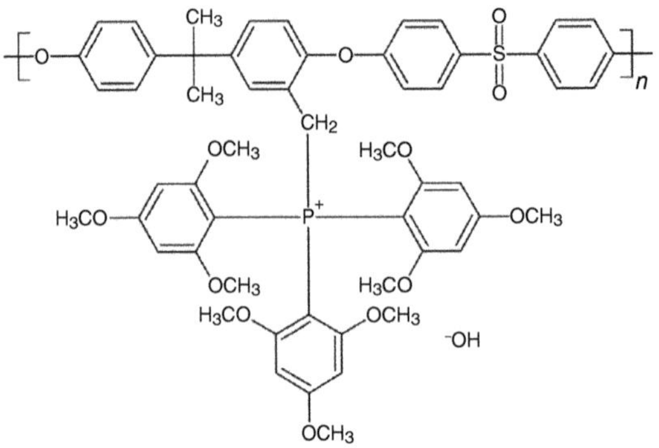

Figure 4.13 Chemical structure of the tri(2,4,6-trimethoxyphenyl) polysulfone-methylene quaternary phosphonium hydroxide (TPQPOH). Source: © Wiley-VCH Verlag GmbH & Co. KGaA, Weinheim, 2009 [61].

4.4.2 Homogenous Membranes Based on Poly(styrene)s

Zeng et al. [63] synthesized AEMs using polystyrene-block-poly(ethylene-ran-butyl)-block-polystyrene (SEBS) as the starting material (Figure 4.14). Chloromethyl groups were introduced into the SEBS by using paraformaldehyde and hydrochloric acid in the presence of $ZnCl_2$ catalyst [63]. The reaction of chloromethyl groups with trimethylamine gave the quaternized SEBS which was converted to an AEM by immersion in 1 M KOH solution [63]. This kind of AEMs exhibited low hydroxide conductivity (9.37 mS cm^{-1}) in deionized water at 80 °C due to the low IEC [63].

Mohanty et al. [65] reported AEMs based on the chemically stable and elastomeric triblock copolymer, polystyrene-*b*-poly(ethylene-*co*-butylene)-*b*-polystyrene (SEBS), which demonstrated high hydroxide conductivity and chemical stability (Figure 4.15). The SEBS triblock copolymer was functionalized via iridium-catalyzed aromatic C–H borylation and palladium-catalyzed Suzuki cross-coupling. This kind of AEM exhibited hydroxide conductivity as high as 45 mS cm^{-1} in deionized water at 30 °C with IEC = 2.41 mmol g^{-1} [65]. The H_2–O_2 fuel cell of the MEA based on quaternized SEBS using Pt/C as catalysts exhibited a peak power density of 240 mW cm^{-2} at 80 °C [65]. These membranes showed a high hydroxide conductivity (86.1 mS cm^{-1} at room temperature). Zhu et al. [64] prepared homogenous AEMs based on quaternized diblock copolystyrene containing a fluorinated hydrophobic block [64]. The PS-based AEMs exhibited a hydroxide conductivity up to 86.1 mS cm^{-1} at room temperature due to the enhanced hydrophilic–hydrophobic discrimination between the different polymer segments upon introduction of fluoro- or pentafluoro-substituted benzene groups [64].

4.4.3 Homogenous Membranes Based on Poly(2,6-dimethyl-1,4-phenylene oxide)

PPO is a good polymer membrane backbone candidate for its high thermal, mechanical, and chemical stabilities. Additionally, it is possible to have modifications on

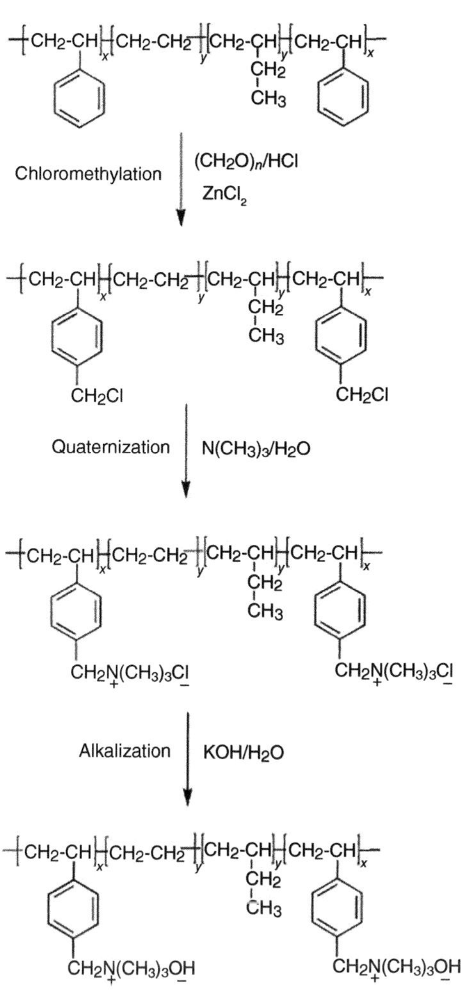

Figure 4.14 Synthetic strategy of SEBS AEMs. Source: © Elsevier, 2010 [63].

both aryl and benzyl positions because of its unique repeat unit structure [74]. For example, reactions such as the free-radical substitution of hydrogen from the benzyl position of PPO; electrophilic substitution on the aryl positions of PPO; metalation of PPO with organometallic compounds; modification and functionalization of PPO from the terminal hydroxyl groups are possible [74].

Yang et al. [71] synthesized poly (2,6-dimethyl-1,4-phenylene oxide)-*b*-poly (vinylbenzyltrimethylammonium) diblock copolymer-based AEMs through growing PVBC blocks from a PPO macroinitiator using nitroxide-mediated polymerization, shown in Figure 4.16. These kind of AEMs exhibited hydroxide conductivity as high as $132\,mS\,cm^{-1}$ in deionized water at 60 °C with IEC = $2.9\,mmol\,g^{-1}$ [71]. In addition, no significant loss of hydroxide conductivity was observed for the AEMs after exposure to 1 M KOH over 13 days at 60 °C [71].

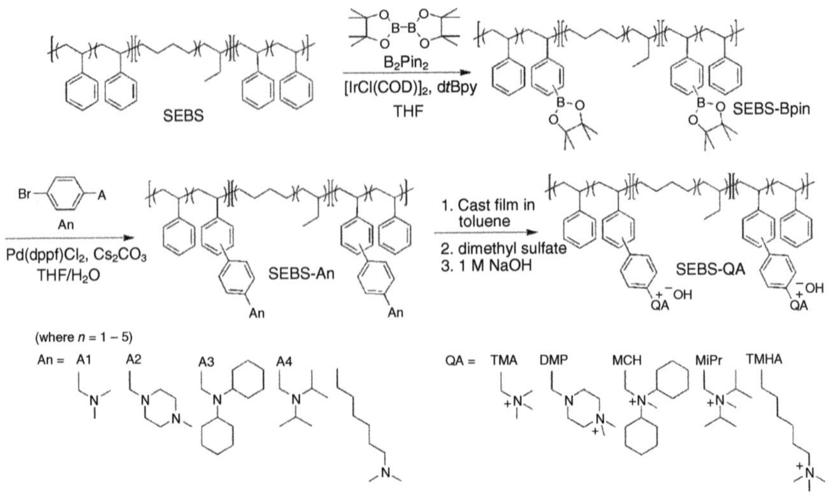

Figure 4.15 Synthesis of quaternized SEBS via iridium-catalyzed aromatic C–H borylation and palladium-catalyzed Suzuki cross-coupling. Source: © American Chemical Society, 2015 [65].

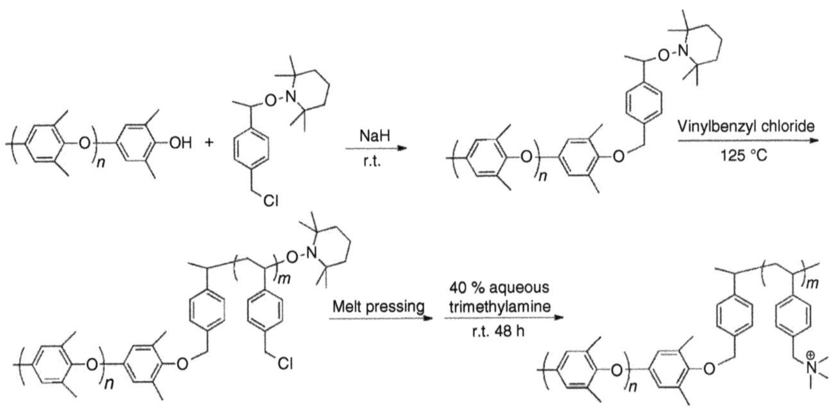

Figure 4.16 Synthetic route for the poly(2,6-dimethyl-1,4-phenylene oxide)-*b*-poly (vinylbenzyltrimethylammonium) diblock copolymer-based AEMs. Source: © American Chemical Society, 2015 [71].

Yang et al. [73] synthesized AEMs based on functionalized PPO with quaternary ammonium side chains via flexible spacers (gQAPPO) by Suzuki–Miyaura coupling reaction, quaternization, and alkalization (Figure 4.17). When the IEC was 1.78 mmol g^{-1}, the gQAPPO exhibited hydroxide conductivity of 0.027 and 0.063 S cm^{-1} at 30 and 70 °C, respectively [73]. The gQAPPO retained more than 90% of their hydroxide conductivities after immersing in 1 M NaOH solution at 60 °C for 168 h [73].

Dang et al. [72] reported AEMs based on PPO modified with heptyltrimethylammonium side chains (PPO-7Q) via lithiation, bromoalkylation, and quaternization (Figure 4.18). Compared to the corresponding benzyltrimethylammonium

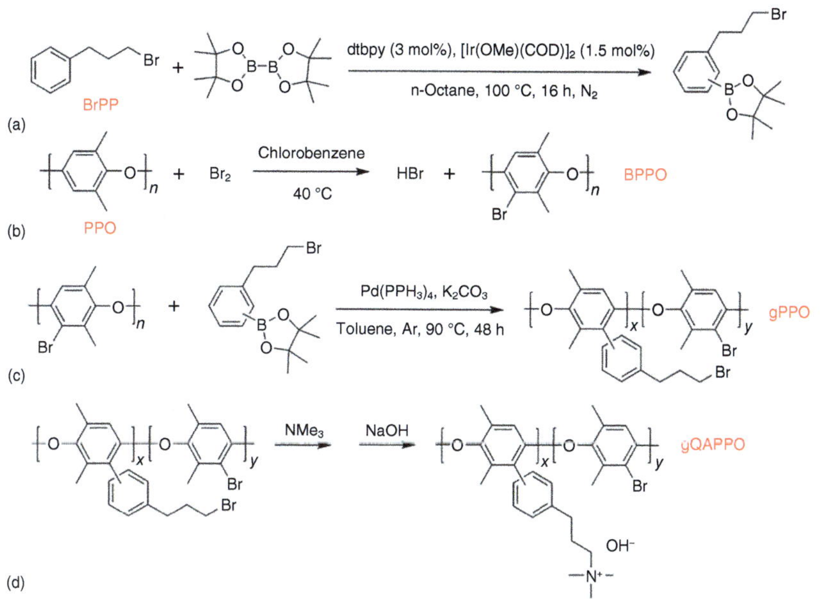

Figure 4.17 Synthetic route of the AEMs containing ammonium groups via flexible spacers (gQAPPO); (a) iridium catalyzed Miyaura reaction; (b) bromination of PPO; (c) Suzuki coupling catalyzed by palladium; (d) quaternization and hydroxide exchange of gPPO membrane. Source: © The Royal Society of Chemistry, 2015 [73]. Reproduced from Yang et al. [73] with permission from the Royal Society of Chemistry.

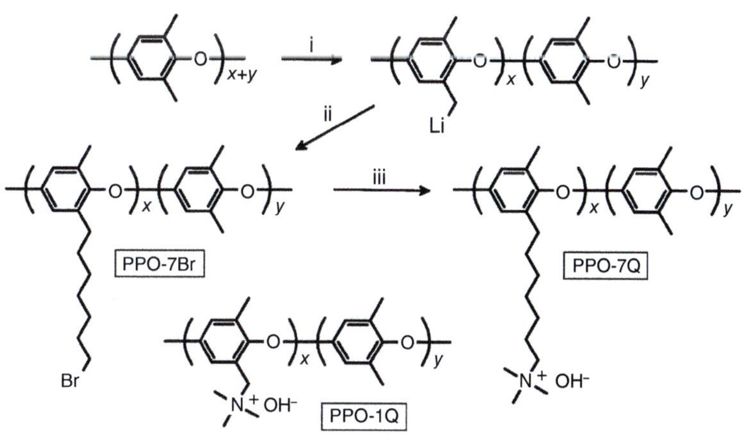

Figure 4.18 Synthetic route for PPO modified with heptyltrimethylammonium side chains (PPO-7Q) via lithiation, bromoalkylation, and quaternization. (i) *n*-BuLi, THF, −78 °C (ii) 1,6-dibromohexane, THF, −78 °C, (iii) trimethylamine, NMP, −78 °C; NaOH. Source: © The Royal Society of Chemistry, 2015 [72]. Reproduced from Dang et al. [72] with permission from the Royal Society of Chemistry.

functionalized polymer, significantly enhanced ionic conductivity and alkaline stability were observed for the PPO-7Q with QA groups via flexible alkyl spacers [72]. The phase-separated PPO-7Q AEMs reached a hydroxide conductivity of 0.035 S cm^{-1} at 20 °C and higher than 0.085 S cm^{-1} at 80 °C in water [72]. After exposure to 1 M NaOH at 80 °C for 196 h, no detectable degradation was observed for the PPO-7Q, suggesting their high performance as AEMs [72]. They had a systematic study including PPO-based AEMs functionalized with 1-methylazepanium, 1-methylpyrrolidinium, 1-methylmorpholinium, quinuclidinium, as well as 1-methyl-, 1,4-dimethyl-, 1,3,5-trimethyl-, and 1,2,6-trimethylpiperidinium [70]. The results demonstrated the quaternary piperidinium and quinuclidinium functionalized AEM materials showed the best overall properties [70].

4.4.4 Fluorene-Containing Homogenous Membranes

AEMs containing fluorenyl groups have been proven to not only have good chemical, thermal, and chemical stability, but these types of AEMs have also been shown to have high hydroxide ion conductivity [75–78].

Tanaka et al. [76] synthesized poly (arylene ether)s containing quaternized ammonia-substituted fluorene groups via block copolycondensation of fluorene-containing oligomers and oligo(arylene ether sulfone ketone)s (Figure 4.19). The IEC of the AEMs could be tunable over a wide range due to the large amount of quaternizable positions on fluorene-containing units [76]. When the IEC was 1.93 mmol g^{-1}, the AEM-containing multiblock exhibited hydroxide conductivity as high as 0.144 S cm^{-1} at 80 °C [76]. Tanaka et al. [76] ascribed the high ionic conductivity to developed hydrophobic/hydrophilic phase separation and interconnected ion transport channels, facilitating the ion transport efficiency. The hydrazine/O_2 fuel cell performance of the MEA exhibited a peak power density of 297 mW cm^{-2} at 80 °C [76].

Lee et al. [77] reported novel fluorene-based AEMs with long alkyl-tethered QA groups on the side chains via $Pd(PPh_3)_4$-catalyzed Suzuki couplings (Figure 4.20). Compared to the aryl ether-containing polymers, degradation of the polyfluorene (PF) backbone in the strong alkaline conditions was largely restrained due to the replacement of C—O bonds on the polymer backbones [77]. The fluorene-based AEMs exhibited hydroxide conductivity up to 0.124 S cm^{-1} at 80 °C with IEC = 3.59 mmol g^{-1} [77]. The ^1H NMR degradation study of the fluorene-based AEMs demonstrated that there was no backbone degradation or ammonium group loss after exposure to 1 M NaOH at 80 °C for 30 days, suggesting the excellent thermo-chemical stability in alkaline conditions [77]. Recently, Xu et al. [78] prepared ether-free PF bearing pendent piperidinium groups (PFPE-Pi) via $Pd(PPh_3)_4$-catalyzed Suzuki couplings. The H_2/O_2 fuel cell performance of the MEA exhibited a peak power density of 661 mW cm^{-2} under 100% RH at 80 °C [78].

4.4.5 Homogenous Membranes Based on Polyolefins

Recently, poly(olefin)-based AEMs has been extensively studied since the polymer backbones contain no heteroatoms [66–69]. Kostalik et al. [69] reported a solvent

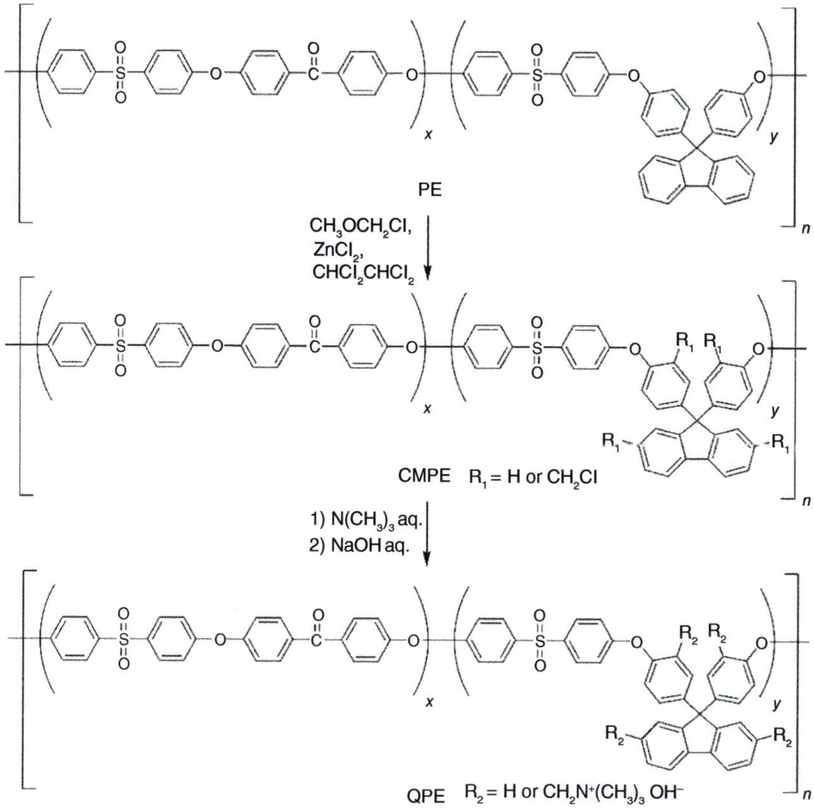

Figure 4.19 Synthetic route of QPE. Source: © American Chemical Society, 2011 [76].

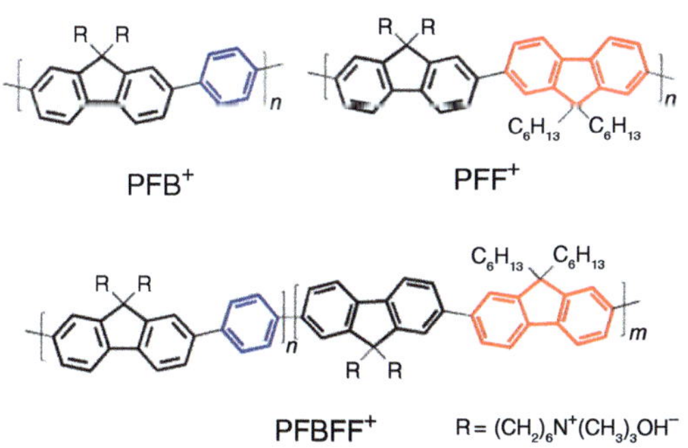

Figure 4.20 Polymer structures of fluorene-based AEMs. Source: © American Chemical Society, 2015 [77].

processable, tetraalkylammonium-functionalized polyethylene-based AEMs through the ring-opening metathesis polymerization (ROMP) [69]. The AEMs was synthesized from tetraalkylammonium-functionalized monomer and cyclooctene by using Grubb's second-generation catalyst. These kind of AEMs exhibited hydroxide conductivity as high as 40 mS cm^{-1} in deionized water at 20 °C with

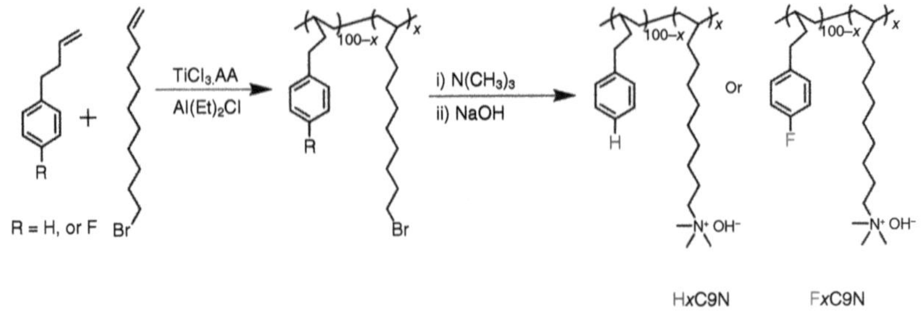

Figure 4.21 Synthetic route for polyolefin-based AEMs using the Ziegler–Natta catalyst.

IEC = 1.29 mmol g^{-1} [69]. In the development of highly chemically stable AEMs, Noonan et al. [68] designed a new kind of highly chemically stable AEMs based on tetrakis(dialkylamino)phosphonium-functionalized polyethylene due to the excellent chemical stability of bulky $[P(N(Me)Cy)_4]^+$ [68]. The tetrakis(dialkylamino) phosphonium-functionalized AEM exhibited a hydroxide conductivity of 22 mS cm^{-1} in deionized water at 22 °C with IEC = 0.67 mmol g^{-1}. No significant loss of hydroxide conductivity was observed for the AEMs after exposure to 15 M KOH for 20 weeks at room temperature [68].

Recently, poly(olefin)-based AEMs prepared by Ziegler–Natta polymerization have become a target because of easy and low-cost preparation. Zhu et al. [39] designed and synthesized poly(olefin)-based AEMs with three cations per side chain and boosted the hydroxide ion conductivities up to 115 mS cm^{-1} under 95% RH at 80 °C. Using these triple-cation poly(olefin) AEMs, they demonstrated a fuel cell with a peak power density of 0.94 W cm^{-2} under H_2/O_2 at 70 °C. To obtain consistently high performance and steady-state operation, Zhu et al. [66] designed fluorine-substituted poly(olefin)-based AEMs (Figure 4.21). Using this new generation of AEMs, they demonstrated a stable and high-performance AEMFC, which was subjected to a constant current density of 600 mA cm^{-2} under H_2/O_2 flow for 120 h [66].

4.4.6 Other Kinds of Homogenous Membranes

Thomas et al. [79] prepared a stable hydroxide-conducting membrane based on benzimidazolium hydroxide via methylation of poly(benzimidazole) (Figure 4.22). Traditionally, the hydroxide form of poly(benzimidazole) was shown to be unstable under alkaline conditions [79]. This polymer was stabilized by steric crowding around the labile benzimidazolium C2 position. The resulting AEMs exhibited a hydroxide conductivity up to 0.013 S cm^{-1} at 20 °C with an IEC = 1.0 mmol g^{-1} [79]. Yang et al. [80] synthesized a new type of AEMs containing the V-shape rigid Tröger's base unit. The resulting AEMs exhibited high hydroxide conductivity at relatively low IEC, as well as a high swelling resistance and chemical stability. It demonstrated a hydroxide conductivity as high as 164.4 mS cm^{-1} at 80 °C with an IEC = 0.82 mmol g^{-1} [80]. Olsson et al. [81] designed and synthesized a series of

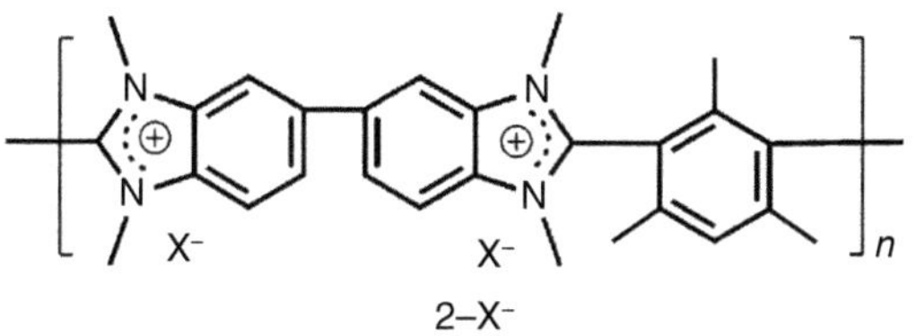

2–X⁻

Figure 4.22 Chemical structure of hydroxide-conducting AEMs based on benzimidazolium hydroxide. Source: © American Chemical Society, 2012 [79].

aliphatic polyelectrolytes based on different *N*-spirocyclic QA cations by radical-initiated cyclo-polymerizations of *N,N*-diallylazacycloalkane quaternary salts. The poly(*N,N*-diallylpiperidinium) demonstrated high alkaline stability, with only minor degradation at 120 °C under alkaline conditions [81]. In addition, the resulting AEMs showed hydroxide conductivity above 100 mS cm^{-1} at 80 °C [81]. Pham et al. [82] prepared *N*-spirocyclic QA ionenes (spiro-ionenes) containing 5- and 6-membered rings fused by central nitrogen cations. The resulting AEMs exhibited excellent thermal and alkaline stability with no degradation observed after 1800 h treatment in 1 M KOD/D_2O at 80 °C [82]. Pham et al. [83] prepared AEMs based on copoly(biphenyl alkylene)s carrying *N*-spirocyclic QA cations, which was synthesized by superacid-mediated polycondensation involving piperidone, biphenyl, and trifluoromethylketones, followed by cyclo-quaternization of the piperidine rings. The resulting AEMs showed both high hydroxide conductivity and chemical stability [83].

It is reported that both interstitial spacers and terminal pendent *n*-alkyl groups enhanced the stability of the tetraalkylammonium, but the interstitial spacers seem to make the greatest contribution to eliminate the benzyl substitution in the benzyl-linked ammonium small molecules and polymers [84]. Inspired by these results, Zhu et al. [75] designed fluorene-containing side chain AEMs (Figure 4.23) in order to obtain *n*-alkyl interstitial spacers tetraalkylammonium cations AEMs. The fluorene side chain containing membranes showed considerably higher hydroxide ion conductivities, up to 176 mS cm^{-1} at 80 °C, compared to that of typical AEMs based

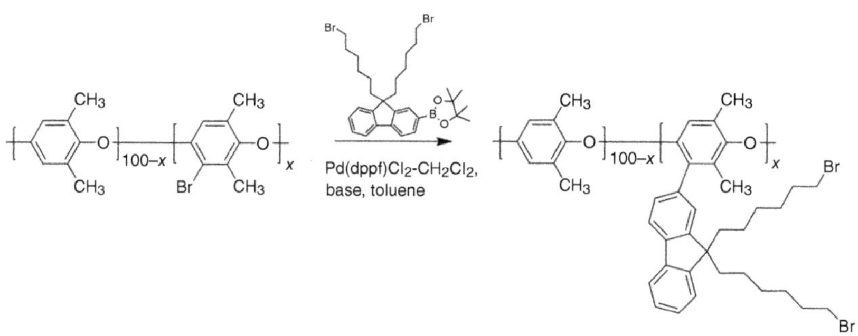

Figure 4.23 Synthetic strategy for the fluorene single-side chain poly(2,6-dimethyl-1, 4-phenylene oxide).

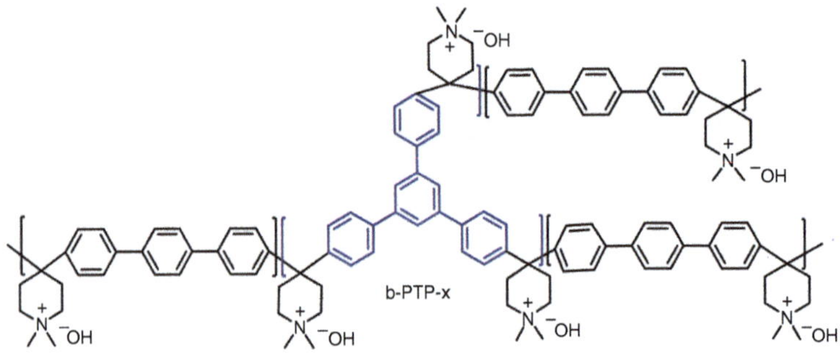

Figure 4.24 Chemical structure of branched poly(terphenyl piperidinium)s. Source: © Wiley-VCH Verlag GmbH & Co. KGaA, Weinheim, 2021 [85].

on the benzyltrimethylammonium motif. The long-term alkaline stability of the AEMs was evaluated by treating the AMEs with argon-saturated 1 M NaOH solution at 80 °C for 1000 h [75]. Only 7.0% hydroxide conductivity degradation in the was AEMs observed after the testing period, which clearly indicates that the fluorene-containing AEMs have excellent stability under alkaline conditions [75].

Wu et al. [85] prepared a branched poly(aryl piperidinium) AEM (Figure 4.24), which exhibited high hydroxide conductivity, low swelling ratio, and desirable mechanical properties. Using these AEMs, they demonstrated a fuel cell with a peak power density of 2.3 W cm^{-2} under H_2/O_2 at 60 °C [85]. The AEMFCs based on the membranes run stably over 500 h, indicating their promising applications in AEMFCs [85]. In order to enhance the mechanical properties of the poly(aryl-*co*-aryl piperidinium) (c-PAP) AEMs, Chen et al. [86] prepared a series of *in situ* styrene-crosslinked c-PAP (x-PAP-PS) membranes, which exhibited hydroxide conductivity (>150 mS cm^{-1} at 90 °C), tensile strength (>80 MPa), and chemical and mechanical stability (~90% remain of ion conductivity and mechanical strength after treatment in 1 M NaOH at 80 °C for 2160 h). Using the x-PAP-PS membrane, they demonstrated a stable AEMFC, which was subjected to a constant current density of 400 mA cm^{-2} under H_2/O_2 flow for 120 h [86].

Research on the homogenous membranes has generated some AEMs which demonstrated high performance in the device, including the high peak power density and long-term durability. Some of the results are very promising for large-scale applications in AEMFCs. In order to achieve the low-cost, durable, commercially viable AEM materials, more researches need to be focused on how to scale up the promising AEMs and manufacture in large quantities.

4.5 Conclusions

For the hybrid membranes, the organic parts typically provide the electrochemical properties, and the inorganic ones ensure the mechanical strength. However, the heterogenous membrane usually demonstrates relatively low hydroxide conductivities, hindering their application on fuel cells. The optimized combination of the polymer backbone and grafts can provide a graft membrane which exhibits both high hydroxide ion conductivity and chemical stability, resulting in high-performance devices. It is well-known that the mechanical flexibility and toughness of AEMs is of great importance to the long-term fuel cell performance where the device is operated under wet–dry cycling conditions. The IPNs demonstrate outstanding mechanical strength and flexibility because of containing both rigid and flexible polymer constituents. Thus, the IPNs materials have drawn great interest because of the excellent combination of good mechanical strength, flexibility, and high ion conductivities. Although they are generally better than hybrid membranes, IPNs are still unsatisfactory in terms of ion conductivity and durability for AEMFC applications. By tuning the structures of polymer backbones and side chains, the homogeneous membranes have shown high ion conductivity, excellent chemical stability, and great performance in devices. The homogenous membrane is regarded as the most promising candidate for AEMFC because of its great cell performance and easy preparation.

References

1 Merle, G., Wessling, M., and Nijmeijer, K. (2011). Anion exchange membranes for alkaline fuel cells: a review. *J Membr Sci* 377 (1–2): 1–35.

2 Fenton, D.E., Parker, J.M., and Wright, P.V. (1973). Complexes of alkali metal ions with poly(ethylene oxide). *Polymer* 14: 589.

3 Armand, M.B., Chabagno, J.M., and Duclot, M.J. (1979). Poly-ethers as solid electrolytes. *Fast Ion Trans Solids* 131–136.

4 Wu, Y., Wu, C., Gong, M., and Xu, T. (2006). New anion exchanger organic–inorganic hybrid materials and membranes from a copolymerof glycidylmethacrylate and γ-methacryloxypropyl trimethoxy silane. *J Appl Polym Sci* 102 (4): 3580–3589.

5 Yang, C.-C. (2007). Synthesis and characterization of the cross-linked PVA/TiO_2 composite polymer membrane for alkaline DMFC. *J Membr Sci* 288 (1–2): 51–60.

6 Wu, Y., Wu, C., Yu, F. et al. (2008). Free-standing anion-exchange PEO–SiO_2 hybrid membranes. *J Membr Sci* 307 (1): 28–36.

7 Wu, Y., Wu, C., Xu, T. et al. (2009). Novel silica/poly(2,6-dimethyl-1,4-phenylene oxide) hybrid anion-exchange membranes for alkaline fuel cells: effect of heat treatment. *J Membr Sci* 338 (1–2): 51–60.

8 Wu, Y., Wu, C., Li, Y. et al. (2010). PVA–silica anion-exchange hybrid membranes prepared through a copolymer crosslinking agent. *J Membr Sci* 350 (1–2): 322–332.

9 Wu, Y., Wu, C., Varcoe, J.R. et al. (2010). Novel silica/poly(2,6-dimethyl-1, 4-phenylene oxide) hybrid anion-exchange membranes for alkaline fuel cells: effect of silica content and the single cell performance. *J Power Sources* 195 (10): 3069–3076.

10 Couture, G., Alaaeddine, A., Boschet, F., and Ameduri, B. (2011). Polymeric materials as anion-exchange membranes for alkaline fuel cells. *Prog Polym Sci* 36 (11): 1521–1557.

11 Liao, X., Ren, L., Chen, D. et al. (2015). Nanocomposite membranes based on quaternized polysulfone and functionalized montmorillonite for anion-exchange membranes. *J Power Sources* 286: 258–263.

12 Liu, L., Tong, C., He, Y. et al. (2015). Enhanced properties of quaternized graphenes reinforced polysulfone based composite anion exchange membranes for alkaline fuel cell. *J Membr Sci* 487: 99–108.

13 Yi, Z., Chaorong, C., Danying, Z., and Hongwei, Z. (2015). Hybrid anion-exchange membranes derived from quaternized polysulfone and functionalized titanium dioxide. *Electrochim Acta* 177: 128–136.

14 Feng, T., Lin, B., Zhang, S. et al. (2016). Imidazolium-based organic–inorganic hybrid anion exchange membranes for fuel cell applications. *J Membr Sci* 508: 7–14.

15 Hu, B., Liu, L., Zhao, Y., and Lü, C. (2016). A facile construction of quaternized polymer brush-grafted graphene modified polysulfone based composite anion exchange membranes with enhanced performance. *RSC Adv* 6 (56): 51057–51067.

16 Li, J., Yan, X., Zhang, Y. et al. (2016). Enhanced hydroxide conductivity of imidazolium functionalized polysulfone anion exchange membrane by doping imidazolium surface-functionalized nanocomposites. *RSC Adv* 6 (63): 58380–58386.

17 Shi, B., Li, Y., Zhang, H. et al. (2016). Tuning the performance of anion exchange membranes by embedding multifunctional nanotubes into a polymer matrix. *J Membr Sci* 498: 242–253.

18 Miao, L., Wang, X., Fu, Y. et al. (2017). Quaternized polyhedral oligomeric silsesquioxanes (QPOSS) modified polysulfone-based composite anion exchange membranes. *Solid State Ionics* 309: 170–179.

19 Danks, T.N., Slade, R.C.T., and Varcoe, J.R. (2003). Alkaline anion-exchange radiation-grafted membranes for possible electrochemical application in fuel cells. *J Mater Chem* 13 (4): 712–721.

20 Liu, H., Yang, S., Wang, S. et al. (2011). Preparation and characterization of radiation-grafted poly (tetrafluoroethylene-*co*-perfluoropropyl vinyl ether) membranes for alkaline anion-exchange membrane fuel cells. *J Membr Sci* 369 (1–2): 277–283.

21 Mamlouk, M., Horsfall, J.A., Williams, C., and Scott, K. (2012). Radiation grafted membranes for superior anion exchange polymer membrane fuel cells performance. *Int J Hydrog Energy* 37 (16): 11912–11920.

22 Ponce-González, J., Whelligan, D.K., Wang, L. et al. (2016). High performance aliphatic-heterocyclic benzyl-quaternary ammonium radiation-grafted anion-exchange membranes. *Energy Environ Sci* 9 (12): 3724–3735.

23 Wang, L., Brink, J.J., Liu, Y. et al. (2017). Non-fluorinated pre-irradiation-grafted (peroxidated) LDPE-based anion-exchange membranes with high performance and stability. *Energy Environ Sci* 10 (10): 2154–2167.

24 Wang, L., Magliocca, E., Cunningham, E.L. et al. (2017). An optimised synthesis of high performance radiation-grafted anion-exchange membranes. *Green Chem* 19 (3): 831–843.

25 Wang, L., Peng, X., Mustain, W.E., and Varcoe, J.R. (2019). Radiation-grafted anion-exchange membranes: the switch from low- to high-density polyethylene leads to remarkably enhanced fuel cell performance. *Energy Environ Sci* 12 (5): 1575–1579.

26 Herman, H., Slade, R.C.T., and Varcoe, J.R. (2003). The radiation-grafting of vinylbenzyl chloride onto poly(hexafluoropropylene-*co*-tetrafluoroethylene) films with subsequent conversion to alkaline anion-exchange membranes: optimisation of the experimental conditions and characterisation. *J Membr Sci* 218 (1–2): 147–163.

27 Li, N., Leng, Y., Hickner, M.A., and Wang, C.Y. (2013). Highly stable, anion conductive, comb-shaped copolymers for alkaline fuel cells. *J Am Chem Soc* 135 (27): 10124–10133.

28 Pan, J., Li, Y., Han, J. et al. (2013). A strategy for disentangling the conductivity-stability dilemma in alkaline polymer electrolytes. *Energy Environ Sci* 6 (10): 2912–2915.

29 Dang, H.-S. and Jannasch, P. (2016). Anion-exchange membranes with polycationic alkyl side chains attached via spacer units. *J Mater Chem A* 4 (43): 17138–17153.

30 Guo, D., Lai, A.N., Lin, C.X. et al. (2016). Imidazolium-functionalized poly(arylene ether sulfone) anion-exchange membranes densely grafted with flexible side chains for fuel cells. *ACS Appl Mater Interfaces* 8 (38): 25279–25288.

31 Lin, C.X., Huang, X.L., Guo, D. et al. (2016). Side-chain-type anion exchange membranes bearing pendant quaternary ammonium groups via flexible spacers for fuel cells. *J Mater Chem A* 4 (36): 13938–13948.

32 Zhu, L., Pan, J., Wang, Y. et al. (2016). Multication side chain anion exchange membranes. *Macromolecules* 49 (3): 815–824.

33 Pan, J., Han, J., Zhu, L., and Hickner, M.A. (2017). Cationic side-chain attachment to poly(phenylene oxide) backbones for chemically stable and conductive anion exchange membranes. *Chem Mater* 29 (12): 5321–5330.

34 Li, L., Lin, C.X., Wang, X.Q. et al. (2018). Highly conductive anion exchange membranes with long flexible multication spacer. *J Membr Sci* 553: 209–217.

35 Pan, Y., Zhang, Q., Yan, X. et al. (2018). Hydrophilic side chain assisting continuous ion-conducting channels for anion exchange membranes. *J Membr Sci* 552: 286–294.

36 Zhu, L., Yu, X., and Hickner, M.A. (2018). Exploring backbone-cation alkyl spacers for multi-cation side chain anion exchange membranes. *J Power Sources* 375: 433–441.

37 Zhu, Y., Ding, L., Liang, X. et al. (2018). Beneficial use of rotatable-spacer side-chains in alkaline anion exchange membranes for fuel cells. *Energy Environ Sci* 11 (12): 3472–3479.

38 Wei, H., Li, Y., Wang, S. et al. (2019). Side-chain-type imidazolium-functionalized anion exchange membranes: the effects of additional hydrophobic side chains and their hydrophobicity. *J Membr Sci* 579: 219–229.

39 Zhu, L., Yu, X., Peng, X. et al. (2019). Poly(olefin)-based anion exchange membranes prepared using Ziegler–Natta polymerization. *Macromolecules* 52 (11): 4030–4041.

40 Ran, J., Wu, L., Lin, X. et al. (2012). Synthesis of soluble copolymers bearing ionic graft for alkaline anion exchange membrane. *RSC Adv* 2 (10): 4250–4257.

41 Lin, C.X., Zhuo, Y.Z., Lai, A.N. et al. (2016). Side-chain-type anion exchange membranes bearing pendent imidazolium-functionalized poly(phenylene oxide) for fuel cells. *J Membr Sci* 513: 206–216.

42 Yang, C., Wang, S., Ma, W. et al. (2016). Highly stable poly(ethylene glycol)-grafted alkaline anion exchange membranes. *J Mater Chem A* 4 (10): 3886–3892.

43 Hu, M., Ding, L., Shehzad, M.A. et al. (2019). Comb-shaped anion exchange membrane with densely grafted short chains or loosely grafted long chains? *J Membr Sci* 585: 150–156.

44 Lin, B., Xu, F., Su, Y. et al. (2019). Facile preparation of anion-exchange membrane based on polystyrene-*b*-polybutadiene-*b*-polystyrene for the application of alkaline fuel cells. *Ind Eng Chem Res* 58 (49): 22299–22305.

45 Zhang, S., Wang, Y., Gao, X. et al. (2020). Enhanced conductivity and stability via comb-shaped polymer anion exchange membrane incorporated with porous polymeric nanospheres. *J Membr Sci* 597: 117750.

46 Pan, J., Zhu, L., Han, J., and Hickner, M.A. (2015). Mechanically tough and chemically stable anion exchange membranes from rigid-flexible semi-interpenetrating networks. *Chem Mater* 27 (19): 6689–6698.

47 Gong, Y., Liao, X., Xu, J. et al. (2016). Novel anion-conducting interpenetrating polymer network of quaternized polysulfone and poly(vinyl alcohol) for alkaline fuel cells. *Int J Hydrog Energy* 41 (13): 5816–5823.

48 Guo, D., Zhuo, Y.Z., Lai, A.N. et al. (2016). Interpenetrating anion exchange membranes using poly(1-vinylimidazole) as bifunctional crosslinker for fuel cells. *J Membr Sci* 518: 295–304.

49 Zeng, L., He, Q., Liao, Y. et al. (2020). Anion exchange membrane based on interpenetrating polymer network with ultrahigh ion conductivity and excellent stability for alkaline fuel cell. *Research* 2020: 4794706.

50 Zeng, L., Liao, Y., Wang, J., and Wei, Z. (2021). Construction of highly efficient ion channel within anion exchange membrane based on interpenetrating polymer network for H_2/Air (CO_2-free) alkaline fuel cell. *J Power Sources* 486: 229377.

51 Wang, J., He, R., and Che, Q. (2011). Anion exchange membranes based on semi-interpenetrating polymer network of quaternized chitosan and polystyrene. *J Colloid Interface Sci* 361 (1): 219–225.

52 He, S.S., Strickler, A.L., and Frank, C.W. (2015). A semi-interpenetrating network approach for dimensionally stabilizing highly-charged anion exchange membranes for alkaline fuel cells. *ChemSusChem* 8 (8): 1472–1483.

53 Chen, C., Chen, B., and Hong, R. (2018). Preparation and properties of alkaline anion exchange membrane with semi-interpenetrating polymer networks based on poly(vinylidene fluoride-*co*-hexafluoropropylene). *J Appl Polym Sci* 135 (5): 45775.

54 Xue, J., Liu, L., Liao, J. et al. (2018). Semi-interpenetrating polymer networks by azide–alkyne cycloaddition as novel anion exchange membranes. *J Mater Chem A* 6 (24): 11317–11326.

55 Ureña, N., Pérez-Prior, M.T., del Rio, C. et al. (2021). New amphiphilic semi-interpenetrating networks based on polysulfone for anion-exchange membrane fuel cells with improved alkaline and mechanical stabilities. *Polymer* 226: 123824.

56 Yang, W., Xu, P., Li, X. et al. (2021). Mechanically robust semi-interpenetrating polymer network via thiol-ene chemistry with enhanced conductivity for anion exchange membranes. *Int J Hydrog Energy* 46 (17): 10377–10388.

57 Yang, Z., Zhang, M., Zhao, Z. et al. (2021). Construction of quaternized polysulfone/polyquaternium-10 anion exchange membrane with semi-interpenetrating network for alkaline fuel cell. *Macromol Mater Eng* 307 (1): 2100539.

58 Mohanty, A.D., Lee, Y.-B., Zhu, L. et al. (2014). Anion exchange fuel cell membranes prepared from C–H borylation and Suzuki coupling reactions. *Macromolecules* 47 (6): 1973–1980.

59 Yan, J. and Hickner, M.A. (2010). Anion exchange membranes by bromination of benzylmethyl-containing poly(sulfone)s. *Macromolecules* 43 (5): 2349–2356.

60 Pan, J., Lu, S., Li, Y. et al. (2010). High-performance alkaline polymer electrolyte for fuel cell applications. *Adv Funct Mater* 20 (2): 312–319.

61 Gu, S., Cai, R., Luo, T. et al. (2009). A soluble and highly conductive ionomer for high-performance hydroxide exchange membrane fuel cells. *Angew Chem Int Ed Eng* 48 (35): 6499–6502.

62 Hibbs, M.R., Fujimoto, C.H., and Cornelius, C.J. (2009). Synthesis and characterization of poly(phenylene)-based anion exchange membranes for alkaline fuel cells. *Macromolecules* 42 (21): 8316–8321.

63 Zeng, Q.H., Liu, Q.L., Broadwell, I. et al. (2010). Anion exchange membranes based on quaternized polystyrene-block-poly(ethylene-ran-butylene)-block-polystyrene for direct methanol alkaline fuel cells. *J Membr Sci* 349 (1–2): 237–243.

64 Zhu, M., Zhang, X., Wang, Y. et al. (2018). Novel anion exchange membranes based on quaternized diblock copolystyrene containing a fluorinated hydrophobic block. *J Membr Sci* 554: 264–273.

65 Mohanty, A.D., Ryu, C.Y., Kim, Y.S., and Bae, C. (2015). Stable elastomeric anion exchange membranes based on quaternary ammonium-tethered polystyrene-*b*-poly(ethylene-*co*-butylene)-*b*-polystyrene triblock copolymers. *Macromolecules* 48 (19): 7085–7095.

66 Zhu, L., Peng, X., Shang, S.L. et al. High performance anion exchange membrane fuel cells enabled by fluoropoly(olefin) membranes. *Adv Funct Mater* 29 (26): 1902059.

67 Zha, Y., Disabb-Miller, M.L., Johnson, Z.D. et al. (2012). Metal-cation-based anion exchange membranes. *J Am Chem Soc* 134 (10): 4493–4496.

68 Noonan, K.J., Hugar, K.M., Kostalik, H.A. et al. (2012). Phosphonium-functionalized polyethylene: a new class of base-stable alkaline anion exchange membranes. *J Am Chem Soc* 134 (44): 18161–18164.

69 Kostalik, H.A., Clark, T.J., Robertson, N.J. et al. (2010). Solvent processable tetraalkylammonium-functionalized polyethylene for use as an alkaline anion exchange membrane. *Macromolecules* 43 (17): 7147–7150.

70 Dang, H.-S. and Jannasch, P. (2017). A comparative study of anion-exchange membranes tethered with different hetero-cycloaliphatic quaternary ammonium hydroxides. *J Mater Chem A* 5 (41): 21965–21978.

71 Yang, Y. and Knauss, D.M. (2015). Poly(2,6-dimethyl-1,4-phenylene oxide)-*b*-poly(vinylbenzyltrimethylammonium) diblock copolymers for highly conductive anion exchange membranes. *Macromolecules* 48 (13): 4471–4480.

72 Dang, H.-S., Weiber, E.A., and Jannasch, P. (2015). Poly(phenylene oxide) functionalized with quaternary ammonium groups via flexible alkyl spacers for high-performance anion exchange membranes. *J Mater Chem A* 3 (10): 5280–5284.

73 Yang, Z., Zhou, J., Wang, S. et al. (2015). A strategy to construct alkali-stable anion exchange membranes bearing ammonium groups via flexible spacers. *J Mater Chem A* 3 (29): 15015–15019.

74 Xu, T., Wu, D., and Wu, L. (2008). Poly(2,6-dimethyl-1,4-phenylene oxide) (PPO)—a versatile starting polymer for proton conductive membranes (PCMs). *Prog Polym Sci* 33 (9): 894–915.

75 Zhu, L., Pan, J., Christensen, C.M. et al. Functionalization of poly(2,6-dimethyl-1, 4-phenylene oxide)s with hindered fluorene side chains for anion exchange membranes. *Macromolecules* 49 (9): 3300–3309.

76 Tanaka, M., Fukasawa, K., Nishino, E. et al. (2011). Anion conductive block poly(arylene ether)s: synthesis, properties, and application in alkaline fuel cells. *J Am Chem Soc* 133 (27): 10646–10654.

77 Lee, W.-H., Mohanty, A.D., and Bae, C. (2015). Fluorene-based hydroxide ion conducting polymers for chemically stable anion exchange membrane fuel cells. *ACS Macro Lett* 4 (4): 453–457.

78 Xu, F., Chen, Y., Lin, B. et al. (2021). Highly durable ether-free polyfluorene-based anion exchange membranes for fuel cell applications. *ACS Macro Lett* 10 (10): 1180–1185.

79 Thomas, O.D., Soo, K.J., Peckham, T.J. et al. (2012). A stable hydroxide-conducting polymer. *J Am Chem Soc* 134 (26): 10753–10756.

80 Yang, Z., Guo, R., Malpass-Evans, R. et al. (2016). Highly conductive anion-exchange membranes from microporous Troger's base polymers. *Angew Chem Int Ed Eng* 55 (38): 11499–11502.

81 Olsson, J.S., Pham, T.H., and Jannasch, P. (2017). Poly(*N,N*-diallylazacycloalkane)s for anion-exchange membranes functionalized with *N*-spirocyclic quaternary ammonium cations. *Macromolecules* 50 (7): 2784–2793.

82 Pham, T.H., Olsson, J.S., and Jannasch, P. (2017). *N*-Spirocyclic quaternary ammonium ionenes for anion-exchange membranes. *J Am Chem Soc* 139 (8): 2888–2891.

83 Pham, T.H., Olsson, J.S., and Jannasch, P. (2018). Poly(arylene alkylene)s with pendant *N*-spirocyclic quaternary ammonium cations for anion exchange membranes. *J Mater Chem A* 6 (34): 16537–16547.

84 Nuñez, S.A., Capparelli, C., and Hickner, M.A. (2016). *N*-Alkyl interstitial spacers and terminal pendants influence the alkaline stability of tetraalkylammonium cations for anion exchange membrane fuel cells. *Chem Mater* 28 (8): 2589–2598.

85 Wu, X., Chen, N., Klok, H.A. et al. (2022). Branched poly(aryl piperidinium) membranes for anion-exchange membrane fuel cells. *Angew Chem Int Ed Eng* 61: e202114892.

86 Chen, N., Hu, C., Wang, H.H. et al. (2021). Chemically & physically stable crosslinked poly(aryl-*co*-aryl piperidinium)s for anion exchange membrane fuel cells. *J Membr Sci* 638: 119685.

5

Proton Exchange Membranes Versus Anion Exchange Membranes

Marilyn M. Xavier[1] *and Suresh Mathew*[2]

[1] *Morning Star Home Science College, Department of Chemistry, Angamaly 683573, Kerala, India*
[2] *Mahatma Gandhi University, School of Chemical Sciences, Kottayam 686560, Kerala, India*

5.1 Introduction

In this century, fuel cell technologies are receiving increasing attention. They are considered to be energy-generating resources with the ability to address the global energy issue without causing environmental harm. With the production of heat and water, fuel cells transform chemical energy found in tiny molecules like hydrogen and methanol straight into electrical energy. Electricity is produced by fuel cells as long as the fuel is available. It has been determined that fuel cell technology is more environmentally friendly since it lessens the geopolitical effects of using nonrenewable fossil fuels. The energy efficiency of a fuel cell is high as 60% in converting electrical energy, highlighting a >90% reduction in the generation of major pollutants. [1]

The working principle of a typical fuel cell can be divided into three steps. (i) Gas fuel such as hydrogen flows into the anode side of the fuel cell, where the platinum catalyst facilitates its oxidation producing hydrogen ions (protons) and electrons, (ii) diffusion of these protons toward the cathode through a membrane which separates the cathode and anode, (iii) reduction of these hydrogen ions combined with oxygen and electron from the assistance of platinum catalyst to form water.

The fuel cells in general are divided into five different categories: (i) polymer electrolyte membrane fuel cells (also called PEFCs), (ii) solid oxide fuel cells (SOFCs), (iii) alkaline fuel cells (AFCs), (iv) phosphoric acid fuel cells (PACs), and (v) molten carbonate fuel cells (MCFCs).

Among these fuel cells, proton-exchange membrane fuel cell (PEMFC) has achieved wide appreciation and development during the past decade. It is commercialized in vehicles, military equipment, portable devices, and so on. But the use of noble metal

Alkaline Anion Exchange Membranes for Fuel Cells: From Tailored Materials to Novel Applications, First Edition. Edited by Jince Thomas, Alex Schechter, Flavio Grynszpan, Bejoy Francis, and Sabu Thomas.

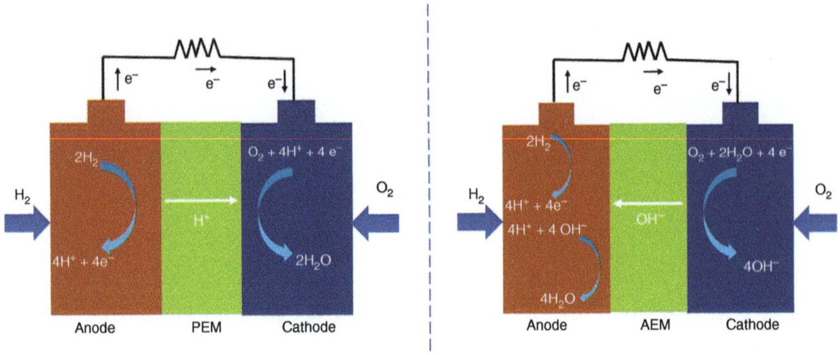

Figure 5.1 Comparison of AAEMFC and PEMFC working principles.

catalysts in PEMFC makes it very costly, and also the electrodes are prone to damage caused by CO or other impurities in the fuel. An alternative to PEMFC is an alkaline anion-exchange membrane fuel cell (AAEMFC) which is currently receiving wide attention, while AAEMFC possesses enhanced electrochemical kinetics of the oxygen reduction reaction (ORR) and could work with a non-precious electrode, which makes it more affordable to use. But the membrane electrode assembly (MEA) where the electrolyte membranes are sandwiched between two bipolar plates, as well as the inflow of the fuel and operation conditions are similar for both PEMFC and AAEMFC (Figure 5.1). In AAEMFC, the ORR is more facile compared to that in PEMFC due to the alkaline conditions with the aid of non-precious ORR catalysts like nickel cobalt and silver or transition metal oxides. Also, the fuel crossover could be more suppressed in an AAEMFC. However, PEMs typically exhibit better ionic conductivity and durability than AEMs and are more commercialized. A sudden resurgence in AAEMFC research and development is as it may provide an inexpensive fuel cell technology.

In a fuel cell MEA, the electrolyte membrane plays an essential role in permeating ionic charges without passing the fuels and electrons through it. Common membrane materials are perfluorinated polymer, non-fluorinated or hydrocarbon polymer, and natural polymer. The overall performance of a fuel cell is decided to a large extent by the mechanical stability, ionic conductivity, power output, and longevity of these electrolyte membranes. Intense research has been going into the designing of specific electrolyte polymeric membranes which could outperform commercial membranes in fuel cell applications. This chapter illustrates some major similarities and dissimilarities between proton-exchange membranes and anion-exchange membranes (AEMs) and how it is affecting the overall performance of corresponding fuel cells.

5.2 Proton Exchange Membrane (PEM)

Proton-exchange membranes (PEMs) or polymer electrolyte membranes are semipermeable membranes that conduct cations, especially protons, and also act as a barrier for the mixing of reactants across the membrane [2]. PEMFCs have high energy

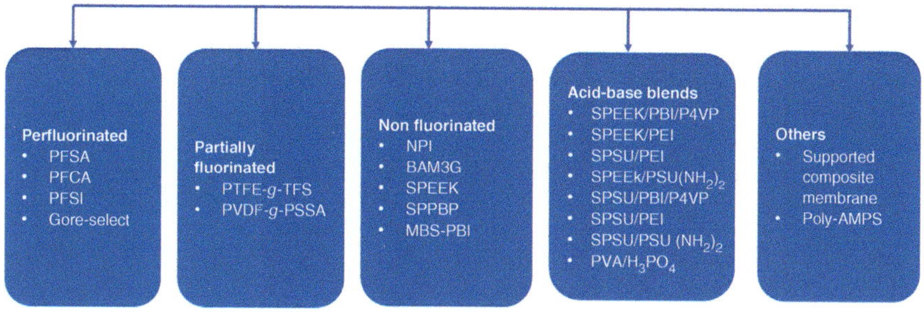

Figure 5.2 Classification of membranes based on materials of synthesis and preparation methods.

and power density, excellent scalability, efficient energy conversion, pollutant-free emission, noiseless operation, and low operation temperature. These advantages make PEMFCs a next-generation device in the field of renewable energy [3].

5.2.1 Classification of PEM Membranes Based on the Materials of Synthesis

Based on the materials used in synthesis, the polymer electrolyte membranes are classified into three main groups: perfluorinated ionomers (or partially perfluorinated), nonfluorinated hydrocarbons (including aliphatic or aromatic structures), and acid–base complexes. In Figure 5.2, the classification of membranes based on materials (perfluorinated, partially fluorinated, and non-fluorinated) and preparation methods (acid–base blends and others) has been shown [4].

5.2.1.1 Perfluorinated Ionomeric Membranes

These membranes have high equivalent weight (EW) and consume high power density [5]. The backbones of these small chain polymers have strong C—F bonds with a highly electronegative fluorine atom and hence possess low polarizability. The reasonable thermal stability, chemical inertness, and enhanced acidity of the sulfonic acid group in CF_2SO_3H make these polymeric membranes an excellent candidate for proton-exchange membranes [6]. These membranes are prepared by polymerizing monomers that contain a moiety that can be converted to either a cationic or an anionic state by additional processing. Dupont developed these membranes with good chemical and thermal stability and named as Nafion (shown in Figure 5.3). Even though many other polymers like Flemion produced by Asahi Glass and Aciplex-S produced by Asahi Chemical were introduced into the market, Nafion is considered superior owing to its high proton conductivity, chemical inertness, and mechanical stability [7].

The Nafion membrane is one of the most widely used PEMs in PEMFCs. Nafion consists of a hydrophobic perfluorinated polyethylene backbone and a highly hydrophilic sulfonic acid-terminated perfluoro vinyl ether pendant [8]. Nafion has high proton conductivity ($0.13\,S\,cm^{-1}$ at 75 °C), 100% relative humidity [RH]), chemical stability, and longevity (>60 000 h) in a fuel cell environment [9]. The sulfonate group

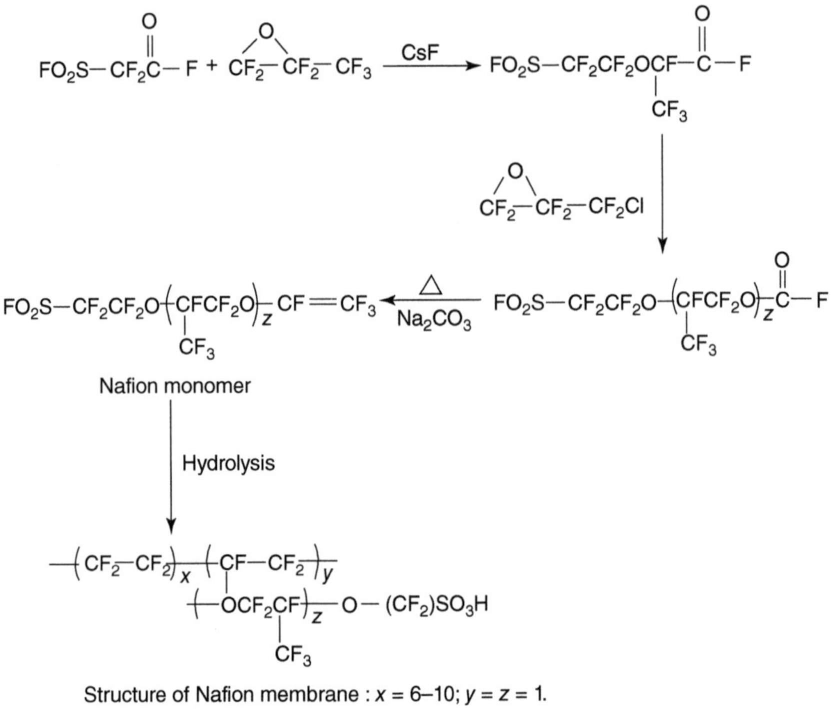

Figure 5.3 Preparation scheme of Nafion and its structure. Source: Reprinted with permission from Kariduraganavar et al. [5].

in Nafion has been chemically grafted into the polytetrafluoroethylene (PTFE) or Teflon backbone. These ionic groups have caused the absorption of a large amount of water by the polymer which in turn leads to the hydration of the polymer. Therefore, the factors affecting the performance of the suitable proton-exchange membrane are the level of hydration and thickness of the membrane which plays an important role in deciding their suitability for their application in the fuel cell [10]. Hence the unique structure of the Nafion membrane renders its long-term stability in oxidative and reductive conditions [11]. Various categories of Nafion membranes are available to cover the market needs for electrolyzers and their various operating conditions. Nafion HP, Nafion 211, Nafion Xl, Nafion 212, Nafion 115, Nafion 117, Nafion 1110 are some of the categories of Nafion membranes currently available in the market. These membranes differs in their thickness, mechanical strength, water uptake capacity, etc. The thickness of the membrane directly affects the pressure difference across the membrane and the operating lifetime. Thicker membranes are more appropriate for very high differential pressure applications or very long operating lifetime, while thin membranes are for fuel cells with negligible differential pressure across the membrane.

5.2.1.2 Partially Fluorinated Hydrocarbon Membranes

The high cost and high methanol permeability of perfluorinated membranes lead to the development of membrane which is impermeable to methanol and more economic as well. Poly(vinylidene fluoride) (PVDF) is a partially fluorinated polymer that has been widely used in batteries as well as in fuel cells [12]. PVDF-g-PSSA membranes prepared by solution grafting technique have shown high proton conductivity, water uptake, and lower oxygen solution ability compared to Nafion membranes [13]. When nano-Al_2O_3 is doped to PVDF before grafting, the resulting PVDF-g-PSSA membranes doped with Al_2O_3 had shown lower methanol permeability [14]. Mordenite-incorporated polyvinyl alcohol (PVA)–polystyrene sulfonic acid (PSSA) blend also exhibits mitigate methanol permeability [15].

5.2.1.3 Non-fluorinated Hydrocarbon Membranes

Non-fluorinated hydrocarbon polymers can be aliphatic or aromatic polymers having an aromatic ring in the backbone or as a pendant group. The hydrocarbon polymers used here are less expensive, easily available, and introduce polar sites around the pendant groups [16]. The polar groups in the polymeric chain facilitate high water absorption over a wide temperature range. Also, proper molecular design is recommended to diminish the decomposition of hydrocarbon polymers [16a].

To achieve high thermal stability for hydrocarbon membranes, either aromatic hydrocarbons are incorporated into the backbone of hydrocarbon polymer or introducing polymers with bulky groups in the backbone [16b, 17]. Due to the presence of inflexible and bulky aromatic groups, polyaromatic hydrocarbons are rigid polymers with high thermal stability with $T_g > 200\,°C$. The aromatic rings in the backbone of these polymers are likely to undergo electrophilic or nucleophilic substitution [18]. Some of the popular main chain polyaromatics are polyether ketones (PEK), polyether sulfones (PESF), polyimides, poly (arylene ether), and polyesters. The extra thermal stability of polyaromatic hydrocarbons as well as environmental friendliness makes them a suitable candidate for proton-exchange membranes. Recently Pemion™ a sulfo-phenylated polyphenylene-based cation-exchange material, based fuel cell has been identified as a next-generation fuel cell for heavy-duty or aeronautic applications [19].

5.2.1.4 Acid–Base Complexes

Membranes based on acid–base complexes possess high proton conductivity at elevated temperatures. This is due to the incorporation of an acid component into the alkaline polymer base. [16b] Here an alkaline polymer is doped with an amphoteric acid which could act as both proton donor and acceptor, which in turn facilitate proton migration [20]. In this type of complex, strong acids like phosphoric acid or sulfuric acid connect with basic functionalities in the polymer backbones through hydrogen bonding or protonation. These strong acids possess unique proton conduction mechanisms through self-ionization and self-dehydration even in their anhydrous state. The basic polymers act as a medium for this acid functionality and increase its dissociation through the interaction which results in enhanced proton conduction [21].

Some of the basic polymers that have been investigated for preparing acid–base electrolytes are PVA [22], Nylon [23], polyethyleneimine (PEI) [24], polyethylene oxide (PEO) [25], polyacrylamide (PAAM) [26], and poly(diallyl dimethyl ammonium)-dihydrogen phosphate, ($PAMA + H_2PO_4$) [27]. But unfortunately, these acid–base complexes failed to perform well at high temperatures. At high temperatures either they lose mechanical stability or proton conductivity. The polybenzimidazole (PBI/H_3PO_4) membrane doped with phosphoric acid appears to be the most effective high-temperature system to date [28]. The ionic crosslinking (electrostatic forces) and hydrogen bonding interactions between acid and base polymers, markedly control the membrane swelling without altering flexibility. Hence these membranes displaying low water absorption, diminished methanol permeability, high proton conductivity, good thermal stability, and high mechanical flexibility and strength were considered excellent membrane for developing low-cost high performing fuel cells [29].

5.2.2 Preparation Methods of PEM

When it comes to the commercialization and marketing of fuel cell technology, the development of cost-effective functional materials for various components is of prime importance. PEM is one of the major contributors to the cost of fuel cells along with noble metal catalysts. Hence researchers always look to fabricate a low-cost but highly efficient PEM. The methods of preparation and fabrication of PEM are generally classified as follows: (i) grafting polymerization method using γ-ray irradiation, (ii) plasma-induced grafting polymerization method, (iii) crosslinking method, (iv) sol–gel method, and (v) direct polymerization of monomers [30].

Graft polymerization employing γ-ray irradiation could initiate a polymerization without any chemical initiators or catalysts. The grafting is initiated by creating active sites on the base polymer matrix by high-energy γ-ray irradiation. The base membranes here act as a hydrophobic host which restrains water absorption and gives mechanical stability to the grafted membrane. Three steps are involved in grafting polymerization: (i) Pre-irradiation step: In this step, the base polymer for example polyethylene tetrafluoroethylene (ETFE) film is pre-irradiated in an inert atmosphere to create active sites and thereby get activated for the grafting process. (ii) Substitution stage: In this stage monomers of our choice are introduced to get grafted onto the active site formed in the base polymer. (iii) In this last stage the grafted polymer is sulfonated in chlorosulfonic acid solution to introduce the sulfonic acid group. [30a]

Plasma-induced grafting polymerization is introduced to prepare ultrathin proton-exchange membranes (in about 1 μm) with low resistance and a sharp reduction in methanol permeability. It is widely used to create miniaturized fuel cells. Plasma-polymerized films display a high degree of cross-linking. Hence energy of plasma is monitored to control the degree of crosslinking, as increased crosslinking may reduce the proton conductivity. The structure of plasma-induced grafted polymer membranes is denser compared to membranes prepared via conventional methods. Hence it prevents methanol crossover through the grafted membrane [31].

The cross-linking method is used to prepare three-dimensional networking in the membrane to reduce the methanol crossover. In this method, a membrane without any charge is charged by crosslinking with a charged organic group. This crosslinking improves the thermal as well as the chemical stability of the proton-exchange membrane [32].

The sol–gel method was employed for the preparation of inorganic–organic hybrid membrane materials. In this method, a metallic alkoxide sol is then introduced into a polymer matrix. Later by the addition of water, these metallic alkoxides get condensed to form metallic nanoparticles linked with the polymer chain to form a gel-like network and ultimately resulted in the dispersion of metal nanoparticles in the polymer matrix [33].

Also, some proton-exchange membranes could be prepared by the traditional method of *in situ* direct polymerization of respective monomers like styrene or divinyl benzene, followed by sulfonation to introduce sulfonic groups into the side chain. To avoid post sulfonation step, sulfonated monomers are also subjected to direct polymerization [34].

5.2.3 Proton Transport Mechanism in PEM

PEM transport of protons across the membrane occurs via the following mechanisms: Grotthus mechanism (hopping mechanism) and vehicular mechanism (diffusion mechanism) [35]. In the Grotthus mechanism, the steady hopping of protons from proton donors to proton acceptors is the prime mode of proton conduction via hopping. The sulfonic acid group present in the side chain of polymeric membranes plays a vital role in this hopping mechanism. When the membranes get hydrated, the sulfonic acid groups dissociate to form transportable proton and sulfonate, anion group. The fluorine atoms in the backbone of Nafion-like membranes converged to form a hydrophobic domain as well as these sulfonate anionic group clusters give rise to a hydrophilic domain. The protons hop from one sulfonate anion to the nearest water molecule or another sulfonate anion. For sulfonic acid membranes other than Nafion-like membranes, the only difference is in the hydrophobic domain where the fluorine backbone is replaced by aromatic hydrocarbon backbones. Some commonly used sulfonated proton-exchange membranes were sulfonated poly(ether ether ketone) (SPEEK) and sulfonated polysulfone (SPSU) [36].

In vehicular mechanism proton in the form of hydronium cation or H_3O^+ diffuses through the aqueous medium in response to the external electrochemical difference. Due to the electroosmotic drag, each protonated water molecule carries two or more water molecules along with them across the membrane [37]. Transportable protons associate with water molecules to form various cationic species like hydronium ions, Zundel cations, and Eigen cations, which are easily diffused into the hydrophilic domains for transportation [38]. But when it comes to composite membranes, the transport mechanism is entirely different and much more complex.

5.2.4 Current State of Art of PEM

Even though Nafion remains one of the most widely commercialized PEM, poor proton conductivity at elevated temperatures restricted its application in many fields [39]. This is due to the restriction of water channels to conduct protons due to dehydration at elevated temperatures [40]. Based on the operating temperatures PEMs could be classified as low-temperature PEMs (functional in the range 60–80 °C) and high-temperature PEMs (functional in the range 100–200 °C). Among them, high-temperature PEMs are in high demand. Hence researchers are more engaged in developing high-temperature PEMs, which could win over Nafion which cannot work beyond a certain temperature. Even the incorporation of additives to pure Nafion had shown improved properties. Low-cost membranes retaining the proton conductivity and mechanical properties of Nafion under low-temperature operating conditions are the primary concern in the development of polymeric electrolyte membranes. When inorganic fillers like phosphor-tungstic acid were incorporated into the Nafion membrane, they facilitate greater water retention capability by reducing the vapor pressure [41].

Good electrochemical and mechanical stability as well as high proton conductivity under low humidity conditions are ideal for an excellent proton-exchange membrane [42]. One of the main strategies to increase the power density of fuel cells is to reduce the thickness of perfluorosulfonic acid (PFSA)-based membranes. The reduction in thickness reduces the water transport pathway and thereby increases the proton conductivity. But their mechanical instability and chances of electrochemical degradation remain a bottleneck for their development [43]. The first-generation MIRAI is one such example, of an ultrathin membrane (10 μm) which even possesses self-humidification properties to avoid anode dryness. In MIRAI they used a porous-medium flow field over the cathode gas diffusion layer (GDL) to ensure good mechanical support for the thin PEM film. [42, 44] When it comes to second-generation MIRAI (GORE-SELECT Membrane [45]), they have achieved a thinner version of PEM, with improved performance and mechanical durability by the addition of cerium salt [46]. This patented work is a promising note for future high-performance PEMFCs. Various modified perfluorinated and partially fluorinated membranes which could facilitate PEMFC operation under low-humidity and high-temperature (>80 °C) conditions were reported [47]. PTFE-reinforced perfluorinated membranes exhibit improved mechanical stability when made into thin films for PEM [44].

Other promising alternatives for perfluorinated proton-exchange membranes were sulfonated hydrocarbon polymers like sulfonated poly (ether sulfone) (SPES) [48], sulfonated poly (ether ketones) [49], sulfonated polyimides (SPIs) [45], and sulfonated polyphenylene-based PEMs [50]. Their high proton conductivity at high temperatures and humidities made them an excellent choice for proton-exchange membranes. When compared to PFSA, some sulfonated membranes have low conductivity at low humidity conditions (<40% RH) [16a]. An exception to this scenario is sulfonated (poly phenylene)-based hydrocarbon membranes [51], phosphoric acid quaternary amine-biphosphate ion pair membranes [52], and interpenetrating network of highly sulfonated poly(phenylene sulfide nitrile) (XESPSN) and random network copolymers [53]. Self-humidifying nano valve membrane coated with hydrophobic surface coating was fabricated to retard water desorption from the membrane, thereby maintaining conductivity in the membrane at high temperatures [54]. An aligned nanostructure of perfluorinated membrane achieved by capillary force lithography has created controlled pathways for proton transport. In this hierarchical structure, pre-patterned poly(dimethylsiloxane) mold is employed to create topographic patterns in Nafion membranes [55].

PFSA-based polymer membranes are expected to rule for the next 5–10 years, and continuous improvements in PEMs are expected to contribute 10–20% to the improvement of power density. However, with substantial development and innovations instability, less expensive non-perfluorinated PEMs are expected to replace PFSA. Furthermore, PEMs capable of operating under low RH conditions may reduce the humidification requirements, thus indirectly improving the power density [56].

5.3 Comparison with AEM

AEMs are a counterpart of PEM, consisting of solid electrolyte used in fuel cells that conducts anionic species especially hydroxide ion across the electrodes. Faster reaction kinetics at the electrodes than in acidic conditions, high electrical efficiency even in the absence of expensive noble metal catalysts, and low operating temperature are the advantages of AFC over other fuel cells. [57] Initially in AFC liquid electrolytes like potassium hydroxide (KOH) were employed in the fuel cell. This involves leakage problems, corrosion, and CO_2 poisoning by forming metal carbonate crystals in the electrolyte retards the hydroxyl ion transport. In addition to that proper maintenance of liquid electrolyte level is required; otherwise, it may lead to either electrode flooding (high liquid electrolyte) or electrode drying conditions (insufficient liquid electrolyte) [58]. When a solid electrolyte membrane replaces the liquid electrolyte, AFC based on solid AEMs initiated numerous types of research in this field. These solid membranes popularize the usage of AFCs, as it retards the negative effect of CO_2.

5.3.1 Materials Used for Preparations

As we have seen in the, Section 5.2 PEM membranes have a polymeric backbone made of fluorinated ionomeric membranes, non-fluorinated hydrocarbon membranes, or acid–base complexes. In all these polyelectrolyte membranes, an acid functionality group is attached to the active sites for the conduction of protons. Generally, sulfonic acid or phosphoric acid groups which could yield an active site for the easy transport of the proton at very low pH is preferred. In the case of AEMs, the polymeric electrolyte membrane should have cationic functional groups. Generally, quaternary ammonium, sulfonium, or phosphonium groups were grafted onto the polymeric backbone. But at very high pH, it is found that phosphonium groups are more stable, compared to the other two. Alkaline degradation of alkaline solid polymer membrane by the nucleophilic attack and Hoffmann elimination reduces mechanical stability and ionic conductivity [59]. As of now, researchers are still in search of a polymer membrane that could withstand high alkalinity and prevent chemical degradation in alkaline conditions like poly(triphenylene)-based membrane [60].

5.3.2 Investigative Methods and Measurement for Ion-Exchange Membranes

This section illustrates various methods of evaluation and characterization of both proton-exchange membranes and AEMs.

5.3.2.1 Ionic Conductivity

The ionic conductivity of PEMs and AEMs is measured using electron impedance spectroscopy. The ionic conductivity, σ, is directly related to the specific resistance of the membrane according to the equation; $\sigma = \frac{l}{RA}$ where R is the resistance, l is the thickness, and A is the area of the membrane. The conductivity of a membrane is measured by keeping a swollen membrane, kept in water for 24 h between two electrodes. The resistance of the membrane is then measured by changing the frequency of the alternating current, and the response of the system to the change in frequency is determined by a spectrometer. The resistance R of the membrane is determined from the Nyquist plot and the conductivity is measured in siemens per centimeter ($S\,cm^{-1}$) using the above equation [61].

We have observed that ionic conductivity increases with temperature, and this led to the idea of high-temperature fuel cells. High-temperature measurements were done in a conductivity cell kept inside a wrap-round resistance heater and feedback temperature controller. The temperature of the membrane is increased with the aid of water vapor, by keeping the conductivity cell just above liquid water in the head-space of a closed container. Before taking the measurements, the membrane is allowed to get saturated with water vapor at the required temperature.

The proton conductivity of the membrane at high temperatures also depends on the glass transition temperature of the polymeric matrix of the membrane. At temperatures below the glass transition temperature, the conductivity obeys the Arrhenius-type law [62], given by the equation

$$\sigma = \frac{A}{T}\exp\left(\frac{Ea}{RT}\right)$$

where A is a constant proportional to the number of charge carriers, Ea is the activation energy, and R is the ideal gas constant. When the temperature is above the glass transition temperature, the segmental motion of the polymeric chains also contributes to ionic conduction. Hence at temperatures above the glass transition temperature, conductivity obeys the Vogel–Tamman–Fulcher (VTF) equation, in which pseudo activation energy related to the segmental motions of the polymer is included. VTF formula for conductivity is [62]

$$\sigma = \frac{A}{\sqrt{T}}\exp\left(-\frac{B}{T - T_0}\right)$$

where B is pseudo activation energy and T_0 is called equilibrium glass transition temperature. This phenomenon is prominently observed for amorphous polymers. Polymeric membranes having rigid structures always follow the Arrhenius-type law throughout the whole temperature range.

When it comes to AEM the charge density due to cationic functional groups present in the polymeric chain, cross-linking density of the membrane, and water content present in the membrane is responsible for the ionic conductivity. As temperature increases the micropores in the membranes open up and facilitate water absorption by the formation of the hydrated ion clusters. The difference in the water uptake between membranes is due to the difference in the wettability of their main chains and side chains where the cationic group is located [63].

5.3.2.2 Water Absorption or Swelling Index

Water content in the membrane is important as ionic conductivity increases with increased water uptake. Water also assists in the mass transport of fuel through the membrane. But increased uptake of water content may affect the polymer microstructure and mechanical properties of the membrane adversely. Hence water uptake of the membranes should not be too high or not too low [64].

The gravimetric method of double weighing is used to measure the water uptake quantitatively. Wet weight (W_{wet}) of the membrane is taken by equilibrating the membrane with water or water vapor at different temperatures, and dry weight (W_{dry}) of the membrane is taken by drying the sample at a temperature above its boiling point for a particular period. Then the total water uptake is calculated by ($W_{uptake}\%$) by the equation [65]

$$W_{uptake} = \frac{W_{wet} - W_{dry}}{W_{dry}} \times 100$$

For higher conductivity, a certain degree of swelling is required, but excessive uptake of water may make the membrane fragile and affect the durability of the fuel cell.

The swelling ratio may be calculated by measuring the thickness of the wet (t_w) and dry membranes (t_d).

$$\text{Swelling ratio} = \frac{t_w - t_d}{t_d} \times 100$$

Another membrane parameter, hydration number could be calculated as,

$$\lambda = \frac{W_{uptake} \times 10}{\text{IEC} \times 18}$$

Temperature, ion-exchange capacity (IEC), and pretreatment of the membrane, the physical state of absorbing water, whether it is in liquid or vapor phase, and the elastic modulus of the membrane are the factors that affect the extent of the water uptake of a membrane.

5.3.2.3 Ion-Exchange Capacity (IEC) of the Membrane

The IEC of the membrane is an important parameter of an ion-exchange membrane. It is the relative concentration of cationic groups per unit mass of the dry polymer. IEC is usually expressed in millimoles per gram or milli-equivalents per gram when the cationic groups involved are monovalent. In both PEM and AEM, the titration method at room temperature is used to determine the IEC.

In PEM, the acidic groups in the membrane are initially exchanged with sodium forms by immersing the membrane in NaCl solutions.

$$R-H+Na^{+} \rightarrow R-Na+H^{+}$$

The exchange H^+ ions in the solution are quantified by titrating with 0.01 N NaOH solutions. IEC values are calculated using the formula, [66]

$$IEC = \frac{\text{Consumed mL of NaOH} \times \text{Molarity of NaOH}}{\text{Weight dried membrane}} \text{ mequiv g}^{-1}$$

When it comes to AEM membranes, acid/base titration or the Mohr method is the most common method to determine IEC.

Acid/base titrations for AEM are performed by soaking the membrane in a strong base (1 M NaOH) to convert AEM to hydroxyl ions, followed by soaking in a strong acid solution of known volume and concentration to convert AEM to the chloride form. Then AEM is removed and rinsed with deionized water, and the resulting dilute HCl is titrated against standard NaOH. [67] Then IEC is calculated using the formula,

$$IEC = \frac{(V_{acid} * C_{acid}) - (V_{base} * C_{base})}{m_d}$$

In the case of acid/base titrations, when the membrane is exposed to ambient air, the CO_2 poisoning of the OH^- groups occurs, leading to the formation of HCO_3^- which in turn leads to an error in calculated IEC. Hence efficient procedures should be adopted while performing acid/base titration to minimize CO_2 poisoning. [68]

In the Mohr method, an AEM is converted to the Cl^- form by soaking in a salt solution (e.g. 1 M NaCl). The AEM is then rinsed and equilibrated in a 0.5 M Na_2SO_4 solution to facilitate the release of Cl^-. Using $AgNO_3$ solution with K_2CrO_4 indicator, the AEM/Na_2SO_4 solution is titrated until the K_2CrO_4 endpoint, which indicates when all the chlorides have been precipitated and now Ag_2CrO_4 forms [69]. The resulting IEC calculated by the Mohr method is

$$IEC = \frac{V_{AgNO_3} * C_{AgNO_3}}{m_d}$$

where m_d is the dry mass of the membrane after drying at 80 °C, for 48 h until the mass remains constant. Since the Mohr method involves carcinogenic hexavalent chromium (CrO_4^{2-}), acid/base titrations are preferred.

To avoid human errors in titrations, ultraviolet–visible (UV–Vis) spectroscopy is employed to determine the NO_3^- ions that exchange with Cl^- ions in an AEM [70].

5.3.2.4 Thermal Stability and Mechanical Strength

The thermal stability of membranes is determined using thermogravimetric analysis (TGA), differential thermal analysis (DTA), and differential scanning calorimetry (DSC). TGA monitors the temperature at which membrane changes occur due to water loss, head group decomposition, and/or polymer decomposition and thereby assesses the thermal stability. The primary degradation of these membranes involves the degradation of the polymeric backbone of membranes, and the secondary degradation involves the degradation of the pendant groups or inorganic compounds inside the membranes at higher temperatures than the primary degradation. DSC and DTA can be used to evaluate the glass transition temperature, the effects of thermal cycling, and changes in polymer crystallinity and cross-linking [55, 71].

A universal testing machine is employed to study the physical properties of the membranes like tensile strength, stress–strain curves, and elongation break. This gives an idea about the mechanical strength of the membrane [72].

5.3.2.5 Durability of the Membranes

Since ion-exchange membranes are the key components in a fuel cell, enhancement in their durability is critical to the lifetime of the fuel cells.

In the case of PEM, membrane degradation is due to the hydrogen peroxide and its intermediate decomposition free radicals like $OH^{\cdot}$ and $HO_2^{\cdot}$ with strong oxidative characteristics, produced during the fuel cell operation. Therefore, the membrane durability was evaluated by both the *ex situ* Fenton test and *in situ* OCV (open circuit voltage) accelerated test. Membrane samples were immersed in 50 ml Fenton solution (3 wt.% hydrogen peroxide solution and 20 ppm Fe^{2+}). The durability tests were carried out at 80 °C for 150 h before fluorine ion characterization, the Pt wire was immersed into the analyte solution to decompose the residual hydrogen peroxide, which was to assure accurate and reproducible results [73].

In the case of AEM, prolonged exposure to highly alkaline solution led to the degradation of AEMs. The alkaline stability is a measure of how the AEM performance changes over time when exposed to high-pH environments. While testing for alkaline stability, AEM is soaked in a high-pH solution (e.g. 1–10 M KOH) at a given temperature (room temperature or elevated temperature) for extended periods and periodically tests the membrane IEC to see how it changes with time. The alkaline stability also depends on the hydration level, as the hydration level increases, reduces the nucleophilicity of hydroxyl ions which in turn improves the alkaline stability. Hence the *ex situ* alkaline stability measurements are inconsistent and may give artificially high alkaline stability which is not congruent with the *in situ* alkaline stability of AEM [74]. So to determine true alkaline stability, Dekel et al. proposed an alternative *ex situ* alkalinity stability testing procedure using nuclear magnetic resonance (NMR) and water-free hydroxide (crown ether/KOH) solution where the water/OH^- ratio (γ) could be controlled to assess alkaline stability at different hydration levels [75].

5.3.3 Water Management

In a PEM fuel cell, water is produced at the cathode, whereas, in an AEM fuel cell, water is produced and consumed on the anode and cathode sides, respectively (Figure 5.1). The amount of water in the membrane has a considerable impact on ionic conductivity; hence, the fuel cell needs a high water content to maintain the membrane's high conductivity. However, as water is a byproduct of electrochemical processes, too much of it prevents gas from moving through the electrode. Additionally, the unequal water distribution between the anode and cathode is made worse by the electro-osmotic drag. For the PEM fuel cell, the anode is drier than the cathode, but for the AEM, the reverse is true.

During the operation of an anion-exchange membrane fuel cell (AEMFC), when the ambient air is used, several anions are present in the cell: carbonate ions, bicarbonate ions, and hydroxyl ions. When AEMFC is operated at high current densities, the most common anion species present across the membrane is the hydroxyl ion [76], generated according to the electrochemical ORR at the cathode. The hydroxyl ion is transported to the anode where hydrogen is oxidized as fuel. In the overall reaction, both PEMFC and AEMFC produce water as a byproduct. However, in contrast to PEMFC technology, in AEMFC water is generated at the anode (twice as much as in a PEMFC, per electron), while at the same time, water is a reactant at the cathode. This distinctive water transport scenario, together with the high alkaline medium in AEMFCs, represents a unique feature of AEMFCs.

In AEMFC, the OH^- ions generated in the cathode travel through the AEM (with the aid of water present in AEM) to the anode where they react with H_2 fuel to produce H_2O [76b]. Hence it is pertinent to control the humidification level and water transport properties of the ion-conducting membranes. Ineffective water management includes water flooding and dehydration. The flooding and dehydration in the anode and cathode may lead to poor cell performance and hence transport of water across the membranes should be regulated effectively. The electro-osmotic drag coefficient, which is the number of water molecules associated with each mobile ion, plays a significant role in the water management of AEMs. The activity of water on both the anode and cathode decides the movement of water inside the membrane through diffusion. [76c] Poor MEA stability and poor cell performance limited the efforts to investigate the movement of water in AEMs. At first, it was postulated that the movement of water in AEMFCs occurs from the anode side to the cathode side, which is the reverse of the process occurring in PEMFCs. But later it was found that the nature of the membrane has a significant influence on water movement. Hydrophilic membranes facilitate the water movement from anode to cathode, while hydrophobic membranes facilitate the opposite movement (cathode to anode) [76c]. Hence careful attention is needed toward nature, thickness, and diffusion resistance of the ion-exchange membranes to eliminate water imbalance in fuel cells [76].

5.3.4 Transport Mechanism

Understanding the anion conduction mechanism in the AEM is of prime importance for making further improvements in the AEM to make it highly efficient and selective to make it more practical in a fuel cell. Similar to proton conductivity in PEM depends on the RH temperature anion conduction also depends on the same. Hence it is assumed that the conduction mechanisms for both proton and anion are the same [77].

The dominant transport mechanisms are the Grotthus mechanism, diffusion, migration, and convection. In the Grotthus mechanism, hydroxide ion diffuses through the hydrogen-bonded network of water molecules through the formation/cleavage of covalent bonds. The movement of the hydrated hydroxyl ion is accompanied by a hyper-coordinating water molecule. The arrival of another electron-donating water molecule leads to hydrogen bond rearrangements, re-orientations, hydrogen ion transfer, and the formation of a fully tetrahedrally coordinated water molecule [78]. Diffusive transport occurs in the presence of a concentration and/or electrical potential gradient. Convective transport across the membrane appears since hydroxides, moving through the membrane, drag water molecules with them through the membrane, thus generating a convective flow of water molecules within the membrane. Surface site hopping of hydroxyl anions occurs on quaternary ammonium groups present on the membrane. This type of transport is thought to be secondary transport across the membrane because the water present in the system acts as a permanent dipole and interacts with the fixed charges of the membrane. This strong coordination of water molecules around the ammonium groups imparts a reduction in the possibility of the ionic species interacting with the ammonium groups on the membrane.

Even though dominant mechanisms for transport through the membrane are the same, there are fundamental discrepancies between PEM and AEM conductivities. They may arise from insufficient OH^- dissociation and solvation, the morphology of the structure, and/or interactions between the OH^- in solution and the quaternary ammonium side chain. Another possible explanation for the discrepancy in ionic conductivity between PEMs and AEMs may result from contact with air during the measurement. When the system is not fully isolated and air penetrates the cell (as might well be possible in experimental setups), OH^- may react with the CO_2 from the air and convert into carbonate/bicarbonate ions, which is a relatively fast process occurring at short timescales (<30–60 min) [52]. This conversion of hydroxyl to carbonate/bicarbonate leads to a significant decrease in ionic conductivities [79].

5.3.5 Catalyst Used in PEMFC and AEMFC

The increased attention on the development of AEMFC is due to its cost-effectiveness as it allows the usage of non-precious metal catalysts when compared to PEMFC. In PEMFC carbon-supported platinum (Pt/C) having a high active surface area is the best choice in today's fuel cells for higher efficiency. Even though carbon-supported platinum catalysts have significantly reduced the use of platinum than when platinum black is in use, the cost of the fuel cell remains high. The research was going on to develop a non-precious catalyst for PEMFC, but the acidic environment in PEMFC may cause corrosion to these metal catalysts which may reduce its efficiency. But none of them has reached the benchmark provided by the Pt systems. In this scenario, AEMFC, with an alkaline environment that could use non-precious metal catalysts, is a great alternative for the PEMFC. Nickel base materials were considered as a replacement for precious metal catalysts owing to their abundance, moderate cost, and durability in highly alkaline conditions [80]. Several transition metals (like Mo, Co, Fe, and Cu) were alloyed with nickel to improve their activity and stability [81]. Au, Ag, as well as Au/Ag alloys, were also considered a potential replacement for Pt, owing to their high activity toward ORR. [82]

5.3.6 MEA Fabrication

The MEA is the core component of a fuel cell and plays a crucial role in its performance. Hence the proper design of MEA is essential as the true potential of the ion-exchange membranes and electrocatalysts is only revealed in an efficient MEA configuration. MEA consists of an ion-exchange membrane, catalyst layers (CLs), and GDLs. While fabricating MEA it is ensured that all active catalyst sites in the CL are accessible to the reactant (H_2 or O_2), protons, and electrons and would facilitate the effective removal of produced water from the CL and GDL. Compared to PEMFC, MEA fabrication methods as well as the materials are in the budding stage due to the novelties in the AEMs. In the case of PEMFC, Nafion, an easily dispersible, highly conducting ionomer, is available to prepare the catalytic ink. But in the case of AEMFC, anion-exchange binders have limited stability and conductivity which limits its MEA performance [83a].

In contrast to PEMFC, there are distinct differences in reaction processes in both anodic and cathodic compartments of AEMFC. In the cathodic layer, an independent outer sphere electron transfer component is prominent in an alkaline environment, while an inner sphere electron transfer process is prominent in an acidic environment. Hence in an AEMFC, a faster kinetics for ORR is proposed with a wide range of non-noble metal electrocatalysts [83b]. Hydrogen oxidation reaction (HOR) activity of some commercial catalysts based on Pt, Pd, and Ir exhibits a decrease in activity with the decrease in solution acidity [83c]. Higher resistance is observed in the HOR activity in the anodic layer of an AEMFC. Investigations on introducing oxyphilic atoms like Ru, Cu, and Ni which could facilitate the OH^- adsorption may help to design a HOR catalyst with lower resistance in an alkaline environment [83d].

In the fabrication of MEA, the catalyst and ionomer are taken in an appropriate ratio and mixed with isopropanol to form a homogeneous catalyst ink. The ink is then coated on the side of an ion-exchange membrane. In all MEAs, the Pt loading content of the anode and cathode was the same, and the ratio of the Pt/C-ionomer was fixed at 20 wt%. Both sides of the catalyst-coated membranes were then sandwiched with GDLs and a Teflon gasket and were finally assembled into single-unit cells [84]. In the case of AEMs, they are easily ruptured through stresses during the fabrication of MEAs due to their weak mechanical strength and durability. Hence, hot-pressing methods that are used in making MEA in PEMFC will not always work for AEMFCs, and other bonding methods are needed to be formulated. An excessive amount of ionomer in the catalyst ink may reduce the porosity of the CL and may resist the transportation of water and oxygen. Hence the ionomer ratio in the catalytic ink, as well as the CL fabrication process needs to be optimized to get abundant mesopores in the CL. The mesoporous nature of the CL is best suited for its excellent performance with high current density. [85]

Unlike in PEMFC, the electrochemical water generated in the anodic compartment needs to diffuse back into the cathodic compartment to facilitate the ORR process [86a]. This back diffusion process should be in such a way that it should not dry up the anodic CL as well as the ion-exchange membranes nor flood the CL with an excess of water molecules. Hence proper distribution and transport of water is an important factor that determines the hydration of MEA avoiding catalyst flooding in AEMFC [86b]. The inlet RH and backpressure on the anode and cathode layers regulate the water management in an MEA and thereby influence the cell performance [87].

Ionomer binders used in CLs in MEA fabrication also play a crucial role in determining the performance of a fuel cell. In constituting PEMFC, an electrode ionomer with significant proton conducting properties as well as superior mechanical and chemical properties is Nafion. Even though the high cost, high gas permeability, and environmental hazardousness slow down its commercialization, Nafion is one of the most extensively studied and optimized ionomers in PEMFC [88]. Hydrocarbon-based PEMs having high mechanical and thermal stabilities are developed to alleviate the disadvantages of Nafion and to complement it [84]. One major problem that arises in the anionic ionomer is the co-adsorption of cation-hydroxide water hindering the diffusion of the hydrogen gas to the catalyst surface and blocking the phenyl group active sites on the catalytic surface, thereby reducing its activity [89]. This issue is tackled by inserting symmetric dimethyl groups into the backbone of the quaternary poly(biphenyl alkylene) ionomer, which increases the hydrogen diffusion coefficient. Another approach to modifying the anionic ionomer to increase its activity is by changing the functional group from alkyl trimethylammonium to alkyl triethylammonium which causes a doubling of the hydrogen limiting current of the new ionomer [90]. Co-adsorption of cation-hydroxide water was therefore deemed as a superior source of AEMFC performance degradation compared to the slow diffusion of hydrogen gas due to phenyl adsorption [85b]. Hence developing anionic ionomer is an emerging field of study.

5.3.7 Fuels Used in Fuel Cells

Hydrogen is considered the most suitable fuel for a fuel cell system as it could provide high energy density. The difficulty in liquefaction and storage of hydrogen gas restricts its usage in a wide range of applications. Currently, commercially available fuel cells are using hydrogen as fuel in an acidic medium [91]. Currently, most of the hydrogen is produced by steam reforming of hydrocarbons. It can also be made by heating coal to roughly 1000 °C in the absence of oxygen, by electrolysis, or by partially burning coal in the presence of steam. Since water is the only byproduct of hydrogen oxidation, it is a highly sought-after fuel. A wide range of AAEMFCs uses hydrogen as fuel and could achieve high power density [92]. Another widely used fuel is easily oxidizable simplest alcohol methanol.

Methanol has a high energy-to-carbon ratio and an energy density of $4820\,\mathrm{Wh\,l^{-1}}$. [93] In both PEMFC and AEMFC fueled by methanol, extra care is taken to avoid catalyst poisoning by CO formed by the incomplete oxidation of methanol. AEMFC may give a high power density compared to a PEMFC. A Tokuyama AEM exhibited a power density of $6.8\,\mathrm{mW\,cm^{-2}}$ at ambient temperature with a Pt/C cathode and

PtRu/C anode catalyst with 1 M methanol and 1 M KOH fuel when compared to a Nafion 117 membrane which produced a power density of 5.2 mW cm^{-2} with the same catalysts and methanol as fuel [94]. The kinetics of the anode reaction when methanol is used in AFCs are slow, and DMFC also suffers from abundant methanol crossover and low performance. [94b]

Besides methanol, ethanol with an energy density of 8030 Wh kg^{-1} was also employed as fuel in both AEMFC and PEMFC [93]. But complete oxidation of ethanol could be achieved only at high temperatures. Hence catalyst capable of completely oxidizing ethanol at low temperatures in an alkaline medium is needed to be found to maximize the use of ethanol as fuel. The rate of ethanol oxidation is lower in an acidic medium. Hence compared to direct-ethanol alkaline-exchange membrane fuel cell (DEAEMFC), poisoning of the catalyst due to incomplete oxidation is more prominent and to be taken care of in direct-ethanol proton-exchange membrane fuel cell (DEPEMFC).

Another choice of fuel that does not contain carbon and hence there is no chance for catalyst poisoning is hydrazine. The theoretical mass-energy density of hydrazine is 2.6 kWh kg^{-1}. But in the hydrazine-driven fuel cell, a noble metal catalyst like Pt is more preferred as it provides the lowest decomposition rate, since it is alkaline in an aqueous solution and expected to be very compatible with an AEM [95]. Out of all the known fuels possible to fuel direct liquid fuel cells, hydrazine is considered the most toxic to humans, as it can damage the lungs, kidneys, and central nervous system. To address these problems, the Daihatsu Motor Company has created a technique for secure hydrazine storage in the form of solid hydrazone, which has no carcinogenic characteristics and is safer due to its solid nature in the case of tank breakage. The hydrazine may be liberated from the hydrazone state by introducing a solvent to the fuel tank [95].

Ethylene glycol, which has high electroactivity in an alkaline environment, as well as non-inflammable, non-volatile glycerol was also employed as fuel in AEMFC. Ethylene glycol has superior performance and power densities in AAEMFCs in comparison to methanol, while the low cost and non-toxicity urged to look for glycerol as another option. Direct glycerol and ethylene glycol fuel cells have not been the subject of a lot of research since they are still early in the research process [96].

Sodium borohydride produces a high open circuit potential and achieves high power densities is an excellent fuel as it could not produce carbon products which could poison the anode catalyst. But the toxic nature of borohydride acts as a barrier to its commercialization. Borohydride could be used only in an alkaline environment as it is unstable in acidic and neutral conditions [97]. Dimethyl ether is considered a promising fuel without NOx and SOx emissions. It is less toxic than methanol and possesses low fuel cross-over due to the low dipole moment, making it a favorable fuel for fuel cells [98]. Another widely accepted fuel is ammonia, which could act as a direct fuel for ammonium fuel cells and an indirect fuel for hydrogen fuel cells. Since liquid ammonia contains 70% more hydrogen than liquid hydrogen and has a 50% greater specific energy density, it makes a great low-temperature fuel [99]. Compared to PEMFC, a wide choice of fuels are suitable for AEMFC.

5.3.8 Fuel Cell Efficiency

It is the efficiency of a process that converts chemical potential energy contained in the fuel into kinetic energy or work. The efficiency of a fuel cell is measured by the ratio of the amount of useful energy put out by the cell (energy output) to the total amount of energy put in (energy input).

$$\text{Energy efficiency} = \frac{\text{Energy output}}{\text{Total energy input}}$$

The load current, temperature, RH, membrane thickness, membrane-active area, electrode active area, corrosion, purity, pressure, and concentration of hydrogen fuel, maintenance of water inside the cell, pressure in the electrode particularly on both sides of the membrane, etc. are some of the factors that determine the fuel cell efficiency.

Life time of a fuel cell is determined by the electrolyte and catalyst employed in it. When the electrolyte becomes corrosive during the transit of H^+, it causes the metal component to degrade and shortens the life of the fuel cell. The lifespan of such fuel cells tends to be decreased when the catalyst utilized at the electrodes becomes exhausted throughout the operation. Majority of low-temperature fuel cells have long life span, whereas the high-temperature fuel cells have poor longevity.

Power density is the power produced per unit volume. Power density varies depending on operating circumstances and is compensated for by the catalyst and electrolyte used in each fuel cell type. The power density of a fuel cell is determined by the efficiency of the used catalyst and the capacity of the electrolyte to swiftly transfer ions from anode to cathode. Excellent power density is needed for fuel cell types that can effectively compete in the energy market, and this is accomplished by careful catalyst selection.

5.4 Conclusion

Massive attempts have been made to find new energy solutions as the world's energy and environmental concerns become more and more urgent. Fuel cells are a novel energy technology, and have the potential to replace traditional fossil fuels due to their tremendous effectiveness and low, or zero, emission. In a fuel cell, ion-exchange membranes are one of the main core components responsible for its efficiency. An ion-exchange membrane controls a fuel cell's performance through its water absorption, ionic conductivity, gas permeability, electrical insulation, and durability.

Among various types of fuel cells, the AFC was the first type to be used in space programs and is still in use. But proton-exchange membrane fuel cells were the most rapidly developed fuel cell in the past decades. It was even commercialized in the form of PEMFC-powered vehicles, mobiles, watches, and military technology. The credit for the rapid commercialization of PEMFC technology goes to tremendous development in the area of proton-exchange membranes. Fluorinated sulfonic acid Nafion membrane is considered a benchmark for proton-exchange membranes owing to its high proton conductivity, durability, and efficiency. It is considered that PEMFC may rule the market for the next 10–20 years. But the high cost and low-temperature endurance limit its applications.

The low cost of implementation compared to PEMFC triggered developments in the field of AFC which was previously meant only for extraterrestrial applications. Liquid electrolyte leakage, CO_2 poisoning of AEM, and low hydroxide ion conductivity slow down its development initially. The introduction of solid electrolyte membranes balanced some of these issues to a great extent. But still, poor hydroxyl ion conductivity and chemical instability remain the largest hurdle to the development of AEM. Some modifications to the existing electrolyte membrane enhance the ionic conductivity, thermal stability, and durability making it a strong opponent for PEM and PEMFC technology. The use of inexpensive metal catalysts as electrodes is one of the real advantages of AEMFC. But still, AEMFC requires innovative engineering solutions regarding the materials and processing schemes to make the conductivity and durability of AEM up to the level of PEM. Also, the identification of suitable OH^- ion conductive polymers for anode and cathode binders, and the identification/optimization of MEA preparation methods are the need of the hour. If AEMFCs are to play the envisioned role in the global energy transition, these issues must be overcome for them to function as a cog in the future hydrogen economy.

References

1 Wang, Y., Chen, K.S., Mishler, J. et al. (2011). A review of polymer electrolyte membrane fuel cells: technology, applications, and needs on fundamental research. *Appl Energy* 88: 981–1007.

2 Devanathan, R. (2008). Recent developments in proton exchange membranes for fuel cells. *Energy Environ Sci* 1: 101–119.

3 Hu, Y., Li, X., Yan, L., and Yue, B. (2017). Improving the overall characteristics of proton exchange membranes via nanophase separation technologies: a progress review. *Fuel Cells* 17: 3–17.

4 Peighambardoust, S.J., Rowshanzamir, S., and Amjadi, M. (2010). Review of the proton exchange membranes for fuel cell applications. *Int J Hydrog Energy* 35: 9349–9384.

5 Kariduraganavar, M., Nagarale, R., Kittur, A., and Kulkarni, S. (2006). Ion-exchange membranes: preparative methods for electrodialysis and fuel cell applications. *Desalination* 197: 225–246.

6 Souzy, R. and Ameduri, B. (2005). Functional fluoropolymers for fuel cell membranes. *Prog Polym Sci* 30: 644–687.

7 Motupally, S., Becker, A.J., and Weidner, J.W. (2000). Diffusion of water in Nafion 115 membranes. *J Electrochem Soc* 147: 3171.

8 (a) Banerjee, S. and Curtin, D.E. (2004). Nafion® perfluorinated membranes in fuel cells. *J Fluor Chem* 125: 1211–1216. (b) Mauritz, K.A. and Moore, R.B. (2004). State of understanding of Nafion. *Chem Rev* 104: 4535–4586.

9 (a) Roziere, J. and Jones, D.J. (2003). Non-fluorinated polymer materials for proton exchange membrane fuel cells. *Annu Rev Mater Res* 33: 503–555. (b) Zatoń, M., Rozière, J., and Jones, D. (2017). Current understanding of chemical degradation mechanisms of perfluorosulfonic acid membranes and their mitigation strategies: a review. *Sustain Energy Fuels* 1: 409–438.

10 Appleby, A. and Foulkes, F. (1989). *Fuel Cell Handbook*, 364. New York: Van Nostrand Reinhold.

11 Panchenko, A., Dilger, H., Möller, E. et al. (2004). In situ EPR investigation of polymer electrolyte membrane degradation in fuel cell applications. *J Power Sources* 127: 325–330.

12 (a) Prakash, G.S., Smart, M.C., Wang, Q.J. et al. (2004). High efficiency direct methanol fuel cell based on poly (styrenesulfonic) acid (PSSA)–poly (vinylidene fluoride)(PVDF) composite membranes. *J Fluor Chem* 125: 1217–1230. (b) Navarra, M., Materazzi, S., Panero, S., and Scrosati, B. (2003). PVdF-based membranes for DMFC applications. *J Electrochem Soc* 150: A1528.

13 Lehtinen, T., Sundholm, G.r., Holmberg, S. et al. (1998). Electrochemical characterization of PVDF-based proton conducting membranes for fuel cells. *Electrochim Acta* 43: 1881–1890.

14 Shen, Y., Qiu, X., Shen, J. et al. (2006). PVDF-g-PSSA and Al_2O_3 composite proton exchange membranes. *J Power Sources* 161: 54–60.

15 Bhat, S., Sahu, A., George, C. et al. (2009). Mordenite-incorporated PVA–PSSA membranes as electrolytes for DMFCs. *J Membr Sci* 340: 73–83.

16 (a) Rikukawa, M. and Sanui, K. (2000). Proton-conducting polymer electrolyte membranes based on hydrocarbon polymers. *Prog Polym Sci* 25: 1463–1502. (b) Smitha, B., Sridhar, S., and Khan, A. (2005). Solid polymer electrolyte membranes for fuel cell applications—a review. *J Membr Sci* 259: 10–26.

17 Higashihara, T., Matsumoto, K., and Ueda, M. (2009). Sulfonated aromatic hydrocarbon polymers as proton exchange membranes for fuel cells. *Polymer* 50: 5341–5357.

18 (a) Wen, S., Gong, C., Tsen, W.C. et al. (2009). Sulfonated poly (ether sulfone) (SPES)/boron phosphate (BPO4) composite membranes for high-temperature proton-exchange membrane fuel cells. *Int J Hydrog Energy* 34: 8982–8991. (b) Lee, J., Li, W., and Manthiram, A. (2009). Poly (arylene ether sulfone) s containing pendant sulfonic acid groups as membrane materials for direct methanol fuel cells. *J Membr Sci* 330: 73–79.

19 Nguyen, H., Lombeck, F., Schwarz, C. et al. (2021). Hydrocarbon-based Pemion™ proton exchange membrane fuel cells with state-of-the-art performance. *Sustain Energy Fuels* 5: 3687–3699.

20 Colomban, P. and Philippe, C. (1992). *Proton Conductors: Solids, Membranes and Gels-Materials and Devices*. Cambridge University Press.

21 Li, Q., He, R., Jensen, J.O., and Bjerrum, N.J. (2003). Approaches and recent development of polymer electrolyte membranes for fuel cells operating above 100 C. *Chem Mater* 15: 4896–4915.

22 Petty Weeks, S., Zupancic, J., and Swedo, J. (1988). Proton conducting interpenetrating polymer networks. *Solid State Ionics* 31: 117–125.

23 Grondin, J., Rodriguez, D., and Lassegues, J. (1995). Proton conducting polymer electrolyte-The Nylon 6–10/H3PO4 blends. *Solid State Ionics* 77: 70–75.

24 Tanaka, R., Yamamoto, H., Shono, A. et al. (2000). Proton conducting behavior in non-crosslinked and crosslinked polyethylenimine with excess phosphoric acid. *Electrochim Acta* 45: 1385–1389.

25 Donoso, P., Gorecki, W., Berthier, C. et al. (1988). NMR, conductivity and neutron scattering investigation of ionic dynamics in the anhydrous polymer protonic conductor PEO $(H_3PO_4)_x$. *Solid State Ionics* 28: 969–974.

26 Stevens, J., Wieczorek, W., Raducha, D., and Jeffrey, K. (1997). Proton conducting gel/H_3PO_4 electrolytes. *Solid State Ionics* 97: 347–358.

27 Bozkurt, A., Ise, M., Kreuer, K. et al. (1999). Proton-conducting polymer electrolytes based on phosphoric acid. *Solid State Ionics* 125: 225–233.

28 Hu, J., Zhang, H., Zhai, Y. et al. (2006). 500 h Continuous aging life test on PBI/H3PO4 high-temperature PEMFC. *Int J Hydrog Energy* 31: 1855–1862.

29 Wu, D., Xu, T., Wu, L., and Wu, Y. (2009). Hybrid acid–base polymer membranes prepared for application in fuel cells. *J Power Sources* 186: 286–292.

30 (a) Chen, J., Asano, M., Yamaki, T., and Yoshida, M. (2006). Preparation and characterization of chemically stable polymer electrolyte membranes by radiation-induced graft copolymerization of four monomers into ETFE films. *J Membr Sci* 269: 194–204. (b) Choi, E.Y., Strathmann, H., Park, J.M., and Moon, S.H. (2006). Characterization of non-uniformly charged ion-exchange membranes prepared by plasma-induced graft polymerization. *J Membr Sci* 268: 165–174. (c) Mosa, J., Durán, A., and Aparicio, M. (2009). Proton conducting sol–gel sulfonated membranes produced from 2-allylphenol, 3-glycidoxypropyl trimethoxysilane and tetraethyl orthosilicate. *J Power Sources* 192: 138–143. (d) Nasef, M.M. and Hegazy, E.S.A. (2004). Preparation and applications of ion exchange membranes by radiation-induced graft copolymerization of polar monomers onto non-polar films. *Prog Polym Sci* 29: 499–561.

31 (a) Brumlik, C.J., Parthasarathy, A., Chen, W.J., and Martin, C.R. (1994). Plasma polymerization of sulfonated fluorochlorocarbon ionomer films. *J Electrochem Soc* 141: 2273. (b) Mex, L., Sussiek, M., and Müller, J. (2003). Plasma polymerized electrolyte membranes and electrodes for miniaturized fuel cells. *Chem Eng Commun* 190: 1085–1095.

32 Krumova, M., Lopez, D., Benavente, R. et al. (2000). Effect of crosslinking on the mechanical and thermal properties of poly (vinyl alcohol). *Polymer* 41: 9265–9272.

33 Sacca, A., Carbone, A., Passalacqua, E. et al. (2005). Nafion–TiO_2 hybrid membranes for medium temperature polymer electrolyte fuel cells (PEFCs). *J Power Sources* 152: 16–21.

34 Xu, T., Woo, J.J., Seo, S.J., and Moon, S.H. (2008). In situ polymerization: a novel route for thermally stable proton-conductive membranes. *J Membr Sci* 325: 209–216.

35 (a) Marx, D. (2006). Proton transfer 200 years after von Grotthuss: insights from ab initio simulations. *ChemPhysChem* 7: 1848–1870. (b) Kreuer, K.D., Rabenau, A., and Weppner, W. (1982). Vehicle mechanism, a new model for the interpretation of the conductivity of fast proton conductors. *Angew Chem Int Ed Engl* 21: 208–209.

36 (a) Zawodzinski, T., Derouin, C., and Radzinski, S. (1993). Water uptake by and transport through Nafion® 117 membranes. *J Electrochem Soc* 140: 1041. (b) Di Noto, V., Zawodzinski, T.A., Herring, A.M. et al. (2012). Polymer electrolytes for a hydrogen economy. *Int J Hydrog Energy* 7: 6120–6131.

37 Deluca, N.W. and Elabd, Y.A. (2006). Polymer electrolyte membranes for the direct methanol fuel cell: a review. *J Polym Sci B Polym Phys* 44: 2201–2225.

38 Petersen, M.K. and Voth, G.A. (2006). Characterization of the solvation and transport of the hydrated proton in the perfluorosulfonic acid membrane nafion. *J Phys Chem B* 110: 18594–18600.

39 (a) Cappadonia, M., Erning, J.W., Niaki, S.M.S., and Stimming, U. (1995). Conductance of Nafion 117 membranes as a function of temperature and water content. *Solid State Ionics* 77: 65–69. (b) Uosaki, K., Okazaki, K., and Kita, H. (1990). Conductivity of Nation membranes at low temperatures. *J Electroanal Chem Interfacial Electrochem* 287: 163–169.

40 (a) Antonucci, P.L., Arico, A.S., Creti, P. et al. (1999). Investigation of a direct methanol fuel cell based on a composite Nafion®-silica electrolyte for high temperature operation. *Solid State Ionics* 125: 431–437. (b) Adjemian, K., Srinivasan, S., Benziger, J., and Bocarsly, A.B. (2002). Investigation of PEMFC operation above 100 C employing perfluorosulfonic acid silicon oxide composite membranes. *J Power Sources* 109: 356–364. (c) Kim, Y.M., Choi, S.H., Lee, H.C. et al. (2004). Organic–inorganic composite membranes as addition of SiO_2 for high temperature-operation in polymer electrolyte membrane fuel cells (PEMFCs). *Electrochim Acta* 49: 4787–4796.

41 Malhotra, S. and Datta, R. (1997). Membrane-supported nonvolatile acidic electrolytes allow higher temperature operation of proton-exchange membrane fuel cells. *J Electrochem Soc* 144: L23.

42 Kraytsberg, A. and Ein Eli, Y. (2014). Review of advanced materials for proton exchange membrane fuel cells. *Energy Fuel* 28: 7303–7330.

43 (a) Kinumoto, T., Inaba, M., Nakayama, Y. et al. (2006). Durability of perfluorinated ionomer membrane against hydrogen peroxide. *J Power Sources* 158: 1222–1228. (b) Zamel, N. and Li, X. (2011). Effect of contaminants on polymer electrolyte membrane fuel cells. *Prog Energy Combust Sci* 37: 292–329.

44 Wang, Y., Diaz, D.F.R., Chen, K.S. et al. (2020). Materials, technological status, and fundamentals of PEM fuel cells–a review. *Mater Today* 32: 178–203.

45 Woo, Y., Oh, S.Y., Kang, Y.S., and Jung, B. (2003). Synthesis and characterization of sulfonated polyimide membranes for direct methanol fuel cell. *J Membr Sci* 220: 31–45.

46 Durante, V.A. and Delaney, W.E. (2011). Highly stable fuel cell membranes and methods of making them. US 7,989,115 B2. Google Patents.

47 Savadogo, O. (2004). Emerging membranes for electrochemical systems: part II. High temperature composite membranes for polymer electrolyte fuel cell (PEFC) applications. *J Power Sources* 127: 135–161.

48 Gahlot, S., Sharma, P.P., Kulshrestha, V., and Jha, P.K. (2014). SGO/SPES-based highly conducting polymer electrolyte membranes for fuel cell application. *ACS Appl Mater Interfaces* 6: 5595–5601.

49 Mikhailenko, S.D., Wang, K., Kaliaguine, S. et al. (2004). Proton conducting membranes based on cross-linked sulfonated poly (ether ether ketone)(SPEEK). *J Membr Sci* 233: 93–99.

50 Miyake, J., Taki, R., Mochizuki, T. et al. (2017). Design of flexible polyphenylene proton-conducting membrane for next-generation fuel cells. *Sci Adv* 3: eaao0476.

51 Kang, N.R., Pham, T.H., and Jannasch, P. (2019). Polyaromatic perfluorophenylsulfonic acids with high radical resistance and proton conductivity. *ACS Macro Lett* 8: 1247–1251.

52 Lee, K.S., Spendelow, J.S., Choe, Y.K. et al. (2016). High-capacity battery cathode prelithiation to offset initial lithium loss. *Nat Energy* 1: 1–7.

53 Lee, S.Y., Kang, N.R., Shin, D.W. et al. (2012). Morphological transformation during cross-linking of a highly sulfonated poly (phenylene sulfide nitrile) random copolymer. *Energy Environ Sci* 5: 9795–9802.

54 Park, C.H., Lee, S.Y., Hwang, D.S. et al. (2016). Nanocrack-regulated self-humidifying membranes. *Nature* 532: 480–483.

55 Lee, H., Seo, H., Kim, S.K., and Bae, I. (2022). Aligned proton transport highway of hierarchically structured proton-exchange membranes constructed via capillary force lithography. *ACS Applied Energy Materials* 5: 6256–6264.

56 Jiao, K., Xuan, J., Du, Q. et al. (2021). Designing the next generation of proton-exchange membrane fuel cells. *Nature* 595: 361–369.

57 Merle, G., Wessling, M., and Nijmeijer, K. (2011). Anion exchange membranes for alkaline fuel cells: a review. *J Membr Sci* 377: 1–35.

58 (a) Gülzow, E. and Schulze, M. (2004). Long-term operation of AFC electrodes with CO_2 containing gases. *J Power Sources* 127: 243–251. (b) Gouérec, P., Poletto, L., Denizot, J. et al. (2004). The evolution of the performance of alkaline fuel cells with circulating electrolyte. *J Power Sources* 129: 193–204.

59 Henkensmeier, D., Najibah, M., Harms, C. et al. (2021). Overview: state-of-the art commercial membranes for anion exchange membrane water electrolysis. *J Electrochem Energy Convers Storage* 18: 024001.

60 Park, E.J., Capuano, C.B., Ayers, K.E., and Bae, C. (2018). Chemically durable polymer electrolytes for solid-state alkaline water electrolysis. *J Power Sources* 375: 367–372.

61 (a) Kim, Y.S., Wang, F., Hickner, M. et al. (2003). Fabrication and characterization of heteropolyacid (H3PW12O40)/directly polymerized sulfonated poly (arylene ether sulfone) copolymer composite membranes for higher temperature fuel cell applications. *J Membr Sci* 212: 263–282. (b) Park, J.S., Choi, J.H., Woo, J.J., and Moon, S.H. (2006). An electrical impedance spectroscopic (EIS) study on transport characteristics of ion-exchange membrane systems. *J Colloid Interface Sci* 300: 655–662.

62 Ostendorf, A., Schönhoff, M., and Cramer, C. (2019). Ionic conductivity of solid polyelectrolyte complexes with varying water content: application of the dynamic structure model. *Phys Chem Chem Phys* 21: 7321–7329.

63 (a) Stoica, D., Ogier, L., Akrour, L. et al. (2007). Anionic membrane based on polyepichlorhydrin matrix for alkaline fuel cell: synthesis, physical and electrochemical properties. *Electrochim Acta* 53: 1596–1603. (b) Slade, R.C. and

Varcoe, J.R. (2005). Investigations of conductivity in FEP-based radiation-grafted alkaline anion-exchange membranes. *Solid State Ionics* 176: 585–597.

64 Kreuer, K.D., Paddison, S.J., Spohr, E., and Schuster, M. (2004). Transport in proton conductors for fuel-cell applications: simulations, elementary reactions, and phenomenology. *Chem Rev* 104: 4637–4678.

65 Ahmad, M., Zaidi, S., and Rahman, S. (2006). Proton conductivity and characterization of novel composite membranes for medium-temperature fuel cells. *Desalination* 193: 387–397.

66 Yagizatli, Y., Ulas, B., Cali, A. et al. (2020). Improved fuel cell properties of Nano-TiO2 doped Poly (Vinylidene fluoride) and phosphonated Poly (Vinyl alcohol) composite blend membranes for PEM fuel cells. *Int J Hydrog Energy* 45: 35130–35138.

67 Wang, T., Jeon, J.Y., Han, J. et al. (2020). Poly (terphenylene) anion exchange membranes with high conductivity and low vanadium permeability for vanadium redox flow batteries (VRFBs). *J Membr Sci* 598: 117665.

68 Wang, F. and Fu, B. (2020). Anion exchange membranes for direct methanol alkaline fuel cells. In: *Direct Methanol Fuel Cell Technology* (ed. K. Dutta), 71–106. Elsevier.

69 Ziv, N., Mondal, A.N., Weissbach, T. et al. (2019). Effect of CO_2 on the properties of anion exchange membranes for fuel cell applications. *J Membr Sci* 586: 140–150.

70 Strasser, D.J., Graziano, B.J., and Knauss, D.M. (2017). Base stable poly (diallylpiperidinium hydroxide) multiblock copolymers for anion exchange membranes. *J Mater Chem A* 5: 9627–9640.

71 Cheng, C., He, X., Huang, S. et al. (2020). Novel self-cross-linked multi-imidazolium cations long flexible side chains triblock copolymer anion exchange membrane based on ROMP-type polybenzonorbornadiene. *Int J Hydrog Energy* 45: 19676–19690.

72 Elumalai, V. and Sangeetha, D. (2018). Preparation of anion exchangeable titanate nanotubes and their effect on anion exchange membrane fuel cell. *Mater Des* 154: 63–72.

73 Zhu, Y., Pei, S., Tang, J. et al. (2013). Enhanced chemical durability of perfluorosulfonic acid membranes through incorporation of terephthalic acid as radical scavenger. *J Membr Sci* 432: 66–72.

74 Li, D., Matanovic, I., Lee, A.S. et al. (2019). Phenyl oxidation impacts the durability of alkaline membrane water electrolyzer. *ACS Appl Mater Interfaces* 11: 9696–9701.

75 Dekel, D.R., Amar, M., Willdorf, S. et al. (2017). Effect of water on the stability of quaternary ammonium groups for anion exchange membrane fuel cell applications. *Chem Mater* 29: 4425–4431.

76 (a) Krewer, U., Weinzierl, C., Ziv, N., and Dekel, D.R. (2018). Impact of carbonation processes in anion exchange membrane fuel cells. *Electrochim Acta* 263: 433–446; (b) Gutru, R., Turtayeva, Z., Xu, F. et al. (2020). A comprehensive review on water management strategies and developments in anion exchange membrane fuel cells. *Int J Hydrog Energy* 45: 19642–19663. (c) Zhang, H., Ohashi, H., Tamaki, T., and Yamaguchi, T. (2013). Water movement in a solid-state alkaline fuel cell affected by the anion-exchange pore-filling membrane properties. *J Phys Chem C* 117:

16791–16801. (d) Wang, X.R., Ma, Y., Gao, J. et al. (2021). Review on water management methods for proton exchange membrane fuel cells. *Int J Hydrog Energy* 46 (22): 12206–12229.

77 (a) Barbir, F. (2005). *PEM Fuel Cells: Theory and Practice*. Amsterdam: Elsevier Academic Press. (b) Grew, K.N. and Chiu, W.K. (2010). A dusty fluid model for predicting hydroxyl anion conductivity in alkaline anion exchange membranes. *J Electrochem Soc* 157: B327.

78 Tuckerman, M.E., Marx, D., and Parrinello, M. (2002). The nature and transport mechanism of hydrated hydroxide ions in aqueous solution. *Nature* 417: 925–929.

79 Yanagi, H. and Fukuta, K. (2008). Anion exchange membrane and ionomer for alkaline membrane fuel cells (AMFCs). *ECS Trans* 16: 257.

80 Davydova, E.S., Mukerjee, S., Jaouen, F., and Dekel, D.R. (2018). Electrocatalysts for hydrogen oxidation reaction in alkaline electrolytes. *ACS Catal* 8: 6665–6690.

81 (a) Kabir, S., Lemire, K., Artyushkova, K. et al. (2017). Platinum group metal-free NiMo hydrogen oxidation catalysts: high performance and durability in alkaline exchange membrane fuel cells. *J Mater Chem A* 5: 24433–24443. (b) Hwang, G.J. and Ohya, H. (1997). Crosslinking of anion exchange membrane by accelerated electron radiation as a separator for the all-vanadium redox flow battery. *J Membr Sci* 132: 55–61. (c) Wang, G., Li, W., Huang, B. et al. (2019). Exploring the composition–activity relation of Ni–Cu binary alloy electrocatalysts for hydrogen oxidation reaction in alkaline media. *ACS Appl Energy Mater* 2: 3160–3165.

82 (a) Lu, F., Zhang, Y., Liu, S. et al. (2017). Surface proton transfer promotes four-electron oxygen reduction on gold nanocrystal surfaces in alkaline solution. *J Am Chem Soc* 139: 7310–7317. (b) Hu, P., Song, Y., Chen, L., and Chen, S. (2015). Electrocatalytic activity of alkyne-functionalized AgAu alloy nanoparticles for oxygen reduction in alkaline media. *Nanoscale* 7: 9627–9636.

83 (a) Sassin, M.B., Garsany, Y., Gould, B.D., and Swider Lyons, K.E. (2017). Fabrication method for laboratory-scale high-performance membrane electrode assemblies for fuel cells. *Anal Chem* 89 (1): 511–518. (b) Ramaswamy, N. and Mukerjee, S. (2011). Influence of inner-and outer-sphere electron transfer mechanisms during electrocatalysis of oxygen reduction in alkaline media. *J Phys Chem C* 115: 18015. (c) Durst, A.S., Simon, C., Hasche, F. et al. (2014). New insights into the electrochemical hydrogen oxidation and evolution reaction mechanism. *Energy Environ Sci* 7: 2255-60. (d) Zheng, J., Sheng, W., Zhuang, Z. et al. (2016). Universal dependence of hydrogen oxidation and evolution reaction activity of platinum-group metals on pH and hydrogen binding energy. *Sci Adv* 2: e1501602.

84 Chae, J.E., Yoo, S.J., Kim, J.Y. et al. (2020). Hydrocarbon-based electrode ionomer for proton exchange membrane fuel cells. *Int J Hydrog Energy* 45: 32856–32864.

85 (a) Mandal, M., Huang, G., Hassan, N.U. et al. (2019). The importance of water transport in high conductivity and high-power alkaline fuel cells. *J Electrochem Soc* 167: 054501. (b) Ferriday, T.B. and Middleton, P.H. (2021). The importance of water transport in high conductivity and high-power alkaline fuel cells. *Int J Hydrog Energy* 46: 18489–18510. (c) Zhang, J., Zhu, W., Huang, T. et al. (2021). Recent insights on catalyst layers for anion exchange membrane fuel cells. *Adv Sci* 8: 2100284.

86 (a) Eriksson, B., Grimler, H., Carlson, A. et al. (2019). Quantifying water transport in anion exchange membrane fuel cells. *Int J Hydrog Energy* 44: 4930. (b) Roy, A., Talarposhti, M.R., Normile, S.J. et al. (2018). Nickel–copper supported on a carbon black hydrogen oxidation catalyst integrated into an anion-exchange membrane fuel cell. *Sustain Energy Fuels* 2: 2268.

87 (a) Kim, T.H., Yi, J.Y., Jung, C.Y. et al. (2017). Solvent effect on the Nafion agglomerate morphology in the catalyst layer of the proton exchange membrane fuel cells. *Int J Hydrog Energy* 42: 478–485. (b) Welch, C., Labouriau, A., Hjelm, R. et al. (2012). Nafion in dilute solvent systems: dispersion or solution? *ACS Macro Lett* 1: 1403–1407.

88 (a) Wang, L.S., Wang, J., Zhao, Y. et al. (2019). High-performance hydroxide exchange membrane fuel cells through optimization of relative humidity, backpressure and catalyst selection. *J Electrochem Soc* 166: F3305. (b) Gao, X., Yu, H., Qin, B. et al. (2019). Enhanced water transport in AEMs based on poly (styrene–ethylene–butylene–styrene) triblock copolymer for high fuel cell performance. *Polym Chem* 10: 1894.

89 Li, D., Chung, H.T., Maurya, S. et al. (2018). Impact of ionomer adsorption on alkaline hydrogen oxidation activity and fuel cell performance. *Curr Opin Electrochem* 12: 189–195.

90 Park, E.J., Maurya, S., Lee, A.S. et al. (2019). How does a small structural change of anode ionomer make a big difference in alkaline membrane fuel cell performance? *J Mater Chem A* 7: 25040–25046.

91 (a) Hong, Z., Wei, Z., and Han, X. (2021). Optimization scheduling control strategy of wind-hydrogen system considering hydrogen production efficiency. *J Energy Storage* 47: 103609. (b) Irfan, M., Razzaq, A., Chupradit, S. et al. (2021). Hydrogen production potential from agricultural biomass in Punjab province of Pakistan. *Int J Hydrog Energy* 47: 2846–2861.

92 (a) Zhang, X., Chu, X., Zhang, M. et al. (2019). Molecularly designed, solvent processable tetraalkylammonium-functionalized fluoropolyolefin for durable anion exchange membrane fuel cells. *J Membr Sci* 574: 212–221. (b) Xiao, Y., Huang, W., Xu, K. et al. (2018). Preparation of anion exchange membrane with branch polyethyleneimine as main skeleton component. *Mater Des* 160: 698–707.

93 Ong, B.C., Kamarudin, S.K., and Basri, S. (2017). Direct liquid fuel cells: a review. *Int J Hydrog Energy* 42: 10142–10157.

94 (a) Kim, J., Momma, T., and Osaka, T. (2009). Cell performance of Pd–Sn catalyst in passive direct methanol alkaline fuel cell using anion exchange membrane. *J Power Sources* 189: 999–1002. (b) Zhao, T.S., Li, Y.S., and Shen, S.Y. (2010). Anion-exchange membrane direct ethanol fuel cells: status and perspective. *Front Energy Power Eng China* 4: 443–458.

95 (a) Kaya, S., Ozok-Arici, O., Kivrak, A. et al. (2022). Benzotiyofen@ Pd as an efficient and stable catalyst for the electrocatalytic oxidation of hydrazine. *Fuel* 328: 125355. (b) Hren, M., Božič, M., Fakin, D. et al. (2021). Alkaline membrane fuel cells: anion exchange membranes and fuels. *Sustain Energy Fuels* 5 (3): 604–637.

96 Nayak, S.P., Ventrapragada, L.K., Rao, A.M., and Kumar, J.K. (2023). Porous gold-curcumin nanocomposite for enhanced electrooxidation of glycerol and ethylene glycol. *Mater Lett* 1 (330): 133212.

97 Oshchepkov, A., Bonnefont, A., Maranzana, G. et al. (2022). Direct borohydride fuel cells: a selected review of their reaction mechanisms, electrocatalysts, and influence of operating parameters on their performance. *Curr Opin Electrochem* 32: 100883.

98 Du, L., Lou, S., Chen, G. et al. (2019). Direct dimethyl ether fuel cells with low platinum-group-metal loading at anode: investigations of operating temperatures and anode Pt/Ru ratios. *J Power Sources* 433: 126690.

99 (a) Ye, M., Sharp, P., Brandon, N., and Kucernak, A. (2022). System-level comparison of ammonia, compressed and liquid hydrogen as fuels for polymer electrolyte fuel cell powered shipping. *Int J Hydrog Energy* 47 (13): 8565–8584. (b) Wang, B., Li, T., Gong, F. et al. (2022). Ammonia as a green energy carrier: electrochemical synthesis and direct ammonia fuel cell-a comprehensive review. *Fuel Process Technol* 235: 107380.

6

Transport and Conductive Mechanisms in Anion Exchange Membranes

Ramato A. Tufa[1], *Misgina T. Tsehaye*[2], *Wenjuan Zhang*[3], *Marco Aquino*[1], *Sergio Santoro*[1], *and Efrem Curcio*[1]

[1] *University of Calabria, Department of Environmental Engineering, Via Pietro Bucci CUBO 44A, Rende, CS, 87036, Italy*
[2] *Flemish Institute for Technological Research (VITO), Separation & Conversion Technology, Boeretang 200, Mol, 2400, Belgium*
[3] *Tianjin Chengjian University, School of Environmental and Municipal Engineering, Tianjin Key Laboratory of Aquatic Science and Technology, Tianjin, 300384, China*

6.1 Introduction

Recent significant research advances in the development of anion exchange membranes (AEMs) drive the reborn interest in the anion exchange membrane fuel cells (AEMFCs). For the most part, AEMFC is similar to the proton exchange membrane fuel cell (PEMFC), with the main difference being that AEMFC utilizes an AEM in which OH^- ion is transported from the cathode to the anode, whereas in PEMFC, H^+ is transported from anode to cathode through the PEM during operations (Figure 6.1) [1]. Unlike PEMFC, AEMFC operates in an alkaline environment which offers advantages like the possible use of non-noble metal catalysts for high energy density and a broad range of fuels like alcohols beyond hydrogen. However, the practical implementation of this technology is still limited by the insufficient OH^- conductivity of the AEMs limiting the fuel efficiency. Moreover, carbonation issues due to the formation of carbonate (CO_3^{2-}) and bicarbonate (HCO_3^-) in the presence of CO_2 result in a substantial performance loss. As such, new designs of AEM with improved transport properties are required, along with a detailed understanding of the transport limitations and mechanisms of the active ion in AEMs.

Despite significant advances in the experimental and theoretical study of AEMs for fuel cells [2–4], the transport mechanism of OH^- and structure–transport efficiency relationships have been less explored. Understanding the mechanisms by

Alkaline Anion Exchange Membranes for Fuel Cells: From Tailored Materials to Novel Applications, First Edition. Edited by Jince Thomas, Alex Schechter, Flavio Grynszpan, Bejoy Francis, and Sabu Thomas.

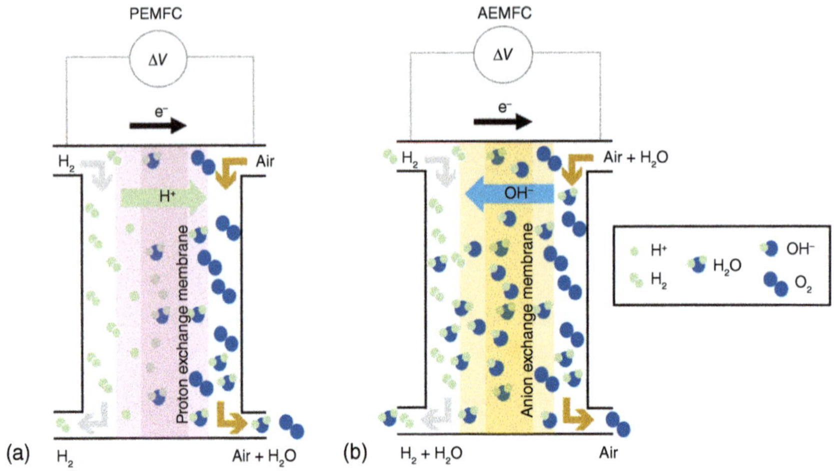

Figure 6.1 (a) Proton exchange membrane fuel cell (PEMFC), and (b) anion exchange membrane fuel cell (AEMFC).

which OH^- diffuses through water channels in AEMs helps to unravel the key correlations between the charge transport efficiency and polymer structure, which allows for thorough investigation and design of high-performance AEMs. In this chapter, we make a systematic analysis of the OH^- ion-conductive properties and the transport mechanism in AEMs supported by the theoretical model developments in the existing literature.

6.2 Transport Mechanisms of Hydroxide Ion in AEMs

At the heart of an alkaline fuel cell lies an AEM which play a key role in the transport of OH^- ions from the cathode to the anode. These AEMs should exhibit good conductive properties for high performance and practical applications of the fuel cell. Thus, it's crucial to understand the transport phenomenon and mechanisms governing the OH^- conductivity of the AEMs.

Early-stage investigations for OH^- transport properties of AEMs were developed, based on the principles and mechanisms for proton exchange membranes (PEMs). It is therefore important to have a good grip on the proton transport mechanisms before illustrating the concepts for OH^- transport.

The H^+ transport mechanism has been debated for a very long while several types of research being carried out to investigate proton transport in an aqueous environment. Compared to OH^- ion, H^+ has a very small ionic radius and strong polarization power which makes it difficult to isolate from its environment under equilibrium conditions. To date, two main H^+ transport mechanisms are considered: (i) the vehicle mechanism and (ii) the Grotthuss mechanism. Other mechanisms like en masse diffusion, migration, surface site hopping along functional side chains in polymeric membranes, and convective processes are also known [3, 5–7]. The vehicular mechanism follows a molecular diffusion process in which hydrated protons are transported through a proton-conducting aqueous channel with water molecules acting as a vehicle. On the other hand, structural diffusion also occurs through the Grotthuss mechanism in which protons are transported across in a subsequent process of hydrogen bond formation and break along with oriental water dipoles [8]. As such, the proton is free and moves by hopping from one oxygen ion to another by rotational diffusion of the OH^- groups after the break of a hydrogen bond (Figure 6.2a–c). Moreover, surface mechanisms or surface site hopping which are based on a process where protons hop between the membrane functional groups driven by strong electrostatic attractions are also known. In another study, the OH^- transport via Grotthuss H^+ hopping was shown to be driven by a dynamic complex with three "fast" and one "slow" H_2O molecules bridging two OH^- groups from the strongly bound remnant water [9]. Overall, the Grotthuss mechanism is often considered the dominant transport mechanism of H^+ in aqueous solutions, in particular, at higher membrane water contents [4, 5, 10].

The dependence of transport mechanisms on external operating factors like temperature is a less understood concept. Zelovich et al. investigated the effect of temperature on the diffusivity of OH^- in AEMs quantized at different hydration

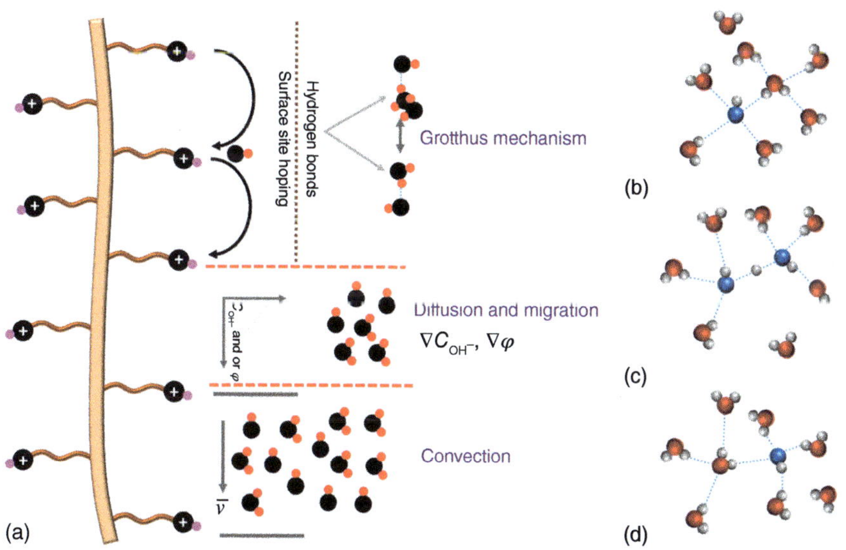

Figure 6.2 (a) Schematic representation of the different transport mechanisms in AEMs: the transport properties are depicted to be the product of any combination of these mechanisms including the Grotthuss mechanism, surface site hopping on the respective membrane quaternary ammonium side chains, diffusion, and migration in the presence of a concentration and/or potential gradient acting. Elaboration of the Grotthuss mechanisms as a dominant transport mechanism in AEMs (b–d); (b) OH^- in a fourfold-coordinated, $OH^-(H_2O)_4$ state; (c) first-solvation-shell H-bond breakage by thermal fluctuations resulting in OH^- in a threefold $OH^-(H_2O)_4$ state, followed by the formation of a weak H-bond between the OH^- hydrogen and neighboring solvent water; (d) migration of OH^- $(H_2O)_3$ to migrate within the H-bond network as a result of complete proton transfer.

levels [11]. They found out that the OH^- diffusion does not increase monotonically with temperature but rather non-monotonic.

The transport mechanisms of H^+ and OH^- ions in aqueous solutions have been depicted to be analogous based on the observed similarity in the dependence of the practical conductivity of both the H^+ and OH^- ions on operating conditions like temperature and relative humidity. As shown in Figure 6.2d, the dominant mechanisms for OH^- ions transport in AEM also follow the Grotthuss mechanism, diffusion, migration, and convection. The OH^- transport in the AEM occurs, however, with some differing specifics in Grotthuss mechanisms. The H_3O^+ is naturally bound by the water hydrogen bonding network and so does not require significant activation and/or solvent rearrangement (Figure 6.2d). On the other hand, OH^- ion tends to rearrange the solvent molecules and shuffle the hydrogen bond network due to stable solvation shells [4, 12]. Diffusive transport of OH^- ions is driven by concentration and/or electrical potential gradient, whereas convective transport across the membrane involves the co-transport of OH^- ions across the membrane along with the water molecules. OH^- ions can also be transported by surface site hopping mechanisms, but this has less significance due to the competitive effect of the strong interaction of water molecules and the hydrophilic membrane functionalities [4].

So far, the relative importance of the different transport mechanisms governing the OH^- transport in AEMs is still not clear among the scientific community, requiring further research to understand the detailed OH^- ion transport mechanisms along with membrane development. Chiu et al. developed a model to elucidate the dominant transfer mechanism along with the factors limiting OH^- conduction in AEMs [4]. Considering PEM and AEM systems, there exists a large deviation of the bulk transport coefficients ions and membrane conductivities. For instance, the relative transport coefficients incorporating en masse diffusion/migration and Grotthuss are about $D_{H^+} = 9.3 \times 10^{-9}\ m^2\ s^{-1}$ and $D_{OH^-} = 5.3 \times 10^{-9}\ m^2\ s^{-1}$ in liquid water at 25 °C are comparable, whereas the conductivities of 0.078, 0.018 S cm^{-1} were reported for Nafion 115 and AEM (based on fully fluorinated poly(tetrafluoroethylene-*co*-hexafluoropropylene) – FEP), respectively. This presents a considerable deviation from the bulk transport coefficients which could be attributed to the intrinsic difference in membrane structure and properties which vary depending on operating/environmental conditions like the relative humidities and/or water uptake coefficients (Figure 6.3a), insufficient OH^- dissociation and solvation, interactions between the OH^- in solution, and the quaternary ammonium side chain, etc. [14]. Mathematically, the water uptake coefficient is described by the Brunauer–Emmett–Teller (BET) equation, which is used to fit experimentally reported water uptake coefficient data with the variations in relative humidity. The standard form of the BET is provided in [4]. Carbonation of OH^- by CO_2, a phenomenon well recognized as a limiting factor for OH^- conductivities, is also another cause for the difference in membrane transport properties. Moreover, a systematic study by Varcoe indicated significantly low conductivities of AEMs based on quaternary ammonium functionalities at low humidity of the surrounding static atmospheres, along with 2–3 times

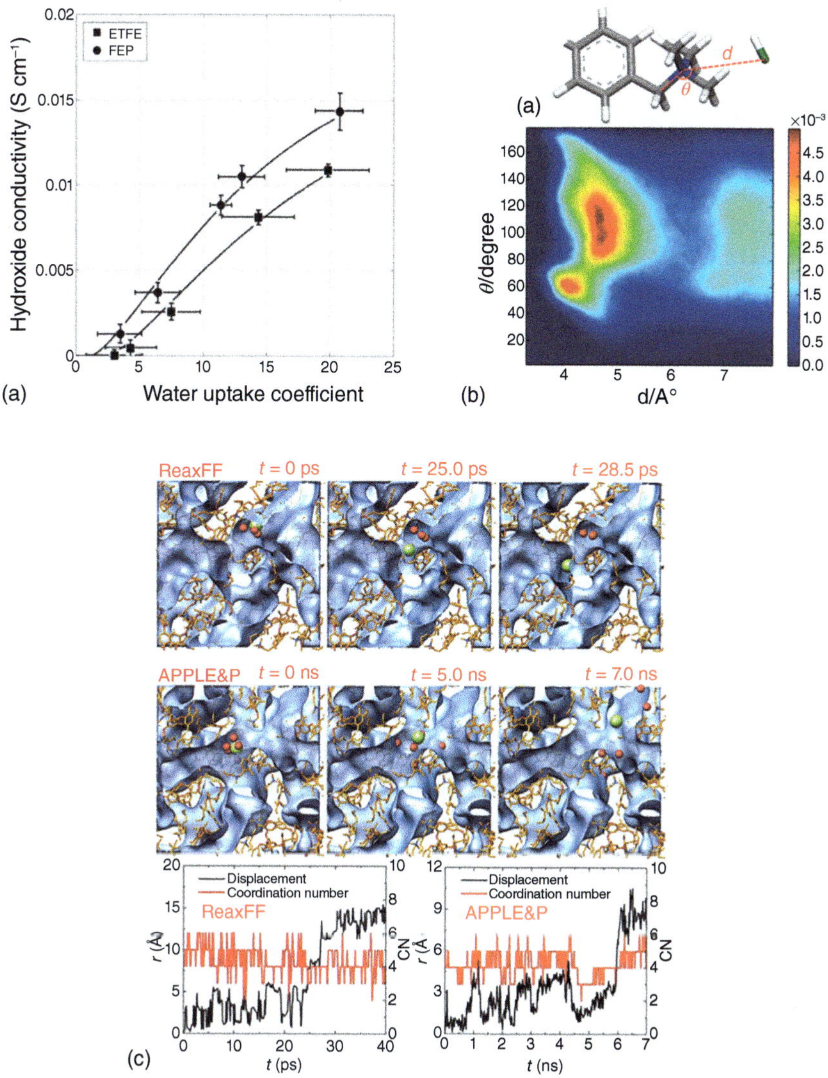

Figure 6.3 (a) Comparison of the OH^- conductivities of AEM-ETFE and AEM-FEP determined by DFM (lines) and experiments (symbol) as a function of water uptake coefficient. Source: Copyright 2010 IOP science. Reproduced with permission from Grew and Chiu [4]. (b) Joint probability density distributions of the N–O* distance (d) and the CCH_2–N–O* angle (θ) for the PVBTMA system at 300 K. Source: Copyright 2016 American Chemical Society. Reproduced with permission from Chen et al. [13]. (c) Diffusion of OH^- through the pocket of a water channel in PPO-based AEM: the green sphere corresponds to the O in OH^- and the red spheres correspond to the coordinated water oxygen at the starting point. The simulation results are displayed in the top three panels, whereas the tracking of displacement from original point and number of water molecules coordinating OH^- within its first hydration shell is shown in the bottom panels. Source: Copyright 2018 American Chemical Society. Reproduced with permission from Dong et al. [10].

higher activation energies for OH^- compared to H^+ conduction in acidic Nafion-115 at all humidities [14].

6.3 AEM Structure–Transport Efficiency Relationships

Understanding the OH^- transport mechanisms as well as the AEM structure–transport efficiency relationship is crucial the design of high-performance AEMs. Literature focusing on this area is still scarce with the topic requiring further research advance to set the benchmarks for designing highly conductive and stable AEMs. For instance, a broad range of AEMs with varying properties like water uptake and conductivity can be prepared by controlling the degree of functionalization. In aqueous systems, the functional groups form a phase-separated domain which can promote high transport rates with slow water diffusion but high water binding. A range of functional groups has been used for the synthesis of AEMs with different physicochemical and electrochemical properties [3, 15, 16]. Typically, quaternary ammonium functional groups have been widely used in synthesizing membranes for a broad range of applications, but the OH^- conductivity is limited due to the weak basicity of such functional groups ($pK_b \sim 4$) [16–18]. The nature and chemistry of the AEM backbone are also highly determinant for enhancing the OH^- transport properties. For example, the use of block copolymer analogs to modify the microphase separation of AEMs has been shown to enhance the OH^- conductivity [19, 20]. Moreover, engineering the side-chain configurations at controlled properties of the spacer units between the functional groups and the polymer backbone can also be used to improve the OH^- conductive properties of AEMs [21]. A molecular dynamics (MD) model developed by Chen et al. on poly(vinyl benzyltrimethylammonium) (PVBTMA) system, depicts a relatively even separation of rigid side chains in AEMs forming a continuous overlapping region for OH^- transport consisting of a hydration shell of a tethered cationic group [13]. The OH^- was observed to have a strong preference for this overlapping region (Figure 6.3b) for conduction with a substantial contribution from both standard diffusion (vehicular) and Grotthuss mechanisms. A substantial contribution from the vehicular mechanism as compared to the Grotthuss mechanism was observed with about only 20% contribution of the Grotthuss mechanism to the total diffusivity of OH^- and the charge transport primarily relying on OH^- vehicular mechanism. On the other hand, Zhang et al. performed atomic-scale simulations to correlate the impact of membrane structural and dynamical properties on ion transport. Their results show that the Grotthuss mechanism makes the proton transport through water channels less dependent on the membrane morphology at high water uptake, and the AEMs with a distinct water channel and enhanced nano-phase separation show a relatively high OH^- transport efficiency [22]. A recent study by Dong et al. on the OH^- charge transport in poly(*p*-phenylene oxide) (PPO-based) AEMs indicate that in nonblock AEMs exhibiting a network of sub-nanometer-wide water channels, the vehicular mechanism for transport requires the OH^- to partially dehydrate which is thermodynamically unfavorable [10]. The Grotthuss hopping was observed to dominate the transport mechanism for OH^- transport through narrow, sub-nanometer size percolating water channels (Figure 6.3c). Other chemistries of AEMs have also been reported like ion-solvating membranes which operate in the presence of an aqueous alkaline electrolyte for enhanced conductivity, although such materials do have intrinsic OH^- conducting properties [23]. Such membranes with different structural and dynamic properties as compared to the classical AEMs also lack a detailed investigation and understanding of the relative significance of the different transport mechanisms calling for further research.

Overall, MD simulations allow a deeper study of the microstructure and ions transport process in a membrane that is not feasible by experimental investigations. Density functional theory (DFT) and MD calculations was used to investigate the impact of membrane water uptake and its ion exchange capacity on OH^- diffusion in alkaline anion exchange membranes (AAEMs).[24]. The study was done on two poly(aryl ether ether ketone) (PEEK)-based membranes in which the side chains contain one or two quaternary ammonium groups. They found similar self-diffusion coefficients in SQ and GQ, which cannot be corrected with the increased conductivity, rather the higher ionic conductivity of GQ was linked to the enhanced ionic conductivity. As such, the fairly identical self-diffusion coefficients had no significant impact on ionic conductivities which lays a prospect for the side chain chemistry in AAEMs and OH^- transport.

6.4 Ion Conductivity Measurement

The performance of an AEMFC is highly influenced by its electrochemical properties, among them: ion conductivity of AEMs [25–29]. Different methods can be employed to measure ionic conductivity: (a) direct current (DC) method or (ii) the alternating current (AC) method. The DC measurements are usually carried out by applying a constant current to a system (an AEM with identical solutions at both sides) and monitoring the voltage drop across the system. The resistance (the reciprocal of ion conductivity) of the system, which is constituted by the resistance of the membrane and interfacial layers, can be obtained according to Ohm's law. In this case, the pure membrane resistance cannot be discriminated from the interfacial layer resistance. Długołęcki et al. [30] applied DC to investigate the effects of solution concentration on the resistance of ion exchange membranes and found that when the NaCl concentration was lower than 0.1 M, the resistance for the membrane and interfacial layers was significantly increased, but the contribution of the pure membrane resistance could not be evaluated. Ziv and Dekel [31] reported a novel, practical, and *ex situ* method for the determination of the true hydroxide conductivity in AEMs with an external electric current across the membrane by a four-probe experimental setting. The applied current forced the electrochemical reactions in the AEMs with the formation of OH^- at one electrode and the transformation of CO_2 from HCO_3^-/CO_3^{2-} at the other electrode. The OH^- conductivity of AEMs could be obtained when all the anions in the membrane were replaced by OH^-.

The AC method, which is also known as electrochemical impedance spectroscopy (EIS), is operated by inputting a small electrical stimulation signal (alternating current or voltage) into the system and obtaining the response signals (voltage/current value and phase shift between the input and response signals) from the reference electrodes adjacent to the test membrane. Therefore, the EIS data are constituted by the Nyquist plots and/or the Bode formats. The Nyquist plots are consisted by a locus of points which represent different measurement frequencies, and the phase angle and modulus as a function of frequency are displayed in the Bode formats. The EIS data are usually interpreted by equivalent circuits with complex nonlinear-least-squares fitting. The fitting results involve the contribution of each component in the test system, i.e. the resistance and capacitance of the pure membrane, electrical double layer, and diffusion boundary layer. EIS is regarded as a noninvasive technology and could dynamically determine the properties of ion exchange membranes in different operating conditions and diverse electrolyte solutions. Zhang et al. [29] compared the ion exchange membrane conductivity obtained from DC and EIS (experimental setting and the equivalent circuit for the EIS fitting as shown in Figure 6.4). It was confirmed that the overall conductivity of membrane systems determined by DC was proportional to the pure membrane conductivity from EIS, and the overall resistance determined by the DC equaled the sum of pure membrane resistance, electrical double layer resistance, and diffusion boundary layer resistance determined by EIS. It was shown by EIS that there were structural changes in the two AEMs, which were caused by a neutral form resulting from the degradation reaction of amino groups (weak basicity) with OH^- in solutions [25]. Reshetenko et al. [32] used DC and EIS to analyze the performance of AEMFCs at different operating conditions with a variation of gas humidification and oxygen partial pressure. It was found that operation with air led to possible electrolyte poisoning by CO_2 and cathode mass transfer losses, which resulted in lower performance. Diffusion limitations in cathode mass transfer contributed to the formation of low-frequency capacitive loop in EIS data.

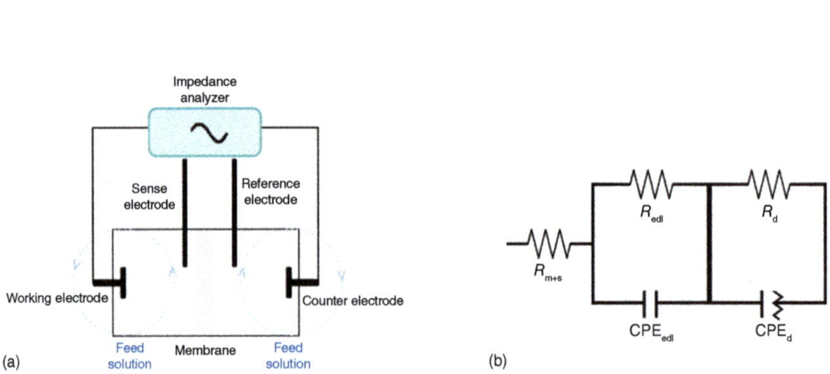

Figure 6.4 (a) Schematic view of the experimental setting for DC and EIS measurement and (b) the equivalent circuit for the EIS fitting. R_{m+s} is the resistance of membrane and solution, R_{edl} and R_d are the resistance of EDL and DBL, respectively, and CPE_{edl} and CPE_d are the constant phase element of EDL and DBL, respectively.

6.5 Carbonation Process in AEMs

6.5.1 Elucidating the Dynamics of Carbonation

High-performance AEMFC requires AEMs with adequate OH^- ion conductivity. Furthermore, to correlate AEMFC performance with AEM ionic conductivity, it is necessary to first understand the true OH^- ion conductivity of AEMs, as well as the factors that influence it. In practice, however, measuring and knowing the actual OH^- ion conductivity of an AEM is difficult due to the fast reaction of OH^- with the carbon dioxide (CO_2) in the air (about 400 ppm CO_2 concentration). As a result, when air is utilized as a source of oxidant instead of oxygen, the OH^- ions react swiftly with atmospheric CO_2 to form larger and less mobile bicarbonate (HCO_3^-) and carbonate (CO_3^{2-}) anions, via acid–base reactions, as shown in Eqs. (6.1) and (6.2) [33]:

$$OH^- + CO_2 \leftrightarrow HCO_3^- \tag{6.1}$$

$$HCO_3^- + OH^- \leftrightarrow CO_3^{2-} + H_2O \tag{6.2}$$

The time it takes for the OH^- ion in the membrane to be completely replaced with HCO_3^- and/or CO_3^{2-} has been estimated to be anywhere between a few minutes and a few hours (Table 6.1). In reality, the majority of carbonation occurs during the first few minutes, with nearly complete conversion occurring within a few hours [35, 36].

6.5.2 Impact of Carbonation on AEM and AEMFC

Carbon dioxide reactions with OH^- ions, which are the dominant charge carriers in AEMFC, have significant impact on both membrane and AEMFC performance [37]. Carbonation is not only limited to AEMFC but also various other technologies like

Table 6.1 Time for complete replacement of OH^- ion with bicarbonate in AEMs.

Membrane	Exposure to CO_2	Temperature	Time until OH^- is completely replaced[a]
A201 (Tokuyama)	Immersed in water in contact with ambient air	20 °C	30 min
A201 (Tokuyama)	Exposed directly to ambient air	Room temperature	6 h
ETFE-based radiation-grafted AEM	Exposed directly to ambient air	—	120 min
FAA-3 (Fumatech)	Exposed directly to ambient air	—	~100 min (less than 10%)
(PTFE-based) PFAEMGen2	Exposed directly to humidified ambient air (95% relative humidity)	30 °C	15 min

[a]Determined by ion exchanging to Cl^- ion and titration using the Warder method.

the recently reported CO_2 pumping/crossover issue in electrochemical CO_2 flow cells [38], where bipolar membranes are reported to be a viable solution [39, 40]. Following the carbonation of OH^- ions in AEMFC, a number of consequences have been reported. One of them is the difficulty in reproducing air exposed reported OH^- ion conductivity values of AEMs, measured with existing measurement techniques [31]. As a result, comparing and analyzing the OH^- ion conductivity of different air exposed AEMs under the same conditions is difficult, if not impossible. The reported values could be the conductivity of the OH^-, HCO_3^-, and CO_3^{2-} anions rather than pure OH^- ion [41]. Indeed, partial/complete carbonation of OH^- ions during conductivity measurement is one of the reasons why OH^- ion conductivity of AEM is reported to be lower than that of H^+ conductivity of PEMs [42]. In this case, proper CO_2 removal precludes this problem.

The CO_3^{2-} and HCO_3^- ions conductivities of AEM are 2.2 and 5.4 times lower than the values measured for OH^- conductivity [43, 44]. This is primarily due to the different molar conductivities of the different anions (CO_3^{2-} = 13.9 and OH^- = 19.9 mS m^2 mol^{-1}) as shown in Table 6.2. Therefore, the ohmic losses in the cell will undoubtedly increase since the main charge carriers are anions with lower mobilities (compared to OH^- ions) [34]. However, this is not the only possible way that the carbonation process can cause a voltage drop. The second issue is that the CO_3^{2-} and HCO_3^- ions are not consumed at the anode electrode, so they accumulate and cause a pH drop. This decrease in the anode pH leads to a Nernstian increase in the anode potential, decreasing overall cell voltage [47]. As a result, the accumulation of those carbonated anions on the anode could result in kinetic and thermodynamic losses [35, 48].

Figure 6.5a–c depicts an AEMFC operating on (i) pure O_2 and (ii) ambient air [49]. In general, the significant decrease in ionic conductivity of AEM and/or ionomer after carbonation processes is estimated to reduce overall AEMFC performance by half [51, 52]. As shown in Figure 6.5c, an initial loss of 50% of power observed when operating the AEMFC with ambient air [50].

6.5.3 Strategies to Avoid Carbonation of OH^- Ions

Few strategies for avoiding carbonation of OH^- ions and/or measuring the true OH^- ions conductivity of AEMs have been proposed [42]. The main strategies are

Table 6.2 Ionic Stokes radii, diffusion coefficient, and ionic conductivity of OH^-, CO_3^{2-}, and HCO_3^- at 25 °C.

Species	Ionic Stokes radii (Å)	Diffusion coefficient (10^{-9} m^2 s^{-1}) in water	Molar ionic conductivity (mS m^2 mol^{-1})
OH^-	0.472	5.27	19.91
CO_3^{2-}	1.296	0.96	13.86
HCO_3^-	2.193	0.6	4.45

Source: Varcoe et al. [16], Nightingale [45], and Besha et al. [46].

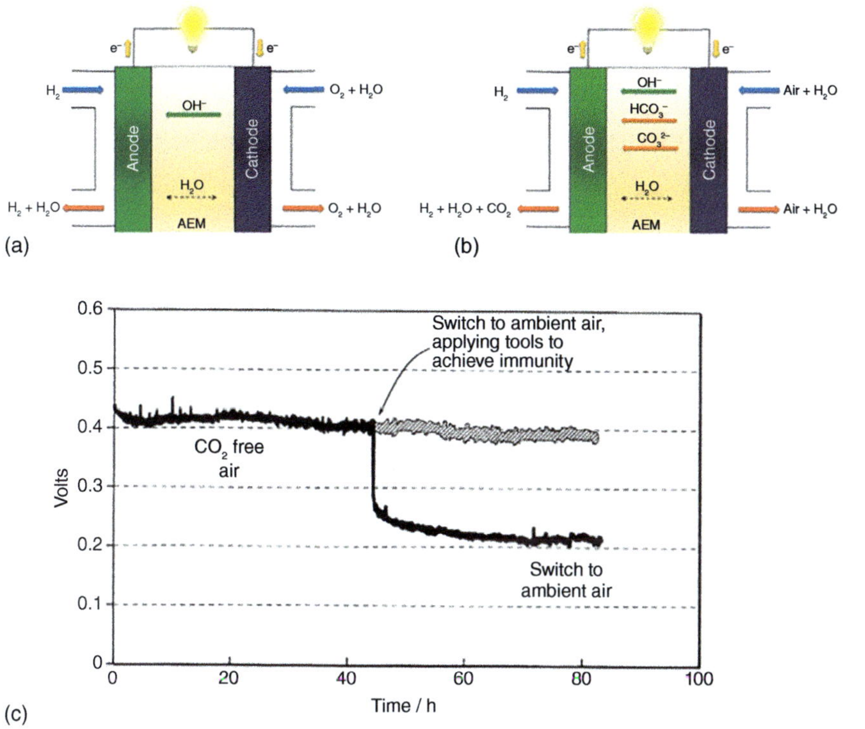

Figure 6.5 Representation of AEMFC operating on (a) pure O_2 and (b) ambient air. Source: Copyright 2018 Elsevier. Reproduced with permission from Krewer et al. [49]. (c) Voltage loss in fuel cell operating with ambient air [50].

discussed in this section: (i) forcing the release of CO_3^{2-} and HCO_3^- as CO_2 gas by passing an electric current through the membrane, (ii) using gas-purged water before or during the experiment, and (iii) using a glovebox containing inert gas.

As discussed in the work of Dekel and coworker [31], the first method proposed to measure the true OH^- ion conductivity of AEMs is forcing the release of the HCO_3^- and CO_3^{2-} anions as CO_2 gas by applying an external electric current through the membrane. Figure 6.6a depicts the experimental setup, Figure 6.6b the processes occurring in the AEM while applying DC, and Figure 6.6c the variation in AEM conductivity before, during, and after applying current. As shown in Figure 6.6a, the samples were exposed to continuous N_2 under testing conditions of constant 90% relative humidity and 40 °C. The corresponding change in ionic conductivity of the AEM over time is shown in Figure 6.6c. The polybenzimidazole (PBI)-based AEM [53] used had a conductivity of 15 mS cm^{-1} without applying current. When a current of 100 μA was applied through the membrane, however, the AEM conductivity increased at an exponential rate due to the electrochemical reactions at the electrodes ($4H_2O + 4e^- \rightarrow 4OH^- + 2H_2$ at the positive electrode) [31]. As shown in Figure 6.6b, the current causes HCO_3^- and/or CO_3^{2-} anions in the

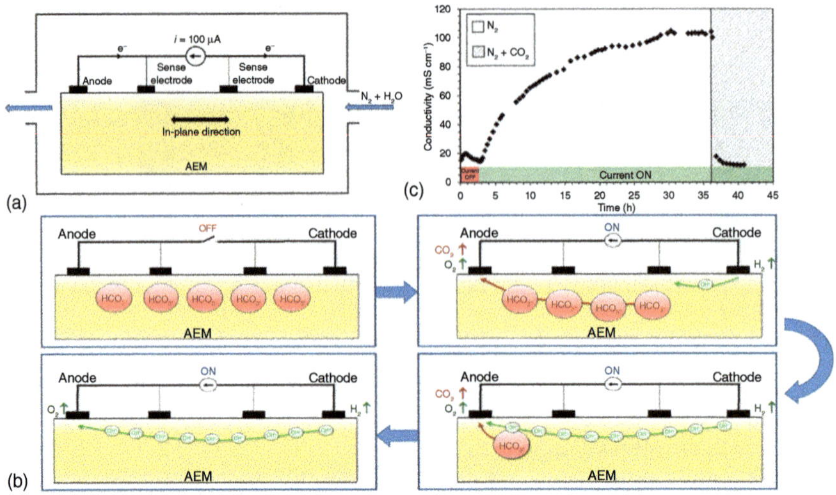

Figure 6.6 Schematic representation of the (a) experimental setup, (b) processes occurring in the AEM while applying the DC, and (c) changes in membrane OH⁻ ion conductivity before, during, and after applying 100 μA. Source: Copyright 2018 Elsevier. Reproduced with permission from Ziv and Dekel [31].

membrane to move to the negative electrode, where CO_2 is released via the reactions shown below:

$$HCO_3^- \leftrightarrow CO_2 + OH^-$$

$$4OH^- \rightarrow 2H_2O + O_2 + 4e^-$$

The experiment was carried out for about 30 h, reaching a maximum conductivity of around 103 mS cm^{-1} (Figure 6.6c). This was reported to be the true OH^- ion conductivity of the membranes as all CO_3^{2-} anions are replaced by the OH^- ions.

Even though this method enables the determination of the actual OH^- ions conductivity of AEMs by mitigating the carbonation problem, it is a time-consuming technique, especially when a screening procedure is required for a large number of membrane samples, at a controlled ion exchange capacity or alkaline stability. Recently, this method was used to monitor the OH^- ion conductivity of AEM in an alkaline stability study [54].

The second method for reducing the negative effect of CO_2 carbonation on the AEMs is to measure the OH^- ion conductivity in a glovebox containing an inert gas that is CO_2-free [55]. In this method, the membranes are first immersed in an aq. NaOH/KOH solution and then washed with pure water to remove excess OH^- ions before being transferred to the conductivity measuring cell. This procedure is complicated, and the presence of a big amount of degassed water inside the glovebox makes it difficult to perform.

The use of gas-purged water during the exchange process [56] or during measurement is another strategy used to combat carbonation and its detrimental consequences. Despite its simplicity, two issues can be raised about this method [31].

First, the measurement of ionic conductivity in water is far from the actual operating conditions for AEMFC membranes. Second, the complete removal of dissolved CO_2 is difficult to achieve, so small amounts of HCO_3^- and CO_3^{2-} anions are likely to remain during the ionic conductivity measurement.

In addition to the measures mentioned above, AEMs with very high OH^- ion conductivities can alleviate the unfavorable influence of carbonation on fuel cell performance [37, 57]. Therefore, novel design routes and membrane materials are required to enhance the ionic conductivity and long-term alkaline stability of AEMs.

6.6 Conclusion and Outlook

AEMs with high hydroxide (OH^-) ion conductivity are required for the practical deployment of AEMFCs with high power density. Understanding the ion transport mechanisms in AEMs, as well as the factors that influence them, is critical for developing high-performance AEMs and AEMFCs. This chapter discussed OH^- ion transport mechanisms, conductivity measurement methods, and carbonation effects on ionic transport in AEMs. Two main H^+ transport mechanisms were described: the vehicle mechanism and the Grotthuss mechanism. In aqueous solutions, the transport mechanisms of H^+ and OH^- are depicted analogous depending on external or operating conditions such as temperature and relative humidity. Although most studies report the Grotthuss mechanism to be the dominant one, there is no collective evidence to support this theory, requiring further research and a detailed understanding of OH^- ion transport mechanisms in aqueous medium in relation to membrane structure–property variations. Elucidating these mechanisms directly supports and assist in identifying the design principles of new AEMs for broad range of technological applications in alkaline medium.

Carbonation effects represent one of the most critical phenomena suppressing OH^- ion conductivity when measurements are performed in open air with exposure to CO_2. This implies that most values reported in the literature are not an exact representation of OH^- conductivity, but include the CO_3^{2-} and HCO_3^{2-} partial ionic conductivities [58]. This has been reported to significantly reduce the performance of AEMFC, and various attempts have been made to avoid carbonation and its negative impacts. For instance, a recently reported method of forcing the release of the HCO_3^- and CO_3^{2-} anions as CO_2 gas by applying an external electric current through the membrane appears to be a promising method. However, it is a time-consuming technique which is not suitable for screening a large number of membrane samples. Various other methods have also been reported for improved OH^- ion conductivity measurement of AEMs. However, there is still a gap in the availability of a simple and accurate method for determining the OH^- ion conductivity of AEMs in a wide range of experimental conditions, such as temperature and relative humidity.

In addition to optimizing ionic conductivity measurement techniques, various membrane design and synthesis strategies are required to improve membrane ion conductivity along with standardized ion conductivity measurement protocols. This includes the use of AEMs with a spacer and/or a block-copolymer structure with a controlled ionic channel size as well as the use of rigid crosslinkers to design fast ion transport channels toward better ionic conductivity and selectivity [59]. These achievements in the combined success of understanding transport mechanisms and design strategies will lay a strong foundation for the advancement of ion exchange membrane technologies.

References

1 Dekel, D. (2014). Alkaline membrane fuel cells, membranes. In: *Encyclopedia of Applied Electrochemistry* (ed. G. Kreysa, K.-I. Ota, and R.F. Savinell), 33–45. New York: Springer New York.

2 Yang, J., Aili, D., Li, Q. et al. (2013). Covalently cross-linked sulfone polybenzimidazole membranes with poly (vinylbenzyl chloride) for fuel cell applications. *ChemSusChem* 6: 275–282.

3 Merle, G., Wessling, M., and Nijmeijer, K. (2011). Anion exchange membranes for alkaline fuel cells: a review. *J Membr Sci* 377: 1–35. https://doi.org/10.1016/j.memsci.2011.04.043.

4 Grew, K.N. and Chiu, W.K.S. (2010). A dusty fluid model for predicting hydroxyl anion conductivity in alkaline anion exchange membranes. *J Electrochem Soc* 157: B327. https://doi.org/10.1149/1.3273200.

5 Kreuer, K.-D., Paddison, S.J., Spohr, E., and Schuster, M.J. (2004). Transport in proton conductors for fuel-cell applications: simulations, elementary reactions, and phenomenology. *Chem Rev* 104: 4637–4678. https://doi.org/10.1021/cr020715f.

6 Weber, A.Z. and Newman, J. (2004). Transport in polymer-electrolyte membranes: III. Model validation in a simple fuel-cell model. *J Electroche Soc* 151: A326.

7 Nagle, J.F. and Morowitz, H.J. (1978). Molecular mechanisms for proton transport in membranes. *Proc Natl Acad Sci U S A* 75: 298–302. https://doi.org/10.1073/pnas.75.1.298.

8 Tuckerman, M.E., Marx, D., and Parrinello, M. (2002). The nature and transport mechanism of hydrated hydroxide ions in aqueous solution. *Nature* 417: 925–929. https://doi.org/10.1038/nature00797.

9 Foglia, F., Berrod, Q., Clancy, A.J. et al. (2022). Disentangling water, ion and polymer dynamics in an anion exchange membrane. *Nat Mater* 21: 555–563. https://doi.org/10.1038/s41563-022-01197-2.

10 Dong, D., Zhang, W., van Duin, A.C.T., and Bedrov, D. (2018). Grotthuss versus vehicular transport of hydroxide in anion-exchange membranes: insight from combined reactive and nonreactive molecular simulations. *J Phys Chem Lett* 9: 825–829. https://doi.org/10.1021/acs.jpclett.8b00004.

11 Zelovich, T., Vogt-Maranto, L., Simari, C. et al. (2022). Non-monotonic temperature dependence of hydroxide ion diffusion in anion exchange membranes. *Chem Mater* 34: 2133–2145. https://doi.org/10.1021/acs.chemmater.1c03594.

12 Marx, D. (2006). Proton transfer 200 years after von Grotthuss: insights from ab initio simulations. *Chemphyschem* 7: 1848–1870.

13 Chen, C., Tse, Y.-L.S., Lindberg, G.E. et al. (2016). Hydroxide solvation and transport in anion exchange membranes. *J Am Chem Soc* 138: 991–1000.

14 Varcoe, J.R. (2007). Investigations of the ex situ ionic conductivities at 30 °C of metal-cation-free quaternary ammonium alkaline anion-exchange membranes in static atmospheres of different relative humidities. *Phys Chem Chem Phys* 9: 1479–1486. https://doi.org/10.1039/B615478F.

15 Miller, H.A., Bouzek, K., Hnat, J. et al. (2020). Green hydrogen from anion exchange membrane water electrolysis: a review of recent developments in critical materials and operating conditions. *Sustain Energy Fuel* 4: 2114–2133.

16 Varcoe, J.R., Atanassov, P., Dekel, D.R. et al. (2014). Anion-exchange membranes in electrochemical energy systems. *Energy Environ Sci* 7: 3135–3191. https://doi.org/10.1039/C4EE01303D.

17 Hibbs, M.R., Hickner, M.A., Alam, T.M. et al. (2008). Transport properties of hydroxide and proton conducting membranes. *Chem Mater* 20: 2566–2573. https://doi.org/10.1021/cm703263n.

18 Hnát, J., Paidar, M., Schauer, J. et al. (2012). Polymer anion-selective membranes for electrolytic splitting of water. Part II: enhancement of ionic conductivity and performance under conditions of alkaline water electrolysis. *J Appl Electrochem* 42: 545–554.

19 Li, Y., Liu, Y., Savage, A.M. et al. (2015). Polyethylene-based block copolymers for anion exchange membranes. 48: 6523–6533.

20 Walker, R., Conrad, R.M., and Grubbs, R.H.J.M. (2009). The living ROMP of trans-cyclooctene. *Macromolecules* 42: 599–605.

21 Dang, H.-S. and Jannasch, P. (2015). Exploring different cationic alkyl side chain designs for enhanced alkaline stability and hydroxide ion conductivity of anion-exchange membranes. *Macromolecules* 48: 5742–5751. https://doi.org/10.1021/acs.macromol.5b01302.

22 Zhang, W., Dong, D., Bedrov, D., and van Duin, A.C.T. (2019). Hydroxide transport and chemical degradation in anion exchange membranes: a combined reactive and non-reactive molecular simulation study. *J Mater Chem A* 7: 5442–5452. https://doi.org/10.1039/C8TA10651G.

23 Kraglund, M.R., Carmo, M., Schiller, G. et al. (2019). Ion-solvating membranes as a new approach towards high rate alkaline electrolyzers. *Energy Environ Sci* 12: 3313–3318. https://doi.org/10.1039/C9EE00832B.

24 Chen, S., Wang, H., Zhang, J. et al. (2020). Effect of side chain on the electrochemical performance of poly (ether ether ketone) based anion-exchange membrane: a molecular dynamics study. *J Membr Sci* 605: 118105. https://doi.org/10.1016/j.memsci.2020.118105.

25 Zhang, W., Cheng, W., Tufa, R.A. et al. (2021). Studies on anion exchange membrane and interface properties by electrochemical impedance spectroscopy: the role of pH. *Membranes* 11: 771. https://doi.org/10.3390/membranes11100771.

26 Fontananova, E., Zhang, W., Nicotera, I. et al. (2014). Probing membrane and interface properties in concentrated electrolyte solutions. *J Membr Sci* 459: 177–189. https://doi.org/10.1016/j.memsci.2014.01.057.

27 Wenjuan Zhang, B.H., Tufa, R.A., Tang, C. et al. (2022). Tracing the impact of stack configuration on interface resistances in reverse electrodialysis by in situ electrochemical impedance spectroscopy. *Front Environ Sci Eng* 4: 46. https://doi.org/10.1007/s11783-021-1480-9.

28 Zhang, W., Ma, J., Wang, P. et al. (2016). Investigations on the interfacial capacitance and the diffusion boundary layer thickness of ion exchange membrane using electrochemical impedance spectroscopy. *J Membr Sci* 502: 37–47. https://doi.org/10.1016/j.memsci.2015.12.007.

29 Zhang, W., Wang, P., Ma, J. et al. (2016). Investigations on electrochemical properties of membrane systems in ion-exchange membrane transport processes by electrochemical impedance spectroscopy and direct current measurements. *Electrochim Acta* 216: 10. https://doi.org/10.1016/j.electacta.2016.09.018.

30 Długołęcki, P., Anet, B., Metz, S.J. et al. (2010). Transport limitations in ion exchange membranes at low salt concentrations. *J Membr Sci* 346: 163–171. https://doi.org/10.1016/j.memsci.2009.09.033.

31 Ziv, N. and Dekel, D.R. (2018). A practical method for measuring the true hydroxide conductivity of anion exchange membranes. *Electrochem Commun* 88: 109–113. https://doi.org/10.1016/j.elecom.2018.01.021.

32 Reshetenko, T., Odgaard, M., Schlueter, D., and Serov, A. (2018). Analysis of alkaline exchange membrane fuel cells performance at different operating conditions using DC and AC methods. *J Power Sources* 375: 185–190. https://doi.org/10.1016/j.jpowsour.2017.11.030.

33 Myles, T.D., Grew, K.N., Peracchio, A.A., and Chiu, W.K.S. (2015). Transient ion exchange of anion exchange membranes exposed to carbon dioxide. *J Power Sources* 296: 225–236. https://doi.org/10.1016/j.jpowsour.2015.07.044.

34 Ziv, N., Mustain, W.E., and Dekel, D.R. (2018). The effect of ambient carbon dioxide on anion-exchange membrane fuel cells. *ChemSusChem* 11: 1136–1150. https://doi.org/10.1002/cssc.201702330.

35 Zheng, Y., Omasta, T.J., Peng, X. et al. (2019). Quantifying and elucidating the effect of CO_2 on the thermodynamics, kinetics and charge transport of AEMFCs. *Energy Environ Sci* 12: 2806–2819. https://doi.org/10.1039/C9EE01334B.

36 Suzuki, S., Muroyama, H., Matsui, T., and Eguchi, K. (2013). Influence of CO_2 dissolution into anion exchange membrane on fuel cell performance. *Electrochim Acta* 88: 552–558. https://doi.org/10.1016/j.electacta.2012.10.105.

37 Mandal, M., Huang, G., Hassan, N.U. et al. (2019). The importance of water transport in high conductivity and high-power alkaline fuel cells. *J Electrochem Soc* 167: 054501. https://doi.org/10.1149/2.0022005jes.

38 Ma, M., Clark, E.L., Therkildsen, K.T. et al. (2020). Insights into the carbon balance for CO_2 electroreduction on Cu using gas diffusion electrode reactor designs. *Energy Environ Sci* 13: 977–985. https://doi.org/10.1039/D0EE00047G.

39 Tufa, R.A., Blommaert, M.A., Chanda, D. et al. (2021). Bipolar membrane and interface materials for electrochemical energy systems. *ACS Appl Energy Mater* 4: 7419–7439. https://doi.org/10.1021/acsaem.1c01140.

40 Blommaert, M.A., Aili, D., Tufa, R.A. et al. (2021). Insights and challenges for applying bipolar membranes in advanced electrochemical energy systems. *ACS Energy Lett* 6: 2539–2548. https://doi.org/10.1021/acsenergylett.1c00618.

41 Robertson, N.J., Kostalik, H.A., Clark, T.J. et al. (2010). Tunable high performance cross-linked alkaline anion exchange membranes for fuel cell applications. *J Am Chem Soc* 132: 3400–3404. https://doi.org/10.1021/ja908638d.

42 Tsehaye, M.T., Alloin, F., and Iojoiu, C. (2019). Prospects for anion-exchange membranes in alkali metal–air batteries. *Energies* 12: 4702.

43 Xue, J., Liu, X., Zhang, J. et al. (2020). Poly(phenylene oxide)s incorporating N-spirocyclic quaternary ammonium cation/cation strings for anion exchange membranes. *J Membr Sci* 595: 117507. https://doi.org/10.1016/j.memsci.2019.117507.

44 Hou, J., Liu, Y., Ge, Q. et al. (2018). Recyclable cross-linked anion exchange membrane for alkaline fuel cell application. *J Power Sources* 375: 404–411. https://doi.org/10.1016/j.jpowsour.2017.06.073.

45 Nightingale, E.R. (1959). Phenomenological theory of ion solvation. effective radii of hydrated ions. *J Phys Chem* 63: 1381–1387. https://doi.org/10.1021/j150579a011.

46 Besha, A.T., Tsehaye, M.T., Aili, D. et al. (2020). Design of monovalent ion selective membranes for reducing the impacts of multivalent ions in reverse electrodialysis. *Membranes* 10: 7.

47 Mustain, W.E., Chatenet, M., Page, M., and Kim, Y.S. (2020). Durability challenges of anion exchange membrane fuel cells. *Energy Environ Sci* 13: 2805–2838. https://doi.org/10.1039/D0EE01133A.

48 Gerhardt, M.R., Pant, L.M., and Weber, A.Z. (2019). Along-the-channel impacts of water management and carbon-dioxide contamination in hydroxide-exchange-membrane fuel cells: a modeling study. *J Electrochem Soc* 166: F3180–F3192. https://doi.org/10.1149/2.0171907jes.

49 Krewer, U., Weinzierl, C., Ziv, N., and Dekel, D.R. (2018). Impact of carbonation processes in anion exchange membrane fuel cells. *Electrochim Acta* 263: 433–446. https://doi.org/10.1016/j.electacta.2017.12.093.

50 Gottesfeld, S. (2010). Systems and methods of securing immunity to air CO_2 in alkaline fuel cells. *US Patent* US20110212370A20110212371.

51 Piana, M., Boccia, M., Filpi, A. et al. (2010). H_2/air alkaline membrane fuel cell performance and durability, using novel ionomer and non-platinum group metal cathode catalyst. *J Power Sources* 195: 5875–5881. https://doi.org/10.1016/j.jpowsour.2009.12.085.

52 Gottesfeld, S., Dekel, D.R., Page, M. et al. (2018). Anion exchange membrane fuel cells: current status and remaining challenges. *J Power Sources* 375: 170–184. https://doi.org/10.1016/j.jpowsour.2017.08.010.

53 Wright, A.G. and Holdcroft, S. (2014). Hydroxide-stable ionenes. *ACS Macro Lett* 3: 444–447. https://doi.org/10.1021/mz500168d.

54 Haj-Bsoul, S., Varcoe, J.R., and Dekel, D.R. (2022). Measuring the alkaline stability of anion-exchange membranes. *J Electroanal Chem* 908: 116112. https://doi.org/10.1016/j.jelechem.2022.116112.

55 Ponce-González, J., Whelligan, D.K., Wang, L. et al. (2016). High performance aliphatic-heterocyclic benzyl-quaternary ammonium radiation-grafted anion-exchange membranes. *Energy Environ Sci* 9: 3724–3735. https://doi.org/10.1039/C6EE01958G.

56 Wright, A.G., Fan, J., Britton, B. et al. (2016). Hexamethyl-p-terphenyl poly(benzimidazolium): a universal hydroxide-conducting polymer for energy conversion devices. *Energy Environ Sci* 9: 2130–2142. https://doi.org/10.1039/C6EE00656F.

57 Aili, D., Yang, J., Jankova, K. et al. (2020). From polybenzimidazoles to polybenzimidazoliums and polybenzimidazolides. *J Mater Chem A* 8: 12854–12886. https://doi.org/10.1039/D0TA01788D.

58 Tang, H., Aili, D., Tufa, R.A. et al. (2022). Anion conductivity of cation exchange membranes in aqueous supporting electrolytes. *Solid State Ionics* 383: 115984. https://doi.org/10.1016/j.ssi.2022.115984.

59 Hu, C., Deng, X., Dong, X. et al. (2021). Rigid crosslinkers towards constructing highly-efficient ion transport channels in anion exchange membranes. *J Membr Sci* 619: 118806. https://doi.org/10.1016/j.memsci.2020.118806.

7

Anion Exchange Membranes Based on Quaternary Ammonium Cations and Modified Quaternary Ammonium Cations

Vijayalekshmi Vijayakumar and Sang Yong Nam

Gyeongsang National University, Research Institute for Green Energy Convergence Technology, Department of Materials Engineering and Convergence Technology, Jinju-Daero, Jinju-Si, 52828, Republic of Korea

7.1 Introduction

The growing concern about energy consumption due to our modern life style and society needs motivates the scientific community to search for alternative energy sources different from fossil fuels [1]. Polymer electrolyte membrane fuel cells convert chemical energy stored in fuels (hydrogen, methanol, natural gas, biogas, etc.) directly to electric energy and heat. Compared to the conventional proton exchange membrane fuel cells, anion exchange membrane fuel cells (AEMFCs) have earned increasing attention due to their unique advantages including faster reaction kinetics especially at the oxygen reduction reaction (ORR) , thereby providing low activation losses, and the extensive electro-catalyst choice such as Ag, Co, Ni, and Fe. As a core component of AEMFCs, anion exchange membranes (AEMs) function as ion transport medium, fuel/oxidant separator, electronic insulator, and electrode support. AEMs directly affect the power output and lifetime of a fuel cell [2–4].

The durability and practical applications of AEMFCs are associated with the conductivity, chemical stability, and dimensional stability of their AEM [5]. The degradation of the cationic group is the main reason for AEM poor alkaline stability. The strong nucleophilic and basic working conditions promote multiple degradation reactions of functional organic cations including Hoffmann elimination, S_N2 nucleophilic substitution, ylide formation, and rearrangement reactions in AEMs at elevated temperatures, resulting in the loss of ion conductivity [6]. Increasing the number of ionic groups can increase the hydroxide ion conductivity to some extent, but this is accompanied by detrimental excessive water uptake and swelling, resulting

Alkaline Anion Exchange Membranes for Fuel Cells: From Tailored Materials to Novel Applications, First Edition. Edited by Jince Thomas, Alex Schechter, Flavio Grynszpan, Bejoy Francis, and Sabu Thomas.

in the deterioration of mechanical strength as well as chemical stability of AEMs. The higher ion exchange capacity is associated to a greater number of ionic groups and hydrophilic regions within the matrix, that increases water uptake and swelling of the membrane. Addition of water generally hydrates the OH^- ions, generating a steric shell around hydroxide and supporting the stability. However, excessive water uptake and swelling creates excess solvation that lowers the stiffness, resulting in the deterioration of mechanical and chemical stability of AEMs [7]. With the aim of developing highly stable AEMs with high power density and long-term durability, numerous novel nitrogen-based cations have been investigated. Among them, optimization of quaternary ammonium (QA) groups has been used as the key element in many studies over the last two decades. In this chapter, various types of QA cations and their applicability to improve AEM stability and electrochemical properties as well as their effect on carbonation issues in AEMFCs are discussed summarizing the past 6 years of research (2016–2022).

7.1.1 Background of AEMFC Invention

Liquid electrolyte-based alkaline fuel cells were originally used for space applications. However, the main drawbacks associated with the hydroxyl ion-containing liquid electrolyte, the carbonation phenomena, and the formation of leaks during extended period of operation limit its practical application. Furthermore, carbonate salt contained in the liquid electrolyte occasionally blocks the pores of the catalyst electrode and limits the fuel cell's efficiency. A number of research works are being carried out on solid alkaline conducting electrolyte polymers, which are less susceptible to carbonation and overcome crossover issues while maintaining all advantages of alkaline fuel cells. The first AEMs from polychloropropene cross-linked by divinylbenzene followed by QA functionalization via trimethylamine (TMA) were developed by Japanese scientists from the Tokuyama Soda Company [8,9]. Since then, AEMFCs have become a point of interest for researchers. The main components of a typical AEMFC include the anode, cathode, membrane electrode assembly (MEA), and a gasket providing the seal around the MEA to prevent fuel leakage. The MEA is fabricated by assembling an anode gas diffusion layer (GDL), followed by the anode catalyst layer (CL), AEM, cathode CL, and a cathode GDL. The GDLs in both electrodes are composed of two layers, a backing layer of carbon paper or carbon cloth and a microporous layer which consists of a mixture of carbon powder and hydrophobic polymer (typically polytetrafluoroethylene [PTFE]). The CLs are generally made of electrocatalysts and an ionomer, resulting in the formation of triple-phase boundaries for redox reactions. The main role of a GDL is to provide support for both anode and cathode CLs, distribute the reactants uniformly, and transport electrons to the current collector. The working principle of AEMFCs is explained elsewhere in many publications and is out of scope of this chapter as this review is mainly focused on the QA-based AEMs [4]. Even though the AEMFCs offer fuel flexibility, low oxidation–reduction over-potential and less fuel crossover, as well as the elimination of acid corrosion issues, major challenges associated with the AEMs, such as dimensional instability, low conductivity, chemical instability in basic environments at high temperatures, insufficient lifetime, and power efficiencies have strictly hampered their commercial success so far [10, 11]. Therefore, extensive efforts have been devoted to developing high-performance membrane materials. An ideal AEM should possess strong electrochemical performance, which can be quantified by its ionic conductivity, ion exchange capacity, alkaline stability, and power density as well as excellent mechanical and thermal stability to play an efficient role in separating oxidants and fuel as well as transporting hydroxyl ions in AEMFCs.

7.2 Quaternary Ammonium (QA)-Based AEMs – Recent Developments and Performances

In fact, developing membrane structures with QA as cationic species is an effective way to solve the conductivity and stability limitations of AEMs. Recently, a series of AEMs based on QA cationic groups have been developed owing to their facile chemical modification, high hydroxide ion conductivity, and low cost, but they tend to be vulnerable to strong nucleophilic attack by hydroxyl ions (S_N2 substitution) and can also degrade through Hoffmann elimination (E1 and E2), ylide formation, and rearrangement reactions, especially in high pH condition and at elevated temperatures. The second-order S_N2 substitution leads mainly to the formation of an alcohol, whereas elimination reactions result in the formation of double bond (Figure 7.1) [12]. It is difficult to prevent the S_N2 substitution on a benzylic derivative which depends mainly on the strength of the nucleophile and on the concentration of both the benzylic ammonium moiety and nucleophile. On the other hand, Hoffmann elimination (E2) can occur only in the presence of a hydrogen atom near to the ammonium group in β-position [13]. The design strategies for stable QA groups are very challenging because of the complexity of group types, degradation mechanism, operation environments, and so forth. Numerous approaches have been made to reduce the weakness of the ammonium group toward nucleophilic substitution and degradation. Scientific studies in recent years have been gradually focused on the various molecular design strategies, steric hindrance effects, spirocyclic structures, etc. Tethered alkyl chain having no β-hydrogens has also been proved to avoid the elimination reaction pathways. Moreover, increasing the alkylene spacer length between the backbone of an aromatic polymer and the nitrogen atom of the ammonium group is reported to be a practical method to enhance the chemical stability of membranes to a great extent in an alkaline environment. On the other hand, ion aggregation in microphase-separated structures of numerous well-defined polymers also boosted the ion migration in membranes [6, 10, 14, 15]. Various studies conducted in the last few years and their performances are explained below in detail.

Pendent QA groups on the side chain have been reported to be a promising method to improve the performance of AEMs. A wide range of polymer backbones including poly(ether ketone), poly(phenylene oxide), poly(phenylene), and poly (ether sulfones) have been used for the preparation of AEMs by introducing QA

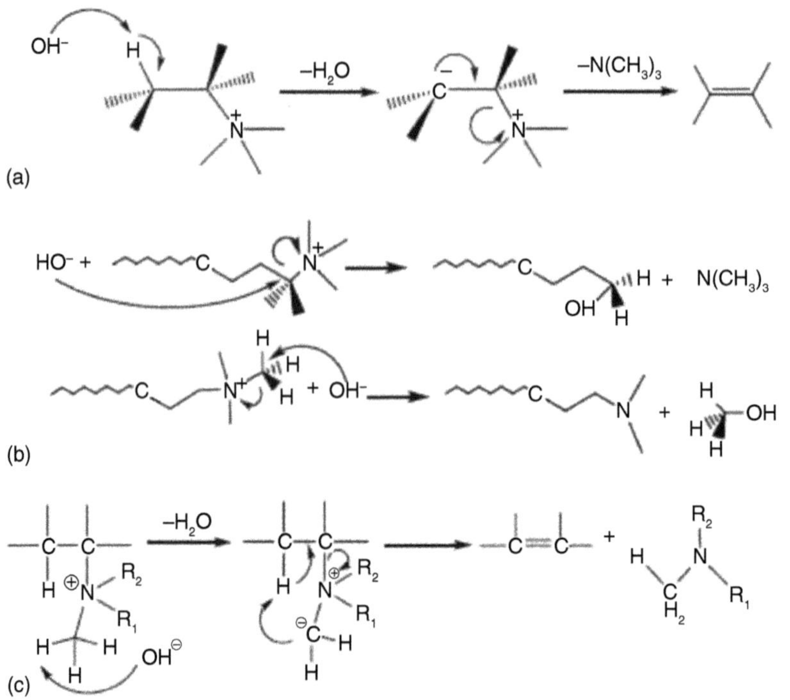

Figure 7.1 Degradation of a quaternary ammonium group by (a) Hoffmann elimination (b), nucleophilic substitution (c), and E1 elimination reaction. Source: Vijayakumar and Nam [12]. Reproduced with permission from Elsevier.

cations into the backbone [16, 17]. Benzyltrimethylammonium (BTMA) is one of the most commonly used cationic group of AEMs that fill out a lot of requirements, but it is liable to degradation under alkaline conditions. The main BTMA degradation routes are direct nucleophilic substitution (S_N2) at benzylic carbon causing tertiary amine formation, substitution at α-carbon producing benzylamine, and *N*-ylide formation through α-hydrogen abstraction from neighboring carbons, followed by Hoffmann elimination at β-hydrogen which gives a tertiary amine. Furthermore, cationic groups close to the polymer backbone will also induce chain scission. Separating the QA cations from the polymer backbone via extending flexible spacers of the alkyl side chain pendent to the nitrogen-centered cation results in side-chain type or comb-shaped AEMs with excellent alkaline stability. The resonances emerging from hydrogen atoms of alkyl chain leading as a barrier to rapid decomposition of QA cations in alkaline solutions. Poly(arylene ether ketone) (PAEK) with pendent QA groups having different alkyl chains, TMA, triethylamine (TEA), and tripropylamine (TPA) were reported by Tuan and Kim. The conductivity and water uptake was higher in PAEK–TMA than in PAEK–TEA and PAEK–TPA membranes, but mechanical strength was vice versa due to the difference in steric hindrance and ionic cluster formation [8]. The effect of various ammonium head groups including dimethylbutyl (DMB), dimethylhexyl (DMH), dibutylmethyl (DB), and heterocyclic *N*-methylpiperidinium (Pip) on partially fluorinated AEMs studied by Mahmoud and Miyatake showed the most balanced properties of high conductivity, low water absorption, alkaline stability, as well as mechanical stability for Pip-based membranes. The Pip-based membrane achieved an ultimate power density of 232 $mW\,cm^{-2}$ and durability for 240 h at a constant current density of 100 $mA\,cm^{-2}$ with an average decay of 1.29 $mV\,h^{-1}$ [18]. PAEK with iodobenzene pendent functionalized with tetra QA carbazole exhibited good thermal and dimensional stability as well as chemical stability (90% conductivity retention after treating in 4 M NaOH solution for a week) [19]. The polymers containing the aryl ether bond generally exhibits poor chemical stability due to the presence of electron-withdrawing groups such as sulfone, ketone, and QA groups. However, polyether sulfones with QA groups at meta position to ether linkage prepared via nucleophilic substitution followed by bromination, quaternization, and anion exchange reaction are reported to be more stable than the polymer with QA groups at ortho position [20].

Side chain type AEMs have been shown to have better chemical stability than those with QA groups located at benzylic position. *m*-Polybenzimidazole (m-PBI) attached with QA groups and alkene as side chains cross-linked by the thiolene reaction exhibited good mechanical, dimensional, and chemical stability [21]. (5-bromopentyl)-Trimethylammonium bromide (BPTMA-Br) grafted PBI also exhibited excellent mechanical stability and conductivity. However, the alkaline stability was not satisfactory (65.3% conductivity retention after 240 h exposure in 1 M KOH at 80 °C) as a result of nucleophilic attack by OH^- ions toward the C2 position of the backbone. Therefore, attention should also be paid to install adjacent bulky groups on the reactive C2 position [22]. Tetraalkyl ammonium functionalized onto fluoropolyolefin also exhibited efficient ion conductivity and alkaline stability, which resulted in impressive device durability during fuel cell application [23]. Long side chain type QA functionalized poly(ether ether ketone) (PEEK) AEMs showed a gradual increment in conductivity with increase in methylhydroquinone, and the ionic conductivity of the membrane in 1 M NaOH at 60 °C after 30 days was around 50% of the initial value. The side chain could enhance the motion of the cations and facilitate hydrophilic/hydrophobic phase separation structure in the membrane, which is more beneficial to the conduction of anions [24]. Hu et al. conducted a study to elucidate the role of the alkyl side chain in poly(aryl piperidinium) copolymers recently. Membranes with alkyl spacers of 0, 1, and 2 showed superior water uptake and ion conductivity, while long chain type with alkyl spacers of 6 and 10 facilitated increase in hydrophobicity and possessed high dimensional stability. The prepared membranes exhibited excellent thermomechanical properties as well as alkaline stability up to 1000 h in 1 M NaOH solution at 80 °C. Short chain type AEMFCs showed a peak power density of about 2.67 $W\,cm^{-2}$ at 80 °C and operation stability under 0.4 $A\,cm^{-2}$ current density for 220 h with a low voltage decline of approximately 77 $\mu V\,h^{-1}$ [25]. The rigid side chain introduced on to poly (2,6-dimethyl-1,4-phenylene oxide) (PPO) by the Friedel–Crafts acylation with 4-fluorobenzoyl chloride and subsequent reaction between phenyl fluoride and secondary amine of 1-methylpiperazine (Figure 7.2) expanded the free volume to construct connected ion transport nano channels, leading to excellent hydroxide conductivity and fuel cell performance as mentioned in Table 7.1 [26].

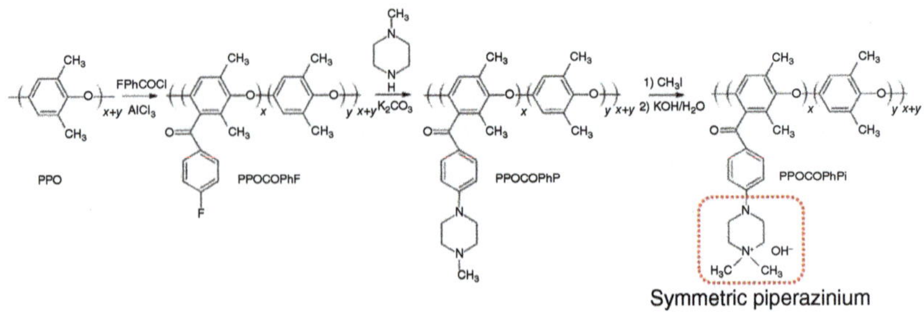

Figure 7.2 Synthesis of PPOCOPhPi (i) aluminum chloride and 4-fluorobenzoyl chloride, 60 °C, 12 h. (ii) 1-Methylpiperazine, K_2CO_3, 100 °C, 48 h (iii) iodomethane, 40 °C, 48 h. Source: Gao et al. [26]. Reproduced with permission from Elsevier.

Table 7.1 Properties of various AEMs and their fuel cell performance.

Membrane designation	Type/specialty of the membrane	Conductivity (mS cm^{-1})	Alkaline stability (% retention)	Power density (mW cm^{-2})	References
PPOCOPhPi-0.55	Rigid side chain, Piperazinium	108.0 (60 °C)	500 h in 1 M NaOH at 60 °C: 99.0%	1210 (80 °C, 100% RH)	[26]
PBP-xQ4	Multication cross-linker	155.0 (80 °C)	1800 h in 2 M NaOH at 80 °C: 91.8%	234 (80 °C, 100% RH)	[27]
PSF-0.1C8-ASD	Rod-coil graft, *N*-spirocyclic	136.2 (80 °C)	720 h in 8 M KOH at 80 °C: 80.0%	850 (80 °C, 100% RH)	[28]
m-PTP-20Q4	Ether free polyaryl polymers, piperidinium	144.2 (80 °C)	1600 h in 2 M NaOH at 80 °C: 87.6%	239 (80 °C)	[29]
POXIH-OH	Cardo polymer, pendent QA	73.6 (80 °C)	1200 h in 1 M NaOH at 80 °C: 90.5%	476 (60 °C)	[30]
QPPO-QPOSS-3	QA composite	87.7 (80 °C)	1200 h in 1 M KOH at RT: 102.0%	288 (60 °C, RH 100%)	[31]
QPPO-QSiO$_2$-3	QA composite	13.6 (80 °C)	1200 h in 1 M KOH at RT: 91.0%	142 (60°C, RH 100%)	[32]
QVT-OH	Pore filling, QA ionic liquid	32.8 (60 °C)	350 h in 1 M KOH at 60 °C: 86.3%	545 (50 °C, RH 100%)	[33]

Table 7.1 (Continued)

Membrane designation	Type/specialty of the membrane	Conductivity (mS cm^{-1})	Alkaline stability (% retention)	Power density (mW cm^{-2})	References
S50P50-P25-OH	QA blend, *N*-spirocyclic	117.0 (90 °C)	500 h in 2 M KOH at 60 °C: 80.0%	—	[34]
PDOTP-*x*	Alkyl chain length	150.0 (80 °C)	1000 h in 1 M NaOH at 80 °C: 97.3%	2670 (80 °C)	[25]
QABNP	Ether free polyaryl polymers, twisted geometry	135.2 (80 °C)	1080 h in 2 M NaOH at 80 °C: 90.0%	1160 (60 °C, RH 100%)	[35]
MXene doped QPSU-PQ-10	QA composite	88.7 (80 °C)	30 d in 1 M NaOH at 60 °C: 80.0%	106 (60 °C, RH 100%)	[36]
s-PDTP	Aryl ether free, pendent ammonium	120.0 (80 °C)	1080 h in 1 M NaOH at 80 °C: 90.0%	1470 (80 °C)	[37]

Recently, a new strategy to improve the performance of AEMs is adopted to build a side chain-type structure with pendent sterically crowded QA cations by tethering terminal alkyne containing QAs, namely, di-cyclohexyl group (TA-CH), di-isopropyl group (TA-iP), and di-*n*-propyl group (TA-*n*P), onto quaternized PPO copolymers through an efficient click reaction (copper catalyzed azide-alkyne cycloaddition reaction, CuAAC). The PPO membrane with di-cyclo group (PPO-CH) presented excellent alkaline stability with 100% conductivity retention after 216 h of aging in 1 M NaOH at 60 °C [15]. Majority of researchers synthesize the cationic QA groups via the reaction of tertiary amine with benzylic chloromethyl groups on the polymer side chain. Chloromethyl methyl ether is an essential raw material in the chloromethylation reaction, which is extremely toxic and strongly carcinogenic. Furthermore, the reaction requires longer reaction time and strict control on temperature. Hence, the search to a new safer, quicker, and controllable method is essential, and many researchers reported that bromination could replace the chloromethylation for the synthesis of QA functionalized polymers. The influence of anchoring QA groups (trimethyl ammonium) on different positions of PPO through the chloromethylation and bromination route on the alkaline stability of AEMs was reported by Becerra-Arciniegas et al. The bromination route demonstrated a higher stability of ionic conductivity, but mechanical stability and reproducibility are better for the chloromethylation route [13].

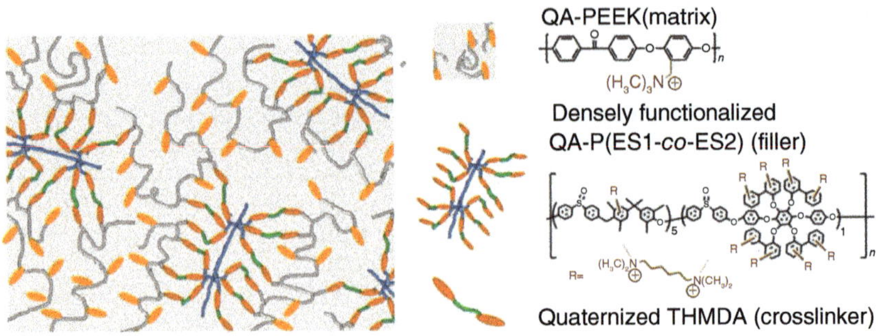

Figure 7.3 Schematic illustration of the blend membrane. Source: Li et al. [16]. Reproduced with permission from Elsevier.

Inspired from the state-of-the-art Nafion polymer electrolyte membranes that possess high conductivity owing to their well-defined hydrophilic/hydrophobic phase separation structure, various AEMs with microphase separation structure were developed and showed improved conductivity and alkaline stability. Optimizing the microphase separation structure is another great challenge to achieve highly efficient ion conducting channels, and various approaches have been pursued recently. Blending QA functionalized polymers aids in the formation of micro-phase separation morphology and favors the formation of ion channels for high-speed ion transfer. Blend membrane of QA functionalized polysulfones and poly(ether ketone) (QA-PEEK) cross-linked with N,N,N',N'-tetramethyl-1,6-hexanedi-amine (TMHDA) reported by Li et al. facilely surmounted the trade-off between ion exchange capacity and physical stability. Illustration of the blend membrane is shown in Figure 7.3. The densely QA functionalized polysulfone acted as "hydroxide ion wires" in the blend membrane, forming straight and efficient ion channels for rapid ion transport. In addition, the strong covalent interaction induced by TMHDA and the steric hindrance by its long aliphatic chain enhanced the physical and chemical stability, respectively, of the membranes. Maximum hydroxide ion conductivity at 90 °C up to 215.4 mS cm^{-1} and power density of single fuel cells up to 137.2 mW cm^{-2} at 60 °C were achieved [16].

High conductivity is essential for an AEM to achieve high power density of fuel cells. Increasing IEC and mobility of hydroxide ions is a direct way to improve conductivity to some extent, but this often comes at the expense of increased swelling and degradation of mechanical stability. The architecture of block copolymer-based membranes with micro-phase separated morphology also helps enhance the ion mobility through the formation of ion-conducting channels, resists dimensional swelling, and improves mechanical stability [38, 39]. Inspired by the superiority of improved alkaline stability introduced by the alkyl-tethered anion head groups and enhanced conductivity through the fabrication of block copolymers, Lin et al. synthesized a triblock copolymer composed of hydrophilic quaternized poly(phenylene oxide) segments and hydrophobic poly(ether sulfone) (PES) segments bearing alkyl-tethered cycloaliphatic QA-head-groups through nucleophilic substitution, Friedel–Crafts acylation, ketone reduction, and Menshutkin reaction. The designed triblock copolymers showed micro phase-separated morphology and well-connected ion domains resulted in a high ionic conductivity of 105.1 mS cm^{-1} at 80 °C. Steric interference offered by the cycloaliphatic structure of cationic groups, and flexible alkyl chains were likely responsible for the robust alkaline stability of membranes for 480 h in 1 M KOH solution at 60 and 80 °C [40]. PES block copolymers containing pendent QA enhances the conductivity at temperature above 40 °C than that of the polymer-bearing imidazolium cations, but the chemical stability was not much satisfactory. The authors have also taken care to avoid the usage of carcinogenic chloromethylation reagents in their study [17]. As a commercially available elastomeric triblock copolymer, poly(styrene-*b*-ethylene-*co*-butylene-*b*-styrene) (SEBS) has been investigated as a promising material with microphase separation structure for highly conductive and stable AEMs. Recently, a series of side chain-type AEMs based on quaternized hexyl bis(quaternary ammonium)-mediated partially cross-linked SEBS have been reported by Xu et al.; these membranes showed exciting conductivity and mechanical as well as chemical stability. A maximum power density of about 315 mW cm^{-2} was also achieved when AEMFCs with this membrane operated at 60 °C and 95% relative humidity (RH). It is no surprise that interionic cluster domains nano-phase morphology with long-range connecting channels was observed by small angle X-ray scattering (SAXS) and atomic force microscopy (AFM) owing to the original triblock structure of SEBS [41].

Introducing multiple QA cations into side chains via alkyl spacers is an effective way to facilitate the formation of well-connected ion conductive channels, but the reaction is complicated and requires strict control. So far, limited studies are reported, and most of them are focused on the Menshutkin reaction between benzyl bromide groups and tertiary amino groups containing multiple QA precursors. However, benzylic QA cation linkages are instable under strong alkaline environments. Zhu et al. reported an initial approach on the synthesis of PPO-based AEMs containing multication side chains from a clickable di-quaternized side-chain precursor via Cu(I)-catalyzed azide–alkyne click chemistry (CuAAC). The precursor was synthesized from 3-bromopropyne and 2-([2-(di-methylamino) ethyl]methylamino) ethanol via nucleophilic substitution and Menshutkin reaction. The membranes have a reasonable alkaline stability, but hydroxide conductivities were not as high as expected, probably due to the absence of efficient and long-range order microphase-separated morphologies that was confirmed by transmission electron microscopy (TEM) and SAXS [42]. Effect of mono and di cations reported by Zhu et al. concluded that olefinic-type AEMs with di-QA cations exhibit better ion transport properties and alkaline stability [43]. Well-connected and clearly phase-separated morphology attained by long flexible multi QA cation side chains in PAEK promoted hydroxide ion conductivity and reasonable alkaline stability to the AEMs [44]. In view of the shortcoming associated with the trade-off between dimensional stability and ion conductivity, a flexible multication cross-linker with five and six alkyl spacers was designed by Chen et al. They have designed a series of tunable multication cross-linked poly(biphenyl piperidinium) (PBP-xQ4) membranes by embedding flexible cationic cross-linkers (1,16-dibromo-5,11-(*N*,*N*-dimethylammonium) hexadecane bromide [DBDMAH])

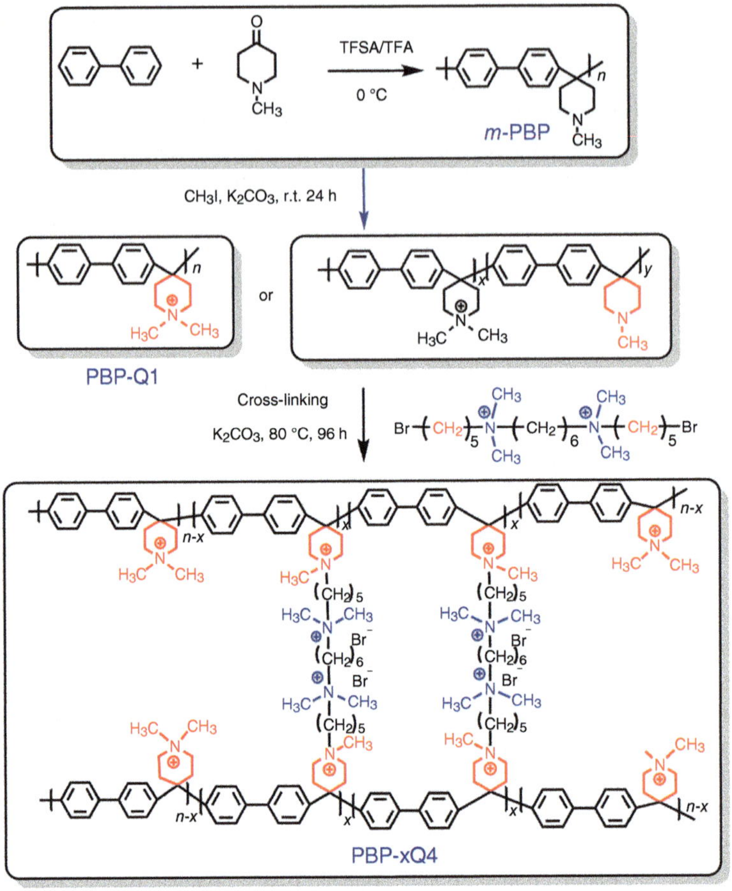

Figure 7.4 Synthesis of the PBP-xQ4 membrane. Source: Chen et al. [27]. Reproduced with permission from Elsevier.

(Figure 7.4). The membranes are foldable to many shapes and exhibit impressive performance of conductivity and durability as mentioned in Table 7.1 [27]. SEBS with multication side chain using 4,4-bipridine as the cross-linker also paves the way for constructing well-defined microphase-separated morphology and high conduction of AEMs [45].

The main degradation routes of QA groups under alkaline conditions are β-hydrogen Hoffmann elimination and direct nucleophilic substitution (S_N2) at α-carbons. These two degradation pathways produce byproducts such as alcohol, tertiary amine, as well as alkene-containing compounds. Utilization of QA groups without β-hydrogen atoms is a more direct and effective way to avoid Hoffmann elimination of QA-type AEMs. For example, Wu et al. found PPO-based AEM with side chain containing dual QA groups without β-hydrogen prepared via CuAAC click reaction showed excellent stability with a high retention of >92% conductivity after immersed in 2 M NaOH at 80 °C for 480 h [46].

The degradation of AEMs including the anion exchange group and polymer backbone is another challenge that has been more closely studied in recent years. Among various cationic head groups such as QA, phosphonium, and imidazolium, QA is regarded as a relatively highly stable cation, however, exhibiting different stabilities in a severe alkaline environment when attached to different polymer backbones [47]. As the stability of the monomer does not correlate to that of the corresponding polymers under the same conditions, stability of QA cations in both monomers and polymer are different and need to be correlated for the design concept of an alkaline stable polymer. In view of that, Zhang et al. developed poly(phenylene-*co*-arylene ether ketone) with five QA groups on a single phenyl unit but exhibited high degradation rate and poor durability than that of the polymer with the same backbone and single QA on the phenyl unit. Moreover, the alkaline stability is excellent in case of the corresponding model monomer than that of the polymer [14]. N1 butyl substituted doubly charged 1-butyl-4-aza-1-azaniabicyclo [2.2.2] octane bromide (BDABCO) functionalized poly(vinylbenzyl chloride) (PVBC)/polybenzimidazole (PBI) cross-linked membranes demonstrated an ionic conductivity of 91.4 $mS\,cm^{-1}$ at 80 °C and a maximum power density of 340 $mW\,cm^{-2}$ at 0.492 V coupling with the membrane resistance less than 0.1 $\Omega\,cm^2$. In their study, BDABCO was used to improve the compatibility among the polymer and QA groups [47].

In the intensive search for AEMs containing base-stable cations, *N*-cyclic QA cations (piperidinium and *N*-spirocyclic QA) have been reported as predominantly long-lived in alkaline environment as the typical tetra alkyl ammonium and benzylic cations. *N*-heterocyclic structure with steric conformation and hindrance effect in QA cations can efficiently alleviate the substitution and elimination reactions. However, tethering of QAs with *N*-spirocylic structures into the polymer backbone is rather complicated, and only a few reports are available. AEMs with pure aliphatic polymer backbone containing *N*-spirocyclic QA prepared via ultraviolet (UV) irradiation using diallyldimethylammonium chloride and diallylpyrrolidinium chloride at room temperature avoiding carcinogenic solvents retain 40% IEC after 30 days of exposure in 1 M KOH solution at 80 °C [48]. Among cyclic QAs, piperidine-based 6-azonia-spiro[5.5]undecane (ASU) and *N,N*-dimethylpiperidinium (DMP), both having spiro-connected six-membered rings, possess exceptional resistance to hydroxide ion attack, even at elevated temperatures. Cross-linked poly(4-(4,4′-biphenyl)-6-azaspiro[5.5] undecane)-*co*-4-(4,4′-biphenyl) piperidinium (PB-ASU-PBP) bromide membranes displayed excellent alkaline stability of 2000 h in 3 M NaOH solution at 80 °C. However, the application of these membranes in fuel cells is still limited by their low fuel cell performance [49]. Xue et al. synthesized a series of PPOs incorporating spirocyclic 3,6-diazaspiro[5.5]undecane (DSU) cation/cation strings by a two-step quaternization strategy. The cyclo quaternization product of DSU was grafted onto the PPO backbone to prepare membranes with single *N*-spirocyclic QA cation, and then second quaternization was conducted on the remaining tertiary piperidine rings to fabricate membranes with DSU cation strings. The presence of tertiary N atoms in the DSU cation resulted in the single-cation membrane degradation through Hoffmann elimination and S_N2 substitution. However, the *N*-methyl group in heterocyclic quaternary piperidinium (Qpip) increases the steric hindrance of DSU cation strings and decreases the degradation rate of spirocyclic DSU cation center by S_N2 substitution that provided the membrane a high conductivity retention and IEC [50]. The incorporation of five-membered and six-membered *N*-spirocyclic QA cations in the phenyl side chain of PEEKs also resulted in good thermal as well as mechanical stability to the MEA fabrication. Conductivity retention of above 74% after 20 days of alkaline exposure in 1 M NaOH at 60 °C showed their better alkaline stability. However, mechanical strength and elongation of the membranes decreased after alkaline treatment suggested the backbone scission of QPEEK [51]. Meanwhile, Qiao et al. developed a spirocyclic QA ionenes containing 5-/6-membered rings based on phenyl and biphenyl which were prepared from tetrakis(bromomethyl) monomers showed a better stability than that of the spirocyclic QA with naphthalene 6-/6-membered rings, indicating the strong electron-withdrawing effect of naphthalene rings [52]. Even though DMP has two anti-planar beta protons, the strong geometric constraints of the six-membered ring make it difficult to reach triangular bipyramidal S_N2 transition state; hence, the chance of elimination and substitution is less and attain good stability. Poly(arylene ether sulfone)s with cyclic tertiary amine groups designed via polycondensation reaction with 4,4′-dichlorodiphenyl sulfone (DCDPS) displayed good stability, which is more when capped with the methoxy group [53]. PES-based AEMs with high-density heterocyclic QA groups prepared by polycondensation of the difluoro monomer with biphenol and bis(4-fluoro phenyl) sulfone owing to their rigid side chain and cations exhibited high hydroxide ion conductivity of 54.1–82 mS cm^{-1} at 80 °C and a power density of 141 mW cm^{-2} with OCV of 1.0 V at 60 °C after assembled in a single cell [54].

Rigid *N*-spirocyclic structure and lack of flexible side chain structural design generally possess poor mechanical and conductive properties to the AEM. Flexible benzyl-free functional side chain containing ether spaced *N*-spirocyclic cation grafted on polysulfone and cross-linking proposed by Zhang et al. endows an excellent alkaline stability as well as trade-off between mechanical stability as well as conductivity to the AEM. The constrained ring conformation of *N*-spirocyclic QA groups elevates the transition state energy of degradation reaction, and ether spacer in side chain helps avoid unstable benzylic structure and diminishes the acidity of β-H, also promoting aggregation of rigid spirocyclic cations and thus improving conductivity significantly. Moreover, cross-linking improved mechanical as well as chemical stability of the membrane [55]. A comparative study on AEMs of PPO with two pendent *N*-cyclic QA groups, six-membered DMP and bis-six membered ASU prepared by Cu(I)-catalyzed "click chemistry" also supported a similar kind of degradation mechanism [56]. Rod-coil graft design for *N*-spirocyclic anion is another strategy that significantly contributed to the reduction of fragility associated with its double-cyclic non-planar structure and conductivity improvement. *n*-Octylamine hydrophobic coil grafted *N*-spirocyclic side chain with polysulfone backbone could break through the low power density bottleneck of the *N*-spirocyclic AEMs and bring *N*-spirocyclic AEMs into the top level among the cycloaliphatic AEMs reported in literatures (Table 7.1) [28]. Partially fluorinated poly(arylene ether) (PAE)-based AEMs bearing pendent *N*-spirocyclic QA synthesized by a facile

Williamson reaction demonstrated improved alkaline stability as well as conductivity. The well-defined microphase-separated structure facilitated the construction of efficient ion transporting pathway and ring conformation of *N*-spirocyclic QA results in an improved stability to the membrane [57]. Porous ionically cross-linked AEM (S50P50-P25-OH) prepared using a blend of pyridine bridged polybenzimidazole (PyPBI) and *N*-spirocyclic QA spiro ionene polymer (SP) resulted in a hydroxide ion conductivity of 117 mS cm^{-1} at 90 °C, and the value retained 80% even after 500 h of alkaline treatment in 2 M KOH at 60 °C [34].

To prevent both polymer and functional groups from base degradation, Yang et al. reported membranes with *N*-alicyclic QA and imidazolium cations with bulky substituents tethered to aryl-ether free polyaromatic backbones. The size of the cation has a significant role in various properties of the membrane. All the membranes exhibited an excellent stability in 2 M NaOH solution at 80 °C for 30 days. A peak power density of 559 mW cm^{-2} at a current density of 2000 mA cm^{-2} at 80 °C under 100% RH with 0.1 MPa back pressure was reported for the H_2/O_2 fuel cell with QA tethered poly((fluorene alkylene)-*co*-(biphenyl alkylene)) (PFBA-QA) membrane [10]. In contrast to *N*-spirocyclic QA, the incorporation of aliphatic monocyclic QA cations such as Qpip is relatively straightforward and less expensive. The ether-free polyfluorene AEMs with alkyl *N*-methyl piperidinium groups were synthesized by Salma and coworkers, and they observed that the alkyl-substituted *N*-methyl piperidinium cationic group is relatively stable in alkaline conditions at 80 °C in 1 M NaOH solution [58]. Configuration with dual pairs of Qpip cations tethered to poly(fluorine alkylene)s via flexible spacers also reached high conductivity and thermal as well as chemical stabilities. Careful evaluation of data revealed that a β-hydrogen in a piperidine ring is nearly eight times less likely to be attacked by hydroxide than a β-hydrogen in an alkyl spacer [59]. Cross-linked PBI/quaternized poly(vinylbenzyl chloride) (QPVC) AEMs with the ether bond free backbone and cyclo aliphatic cation (*N*-methyl piperidinium) possessed well-defined hydrophilic/hydrophobic phase-segregated structure, leading to good alkaline stability and robust mechanical, thermal, as well as conductive properties [60]. Alkaline stability of various functional groups that are linked to the main chain via flexible spacers especially heterocyclic QA in the end of flexible spacers is rarely reported. Trimethylammonium, 1-methylpyrrolidinium, 1-methylpiperidinium, 1-methylimidazolium, and 1,2-dimethylimidazolium grafted to the PES backbone via flexible spacers were analyzed for their electrochemical properties. AEM carrying 1-methylpyrrolidinium and 1-methylpiperidinium displayed the highest stability due to the conformational restrictions imposed by the cyclic structure. Moreover, the former exhibited even more robust alkaline tolerance because substitution was the only active degradation pathways but later undergo both substitution as well as ring opening reaction [61].

Many efforts have been focused recently on designing and synthesizing cationic polymers for improving hydroxide ion conductivity and chemical durability of AEMs. As the long-term durability in strong alkaline environment cannot be guaranteed enough by aryl ether backboned polymers because aryl ether bonds undergo hydrolysis under alkaline conditions, many polymers without aryl ether bonds are highly demanded as the backbone of AEMs. A series of aryl ether free poly(olefin)-based AEMs comprising QA cations were prepared by Qiu and coworkers by introducing triallylmethyl ammonium iodine as both the cross-linker as well as ionic conductor. Considering the excellent chemical and physical property, the single-cell performance reported is not satisfactory, and further optimization on the MEA and fuel cell fabrication is required in their study [62]. Zhang's group developed AEMs with poly(*N*-allylisatin biphenyl)-*co*-poly(alkylene biphenyl)s (PIB-co-PAB) backbone free of C—O bonds by super-acid catalyst Friedel–Crafts polymerization. *N*-spirocyclic QA groups were tethered to the PIB-*co*-PAB polymer backbone by cycloquaternization followed by thiolene click chemistry to solve swelling and achieve mechanically stable high conductive AEMs. *N*-spirocyclic QA groups promote the chemical stability of the prepared membranes and achieved a conductivity retention of 65.19% after storage in 1 M NaOH solution at 80 °C for 1000 h [63]. Wang et al. found an excellent chemical resistance of 3000 h at 80 °C in 1 M NaOH/CD_3OD/D_2O solution for aryl-ether free polystyrene (PS) functionalized with bis-six-membered *N*-cyclic QA (ASU) due to the high ring strain of ASU, which raises the transition state energy of both substitution and elimination reactions. Moreover, the SEBS/PS-ASU blend membrane also displayed 83.2% conductivity retention after 1800 h of exposure in 5 M NaOH at 80 °C [64]. Aryl ether free IEC controlled side-chain type poly(dibenzbeyl-*co*-terphenyl piperidinium) (s-PDTP) copolymer-based membranes synthesized by a bimolecular nucleophilic substitution (S_N2) reaction between 6-bromo-*N,N,N*-trimethylhexanaminium (Br-HTMA) and poly(dibenzyl-*co*-terphenyl *N*-methyl piperidine) (PDTM-50) with pendent ammonium groups exhibited an ion conductivity of >120 mS cm^{-1} at 80 °C and an excellent *ex situ* durability in 1 M NaOH at 80 °C for 1000 h (90% conductivity retention) as well as a peak power density of 1.47 W cm^{-2} in H_2–O_2 and 1.04 W cm^{-2} in H_2–air (CO_2 free) at 80 °C [37].

Ether free polyaryl polymers prepared by superacid catalyzed Friedel–Crafts polymerization owing to their simple synthesis method and high alkali resistance have attracted great interest as an AEM material. However, monomer selection as well as relationship among structural design and properties needs further investigation. In view of that, Long et al. developed a series of free ether poly(aryl piperidinium) membranes with various main chain steric configurations based on meta and para terphenyl and diphenyl/*p*-terphenyl copolymer tethered with stable piperidinium cation. However, their results indicated that the difference in cationic groups has a main influence on the alkaline stability than the polymer chain configuration. An excellent stability of about 1600 h after being treated in 2 M NaOH at 80 °C (80%–90% conductivity retention) was observed for all the membranes [29]. A gradual improvement in steric hindrance is reported for *N*-cyclic cations, with size ranging from small piperidinium to sterically protected *N*-spirocyclic QA. Twisted ether free poly(terphenylene)s backbone tethered with piperidinium resulted in the microphase-separated morphology and a conductivity of 68.7 mS cm^{-1} at 80 °C [2]. Inspired by free volume and side chain benefit, octopus-like side chains grafted onto ether free poly(arylene piperidinium) backbones were prepared by Ma et al. The side chain with bulky rigid β-cyclodextrin (β-CD) as the "octopus head," and long flexible piperidinium ionic liquids as the "arms" contributed to enhanced alkaline stability of membranes [65]. Gao et al. prepared ether free quaterphenyl piperidinium-based AEM with twisted 1,1′-binaphthyl (BN) via a superacid catalyzed Friedel–Crafts polyhydroxy alkylation reaction to ameliorate the performances. The bulky twisted geometry imparted well-connected microphase separation with enhanced ionic conductivity of about 135.25 $mS\,cm^{-1}$ at 80 °C and exceptional alkaline stability in 2 M NaOH at 80 °C for 1080 h (90% conductivity retention). Moreover, the single cell assembled with the membrane possessed a maximum power density of 1.16 $W\,cm^{-2}$ at a current density of 2.37 $A\,cm^{-2}$ [35].

Among the promising polymeric structures, the Cardo structure in which polymers containing cyclic side groups in the repeating units has become a preferable subject for AEMs recently. Cardo polyelectrolytes based on poly(oxoindolebiphenylylene) with pendent cyclic QA cations of different chain lengths synthesized by the superacid-catalyzed polyhydroxyalkylation of bromoalkyl-substituted oxoindoles with biphenyl and subsequent quaternization owing to the presence of flexible side chains and Cardo structure facilitated the formation of micro phase separation and connected ionic domains and hence promoted stability and conductivity. Increasing side chain length improved the scale of micro phase separation and solved trade-off between conductivity and alkaline stability to a certain limit [30].

Various factors such as IEC, micro-phase separation morphology, as well as effective ion diffusion coefficient could affect the electrochemical performance of the membrane. Therefore, more investigation is required to find the mechanism of change in ionic conductivity and stability of AEMs. Generally, molecular dynamic simulation could provide in-depth information of micro structure and ion transport process in membranes beyond the experiments. Coarse-grained molecular dynamics simulations of QA functionalized PEEK with side chains of two QA groups (GQ) and single QA (SQ) under the same grafting degree were conducted by Chen et al. to investigate the mechanism of stability and conductivity. They found that remarkably identical self-diffusion coefficients in SQ and GQ have little contribution, while the high IEC of GQ results in the improved ionic conductivity of GQ. Furthermore, the radial distribution function (RDF) analysis, a trajectory-based analysis, reveals that more water molecules wrap around the OH^{-1} in GQ leading to higher alkaline stability to the membranes. The work provides an in-depth understanding of the design of graft copolymer-based AEM with QA functional side chains [66].

Room temperature ionic liquids, liquids at room temperature composed of organic cations and organic/inorganic anions owing to their low volatility, good thermal stability, high conductivity, and wide electrochemical window, have displayed curiously beneficial qualities as a potential material for AEM preparation for fuel cell application. However, the poisoning effect on catalyst and leakage are the main drawbacks. Polymerizing ionic liquids helps increase the stability and ionic conductivity, and considerable studies on polyionic liquids for AEMFC applications are available. Ionic liquid with QA and tertiary amine head groups synthesized from 4-vinylbenzylchloride (VBC) and tetramethylethylenediamine (TMEDA) after polymerization at 100 °C in porous polyethylene substrate result in pore-filled membranes with no catalyst poison anymore [33].

Embedding inorganic components into organic polymers can combine the advantages of both polymer and inorganic materials. Considerable research has been devoted to inorganic–organic hybrid membranes to achieve high performance of AEMs, and a variety of inorganic materials including graphene oxide (GO), silica (SiO_2), layered double hydroxide (LDH), and carbon nanotube (CNT) have been widely applied. However, the properties were always limited by the dosage of the inorganic materials as the high level of the filler will block the ion conductivity as well as mechanical stability of the membranes. More effective strategies are desired to apply inorganic materials in AEMs. Functionalization of inorganic materials with cationic groups is an effective method used widely. *N*-spirocyclic ammonium (6-azonia-spiro[5.5]undecane groups) functionalized graphene oxide (ASU-GO) incorporated piperidine functionalized PPO (PIPPO) exhibited a maximum conductivity of 73.7 mS cm^{-1} at 80 °C and conductivity retention of about 90% even after alkaline immersion in 1 M KOH at 80 °C for 700 h [67]. Doping 5-mercaptotetrazole modified graphene oxide (QGO) in quaternized PPO (QPPO) resulted in a highest ionic conductivity of 75 mS cm^{-1} at 80 °C [68]. Several researchers have reported the application of glycidyl trimethyl ammonium chloride into the nanomaterials to improve fuel cell performances. Our previous works showed that glycidyl trimethyl ammonium chloride has good alkaline stability and high ion conductivity. QA functionalized silica (QSiO2) and polyhedral oligomeric silsesquioxane (QPOSS) in QA functionalized PPO exhibit high conductivity as well as outstanding alkaline stability to the membranes (Table 7.1). The high stability is ascribed to the steric hindrance offered by the long glycidyl trimethyl ammonium chain. The influence of various parameters such as thermal and mechanical stability, IEC, and fuel cell performance has also been reported in the work [31, 32]. Two types of 2D MXenes (LiF-Ti_3C_2Tx and NH_4HF_2-Ti_3C_2Tx) prepared by selective etching of Ti_3AlC_2 with LiF/HCl and NH_4HF_2 aqueous solutions respectively introduced into quaternized polysulfone/polyquaternium (QPSU/PQ) interpenetrating network resulted in an obvious nanoscale microphase separation morphology with excellent thermomechanical as well as electrochemical properties. A maximum conductivity of 88.76 mScm^{-1} at 80°C and power density of 88.76 mWcm^{-2} at 80°C and 106.28 mWcm^{-2} at 60 °C were observed for the composite AEMs [36]. Properties of various AEMs with better stability and cell performances discussed in the above section are summarized in Table 7.1.

7.3 Other Factors Affecting Performance of Fuel Cells

It is worth mentioning that the performance of fuel cells is pertinent to the MEA fabrication, ionomers, catalyst, and operating conditions such as temperature, humidity, and gas flow rate. Therefore, it is difficult to directly compare the cell performance of various AEMs measured in different conditions published in various studies. Water management has a significant role in fuel cell performance as the water moves from the cathode to anode due to electro-osmotic drag resulting in anodic catalyst flooding that can lead to the deterioration of cell performance. Optimizing this aspect can make a great difference to AEMFC performance [69].

It is well known that the deterioration of AEMs and un-optimized MEA fabrication is supposed to be responsible for the insufficient durability of the fuel cell devices. Mechanical stability and thickness of the membranes is of high importance as the very thin membranes enable fast water transport and reduce the influence of voltage drop across the membrane. Therefore, the membranes need to be as thin as possible and only as thick as it is necessary to guarantee the mechanical stability under operating conditions. Moreover, hydration level affects ionic conductivity than does temperature in controlling the anion conductivity in water vapor [32, 70]. The catalyst–ionomer interaction should be high to get good adhesion among the electrode and membrane and avoid the crack formation on the electrode that notably impacts on the ultimate cell performance [71]. Optimization of the ionomer and thereby power density is a blooming field of study. AEM carbonation in the CO_2-containing atmosphere also adversely affects the fuel cell performance. An AEM in OH^- forms through acid–base reactions with CO_2, creating carbonate (CO_3^{2-}) and bicarbonate (HCO_3^-) anion that reduces conductivity and hence the overall fuel cell performances. Carbonate (CO_3^{2-}) form is the lead species during fuel cell operation and are weaker nucleophiles than OH^- [69]. Achieving the performance goals outlined by the US DOE every facet such as MEA components and its fabrication, fuel cell working conditions, etc. require optimization.

7.4 Summary and Perspectives

The focus on developing QA-based AEMs has increased momentously in the past decade and accelerated in the period from 2016 to the present with the rise in recorded AEMFC performances. Their availability and ease to convert tertiary amines to QA groups, low cost, and adequate alkaline stability makes them have more potential as a novel cation in AEM fabrication. However, electron-deficient properties determine the degradation via nucleophilic substitution or elimination in highly alkaline conditions, especially at high temperatures. An ideal AEM should possess high conductivity and excellent stability under alkaline conditions to meet AEMFCs' practical application and commercialization. The micromorphology and chemical structure of cations as well as the polymer backbone have a great impact on the properties of AEMs. In this chapter, we provided an up-to-date summary of QA-based AEMs with various chemical structures and morphology favorable for fuel cell applications.

Many design criterions and conclusions including avoiding the benzylammonium structure in AEMs, optimizing the microphase separation structure of AEMs, weakening the OH^- susceptibility of cationic groups and polymer main chain, reducing the number of aryl ether bonds in the polymer backbone, and embedding polyionic liquids and functionalized nano particles have been summarized in this review. The use of long flexible multi cation side chains leads to the solution of trade-off between conductivity and alkaline stability of the membranes. Among the ammonium species, *N*-spirocyclic QA cations possess outstanding alkaline stability that extends its application scope to many commercialized high-performance polymer membranes. The alkaline stability could also be enhanced through the formation of side chain containing multi cations without β-hydrogen. Under highly alkaline environment and high temperatures, an excellent stability is crucial to ensure long-term operation of fuel cells. The minimum commercial demand of long-term stability is >5000 h at 80 °C, which has not been realized yet. In fact, the fabrication of AEMFCs is a complex procedure as there are many parameters such as gas flow rate, catalyst, as well as working conditions that affect the performance of AEMFCs. Therefore, more focus on properties of AEMs under practical working conditions is required.

Acknowledgments

The authors acknowledge the financial assistance received from the National Research Foundation of Korea (NRF), Ministry of Science and ICT (NRF – 2021 M1A2A2038115). This research was also supported by basic science research program funded by Ministry of Education (NRF – 2020R1A6A1A03038697).

References

1 Olvera, L.I., Aldeco-Perez, E., Rico-Zavala, A. et al. (2021). High thermomechanical stability and ion-conductivity of anion exchange membranes based on quaternized modified poly(oxyndoleterphenylene). *Polym Test* 95: 107092.

2 Wang, X., Lin, C., Gao, Y., and Lammertink, R.G.H. (2021). Anion exchange membranes with twisted poly(terphenylene) backbone: effect of the N-cyclic cations. *J Membr Sci* 635: 119525.

3 Kumari, M., Doughlin, J.C., and Dekel, D.R. (2021). Crosslinked quaternary phosphonium-functionalized poly(ether ether ketone) polymer-based anion-exchange membranes. *J Membr Sci* 626: 119167.

4 Hren, M., Bozic, M., Fakin, D. et al. (2021). Alkaline membrane fuel cells: anion exchange membranes and fuels. *Sustain Energy Fuels* 5: 604–637.

5 Chu, X., Liu, J., Miao, S. et al. (2021). Crucial role of side-chain functionality in anion exchange membranes: properties and alkaline fuel cell performance. *J Membr Sci* 625: 119172.

6 Hossain, M.A., Jang, H., Lee, S. et al. (2015). Comparison of properties of anion conductive Parmax membranes containing imidazolium cation and quaternary ammonium. *Int J Hydrog Energy* 40: 1324–1332.

7 Lu, Y., Liu, L., Pu, Y. et al. (2021). Towards performance improved anion exchange membrane: cross-linking with multi-cations oligomer modified graphene oxide. *Int J Hydrog Energy* 46: 23855–23867.

8 Tuan, C.M. and Kim, D. (2016). Anion exchange membranes based on poly(arylene ether ketone) with pendant quaternary ammonium groups for alkaline fuel cell application. *J Membr Sci* 511: 143–150.

9 Couture, G., Alaaeddine, A., Boschet, F., and Ameduri, B. (2011). Polymeric materials as anion exchange membranes for alkaline fuel cells. *Prog Polym Sci* 36: 1521–1557.

10 Yang, K., Li, X., Guo, J. et al. (2020). Preparation and properties of anion exchange membranes with exceptional alkaline stable polymer backbone and cation groups. *J Membr Sci* 596: 117720.

11 Chen, N. and Lee, Y.M. (2021). Anion exchange polyelectrolytes for membranes and ionomers. *Prog Polym Sci* 113: 101345.

12 Vijayakumar, V. and Nam, S.Y. (2019). Recent advancements in applications of alkaline anion exchange membranes for polymer electrolyte fuel cells. *J Ind Eng Chem* 70: 70–86.

13 Becerra-Arciniegas, R.A., Narducci, R., Ercolani, G. et al. (2019). Alkaline stability of model anion exchange membranes based on poly(phenylene oxide) (PPO) with grafted quaternary ammonium groups: influence of the functionalization route. *Polymer* 185: 121931.

14 Zhang, W., Qiu, X., Ueda, M. et al. (2018). Synthesis and properties of poly(phenylene-co-arylene ether ketone)s with five quaternary ammonium groups on a phenyl unit for anion-exchange membranes. *Solid State Ionics* 314: 187–194.

15 Liu, Z., Bai, L., Miao, S. et al. (2021). Structure-property relationship of poly (2, 6-dimethyl-1, 4-phenylene oxide) anion exchange membranes with pendant sterically crowded quaternary ammoniums. *J Membr Sci* 638: 119693.

16 Li, Z., He, X., Jiang, Z. et al. (2017). Enhancing hydroxide conductivity and stability of anion exchange membrane by blending quaternary ammonium functionalized polymers. *Electrochim Acta* 240: 486–494.

17 Mohanty, A.K., Devaraju, S., Kim, N., and Paik, H.J. (2018). Synthesis and characterization of poly(ether sulfone) block copolymers containing pendent quaternary ammonium and imidazolium groups as anion exchange membranes. *Solid State Ionics* 314: 46–56.

18 Mahmoud, A.M.A. and Miyatake, K. (2022). Hghly conductive and alkaline stable partially fluorinated anion exchange membranes for alkaline fuel cells: effect of ammonium head groups. *J Membr Sci* 643: 120072.

19 Li, X., Wang, K., Liu, D. et al. (2020). Poly(arylene ether ketone) with tetra quaternary ammonium carbazole derivative pendant for anion exchange membrane. *Polymer* 195: 122456.

20 Chen, J.C., Chen, P.Y., Chen, H.Y., and Chen, K.H. (2018). Analysis and characterization of an atropisomeric ionomer containing quaternary ammonium groups. *Polymer* 141: 143–153.

21 Abdi, Z.G., Chiu, T.H., Pan, Y.Z., and Chen, J.C. (2020). Anion exchange membranes based on ionic polybenzimidazoles crosslinked by thiol-ene reaction. *React Funct Polym* 156: 104719.

22 Tang, W., Yang, Y., Liu, X. et al. (2021). Long side-chain quaternary ammonium group functionalized polybenzimidazole based anion exchange membranes and their applications. *Electrochim Acta* 391: 138919.

23 Zhang, X., Chu, X., Zhang, M. et al. (2019). Molecularly designed, solvent processable tetraalkylammonium-functionalized fluoropolyolefin for durable anion exchange membrane fuel cells. *J Membr Sci* 574: 212–221.

24 Wang, H., Du, X., Zhang, H. et al. (2021). Synthesis and characterization of long-side-chain type quaternary ammonium-functionalized poly (ether ether ketone) anion exchange membranes. *Int J Hydrog Energy* 46: 8156–8166.

25 Hu, C., Park, J.H., Kim, H.M. et al. (2022). Elucidating the role of alkyl chain in poly(aryl piperidinium) copolymers for anion exchange membrane fuel cells. *J Membr Sci* 647: 120341.

26 Gao, L., Wang, Y., Cui, C. et al. (2019). Anion exchange membranes with "rigid-side-chain" symmetric piperazinium structures for fuel cell exceeding 1.2 $W\,cm^{-2}$ at 60 °C. *J Power Sources* 438: 227021.

27 Chen, N., Lu, C., Li, Y. et al. (2019). Tunable multi-cations-crosslinked poly(arylene piperidinium)-based alkaline membranes with high ion conductivity and durability. *J Membr Sci* 588: 117120.

28 Zhang, Y., Chen, W., Li, T. et al. (2021). A rod-coil grafts strategy for N-spirocyclic functionalized anion exchange membranes with high fuel cell power density. *J Power Sources* 490: 229544.

29 Long, C., Wang, Z., and Zhu, H. (2021). High chemical stability anion exchange membrane based on poly(aryl piperidinium): effect of monomer configuration on membrane properties. *Int J Hydrog Energy* 46: 18524–18533.

30 Ren, R., Zhang, S., Miller, H.A. et al. (2019). Facile preparation of novel cardo poly(oxindolebiphenylylene) with pendent quaternary ammonium by superacid-catalysed polyhydroxyalkylation reaction for anion exchange membranes. *J Membr Sci* 591: 117320.

31 Vijayakumar, V., Son, T.Y., Kim, H.J., and Nam, S.Y. (2019). A facile approach to fabricate poly(2,6-dimethyl-1,4-phenylene oxide) based anion exchange membranes with extended alkaline stability and ion conductivity for fuel cell applications. *J Membr Sci* 591: 117314.

32 Vijayakumar, V., Son, T.Y., Im, K.S. et al. (2021). Anion exchange composite membranes composed of quaternary ammonium-functionalized poly(2,6-dimethyl-1,4-phenylene oxide) and silica for fuel cell application. *ACS Omega* 6: 10168–10179.

33 Xie, F., Shao, Z., Gao, X. et al. (2019). Facile preparation of porefilled membranes based on poly(ionic liquid) with quaternary ammonium and tertiary amine head groups for AEMFCs. *Solid State Ionics* 338: 58–65.

34 Das, A., Sana, B., Bhattacharyya, R. et al. (2022). Crosslinked alkaline anion exchange membrane from N-spirocyclic quaternary ammonium and polybenzimidazole. *ACS Appl Polym Mater* 4: 1523–1534.

35 Gao, W.T., Gao, X.L., Gou, W.W. et al. (2022). High-performance tetracyclic aromaic anion exchange membranes containing twisted binaphthyl for fuel cells. *J Membr Sci* 655: 120578.

36 Yang, Z., Zhang, M., Zhao, Z. et al. (2022). Application of 2D nanomaterial MXene in anion exchange membranes for alkaline fuel cells: improving ionic conductivity and power density. *Int J Hydrog Energy* 47: 18122–18138.

37 Kim, H.M., Hu, C., Wang, H.H. et al. (2022). Impact of side-chains in poly(dibenzyl-co-terphenyl piperidinium) copolymers for anion exchange membrane fuel cells. *J Membr Sci* 644: 120109.

38 Lai, A.N., Zhuo, Y.Z., Hu, P.C. et al. (2020). Enhanced ionic conductivity of anion exchange membranes by grafting flexible ionic strings on multiblock copolymers. *Int J Hydrog Energy* 45: 1998–2008.

39 Nugraha, A.F., Arbi, M.R., Wijaya, F. et al. (2020). Synthesis and characterization of anion-exchange multi-block-copolymer membranes containing highly densified cationic functional groups. *Polymer* 210: 122996.

40 Lin, C.X., Wang, X.Q., Li, L. et al. (2017). Triblock copolymer anion exchange membranes bearing alkyl-tethered cycloaliphatic quaternary ammonium-head-groups for fuel cells. *J Power Sources* 365: 282–292.

41 Al Munsur, A.Z., Hossain, I., Nam, S.Y. et al. (2020). Quaternary ammonium-functionalized hexyl bis(quaternary ammonium)-mediated partially crosslinked SEBSs as highly conductive and stable anion exchange membranes. *Int J Hydrog Energy* 45: 15658–15671.

42 Zhu, Z., Cui, P., Lu, Y. et al. (2019). Facile synthesis of anion conductive poly(2,6-dimethyl-1,4-phenylene oxide)s from a clickable di-quaternized side-chain precursor. *Polymer* 172: 75–82.

43 Pan, J., Zhu, H., Cao, H. et al. (2021). Flexible cationic side chains for enhancing the hydroxide ion conductivity of olefinic-type copolymer-based anion exchange membranes: an experimental and theoretical study. *J Membr Sci* 620: 118794.

44 Gao, X.L., Yang, Q., Wu, H.Y. et al. (2019). Orderly branched anion exchange membranes bearing long flexible multi-cation side chain for alkaline fuel cells. *J Membr Sci* 589: 117247.

45 Zhang, S., Zhu, X., Jin, C., and Hu, H. (2019). Pyridinium-functionalized crosslinked anion exchange membrane based on multication side chain tethered elastomeric triblock poly(styrene-b-(ethylene-co-butylene)-b-styrene). *React Funct Polym* 138: 62–69.

46 Wu, J., Wei, X., Jiang, H., and Zhu, Y. (2021). Synthesis and properties of anion conductive polymers containing dual quaternary ammonium groups without beta-hydrogen via CuAAC click chemistry. *Polymer* 228: 123920.

47 Hao, J., Jiang, Y., Gao, X. et al. (2018). Functionalization of polybenzimidazole-crosslinked poly(vinylbenzyl chloride) with two cyclic quaternary ammonium cations for anion exchange membranes. *J Membr Sci* 548: 1–10.

48 Wang, Z., Parrondo, J., Sankarasubramanian, S. et al. (2022). Alkaline stability of pure aliphatic based anion exchange membranes containing cycloaliphatic quaternary ammonium cations. *J Electrochem Soc* 167: 124504.

49 Zhu, H., Li, Y., Chen, N. et al. (2019). Controllable physical-crosslinking poly(arylene 6-azaspiro[5.5] undecanium) for long-lifetime anion exchange membrane applications. *J Membr Sci* 590: 117307.

50 Xue, J., Liu, X., Zhang, J. et al. (2020). Poly(phenylene oxide)s incorporating N-spirocyclic quaternary ammonium cation/cation strings for anion exchange membranes. *J Membr Sci* 595: 117507.

51 Shang, L., Yao, D., Pang, B., and Zhao, C. (2021). Anion exchange membranes based on poly (ether ether ketone) containing N-spirocyclic quaternary ammonium cations in phenyl side chain. *Int J Hydrog Energy* 46: 19116–19128.

52 Qiao, X., Wang, X., Liu, S. et al. (2021). The alkaline stability and fuel cell performance of poly(N-spirocyclic quaternary ammonium) ionenes as anion exchange membrane. *J Membr Sci* 630: 119325.

53 Wang, F., Xue, B., Zhou, S. et al. (2019). Synthesis and property of novel anion exchange membrane based on poly(aryl ether sulfone)s bearing piperidinium moieties. *J Membr Sci* 591: 117334.

54 Du, S., Li, S., Xie, N. et al. (2022). Development of rigid side-chain poly(ether sulfone)s based anion exchange membrane with ultiple annular quaternary ammonium ion groups for fuel cells. *Polymer* 251: 124919.

55 Zhang, Y., Chen, W., Yan, X. et al. (2020). Ether spaced N-spirocyclic quaternary ammonium functionalized crosslinked polysulfone for high alkaline stable anion exchange membranes. *J Membr Sci* 598: 117650.

56 Chu, X., Liu, L., Huang, Y. et al. (2019). Practical implementation of bis-six-membered N-cyclic quaternary ammonium cations in advanced anion exchange membranes for fuel cells: synthesis and durability. *J Membr Sci* 578: 239–250.

57 Lin, C., Yu, D., Wang, J. et al. (2019). Facile construction of poly(arylene ether) s-based anion exchange membranes bearing pendent N-spirocyclic quaternary ammonium for fuel cells. *Int J Hydrog Energy* 44: 26565–26576.

58 Salma, U. and Nagao, Y. (2020). Alkaline stability of ether bond free fluorene-based anion exchange polymer containing cycloaliphatic quaternary ammonium groups. *Polym Degrad Stab* 179: 109299.

59 Allushi, A., Pham, T.H., and Jannasch, P. (2021). Highly conductive hydroxide exchange membranes containing fluorene-units tethered with dual pairs of quaternary piperidinium cations. *J Membr Sci* 632: 119376.

60 Lin, C., Wang, J., Shen, G. et al. (2019). Construction of crosslinked polybenz imidazole-based anion exchange membranes with ether-bond-free backbone. *J Membr Sci* 590: 117303.

61 Liu, F.H., Lin, C.X., Hu, E.N. et al. (2018). Anion exchange membranes with well-developed conductive channels: effect of the functional groups. *J Membr Sci* 564: 298–307.

62 Xu, F., Qiu, K., Lin, B. et al. (2021). Enhanced performance of poly(olefin)-based anion exchange membranes cross-linked by triallylmethyl ammonium iodine and divinylbenzene. *J Membr Sci* 637: 119629.

63 Zhang, S., Wang, Y., Liu, P. et al. (2020). Photo-cross-linked poly(N-allylisatin biphenyl)-co-poly(alkylene biphenyl)s with pendant N-cyclic quaternary ammonium as anion exchange membranes for direct borohydride/hydrogen peroxide fuel cells. *React Funct Polym* 152: 104576.

64 Wang, X., Sheng, W., Shen, Y. et al. (2019). N-cyclic quaternary ammonium-functionalized anion exchange membrane with improved alkaline stability enabled by aryl-ether free polymer backbones for alkaline fuel cells. *J Membr Sci* 587: 117135.

65 Ma, L., Hussain, M., Li, L. et al. (2021). Octopus-like side chain grafted poly(arylene piperidinium) membranes for fuel cell application. *J Membr Sci* 636: 119529.

66 Chen, S., Wang, H., Zhang, J. et al. (2020). Effect of side chain on the electrochemical performance of poly (ether ether ketone) based anion-exchange membrane: a molecular dynamics study. *J Membr Sci* 605: 118105.

67 Long, C., Lu, C., Li, Y. et al. (2020). N-spirocyclic ammonium-functionalized graphene oxide-based anion exchange membrane for fuel cells. *Int J Hydrog Energy* 45: 19778–19790.

68 Zhang, D., Ye, N., Chen, S. et al. (2020). Enhancing properties of poly(2,6-dimethyl-1, 4-phenylene oxide) based anion exchange membranes with 5-mercaptotetrazole modified graphene oxides. *Renew Energy* 160: 250–260.

69 Luo, X., Rojas-Carbonell, S., Yan, Y., and Kusoglu, A. (2020). Structure-transport relationships of poly(aryl piperidinium) anion-exchange membranes: eeffect of anions and hydration. *J Membr Sci* 598: 117680.

70 Yang, Q., Sun, L.X., Gao, W.T. et al. (2021). Crown ether-based anion exchange membranes with highly efficient dual ion conducting pathways. *J Colloid Interface Sci* 604: 492–499.

71 Park, H.J., Chu, X., Kim, S.P. et al. (2020). Effect of N-cyclic cationic groups in poly(phenylene oxide)-based catalyst ionomer membranes for anion exchange membrane fuel cells. *J Membr Sci* 608: 118183.

8

Guanidinium Cations and Their Derivatives-Based Anion Exchange Membranes

Jifu Zheng[1], Boxin Xue[2], and Suobo Zhang[1]

[1] *Chinese Academy of Sciences, Changchun Institute of Applied Chemistry, Key Laboratory of Polymer Ecomaterials, 5625 Renmin Street, Changchun, 130022, China*
[2] *Chinese Academy of Sciences, Ningbo Institute of Material Technology & Engineering, Institute of Materials Technology, 1219 Zhongguan West Road, Ningbo, 315201, China*

8.1 Introduction

Quaternary ammonium (QA) cations are the most widely studied ionic groups used in anion exchange membranes (AEMs). Due to the existence of various degradation routes including nucleophilic substitution, Hoffmann elimination, and ylide formation under strong alkaline conditions [1], its application in the field of AEMs is largely restricted, making the development of novel cationic groups with excellent alkaline stability become a very important research focus.

Guanidinium is a relatively unexplored organic cationic group with a resonantly delocalized structure: positive charges are distributed over the three nitrogen atoms in a four-atom planar structure [2]. Due to the resonance stability of the π system and the characteristics of "Y-type delocalization," its stability is comparable to that of cyclic aromatic hydrocarbons such as benzene [3]. The structural characteristics of the resonant delocalization give it good thermal and chemical stability. In addition, guanidinium hydroxide is a strong basic ion pair, which can effectively increase the number of dissociated OH^- ions and absorb more water molecules, thereby helping to improve the ion conductivity of the AEM [4]. Moreover, the possibility of introducing six diverse substituents to the guanidinium core boosts its structural tunability and functional designability. Therefore, guanidinium is an ideal ionic group with a great promising potential and applications in AEMs.

Alkaline Anion Exchange Membranes for Fuel Cells: From Tailored Materials to Novel Applications, First Edition. Edited by Jince Thomas, Alex Schechter, Flavio Grynszpan, Bejoy Francis, and Sabu Thomas.

8.2 General Synthetic Method of Various Guanidiniums

The two most common synthetic routes toward various guanidinium cations are shown in Figure 8.1. In Method **1**, alkyl substituted guanidinium can be obtained in one step by a nucleophilic substitution reaction between commercially available *N,N*-dimethylphosgeniminium chloride and a secondary amine [5]. This method was used in the early days to prepare guanidinium ionic liquids, that is, it was often used to make small monomeric molecules or model compounds, but it is not commonly used in the preparation of guanidinium-based AEMs. This may be due to the fact that this method involves the reaction of secondary amine polymers: in polymers containing a large number of secondary amines, the two secondary amine groups will react with a single *N,N*-dimethylphosgeniminium chloride to form a cross-linked structure, resulting in poor controllability of the reaction and thus relatively difficult binding of guanidine compounds to the polymer framework.

The second method for the preparation of guanidinium is the phosphorus oxychloride method or the oxalyl chloride method, using a commercially available symmetrical tetraalkylurea as the starting material [2]. In this synthetic route, primary amines are usually reacted with the Vilsmeier salt to prepare pentaalkylguanidine, followed by quaternization to obtain guanidinium. Since a large number of primary amines are commercially available, it is easy to design and adjust the structure of pentaalkylguanidine to obtain guanidinium with different structures. In addition, guanidinium can also be directly obtained by the reaction of secondary amine and Vilsmeier salt [6, 7]. While preparing guanidine using Method **2**, the following points need to be noted: first, since the Vilsmeier salt decomposes easily in water, it is necessary to remove all traces of water from the reaction system and solvent. The solvents used in this reaction are usually nonpolar, such as toluene and chlorobenzene. Second, the reaction of the primary or secondary amine with the Vilsmeier salt is violent and exothermic, and the process of slowly adding the primary or secondary amine into the Vilsmeier salt solution needs to be performed in an ice-water

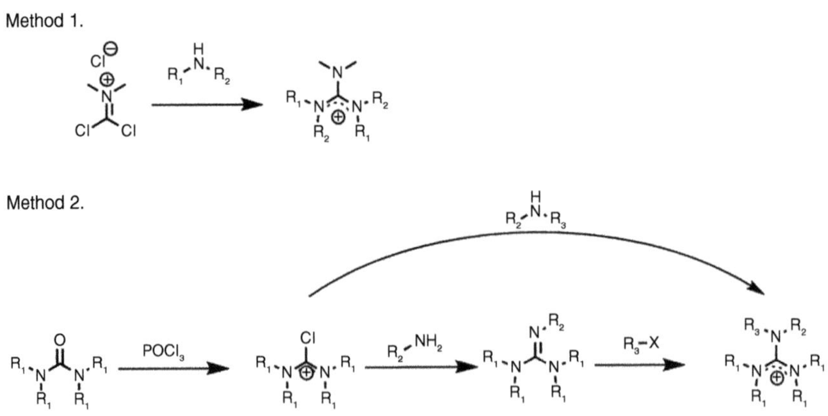

Figure 8.1 The common synthesis methods of guanidinium cations. Source: Jifu Zheng.

bath. Third, when purifying a pentaalkylguanidine, deprotonation with a concentrated NaOH solution is necessary, and finally distillation at reduced pressure is used to purify the pentaalkylguanidine. Method **2** is the most widely used method to prepare guanidinium cations. This method is simple, efficient, and, most importantly, allows for a good control of the six substituents, which can greatly enrich the structural types of guanidinium substituents.

8.3 Degradation Mechanism and Alkaline Stability of Guanidinium Cations

To improve the stability of ionic groups in AEMs, it is important to understand the degradation mechanism of such ionic groups. Compared to the widely studied QA and imidazolium, the degradation process of guanidinium has been less studied. Kim's group [8] analyzed the degradation route of pentamethyl-aryl guanidinium through theoretical calculations. They supposed that in the degradation process of aryl guanidinium, the hydroxide ion first attacked the central carbon atom of the guanidinium to form an intermediate, which then decomposed to give the corresponding tetramethyl urea and dimethyl amine. Mohanty and Bae [9] quantitatively studied the degradation process of a variety of small-molecule model compounds including pentamethyl-benzyl guanidinium by nuclear magnetic resonance (NMR). They found that the benzyl guanidinium rapidly degraded at 60 °C, while benzyltrimethyl ammonium ions maintained long-term stability. In addition, they reported that the guanidinium degradation products might undergo further rearrangements and other side reactions, so the degradation products of guanidinium could not be accurately confirmed, but they speculated on the presence of benzyl alcohol molecules in the degradation mixture based on spectral analysis. Similarly, Li et al. [10] studied the degradation processes of three typical guanidinium cations including pentamethyl-benzyl guanidinium, pentamethyl-aryl guanidinium, and hexamethyl guanidinium through theoretical calculations. According to their calculations, Step 2 of the intermediate decomposition process is thought to be the decisive step in the degradation of the guanidinium (Figure 8.2). At the same time, they speculated that guanidinium has poor alkaline stability since calculations showed that degradation of guanidinium has a lower activation energy barrier. In addition, by

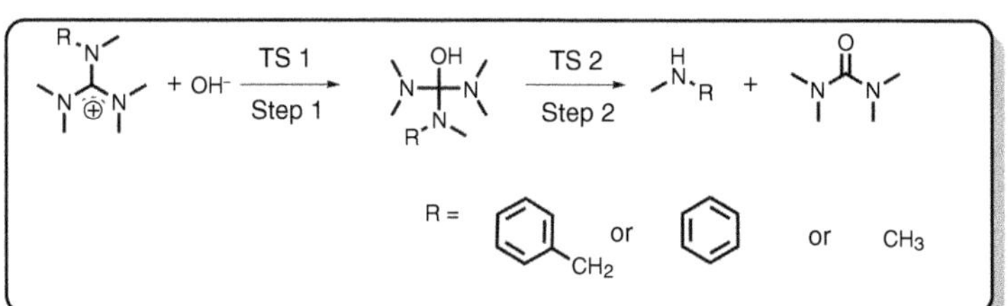

Figure 8.2 The degradation mechanism of guanidinium in alkaline solution [10]. Source: Jifu Zheng.

comparing the activation energy barriers of pentamethyl-benzyl guanidinium, pentamethyl-aryl guanidinium, and hexamethyl guanidinium in the two stages of the degradation process, they found that hexamethyl guanidinium has the highest activation energy, pentamethyl-benzyl guanidinium has moderate activation energy, and pentamethyl-aryl guanidinium has the lowest activation energy. Therefore, they believed that pentamethyl-aryl guanidinium is more susceptible to hydroxide attack than hexamethyl guanidinium and pentamethyl-benzyl guanidinium. In addition, the frontier molecular orbital is an important parameter of chemical reactions, and the lowest unoccupied molecular orbital energy (LUMO) is usually used to measure the ability of a molecule to accept electrons [11, 12]. Therefore, the LUMO energy can be used to illustrate the relative stability of the guanidiniums. The relative order of the LUMO energies of the three guanidiniums is consistent with the activation energy barrier, which provides support that hexamethyl guanidinium has the best alkaline stability, followed by pentamethyl-benzyl guanidinium, and pentamethyl-aryl guanidinium is the least stable. The possible reason is that the methyl group has an electron-donating effect, which enhances the electron density of hexamethyl guanidinium, while the aryl group has an electron-withdrawing effect. The benzyl group has a conjugation effect, which makes benzyl guanidinium more alkaline stable than the aryl guanidinium.

These results confirm that the substituent type of guanidinium strongly influences the corresponding alkaline stability. Our research group [13] analyzed the degradation products of four benzyl guanidiniums containing aliphatic chains of different lengths ($n = 1, 6$, and 12) through experimental methods. Using heteronuclear multiple bond correlation (HMBC) NMR spectra and mass spectroscopy analysis of the degradation products (taking QG6 as an example), we found that the guanidinium was more easily degraded to urea and small amines, while no benzyl alcohol molecules could be found in the degradation mixture (Figure 8.3). This demonstrated that the degradation process of guanidinium is different from that of the benzyl trimethylammonium cation. The degradation reaction site of guanidinium occurs only on the central carbon atom, but not on any other carbon atom, which may be directly related to the unique delocalized resonance structure of guanidinium.

In 2019, our research group [14] prepared more guanidinium model compounds (Figure 8.4) by further tuning the substituent groups and systematically studied the relationship between the guanidinium structure and the alkaline stability by ^{1}H NMR analysis. First, M4, M5, and commercially available benzyl trimethylammonium chloride (BTMA) were taken as examples to describe the degradation process of guanidiniums and QA. As shown in Figure 8.5, their structures contained benzyl groups, which was the most common building unit of the reported AEMs. It was noted that no proton resonances ascribed to olefin or alcohol could be found in the ^{1}H NMR spectra, indicating that no β-H elimination or nucleophilic substitution occurred during the degradation process of M4 and M5. These results demonstrated that the degradation mechanism of guanidinium is different from that of BTMA, where benzyl alcohol is one of its degradation products. As for guanidiniums, the degradation reaction occurred in a nucleophilic addition–elimination way: first, the

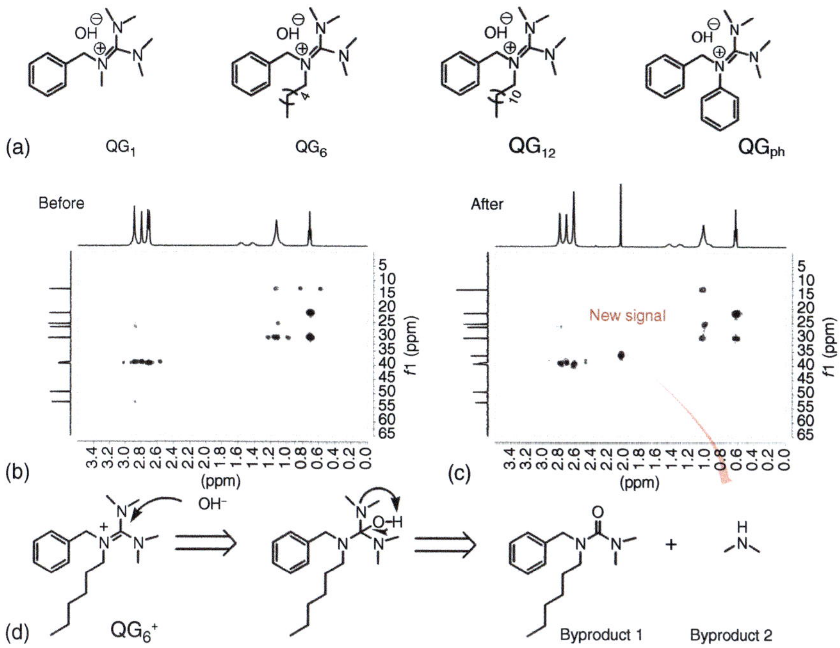

Figure 8.3 (a) Model compounds (QGx) of guanidinium cations. (b) HMBC spectra of QG6 before the alkaline stability test. (c) HMBC spectra of QG6 after the alkaline stability test (sample including QG6, byproduct 1 and byproduct 2). (d) The degradation route of QG6 [13]. Source: Jifu Zheng.

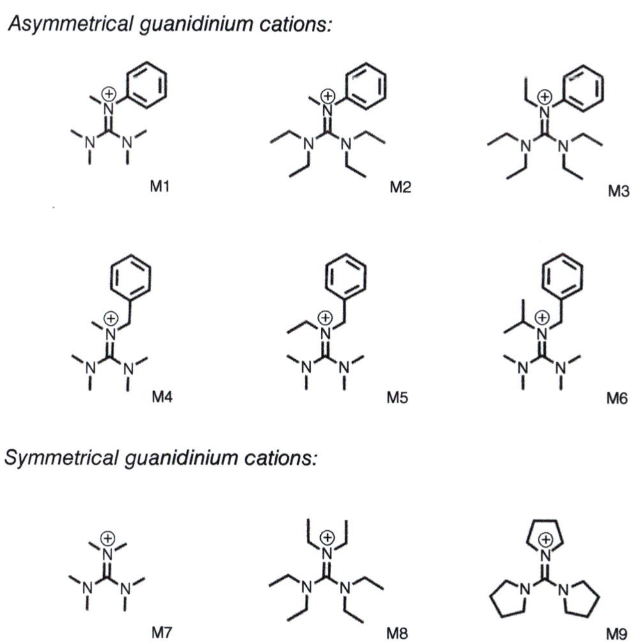

Figure 8.4 Model compounds (M1–M9) of guanidinium cations [14]. Source: Jifu Zheng.

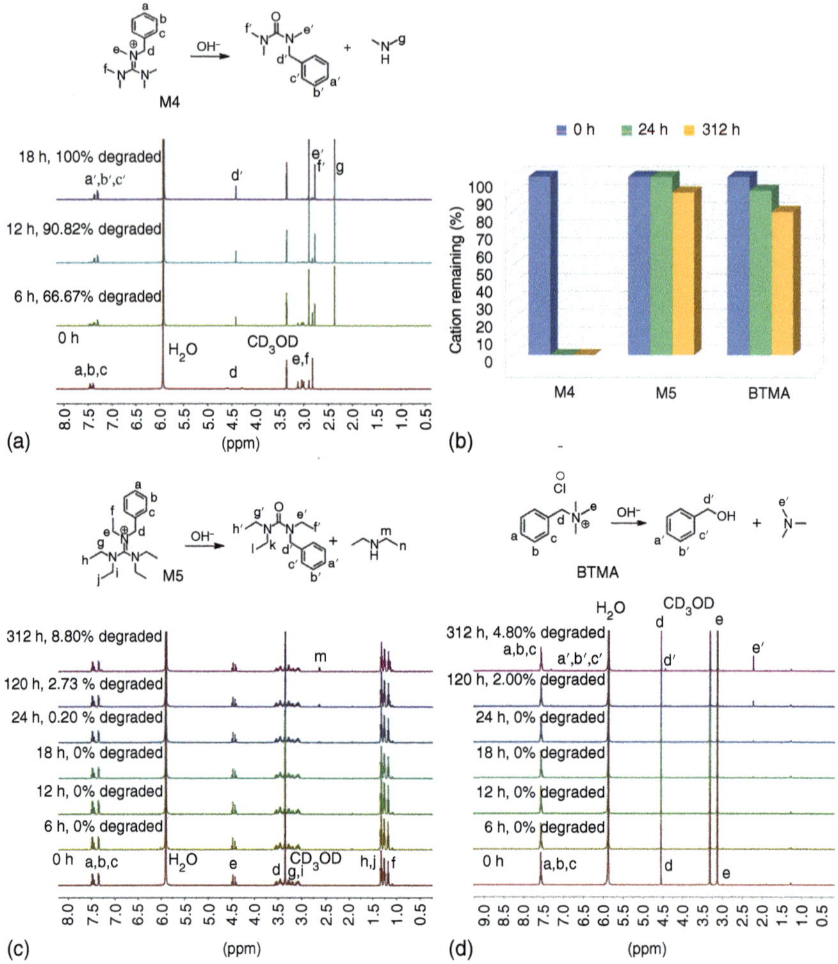

Figure 8.5 (a) The alkaline stability of M4. (b) Remaining cations of model compounds after different immersion times. (c) The alkaline stability of M5. (d) The alkaline stability of BTMA [14]. Source: Jifu Zheng.

center C atom was attacked by OH^- and yielded an intermediate (addition reaction), then one of the N atoms was protonated, and the elimination reaction occurred to form corresponding degradation products including urea and amine. The remaining cations of M4, M5, and BTMA after different immersion times are compared in Figure 8.5b. The alkaline stability sequence was M5 > BTMA > M4, indicating that tuning the substituents of guanidiniums achieves higher stability than typical QA under the same conditions.

Similarly, the degradation process of other guanidiniums was studied in the same method. The resulting cations from each guanidinium model compound in the solution within a certain immersion time are summarized in Figure 8.6a. Based on this results, the relative stability of these guanidinium cations was obtained. Guanidinium

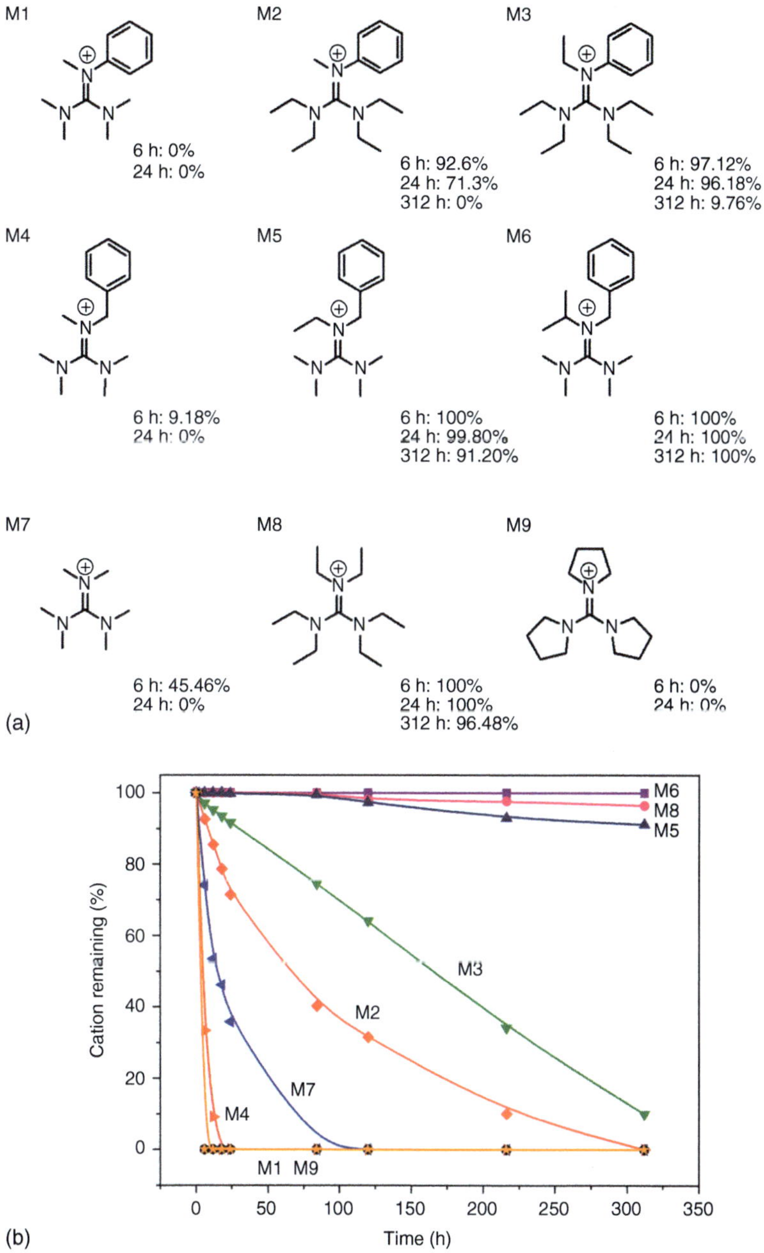

Figure 8.6 (a) Resulting cations of the model compounds after different immersion times. (b) Percentage of remaining cations over time (the counterions were omitted) [14]. Source: Jifu Zheng.

model compounds M1, M2, and M3 all contained a phenyl group at the N2 position, and their alkaline stability followed the sequence of M3 > M2 > M1, showing that when the methyl group was changed to the ethyl group, the guanidiniums presented better stability.

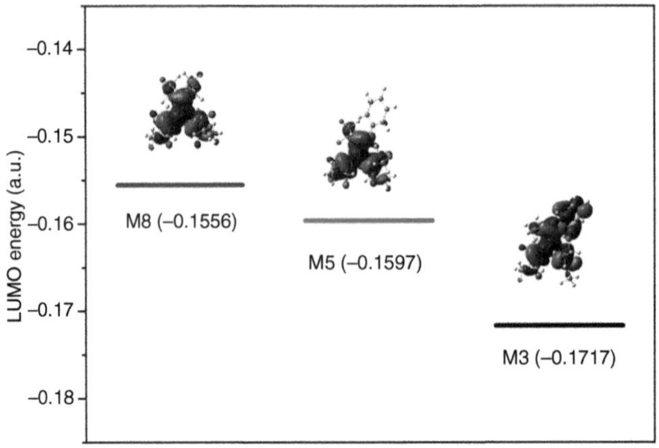

Figure 8.7 LUMO energies of M3, M5, and M8 [14]. Source: Jifu Zheng.

Guanidinium model compounds M4, M5, and M6 all contained a benzyl substituent, and their sequence of alkaline stability was M6 > M5 > M4. Especially, when the isopropyl group was introduced in M6, the guanidinium presented excellent stability (no degradation could be found even after 312 h). Based on the above results, we speculated isopropyl and ethyl groups had greater steric hindrance than methyl groups, which could effectively protect guanidinium from OH^-, so these "protected" guanidinium had better alkaline stability.

As for M3, M5, and M8, besides five ethyl substituents, the sixth substituent was the phenyl, benzyl, and ethyl group, respectively. The sequence of their alkaline stability was M8 > M5 > M3. This was consistent with the LUMO energy calculation results as shown in Figure 8.7, of which the order of LUMO energy was M8 (−0.1556) > M5 (−0.1597) > M3 (−0.1717), and as reported, the LUMO energy was a referable parameter to state the relative ability of cations. The lower the LUMO energy of the cations, the easier they could be attacked by OH^- [9–12]. Hence, the influence sequence of the substituents was ethyl > benzyl > phenyl. This result was different with those of some reported work in which they claimed the benzyl guanidiniums might undergo faster degradation reaction than phenyl guanidiniums because of the nucleophilic attack by OH^- at the benzylic position [15]. But in fact, in our study, no nucleophilic substitution was found for benzyl guanidiniums during the degradation process as described above, and benzyl guanidiniums showed much better alkaline stability than phenyl guanidiniums. This could be due to the higher LUMO energy value and electron-donating characteristics (a proxy for LUMO) of the benzyl group.

Among the symmetrical guanidiniums of M7–M9, M8 containing ethyl substituents exhibited the best alkaline stability, leaving more than 96% cations remaining after 312 h at 60 °C. M7 with six methyl substituents displayed weaker stability than M8; in this case only 45% of the resulting cation was observed after 6 h. Cyclobutyl-substituted M9 showed the lowest alkaline stability, and no cation remained even aging for 6 h. Figure 8.8a,b showed the single-crystal structures of M8 and M9,

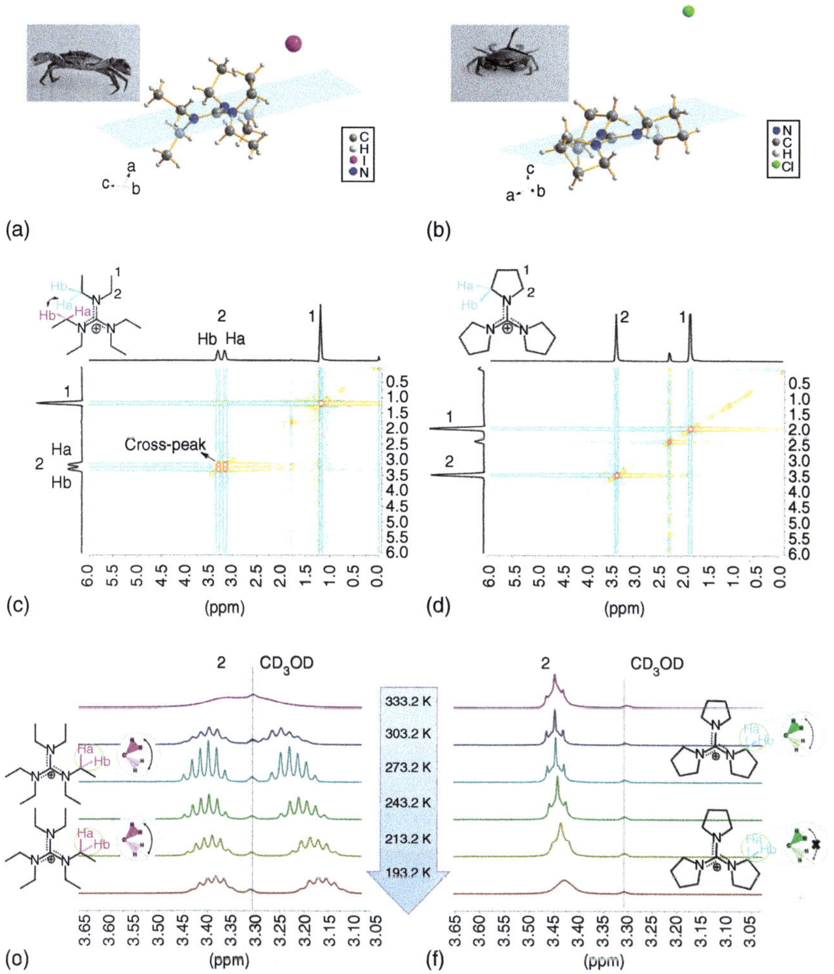

Figure 8.8 (a) The single-crystal structure of M8. (b) The single-crystal structure of M9. (c) NEOSY spectra of M8 (solvent: $CDCl_3$). (d) NEOSY spectra of M9 (solvent: $CDCl_3$). (e) Partial variable-temperature ^1H NMR spectra of M8 (solvent: CD_3OD). (f) Partial variable-temperature ^1H NMR spectra of M9 (solvent: CD_3OD). Source: Xue et al. [14]/Royal Society of Chemistry.

respectively, where the coplanar central C atom and the three N atoms could be clearly seen. M8 with the *N,N*-diethyl substituents resembled a crab with free legs, and the pyrrolidine substituents in M9 were also large but with fewer degrees of freedom, making M9 look like a tied crab. Moreover, as shown in Figure 8.8c,d, there existed cross-peaks correlating Ha and Hb in the 2D ^1H–^1H correlation nuclear overhauser effect spectroscopy (NOESY) of M8; in contrast, no extra correlation signal could be found for M9. These results suggested that the ethyl groups in M8 underwent free rotation around C—N bonds, leading to the strong interplay between the neighboring ethyl groups, while the ring structure of M9 limited the rotation of the cyclobutyl groups. The variable-temperature ^1H NMR spectra in Figure 8.8e,f further identified the difference in rotation contributing to different steric hindrances between M8 and M9. For M8, a broad single H signal could be observed at 333.2 K, demonstrating that the ethyl groups moved faster than the frequency difference between the two sites. At lower temperatures, the peak splits into two multiple H signals, indicating a decrease in the moving speed. But the ethyl groups still moved at a certain speed even at 193.2 K, which was different from M9. Therefore, the low degree of freedom of the ring substitution structure led to small steric hindrance, which might be responsible for the poor stability of M9 in alkaline solutions.

Briefly, the percentage of guanidinium cations remaining in the alkaline solution over time is summarized in Figure 8.6b. It could be seen that the stability of the model compounds followed the sequences: M6 > M8 > M5 > M3 > M2 > M7 > M4 > M1, M9. This suggested the order of substituents contributing to alkaline stability were isopropyl > ethyl > benzyl > phenyl, and ethyl > methyl > cyclobutyl. Guanidiniums with free and bulky alkyl substituents showed excellent resistance to alkaline conditions, demonstrating that the steric effect played a key role in protecting guanidiniums from the attack of OH^-, i.e. the larger steric hindrance was, the more stable guanidinium might be obtained. This steric effect was beneficial to prepare highly stable guanidinium-based AEMs.

In conclusion, current studies on the degradation mechanism and alkaline stability of guanidiniums are still relatively limited, and most of them are in their initial stages. It remains to be further verified whether guanidiniums are suitable ionic groups by introducing them into the AEMs.

8.4 Preparation of Guanidinium and Their Derivative-Based AEMs

In recent years, various methods have been explored to prepare guanidinium functionalized polymers: (i) Nucleophilic substitution reaction between halomethylated polymers and pentaalkylguanidine [4, 16, 17]. (ii) Grafting of polymers containing primary amines (or secondary amines) with Vilsmeier salt [18]. (iii) Nucleophilic substitution reaction of polymers containing activated fluorinated benzene and tetraalkylguanidine and further quaternarization [9, 13, 15]. (iv) Different polymerization reactions using guanidine (or guanidinium) functional monomers [14, 19, 20]. (v) Directly polycondensation reaction of guanidinium hydrochloride and diamine [21]. The researchers used the above methods to design and prepare guanidinium AEMs with different structures. Guanidinium and their derivative-based AEMs are divided into the following categories based on the different types of the substituents in the reported guanidinium structure.

8.4.1 Benzyl-guanidinium AEMs

In 2010, our research group [4] optimized the chloromethylation conditions of poly(aryl ether sulfone) (including solid content, temperature, and time and catalyst content), and chloromethylated poly(aryl ether sulfone)s with different degrees of substitution were obtained, and then chloromethylated poly(aryl ether sulfone) and pentamethyl guanidine underwent a nucleophilic substitution reaction, and a series of guanidinium-based poly(aryl ether sulfone) AEMs PSGOH were successfully prepared (Figure 8.9a). The OH^- conductivity of the PSGOH-1.2 membrane with ion exchange capacity (IEC) of 1.89 mmol g^{-1} at room temperature was 0.045 S cm^{-1}, which was significantly higher than that of QA-based AEMs with similar IEC (IEC = 1.85 mol g^{-1}, 0.029 S cm^{-1}). The strong alkalinity of the guanidinium tended to significantly increase the numbers of dissociated hydroxide ions and absorbed water molecules, thus effectively improving the conductivity of the membrane material. In addition, the PSGOH AEMs with resonance structure also had excellent mechanical properties and thermal stability (the decomposition temperature was higher than 185 °C). The conductivity of PSGOH remained unchanged after 48 h of immersion in 1 M NaOH at 60 °C, indicating that these membranes were well alkaline stable.

Later, Xu and coworkers [16] selected polyphenylene oxide (PPO) as the polymer backbone, by bromomethylation with liquid bromine, and brominated poly(2,6-dimethyl-phenylene oxide) (BPPO) with different degrees of bromination was obtained, and then guanidinium-based AEM guanidinium-functionalized poly(2,6-dimethyl-1,4-phenylene oxide) (GPPO) was prepared by quaternization of pentamethylguanidine and BPPO (Figure 8.9b). This type of AEM had an excellent conductivity of up to 0.071 S cm^{-1} at room temperature. The peak power density of the obtained membranes in the H_2/O_2 fuel cell was 16 mW cm^{-2}. Again, using PPO as the polymer skeleton, BPPO/G was prepared by Bai's research group [17] and then a 2-benzyl-1,1,3,3-tetramethylguanidine (BTMG) was synthesized to react with BPPO; finally, the cross-linked benzyl guanidinium AEM (Figure 8.9b) was obtained by solution casting and heat treatment. Due to the presence of the cross-linked structure, the swelling rate of the membrane was less than 15% at 60 °C, showing very good dimensional stability. In addition, the cross-linked guanidinium AEM also showed good alkaline stability (there was only a small loss of conductivity after immersing in 60 °C, 1 M $NaOH/D_2O$ solution for 2 days). Sherazi et al. [23] reported a radiation-induced method to graft vinyl benzyl chloride onto ultra-high molecular weight polyethylene, followed by hot pressing to prepare a film; finally, guanidinium-based AEMs with an all-carbon skeleton structure were successfully prepared by the quaternization reaction of 1,1,3,3-tetramethyl-2-*n*-butyl (TMBG) and benzyl chloride. Compared to other radiation-grafted AEMs, the obtained guanidinium-based AEMs had higher conductivity under the condition of low grafting degree and low IEC. In addition, the obtained membrane materials still had a lower swelling rate even at high temperatures, indicating good dimensional stability. The methanol permeability of the prepared membrane materials with grafting degrees of 8.5 and 12.5% was 1.5×10^{-8} and 2.72×10^{-8} cm^2 s^{-1}, respectively, which were far lower than Nafion's methanol permeability level, indicating they had good application prospects in the field of alkaline fuel cells.

In general, the IEC is set to higher values in order to obtain sufficient ion conduction capacity, which often results in a high swelling ratio (SR), leading to low

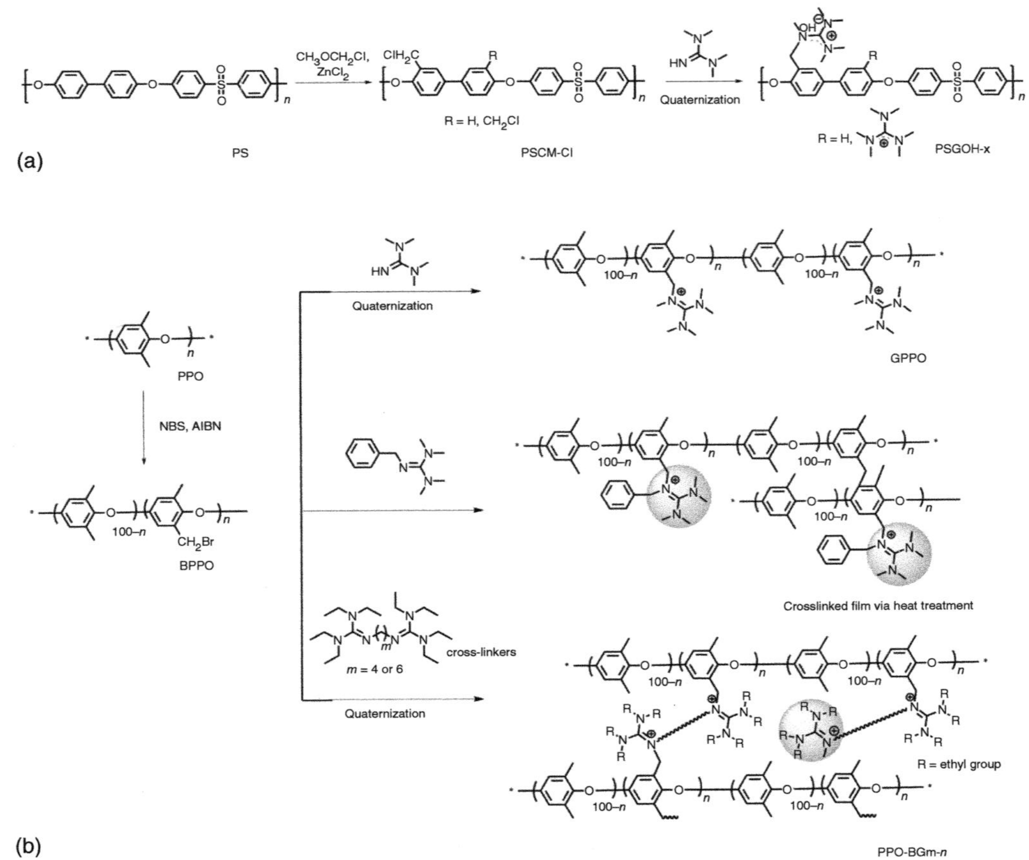

Figure 8.9 (a) Synthetic route of PSGOH-x [4]. (b) Synthetic procedure of GPPO and cross-linked membranes [16, 17, 22]. Source: Jifu Zheng.

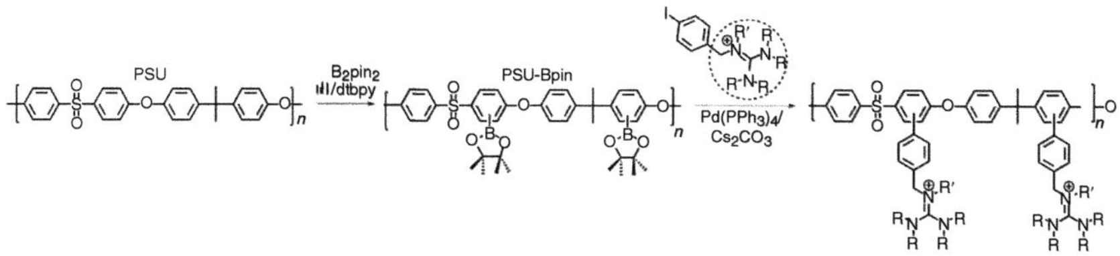

Figure 8.10 The synthetic route of PSU-P-EG and PSU-P-PG AEMs (the counterions were omitted) [14]. Source: Jifu Zheng.

dimensional stability and serious deterioration of mechanical properties. To overcome the above shortcomings, bi-guanidinium cross-linked AEMs (PPO-BGm-*n*) were designed and prepared by Menshutkin reactions using BPPO and bi-guanidine as cross-linkers, and then the remaining guanidines were further quaternized via bromoethane (Figure 8.9b) [22]. The prepared membranes exhibited remarkable dimensional stability (<15%) even at 80 °C because of the introduction of cross-linking structures. Alkaline stability test results showed that the bi-guanidinium cross-linked AEMs had excellent resistance to alkaline solutions, and more than 90% of cations could be found remaining after soaking in 1 mol L^{-1} NaOH solution at 60 °C for 30 days.

It has been reported that when the cation groups are located on the neighboring site of Ar—O bonds, the polymers containing aryl ether (Ar—O) such as polysulfone (PSU) may experience the cleavage of its Ar—O bonds via nucleophilic aromatic substitution (SNAr) of OH^- [24–26]. To address these issues, it is an ideal strategy to select a suitable spacer between the main chain and the positive ion to reduce the influence of the positive ion on the Ar—O bond [14]. In 2019, our research group prepared the boronated PSU precursor PSU-Bpin through the iridium-catalyzed boronation reaction and then performed the Suzuki–Miyaura coupling reaction with the pre-prepared benzylguanidinium-aryl iodide monomer. Two novel side-chain-type guanidinium functionalized AEMs were successfully prepared (Figure 8.10). The two-step highly active transition-metal catalyzed reaction avoided the use of toxic methyl chloromethyl ether and ensured an efficient progress of the grafting reaction. Alkaline stability test results revealed that PSU-P-PG with a large steric hindrance exhibited excellent resistance to alkaline solutions. Practically no degradation of ionic groups or hydroxide conductivity was observed after immersing in 1 mol L^{-1} NaOH at 60 °C for 30 days. At the same time, no breakage of the aromatic ether bond was observed, indicating that the phenyl group between the main chain and the cation acted as a spacer group to reduce the influence of guanidinium on the main chain.

8.4.2 Alkyl-guanidinium AEMs

Our research group [18] prepared a side-chain-type hexaalkyl guanidinium-based poly(aryl ether sulfone) ionomer. The specific experimental process was listed as follows (Figure 8.11a): a phenolphthalein-type poly(ether sulfone) (PES) containing secondary amine was prepared through a typical nucleophilic substitution polycondensation reaction and the reaction of the Vilsmeier salt and the amine group on the polymer side chain. Due to the high conversion rate of the grafting reaction, a high yield of side-chain type alkyl-guanidinium-based AEM PES-GOH was obtained. Experimental studies showed that compared with traditional QA salt-based AEMs, PES-GOH had excellent conductivity (the conductivity of PES-GOH with an IEC of 1.39 $mmol\,g^{-1}$ at 60 °C reached 0.042 $S\,cm^{-1}$), and because hexaalkyl guanidinium had a larger space volume and hydrophilicity, a good phase separation structure could be observed through transmission electron microscopy (TEM).

Qu et al. [19] prepared a biguanidinium-bridged polyhemisesquioxane guanidinium-based composite membrane material BG-BPS/PTFE (polytetrafluoroethylene) through an *in situ* sol–gel reaction (Figure 8.11b). It showed that a three-dimensional network structure formed by the hydrolysis and condensation of sesquisiloxane could effectively inhibit excessive swelling of the membrane material. At the same time, BG-BPS/PTFE had high conductivity and excellent alkaline stability. The conductivity remained constant after 120 h of soaking in a 3 M NaOH solution at 40 °C. In addition, at 40 °C and a current density of 700 $mA\,cm^{-2}$, its direct borohydride fuel cell (DBFC) achieved a peak power density of 321 $mW\,cm^{-2}$. The cell was able to discharge continuously for 50 h at a voltage of 0.76 V and a current density of 200 $mA\,cm^{-2}$, showing very excellent cell performance.

In 2022, our research group [27] reported the combination of the hydrophobic guanidinium salt with ether-free poly(phenyl-alkane)s backbones through a simple and efficient route, and a series of AEMs G-PPTPT-n:m containing novel alkyl-guanidinium salts were successfully prepared (Figure 8.11c). The AEMs showed ultra-low water uptake (WU) and SR (80 °C, WU < 12%, SR < 3%), demonstrating that the introduction of the large steric guanidinium salt was effective in improving the dimensional stability of the AEMs. The ether-free polymer skeleton endowed these AEMs with excellent stability: even at high temperatures under a strong alkaline environment, they could still maintain good stability (at 60 °C and in 1 M KOH solution, kept for 300 days). In addition, the AEMs exhibited excellent thermal stability, with the decomposition temperatures of the guanidinium group and the backbone being 350 and 450 °C, respectively.

8.4.3 Aryl-guanidinium AEMs

Kim's group [15] reported a controllable and effective fluorophenyl-amine reaction, through which guanidinium could be introduced into the polymer backbone: fluorophenyl poly(arylene ether ketone) (PAES) activated by sulfone groups underwent a nucleophilic substitution reaction with tetramethylguanidine (TMG), followed by quaternization, and finally the arylguanidinium-type AEM M-PAES-TMG was successfully prepared (Figure 8.12a). The conductivity of M-PAES-TMG decreased from 0.036 to 0.031 $S\,cm^{-1}$ after 72 h of immersion in a 0.5 $mol\,L^{-1}$ NaOH solution at 80 °C. At the same time, the alkaline fuel cell showed excellent cell performance with M-PAES-TMG as the electrode ionomer under the conditions of Pt as the

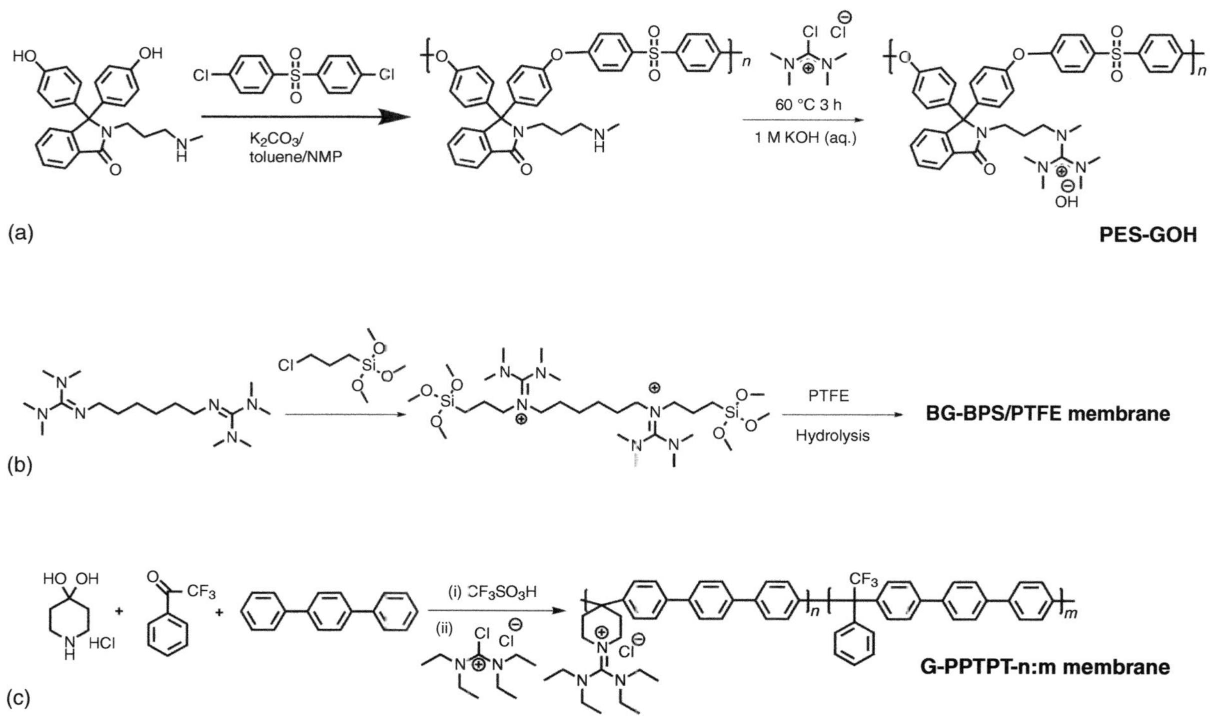

Figure 8.11 (a) Synthetic route of poly(aryl ether sulfone) containing hexaalkylguanidinium groups (PES-G-OH) [18]. (b) Synthetic route for the BG-BPS/PTFE membrane [19]. (c) Synthetic route for the G-PPTPT-n:m membrane [27]. Source: Jifu Zheng.

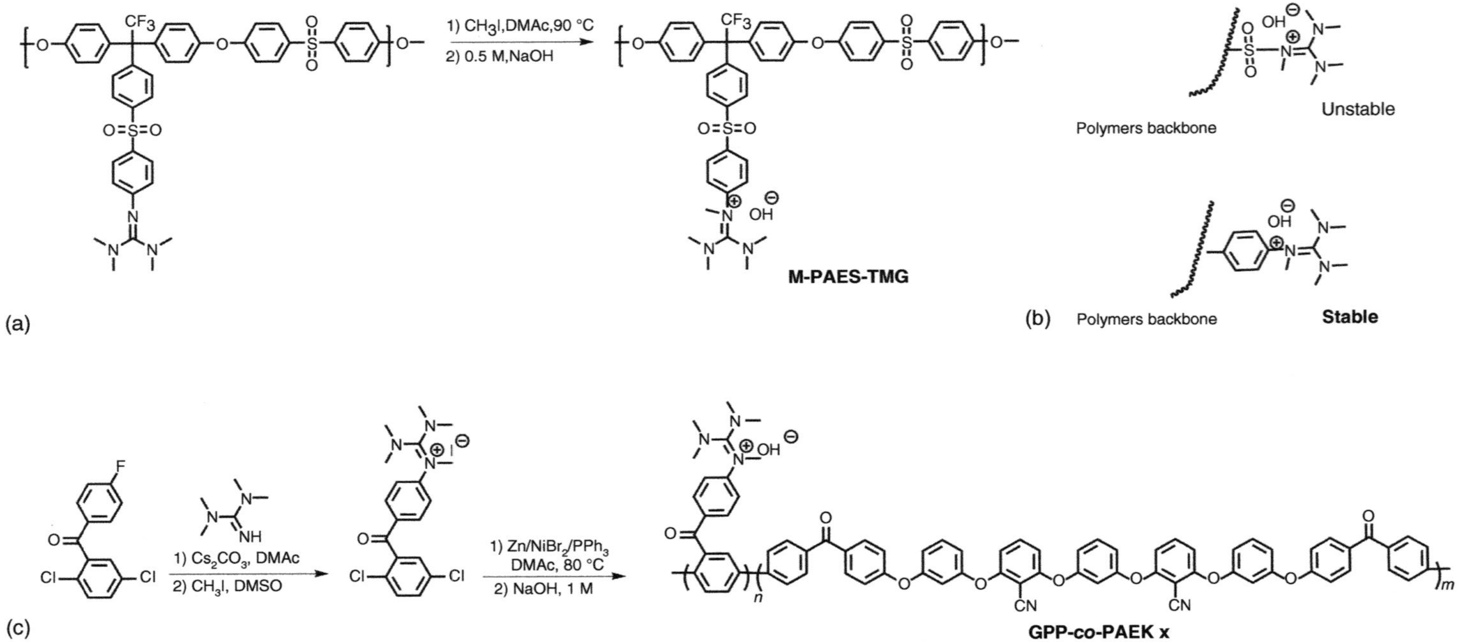

Figure 8.12 (a) Synthetic route of phenyl-guanidinium-functionalized poly(arylene ether sulfone) [15]. (b) Sulfone-pentamethylguanidinium and phenyl-pentamethylguanidinium functionalized ionomers [8]. (c) The synthetic procedure of GPP-*co*-PAEK x copolymers [13]. Source: Jifu Zheng.

catalyst and H_2/O_2 as the reaction raw material (the obtained peak power density at 80 °C is 200 mW cm^{-2}).

Later, using a similar method, a resonance stable aryl guanidinium-based perfluoro-ionomer was successfully prepared (Figure 8.12b) [8]. A comparative study with the sulfone guanidinium-based ionomer found that after soaking at 80 °C in 0.5 M NaOH for 24 h, the sulfone guanidinium ionomer was almost completely degraded, while the aryl guanidinium ionomer did not degrade after 72 h under the same conditions, indicating the stability of the ionomer largely depended on the structure of the adjacent group connected to the cation. In addition, the alkaline fuel cell with the aryl guanidinium polymer as the catalytic layer ionomer had a peak power density of 466 mW cm^{-2} at 80 °C with H_2/O_2 as the gas raw material, indicating that the aryl guanidinium AEM had excellent cell performance. It might be due to improved interaction and compatibility between the electrode binder and catalyst surface.

The fluorophenyl-amine reaction showed high reactivity, which was a good solution to the grafting degree problem. However, both of the above preparation methods required the use of alkylation reagents to further QA to obtain the target polymer, which inevitably led to the problem of incomplete conversion. In 2017, novel aryl guanidinium-based AEMs (GPP-*co*-PAEK x) with controlled microblock structures were prepared by our group via a fluorophenyl-amine reaction, methylation, and nickel (0)-catalyzed coupling (Figure 8.12c). This controllable and efficient method overcomes the difficulties associated with the post-quaternization method in controlling the degree of incorporation and location of the functional groups, and also avoids using hygroscopic Vilsmeier salts. In addition, this method also omits the halomethylation process of the conventional method and provides copolymers with well-defined chemical structures. Under a similar IEC, the micro-block type GPP-*co*-PAEK had a lower water absorption rate (less than 24% at 60 °C) than the reported random guanidinium AEM. The environmentally friendly and ionized monomer polymerization strategy avoids the use of wet-sensitive reagents, which not only provides good control of the IEC and the location of ionic groups in the copolymers but also addresses the potential problem of incomplete post-functionalization.

8.4.4 Other Guanidinium-Based AEMs

Sajjada et al. [21] reported a main chain guanidinium polymer which was prepared by the polycondensation of guanidinium hydrochloride and diamine. Then, the guanidinium polymer was locked into porous PTFE by adding a cross-linking agent. Finally, chemically stable and high-strength composite AEMs were obtained. This synthetic route is very simple and does not require any toxic chloromethylating agent or complicated post-treatment process. Cross-linking anchors the guanidinium cations in the pores of PTFE. Therefore, reinforced AEMs have excellent mechanical properties and dimensional stability and can be processed into very thin films. The measured conductivity was as high as 0.0847 S cm^{-1} at 20 °C, which was comparable to that of the commercial proton exchange membrane Nafion. In addition, the structure of the composite guanidinium-based AEM was unchanged after 50 h treatment in 5 M KOH solution at 55 °C. Compared to the commercialized ammonium-based A201 membrane, it showed obvious advantages in stability.

Ding and coworkers [20] reported that the properties of comb-shaped guanidinium-based AEM could be tuned facilely by the spacer types and lengths. Due to the appropriate water absorption and the formation of connected ion transport channels in the structure, the three carbon atom (3C)-linked guanidinium AEM showed higher conductivity than the 1C- and 9C-linked guanidinium AEM. In addition, the conductivity of the guanidinium AEM with two ethylene oxide segments was more prominent, which could be attributed to the strong hydrophilicity of the ethylene oxide segments and connected ion channels. The thermal decomposition temperature of all guanidinium AEMs was as high as 200 °C, which further confirmed that the guanidinium AEM had excellent thermal stability. However, during the chemical stability test, the membrane material degraded rapidly after being immersed in a solution of 1 M NaOH for 72 h at 60 °C. The analysis was carried out because the guanidinium substituted by hydrogen atoms was more likely to be attacked by hydroxide radicals, leading to the deterioration of the alkaline stability of the membrane material.

Imidazole-guanidinium is a derivative of guanidinium. Different from the conventional guanidinium, the primary amine used in the preparation process of Zhang and coworkers was an imidazole primary amine, and the resulting imidazole guanidine was used as a cross-linking molecule [28]. A two-site reaction between imidazole guanidine and chloromethylated PSU realizes the preparation of cross-linked membranes. Compared to the non-cross-linked AEM, the imidazole–guanidinium cross-linked membrane exhibits better alkaline resistance due to the enhanced resonance effect and the construction of a cross-linked structure by imidazole guanidine. (After 10 days of treatment in 3 M NaOH solution, the conductivity and chemical structure of the membrane remained almost unchanged.) The maximum power density for the H_2/O_2 fuel cell of the prepared membrane was 39 mW cm^{-2} at 50 °C.

Recently, Tang and coworkers [29] developed a novel guanidinium-functionalized covalent organic framework (COF) by grafting guanidinium onto the channel walls of COF via the Williamson reaction, followed by doping the COF into GPPO to prepare composite membranes. The guanidiniums grafted into the COF nanochannels play the role of the "active sites" in the membranes to enhance the migration rate of the hydroxide ions and thus improve the conductivity. The resulting composite membrane with 5 wt.% guanidinium-functionalized-COF doping has a hydroxide conductivity of up to 89.93 mS cm^{-1} at 80 °C under hydrated conditions, 61% higher than that of the pristine GPPO membrane. At the same time, the single-cell performance of the composite membrane reaches a maximum power density of 111.3 mW cm^{-1} at 60 °C and 100% relative humidity (RH).

Compared to QA-based AEMs, the study for guanidinium-based AEMs is relatively late. This can be attributed to the following reasons: on the one hand, there are still very few types of guanidinium cations and their derivative-based AEMs due to the limitations of the synthesis methods. On the other hand, the application of guanidinium cations is not very in-depth, especially in the field of the fuel cells, due to the lack of systematic studies on the structure–activity relationship between the chemical structure and the stability of guanidinium cations. So far, only a few research groups use guanidinium-based AEMs as membrane materials and catalytic layer ionomer (or binder) to prepare the membrane electrode assembly (MEA), and further studied their applications in the H_2/O_2 fuel cell and DBFC. The conductivity, alkaline stability, and fuel cell efficiency of some guanidinium functionalized AEMs with other cation-based AEMs such as imidazolium, phosphonium, and QA are summarized in Table 8.1. Compared to other functional group-based AEMs, especially the QA-type AEMs, the values of the maximum power density of the guanidinium functionalized AEMs are lower. As is well-known, enhancing conductivity is an

Table 8.1 Comparison of conductivity, alkaline stability, and fuel cell efficiency of guanidinium functionalized AEMs with other cation-based AEMs.

Membrane	Ionic group	Conductivity @ 80 °C (mS cm^{-1})	Alkaline stability	Maximum power density	References
GPPO-0.54	Benzyl-guanidinium	71	Conductivity remained unchanged in 1 M KOH solution at 25 °C for 192 h	16 mW cm^{-1} at 50 °C	[16]
M-Nafion-FA-TMG	Aryl-guanidinium	20	Structure remained unchanged in 0.5 M NaOH solution at 80 °C for 72 h	—	[8]
IL-COF/GPPO-5	Alkyl-guanidinium	89.93	90.31% conductivity remained in 2 M NaOH solution at 60 °C for 336 h	111.3 mW cm^{-1} at 60 °C	[29]
P-MeIM-tPb11	Imidazolium	83	Conductivity remained unchanged in 5 M NaOH solution at 80 °C for 1500 h	270 mW cm^{-1} at 80 °C	[30]
QPPEEK-PEG 20	Phosphonium	102.13	85% conductivity remained in 1 M KOH solution at 80 °C for 400 h	154 mW cm^{-1} at 60 °C	[31]
PAP-TP-85	Quaternary ammonium	~160	97% IEC remained in 1 M KOH solution at 80 °C for 2000 h	920 mW cm^{-1} at 95 °C	[32]

effective way to improve the performance of fuel cells, so further efforts are required to optimize membrane chemical structure (or nanostructure) to improve its comprehensive performance.

8.5 Prospect

As a kind of resonance delocalized organic cation, guanidinium has great application potential in the field of the alkaline fuel cell polymer electrolyte (membranes and ionomers) because of its excellent thermal and chemical stability. Some of the guanidinium-based AEMs reported so far have made certain progress. For example, some guanidinium-based AEMs have excellent conductivity, and some show acceptable cell performance in alkaline fuel cell applications. Due to the limited types of guanidinium studied at this stage, researchers still lack a clear understanding of the specific degradation process of guanidinium and the relationship between structure and stability, but the feasibility of using guanidinium as an ionic group is still a subject worth exploring. Therefore, in the future, it is necessary to explore more substituted forms of guanidinium, combining experiments and theoretical calculations, and to systematically evaluate the influence of electronic effects, steric effects, molecular configurations, and other factors on its alkaline stability, as well as to clarify the degradation mechanism of guanidinium. This is crucial for the design of future guanidinium-based AEMs with excellent alkaline stability. In addition to the influence of the guanidinium group, the polymer backbone structure also plays a key role in the performance of guanidinium AEMs. Therefore, polymer backbones of different structures will be selected to prepare guanidinium-type AEMs with various structures (all-carbon or ether-free, etc.) to obtain excellent performance.

References

1 Marino, M.G. and Kreuer, K.D. (2015). Alkaline stability of quaternary ammonium cations for alkaline fuel cell membranes and ionic liquids. *ChemSusChem* 8: 513–523.

2 Duan, H.F., Xu, G., Li, S.H. et al. (2006). Progress in the research of guanidinium ionic liquids. *Chin J Org Chem* 10: 1335–1342.

3 Even, U. and Jortner, J. (1982). Spectroscopy of large molecules in supersonic expansions: isolated ultracold porphyrins. In: *Intramolecular Dynamics*, vol. 15 (ed. J. Jortner and B. Pullman), 227. Netherlands. https://doi.org/10.1007/978-94-009-7927-7: Springer.

4 Wang, J.H., Li, S.H., and Zhang, S.B. (2010). Novel hydroxide-conducting polyelectrolyte composed of an poly(arylene ether sulfone) containing pendant quaternary guanidinium groups for alkaline fuel cell applications. *Macromolecules* 43: 3890–3896.

5 Mateus, N.M.M., Branco, L.C., Lourenço, N.M.Y., and Afonso, C.A.M. (2003). Synthesis and properties of tetra-alkyl-dimethylguanidinium salts as a potential new generation of ionic liquids. *Green Chem* 5: 347–352.

6 Li, S.H., Lin, Y.J., Cao, J., and Zhang, S.B. (2006). Guanidine/Pd(OAc)$_2$-catalyzed room temperature Suzuki cross-coupling reaction in aqueous media under aerobic conditions. *J Org Chem* 11: 4067–4072.

7 Wang, J.H. (2010). Synthesis and properties of quaternized poly(arylene ether)s as anion exchange membranes (in Chinese). PhD dissertation. Changchun Institute of Applied Chemistry, Chinese Academy of Sciences, Changchun.

8 Kim, D.S., Fujimoto, C.H., Hibbs, M.R. et al. (2013). Resonance stabilized perfluorinated ionomers for alkaline membrane fuel cells. *Macromolecules* 46: 7826–7833.

9 Mohanty, A.D. and Bae, C. (2014). Mechanistic analysis of ammonium cation stability for alkaline exchange membrane fuel cells. *J Mater Chem A* 2: 17314–17320.

10 Li, W.W., Wang, S.B., Zhang, X.F. et al. (2014). Degradation of guanidinium-functionalized anion exchange membrane during alkaline environment. *Int J Hydrog Energy* 39: 13710–13717.

11 Fukui, K. (1982). Role of frontier orbitals in chemical reactions. *Science* 218: 747–754.

12 Pernpointner, M. and Hashmi, A.S.K. (2009). Fully relativistic, comparative investigation of gold and platinum alkyne complexes of relevance for the catalysis of nucleophilic additions to alkynes. *J Chem Theory Comput* 5: 2717–2725.

13 Xue, B.X., Dong, X., Li, Y.C. et al. (2017). Synthesis of novel guanidinium-based anion-exchange membranes with controlled microblock structures. *J Membr Sci* 537: 151–159.

14 Xue, B.X., Wang, F., Zheng, J.F. et al. (2019). Highly stable polysulfone anion exchange membranes incorporated with bulky alkyl substituted guanidinium cations. *Mol Syst Des Eng* 4: 1039–1047.

15 Kim, D.S., Labouriau, A., Guiver, M.D., and Kim, Y.S. (2011). Guanidinium-functionalized anion exchange polymer electrolytes via activated fluorophenyl-amine reaction. *Chem Mater* 23: 3795–3797.

16 Lin, X.C., Wu, L., Liu, Y.B. et al. (2012). Alkali resistant and conductive guanidinium-based anion-exchange membranes for alkaline polymer electrolyte fuel cells. *J Power Sources* 217: 373–380.

17 Liu, L., Li, Q., Dai, J.W. et al. (2014). A facile strategy for the synthesis of guanidinium-functionalized polymer as alkaline anion exchange membrane with improved alkaline stability. *J Membr Sci* 453: 52–60.

18 Zhang, Q., Li, S.H., and Zhang, S.B. (2010). A novel guanidinium grafted poly (aryl ether sulfone) for high-performance hydroxide exchange membranes. *Chem Commun* 46: 7495–7497.

19 Qu, C., Zhang, H.M., Zhang, F.X., and Liu, B. (2012). High-performance anion exchange membrane based on bi-guanidinium bridged polysilsesquioxane for alkaline fuel cell application. *J Mater Chem* 22: 8203–8207.

20 Chen, Y., Tao, Y.P., Wang, J.L. et al. (2017). Comb-shaped guanidinium functionalized poly(ether sulfone)s for anion exchange membranes: effects of the spacer types and lengths. *J Polym Sci A Polym Chem* 55: 1313–1132.

21 Sajjada, S.D., Hong, Y., and Liu, F.Q. (2014). Synthesis of guanidinium-based anion exchange membranes and their stability assessment. *Polym Adv Technol* 25: 108–116.

22 Xue, B.X., Wang, Q., Zheng, J.F. et al. (2020). Bi-guanidinium-based crosslinked anion exchange membranes: synthesis, characterization, and properties. *J Membr Sci* 601: 117923.

23 Sherazi, T.A., Zahoor, S., Raza, R. et al. (2015). Guanidine functionalized radiation induced grafted anion-exchange membranes for solid alkaline fuel cells. *Int J Hydrog Energy* 40: 786–796.

24 Arges, C.G. and Ramani, V. (2013). Two-dimensional NMR spectroscopy reveals cation-triggered backbone degradation in polysulfone-based anion exchange membranes. *PNAS* 110 (7): 2490–2495.

25 Arges, C.G., Wang, L.H., Parrondo, J., and Ramani, V. (2013). Best practices for investigating anion exchange membrane suitability for alkaline electrochemical devices: case study using quaternary ammonium poly(2,6-dimethyl 1,4-phenylene) oxide anion exchange membranes. *J Electrochem Soc* 160: F1258–F1274.

26 Mohanty, A.D., Tignor, S.E., Krause, J.A. et al. (2016). Systematic alkaline stability study of polymer backbones for anion exchange membrane applications. *Macromolecules* 49: 3361–3372.

27 Wang, Q., Huang, L., Zheng, J.F. et al. (2022). Design, synthesis and characterization of anion exchange membranes containing guanidinium salts with ultrahigh dimensional stability. *J Membr Sci* 643: 120008.

28 Cheng, J., Yang, G.Q., Zhang, K.B. et al. (2016). Guanidimidazole-quanternized and cross-linked alkaline polymer electrolyte membrane for fuel cell application. *J Membr Sci* 501: 100–108.

29 Zhang, N.X., Li, P., Li, X., and Tang, S.K. (2022). Anion exchange composite membrane based on ionic liquid-grafted covalent organic framework for fuel cells. *Int J Hydrog Energy* 47: 29481–29494.

30 Li, X.F., Wang, Z.M., Chen, Y.H. et al. (2022). Imidazolium-based AEMs with high dimensional and alkaline-resistance stabilities for extended temperature range of alkaline fuel cells. *J Membr Sci* 670: 121352.

31 Kumari, M., Douglin, J.C., and Dekel, D.R. (2021). Crosslinked quaternary phosphonium-functionalized poly(ether ether ketone) polymer-based anion-exchange membranes. *J Membr Sci* 626: 119167.

32 Wang, J.H., Zhao, Y., Setzler, B.P. et al. (2019). Poly(aryl piperidinium) membranes and ionomers for hydroxide exchange membrane fuel cells. *Nat Energy* 4: 392–398.

9

Anion Exchange Membranes Based on Imidazolium and Triazolium Cations

Do-Hyeong Kim, Vijayalekshmi Vijayakumar, and Sang Yong Nam

Gyeongsang National University, Research Institute for Green Energy Convergence Technology, Department of Materials Engineering and Convergence Technology, Jinju-Daero, Jinju-Si, 52828, Republic of Korea

9.1 Introduction

Fuel cells are the most promising environmentally friendly power sources that can potentially replace the existing traditional fossil fuels. Owing to the upsurge in energy demand and mass consumption, the need for clean and efficient energy increases, and significant research efforts are being carried out internationally toward the development of high-performance fuel cells [1, 2]. In particular, polymer electrolyte membrane fuel cells, owing to their high power density and energy conversion efficiencies, as well as their low pollutant formation and working temperature, have attracted much research interest. As one of the core components, the ion exchange polymer electrolyte membrane has a significant influence on cell performance and is nowadays a hot research topic [3–5]. Nafion, a perfluorosulfonic acid membrane, is regularly used as a commercially available membrane in proton exchange membrane fuel cells (PEMFCs) due to various inherent advantages such as high conductivity, excellent mechanical, thermal, and chemical stability. [4, 6, 7]. However, the high cost and non-eco-friendly nature associated with the fluoropolymer, the use of expensive platinum catalysts and its carbon monoxide poisoning at low temperatures, slow redox kinetics, the limited working life of proton exchange membranes, high fuel permeability, and complex water management limit the commercialization of PEMFCs [1, 8–10]. Therefore, anion exchange membrane (AEM) fuel cells comprising the advantages of both an alkaline fuel cell and a solid polymer electrolyte membrane have recently been favored over PEMFCs.

Anion exchange membrane fuel cells (AEMFCs) have a great advantage in terms of system cost owing to their enhanced redox kinetics and the use of cheaper

Alkaline Anion Exchange Membranes for Fuel Cells: From Tailored Materials to Novel Applications, First Edition. Edited by Jince Thomas, Alex Schechter, Flavio Grynszpan, Bejoy Francis, and Sabu Thomas.

non-noble metal catalysts compared to PEMFCs. Membrane electrode assembly (MEA), gas diffusion layers, flow field plates, and current collectors are the general components of a typical AEMFC. MEA, composed of AEMs as well as catalyst-loaded electrodes, acts as the power generation device for the hydrogen oxidation reaction (HOR) and oxygen reduction reaction (ORR). As the key component of AEMFC, AEMs play an important role as fuel/oxidant separators, hydroxide ion transporting medium from cathode to anode, as well as an electronic insulator [11]. The durability and power density of MEAs are primarily determined by the nature of the AEM. An ideal AEM must have high hydroxide ion conductivity, excellent thermal, dimensional, and mechanical stability, and high alkali resistance. The main challenging issue in the commercialization of AEMFC is associated with the trade-off between conductivity and dimensional stability of AEM. The conductivity and chemical stability of AEMs mainly depend on the polymer backbone, cationic groups, and interactions among polymer chains. The polymer chain with unstable sites and cationic groups is prone to degradation by hydroxide ions due to its high nucleophilicity and basicity. To circumvent these drawbacks, many research efforts have been invested in the modification of polymer architecture as well as the chemical structure of the cationic group . Quaternary ammonium (QA), the most commonly used cation is unstable and liable to strong nucleophilic attack/Hoffmann elimination by hydroxyl ions, especially at high pH conditions and elevated temperatures [12, 13]. Therefore, AEMs based on other cationic functional groups such as guanidinium [14–18], imidazolium (IM) [3, 19–21], pyridinium [22–25], tertiary sulfonium [25, 26], spirocyclic QA [27–29], and phosphonium [30, 31] have been proposed over the past few decades to replace QA cations to achieve enhanced stability toward strong bases and high ion conductivity. Among them, the IM-based AEMs are progressively attracting much attention next to the QA-ion-based AEMs. In addition, triazolium cation-based AEMs, owing to their balance between stability and conductivity also have great potential to be used as a promising polymer electrolyte membrane for AEMFCs but only a few studies have been conducted so far [32–34]. It is clear from Figure 9.1, which shows the number of publications categorized based on the cation group over the last nine years that, the trend to use triazole cations has increased over the last five years. In this book chapter, we will discuss the molecular design, synthesis strategies, and performance of AEMs based on IM and triazolium cations from 2013 to 2023. This chapter also summarizes various strategies implemented to improve the alkaline stability of AEMs based on IM and triazolium species.

9.2 AEMs Based on Imidazolium Cations

Cations based on derivatives containing π-conjugated systems have attracted much interest due to the delocalization of the positive charge, which diminishes their electrophilicity and makes them less susceptible to nucleophilic attack. The presence of electron-delocalized π-conjugated structure and steric hindrance makes IM-based cations relatively stable and has been studied recently by several research groups.

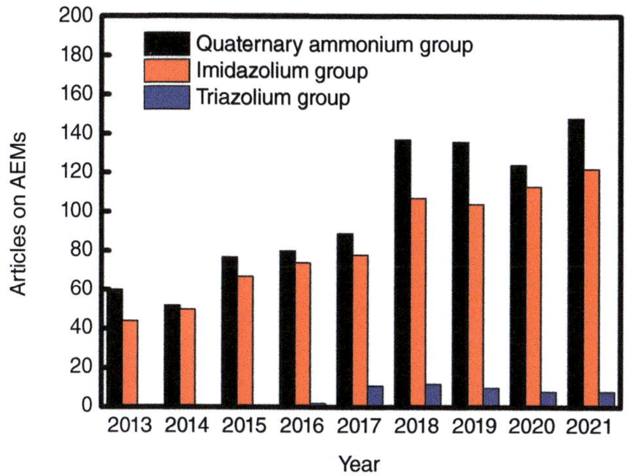

Figure 9.1 Number of research articles published on QA, IM, and triazolium-based AEMs from 2013 to 2021.

However, nucleophilic attack of hydroxide ions on the imidazole rings and membrane degradation at high pH and temperature conditions were also observed, and alkaline stability remains an open topic. The main degradation pathways are illustrated in Figure 9.2 [35]. To solve this issue and improve the alkaline stability of IM cations, various approaches, such as functionalization of various substitutes, use of ionic liquids (ILs), and incorporation of side chains, have been adopted. Incorporation of side chain could promote cations' motion and thereby the formation of hydrophobic/hydrophilic phase separation in membranes, which is more conducive to anion conduction. The incorporation of multiple cations in each side chain could also enhance the alkaline stability and conductivity of AEMs. Control of membrane morphologies via the positioning of IM groups such as clustered, side-chain type, graft/comb-shaped, block, and copolymer architectures to improve AEM performances is also discussed in this chapter.

9.2.1 AEMs Based on Imidazolium-type Ionic Liquids

Active research on IM -based AEMs started in the year 2010 [10, 36–39]. The application of ILs as electrolyte material in membranes has received significant interest owing to their unique properties. IL is an organic salt that dissolves at or below room temperature and has many advantages, such as low melting point, low volatility, high ionic conductivity, wide electrochemical window, and excellent thermal and chemical stability. ILs also have the ideal property of good cosolvent miscibility. Aiming to explore stable and highly conductive AEMs, IM-type ILs have been used in various studies. Zhu et al. have developed a blend membrane of polyvinyl alcohol (PVA) and 1-ethyl-3-methylimidazolium hydroxide ([Bmim][OH]) (Figure 9.3a). An increase in ionic conductivity was noticed with the increase in IM-based IL content, and a maximum value of $1.96 \times 10^{-2}\,S\,cm^{-1}$ was observed at [Bmim] OH to PVA

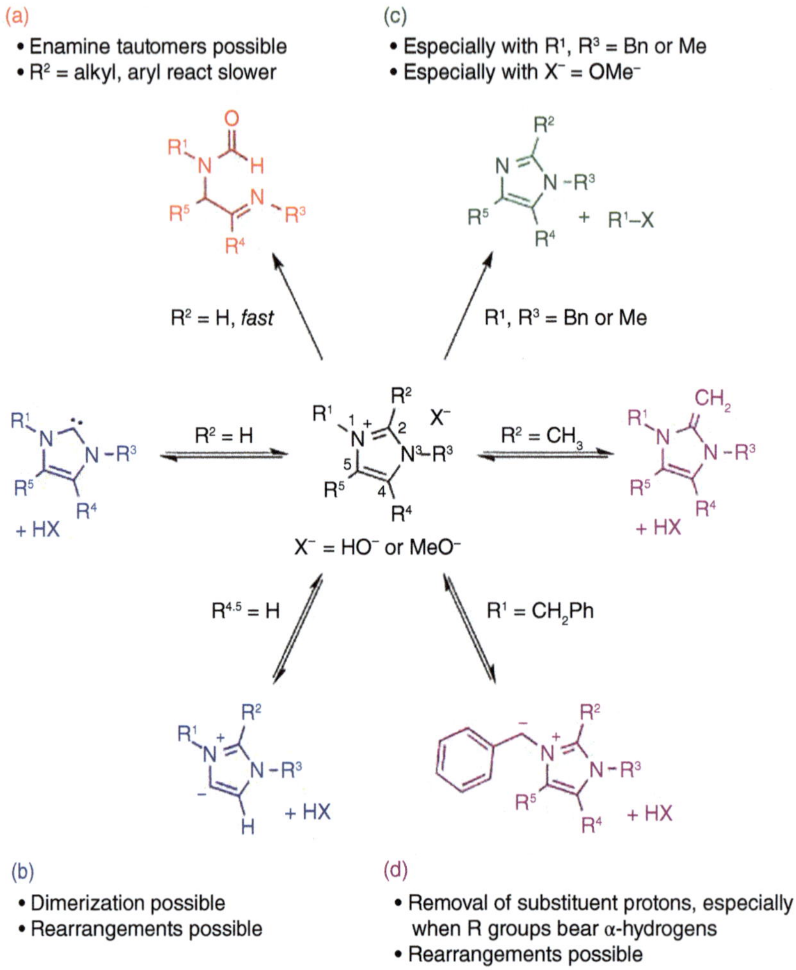

Figure 9.2 Degradation pathways of imidazolium cations under alkaline conditions [35], (a) Ring-opening reactions, (b) heterocycle deprotonations, (c) S_N2 reactions, and (d) substituent deprotonations. Source: Reproduced with permission from American Chemical Society.

ratio of 2. However, the alkaline stability of the membranes was not revealed in their report [38]. Recently, Wang et al. conducted studies on the synthesis of geminal-IM-type ionic liquids ([DimL][OH] ILs) and their application to the production of AEMs. Due to their dual-core structure, geminal-imidazolium-type IL is more mechanically and chemically stable than single-core ILs. Furthermore, the resonance effect of the π-conjugated structure of the five-membered heterocyclic ring in IM together with the presence of steric hindrance could minimize its interaction with hydroxides and contribute to alkaline stability. AEM developed using pyridine-functionalized PVA doped with genomic-IM-type IL followed by cross-linking using glutaraldehyde (GA) (PVA-FP/[DimL][OH]) (Figure 9.3b) attained an

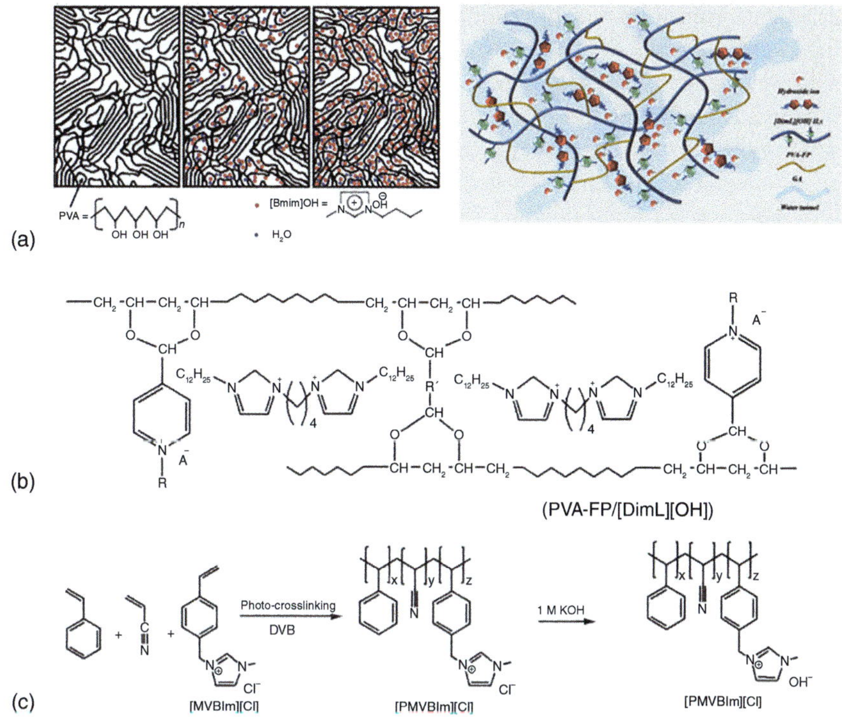

Figure 9.3 Schematic representation of AEMs based on IM-type ionic liquids: (a) PVA/[Bmim][OH], (b) PVA-FP/[DimL][OH], and (c) [PMVBIm][OH] membranes [38–40]. Source: Reproduced with permission from Elsevier.

OH^- conductivity of 5.8×10^{-2} S cm^{-1} at 70 °C. In addition, the reduction rate of the ionic conductivity was only 4.6% after 240 h of exposure to a 1 M KOH solution [39].

IM derivatives with various substitutes can be easily attained due to the unique chemical structure and excellent design possibilities of the imidazole ring. N-substitutes can easily be obtained by nucleophilic substitution reaction between the lone pair electron located at nitrogen atom with halogenoalkane, but the synthesis of C-substituted functional group is more complex and difficult than the former one. Si et al. have focused on the C2 and N3 substitutions of IM cations to analyze the stability of AEMs. The highly stable C2-(methyl) and N3-(butyl) substituted IM cation grafted onto poly(arylene ether sulfone) (PAES) possessed a high ionic conductivity of more than 0.01 S cm^{-1} at room temperature and alkaline stability of about 480 h in a 1 M KOH solution at 60 °C that confirmed the research feasibility of IM-based AEMs [36]. Qiu et al. developed an IM-based copolymer AEM through *in situ* cross-linking of styrene, acrylonitrile, and 1-methyl-3-(4-vinyl benzyl) imidazolium chloride ([MVBIm][Cl]) (Figure 9.3c) and compared with the QA-based AEM prepared in the same way using *N,N,N*-trimethyl-1-(4-vinyl phenyl) methanaminium chloride ([TMVPMA][Cl]). Alkaline IM-type IL-functionalized AEM exhibited excellent alkaline stability up to 1000 h in 1 M KOH at 60 °C without obvious loss in ionic conductivity, whereas QA-based AEM degraded at high pH conditions due to nucleophilic substitution on QA cations. The resonance effect of the conjugated IM rings facilitated a reduction in the positive charge density of cation that weakened the hydroxide-ion interaction in the membrane [40].

As there is no driving force for ion aggregation or phase separation in postfunctionalization, i.e. ionic functionality introduced after membrane casting, the formation of ion-conducting channels is difficult. In contrast, poly(ionic liquids), also called polymerized ionic liquids (PILs) generally synthesized via direct radical polymerization of IL monomers and/or post-polymerization modification, typically through quaternization of alkyl halide polymers with imidazole acted as a new solution to enhance ion conduction. Ionic polymer membranes prepared by casting a mixture of bromomethylated poly(2,6-dimethyl-1,4-phenylene oxide) (BPPO) and 1-methylimidazole displayed a nanophase-separated morphology and enhanced short-term thermal and chemical stability than classical QA-type AEMs [37]. Polymerizable ILs have the advantages of –C=C– sites for free radical copolymerization as well as the cationic group for ionic conduction. AEMs based on 1-vinyl-3-methylimidazolium iodide and 1-vinyl-3-butyl imidazolium bromide synthesized by copolymerization showed an ion conductivity of 0.0226 S cm^{-1} at 30 °C and excellent thermostability as well as chemical stability for 120 h at 60 °C in 10 M NaOH solution. H_2/O_2 fuel cell fabricated using this membrane displayed a peak power density of 116 mW cm^{-2} at a current density of 220 mA cm^{-2} at 60 °C [2]. AEM with IM salt functional group prepared by copolymerization of 1-allyl-3-methylimidazolium chloride (AmimCl) ILs with alkyl acrylates via free radical polymerization has superior thermal and chemical stability (6 M NaOH solution at 80 °C for 120 h) than alkyl QA-functionalized polymer membranes [12].

9.2.2 Imidazole Containing Polymers and Composites

The structure and morphology of the polymer, as well as the type, degree, and position of the cationic species, have a significant impact on the AEM properties. Direct introduction of hydroxide conducting IM groups onto various high-performance polymers such as polysulfone, poly(ether ether ketone), polyether sulfone (PES) cardo, cardo polyether ketone, and poly (phthalazinone) has been successfully utilized for AEM materials by many researchers. Table 9.1 shows the performances of various AEMs incorporating IM groups into several polymer backbones. Among them, polysulfone and PES are the polymer backbones most frequently used as AEM because of their high thermal stability, excellent chemical resistance as well as mechanical properties [41, 42]. Rao et al. synthesized a multi-block PES functionalized with ethyl imidazolium by nucleophilic aromatic substitution from F-terminal oligomers and OH-terminal oligomers followed by bromination and ethyl imidazolium incorporation at the benzyl position. The developed membrane exhibited excellent ion conductivity by forming well-defined phase-separated morphology. In addition, the fluorenylidene biphenylene moiety has also been introduced in the membranes to increase their mechanical as well as hydrolytic stability [43]. 1-Methyl imidazolium functionalization on poly(arylene ether ketone)s and poly(arylene ether ketone sulfone)s containing fluorine group displayed a maximum hydroxide ion conductivity

Table 9.1 AEMs with imidazole functional groups reported in various literatures.

Membrane	Backbone	Modification/ functional group	Ionic conductivity ($S\,cm^{-1}$)	IEC ($mmol\,g^{-1}$)	Alkaline stability	Power density of fuel cell ($mW\,cm^{-2}$)	References
PSf-Im OH 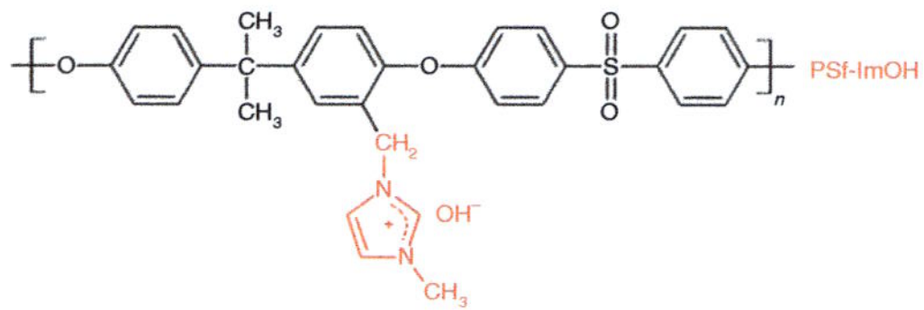	Polysulfone	1-Methylimidazolium	0.053 at 20 °C	2.19	—	—	[41]
PES-MeIm OH 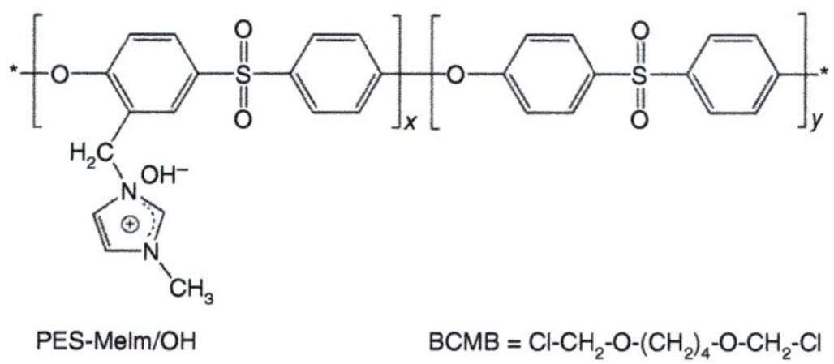	Poly(ether sulfone)	1-Methylimidazolium	0.014 at 20 °C	1.54	—	29.5 (45 °C)	[42]
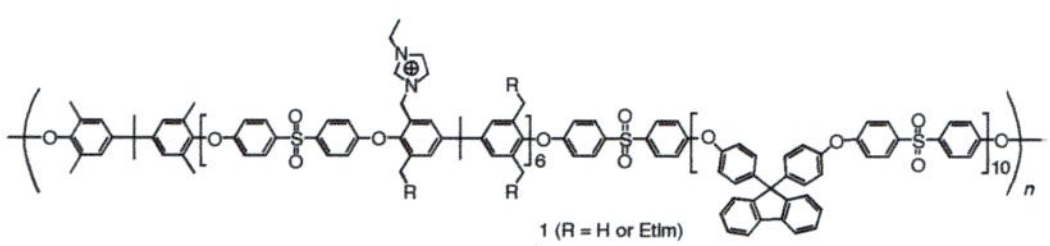	Poly(arylene ether sulfone) block copolymer	Ethyl imidazolium	0.03 at 20 °C	1.45	2 M NaOH, 60 °C, 168 h	—	[43]

(Continued)

Membrane	Backbone	Modification/ functional group	Ionic conductivity ($S cm^{-1}$)	IEC ($mmol g^{-1}$)	Alkaline stability	Power density of fuel cell ($mW cm^{-2}$)	References
HMT-PMPI 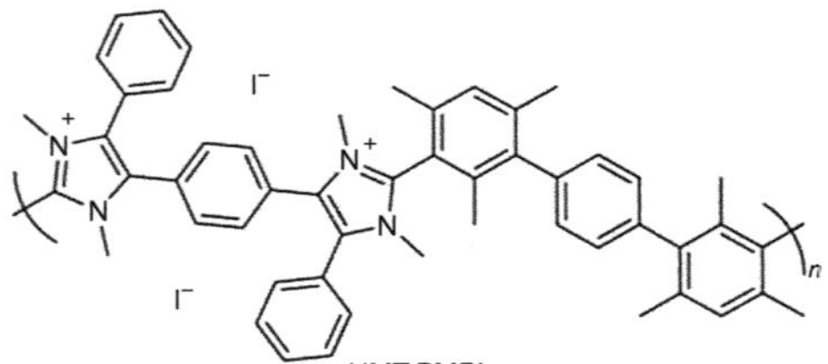	Poly(arylene ether sulfone) block copolymer	1-Methylimidazolium	0.28 at 20 °C	2.61	10 M KOH, 100 °C, 168 h	818 (80 °C)	[44]
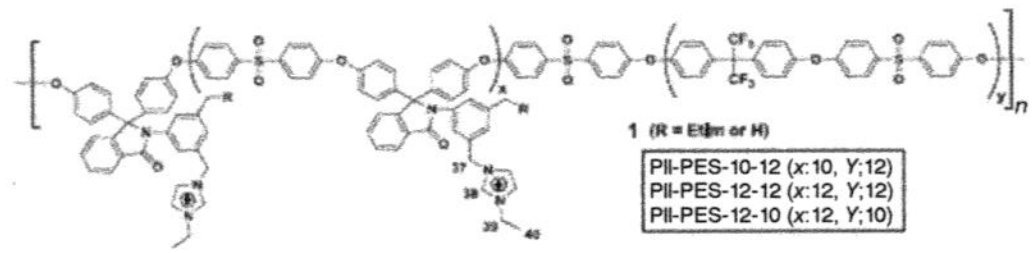	Phenolphthalein-based cardo poly(arylene ether sulfone) block copolymers	Ethyl imidazolium	0.03 at 20 °C, 0.1 at 80 °C	1.29	2 M NaOH, 60 °C, 192 h	—	[45]
PAES-Im-34 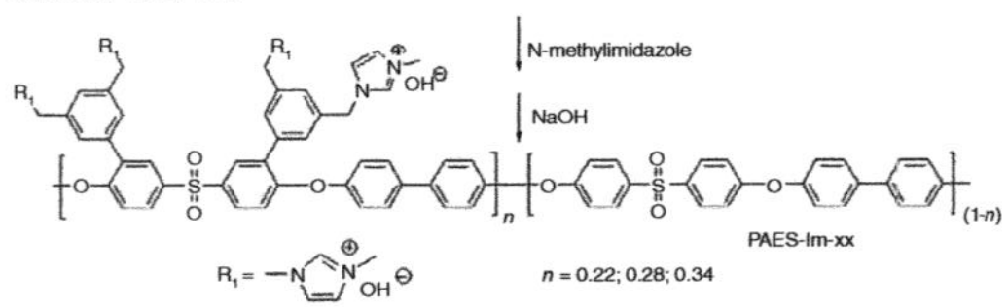	Side-chain-type poly(arylene ether sulfone)s containing imidazolium groups on pendant phenyl rings	*N*-methylimidazolium	0.035 at 80 °C	1.92	4 M NaOH, 80 °C, 336 h	—	[46]

Im-PES-40 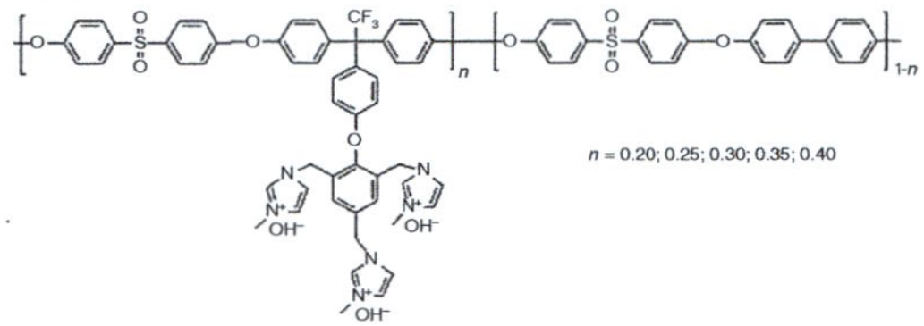	Side-chain-type poly(arylene ether sulfone)s with multiple cation	1-Methyl imidazolium	0.028 at 20 °C 0.070 at 80 °C	1.86	4 M NaOH, 80 °C, 336 h	—	[47]
Cross-Im-PPO-75 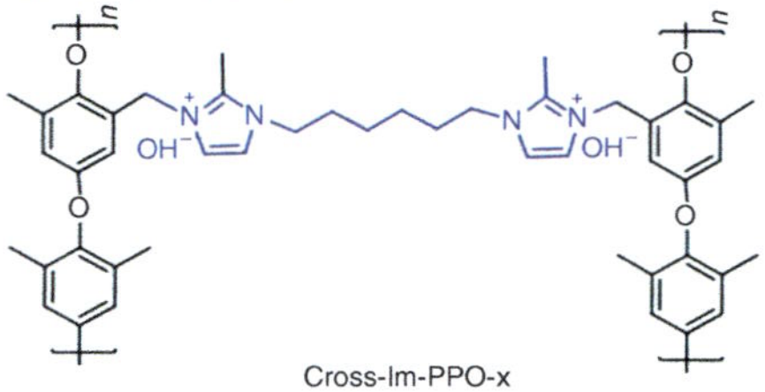	Cross-linked and long-chain pendant bis-imidazolium based AEM	1,6-Bis(2-methylimidazol-1-yl) hexane	0.052 at 90 °C	3.39	2 M KOH, 80 °C, 240 h	212.45 (60 °C)	[48]
PPO-TAM-45 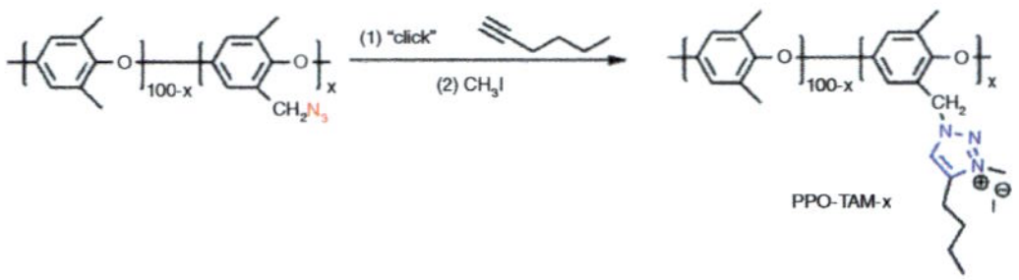	Poly(2,6-dimethyl-1,4-phenylene oxide) containing 1,2,3-triazolium	1,2,3-Triazolium	0.015 at RT	1.05	1 M NaOH, 80 °C, 50 h	—	[49]
X40Y10 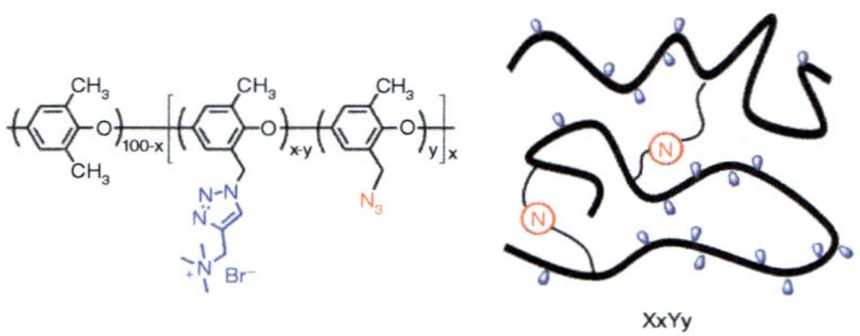	Self-cross-linked side chain type poly(2,6-dimethyl-1,4-phenylene oxide) XxYy copolymer	1,2,3-Triazolium	0.038 at RT	1.68	10 M NaOH, 80 °C, 400 h	—	[50]

of $38.6 \times 10^{-3}\,S\,cm^{-1}$ at 80 °C and a broad basic operation window at room temperature from 1 M to 5 M NaOH for 24 h [8]. Quaternization of bromomethylated fluorinated poly(aryl ether oxadiazole)s with *N*-methyl imidazole as an aminating reagent also demonstrated a better performance than the corresponding QA-based AEMs [10].

It is well known that polyvinylimidazolium cations of grafted polymers are prone to degradation via β-elimination because of the presence of very acidic protons at β-carbons from two IM nitrogen atoms in adjacent vinyl imidazolium units and hydrolytic ring opening reactions. Nucleophilic substitution at the C2 atom position on the imidazolium ring is the other dominant degradation pathway in the presence of hydroxide ions. Various approaches to prevent these types of degradation routes have been published. For instance, Yoshimura et al. have introduced a new graft type of AEM composed of poly(dimethyl-vinyl imidazole-*co*-styrene) copolymer chain grafted onto ETFE films via radiation-induced grafting. Copolymers of IM and styrene units acted as the spacer and reduced the sequence of adjacent vinyl imidazolium units, effectively leading to a significant improvement in stability. Well-connected ionic channels are also confirmed using small-angle X-ray scattering (SAXS) [3].

Both substituents, as well as substitution positions, have a significant effect on the alkaline stability of the AEMs. The performance of AEMs comprised of IM cations with different substitution positions was compared by Si et al. (Figure 9.4). IM cations with butyl substituents at N1, C2, and N3 positions, 1-butyl-2,3-dimethyl imidazolium ([N1-BDMIm]+), 2-butyl-1,3-dimethyl imidazolium ([C2-BDMIm]+), and 3-butyl-1,2-dimethyl imidazolium ([N3-BDMIm]+), respectively, were used to synthesize AEMs in their study. The ^{1}H NMR results of the membranes after exposure to 4 M KOH solution at 80 °C for various intervals showed that the alkaline stability of IM cations with strong electron-donating groups (butyl) at the N1 and N3 positions is better than that at the C2 position. Furthermore, the studies on the effect of different C2 substitutions on the chemical stability of IM cations (1-butyl-2,3-dimethylimidazoliumbromide ([N1-BDMIm][Br]), 2-ethyl-1-butyl-3-methylimidazolium bromide ([C2-EBMIm][Br]), 2,1-dibutyl-3-methylimidazolium bromide ([C2-BBMIm][Br], and 2-hydroxymethyl-1-butyl-3-methylimidazolium bromide ([C2-HMBMIm][Br])) and their corresponding AEMs also revealed that $[C2\text{-}BBMIm]^{+}$-based membrane possessed higher chemical stability than the others [51]. Several researchers have explored the effect of C2 substitution on the alkaline stability of AEMs. Poly(arylene-imidazoliums) synthesized by microwave polycondensation of dialdehyde with bibenzyl and alkylation, sterically protected around C2 position is stable in 10 M KOH solution at 100 °C and attained a peak power density of $818\,mW\,cm^{-2}$ at 80 °C [44]. The effect of C2 substituted (with methyl, isopropyl, or

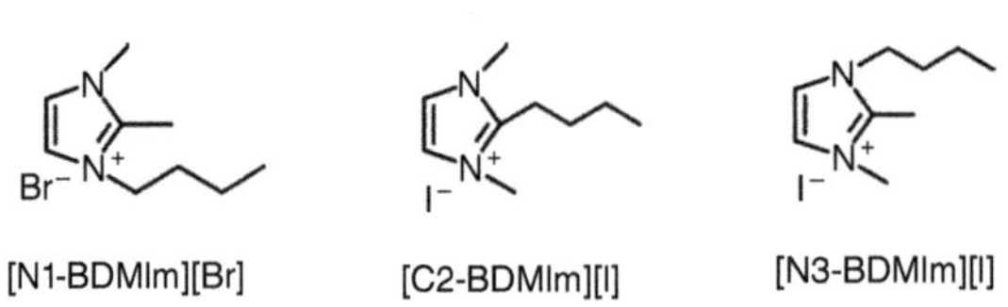

Figure 9.4 Molecular structure of imidazolium cations with different substitution positions [51]. Source: Reproduced with permission from American Chemical Society.

phenyl group) IM salts on alkaline stability of polymer membrane investigated by ^{1}H and ^{13}C NMR spectroscopy suggested that the C2 substitution, owing to the steric hindrance and the σ–π hyper conjugative effect between C—H (σ-bond) of methyl group and π-conjugated imidazole ring could enhance the stability and the order of stability of substituent was found to be methyl > isopropyl > phenyl group [52]. Andrew et al. synthesized a novel alkaline membrane by introducing steric hindrance around the C2-position using methylation of poly[2,2′-(2,2″,4,4″,6,6″-hexamethyl-*p*-terphenyl-3,3″-diyl)-5,5′-bibenzimidazole] (HMT-PMBI). MEA was prepared by applying catalyst ink on both sides of the membrane after fixing to a vacuum table at 120 °C using an ultra-sonicating spray coater to create 5 cm^2 active area with a catalyst loading of 0.4/0.4 mg Pt cm^{-2}. The catalyst ink contained 1 wt% carbon-supported Pt catalyst (46.4 wt% Pt supported on graphitized C) in 3 : 1 (wt/wt) methanol/water ratio. The membrane exhibited a high power density of around 70 mW cm^{-2} for fuel cell after an initial operation of 119 h at 60 °C. No apparent chemical degradation after being subjected to 6 M NaOH at room temperature for seven days represented the exceptional *ex situ* stability of the membranes [53]. Yang et al. evaluated the impact of C2 and N3 substitution with different substituent groups (1-methylimidazole, 1-phenyl imidazole, 1,2-dimethylimidazole, 1-butyl-2-methyl-imidazole, 1-decyl-2-methylimidazole, 1-benzyl-2-methyl-imidazole) on the various properties of poly(2,6-dimethyl-1,4-phenylene oxide) membranes. They observed improved stability (1 M KOH at 60 °C for 180 h) and a maximum conductivity of 42.5×10^{-3} S cm^{-1} at 80 °C for membranes based on C2-methyl as well as N3-butyl substituted IM pendants [54]. The effect of C2,4,5 substituents (including hydrogen, methyl, and phenyl groups) on chemical stability of polyimidazolium/PVA-based AEMs suggested the highest stability for 1,3-dihexyl-4,5-dimethyl-2-phenyl-imidazolium cation than 1,3-dihexyl-imidazolium, 1,3-dihexyl-2-methyl-imidazolium, 1,3-dihexyl-2,4,5-trimethyl-imidazolium, and 1,3-dihexyl-2,4,5-triphenyl-imidazolium cations [55]. To understand the influences of C4/C5-position substitutes on the alkaline stability of IM cations, Xiaofeng et al. embedded methylimidazolium (MeIM) monomer to prepare ether-free branched poly (methyl-imidazolium triphenylbenzene) (P-MeIM-tPb) by superacid-catalytic method. Excellent ion exchange capacity with tensile properties and dimensional stability against swelling were exhibited by the synthesized AEMs. The conductivity was preserved by 96% after soaking for 1000 h in 5 NaOH at 80 °C. Alkaline-resisted membrane with branched structure expanded the operating temperature range of fuel cell to 90 °C and maintained long-term durability [56]. In a recent report from Hugar et al., the effect of various substituents at different locations of IM on the performance of AEMs was studied. They reported slightly higher stability for IM with C4,5 methyl groups than the stability relative to phenyl groups. It was inferred that the substitution at C2 position and 2,6-dimethyl phenyl substituents are more effective, and the use of alkyl substituents on the nitrogen, especially *n*-butyl groups, could prevent the degradation better than benzyl or methyl groups. More than 99% of cations that remain after 30 days of exposure in 5 M KOH/CD_3OH at 80 °C displayed the highest stability reported to that date of their AEMs [35].

Focusing on several challenges associated with incomplete postfunctionalization, especially when grafting large sterically hindered IM onto polymer precursors, Boxin et al. developed a novel super-acid-catalyzed polymerization method. A novel IM-ionized monomer PhIM [PF6] with its N1/N3 position modified by propylbenzene group was copolymerized with 1,1,1-trifluoroacetone under superacid-catalyzed conditions to form functionalized poly(arylene-imidazolium) (poly-PhIM [OH]) polymers without further functionalization. Both random as well as block copolymers were synthesized by using biphenyl as a comonomer and varying the feeding sequence. These polymer-based AEMs exhibited excellent resistance to alkaline solutions. No evident degradation from ^{1}H NMR after treatment with 10 M NaOH at 80 °C for 100 days [57].

Acid–base blending benefiting from strong electrostatic interactions can afford more compact polymer chain packing and thereby stability for the membranes. Blending basic imidazolium-functionalized poly(ether ether ketone) (ImPEEK) and acid-sulfonated PEEKs (SPEEK) via an *in situ* Menshutkin/cross-linking method showed an ion exchange capacity (IEC) up to 3.15 mmol g^{-1} and consequently conductivity up to 31.59×10^{-3} S cm^{-1} at 30 °C [5]. Blends of bromomethylated poly(2,6-dimethyl-1,4-phenylene oxide) (PPO) and polybenzimidazole (PBI) with sulfonic components functionalized with 1-methylimidazole (MIm) and 1-ethyl-2-methylimidazole (EMIm) were also reported to have good chemical stability [21].

Side chain-based AEMs have shown a huge potential for improving alkaline stability, as reported by Wang et al. (PAES)s containing IM groups on pendant phenyl rings had a long-term stability of 336 h in 4 M NaOH at 80 °C, which is higher than the corresponding QA-based AEMs. No apparent deterioration in IEC and conductivity values after strong base treatment was observed. These membranes also showed excellent dimensional stability, with a swelling ratio of only 4.4–7.9% at 80 °C [46]. PAES with a triimidazolium cation in a pendant phenoxy phenyl spacer also displayed excellent dimensional stability, hydroxide conductivity, as well as improved alkali stability. The retention rate of IEC and conductivity were 78.5–83.3% and 75.5–80.6%, respectively, after being treated in strongly alkaline conditions of 4 M NaOH at 80 °C for 336 h. The viability of phenoxy-phenyl spacer with multiple IM cations in this work suggests a facile route to prepare highly conductive and alkali-stable AEM materials [47]. Increasing the chain length of the pendant alkyl IM group can facilitate the hydrophilic–hydrophobic phase separation and thereby ionic cluster formation in membranes. Rao et al. have reported the effect of alkyl chain length on various properties of IM-functionalized PAES block copolymers by changing the chain length from C_2 to C_{16}. Long alkyl chains with C_6 and greater in the comb-shaped system resulted in high ionic conductivity and alkaline stability up to 500 h in 2 M NaOH at 60 °C [19].

Increasing the concentration of basic functional groups, the general method to improve conductivity, also resulted in degradation of dimensional stability by excess swelling. Therefore, cross-linking using imidazole compounds has been used as a facile strategy to improve the performances of AEMs. Functionalization with methyl-imidazole and 1,4-bis(imidazolyl)butane acted both as cross-linkers as well as conductive sites in bromomethylated PPO that resulted in high ion exchange capacity without compromising on mechanical robustness [20]. 1,6-bis(2-methyl imidazol-1-yl)hexane and bis-imidazolium salts grafted onto PPO as a long, flexible cross-linker and long-chain pendant, respectively, also showed an improvement in stability as well as mechanical properties of the AEM. Well-developed hydrophilic/hydrophobic phase-separated morphology formed after bis-imidazolium cross-linking facilitated hydroxyl ion mobility as well as conductivity to the membrane and showed a maximum power density of 212.45 $mW\,cm^{-2}$ indicating their potential suitability for fuel cell applications [48]. The flexible side chains with a cross-linked topological structure designed from norbornene copolymer backbone having epoxy group-functionalized side chain and IM-terminated flexible side chain resulted in well-defined microphase separation and a hydroxide ion conductivity of $90 \times 10^{-3}\,S\,cm^{-1}$ at 80 °C. It was noted that the conductivity only decreased by 25% after 10 days of exposure in 2 M NaOH at 60 °C and resulted in a peak power density of 118.4 $mW\,cm^{-2}$ for H_2/O_2 single cell [58]. Cross-linked vinyl polymer-based AEM using 1,3-diallyl-2-methyl imidazolium bromide ([DAMIm][Br]) as cross-linker exhibited a hydroxide ion conductivity above $10^{-2}\,S\,cm^{-1}$ and chemical stability of in 1 M KOH at 60 °C as a result of the resonant effect of conjugated imidazole ring and hyper conjugative effect between C—H (σ bond) of methyl group as well as π-conjugated imidazole ring also suggested a feasible approach for synthesis and practical application of AEMs for fuel cells [59]. Cross-linked membrane synthesized by copolycondensation and IM functionalization between phenolphthalein-based poly(arylene ether ketone) and 1-vinyl imidazole demonstrated a high ionic conductivity of 0.0836 $S\,cm^{-1}$ and excellent dimensional as well as chemical stability. After 400 h of immersion in 1 M NaOH at 40 °C, the conductivity for cross-linked membranes was significantly maintained (82.4%) [60]. Generally, cross-linked membranes displayed a reduction in water absorption and hence conductivity due to their more compact polymer architecture. To study the effect of cross-linker structure on the performance of AEM, Firouz Tadavani and his coworkers prepared a series of membranes based on poly(benzimidazolium-imide) containing triazolium groups using "click chemistry." A hydrophilic aromatic diazide and a hydrophobic aliphatic diazide were used as cross-linkers and found better long-term chemical stability than the uncross-linked membranes. Furthermore, the hydrophilic cross-linker contributed to better water uptake and conductivity, but the chemical stability increased in presence of hydrophobic one [61]. A base-stable and mechanically strong AEM was prepared by Wei et al. by subjecting IM-fused cyclooctane monomer to ring-opening metathesis polymerization conditions. The resultant AEM showed a conductivity of $59 \times 10^{-3}\,S\,cm^{-1}$ at 50 °C and stability in 1 M KOH solution at 80 °C for 30 days (minor conductivity drop, $<3 \times 10^{-3}\,S\,cm^{-1}$) [62]. Tao et al. consecutively stated that, simple and easily available IM structural isomer, pyrazolium cation showed excellent alkaline stability in 5 M NaOH at 80 °C for at least 240 h, which is substantially more stable than IM. The degradation energy barriers calculated by density functional theory showed AEM based on (arylene alkylene) architecture modified with pyrazolium retained 80% of its initial conductivity after 1000 h of treatment in 1 M NaOH at 80 °C. Furthermore, high ion conductivity of

120×10^{-3} S cm^{-1} at 80 °C and a peak power density of 502 mW cm^{-2} after assembled into H_2–O_2 fuel proving its good potential in AEMFC application [63].

The effect of mono- or di-cationic functional groups on the physicochemical, electrochemical, as well as alkaline stability of polyolefin-based AEMs using QA, pyrrolidinium, and IM cations was studied via experimental as well as theoretical methods. 3-Butyl-2-methyl-1-(4-vinylbenzyl) imidazolium chloride ([BMVBIm][Cl]), 3-butyl-2-methyl-1-(6-(2-methyl-1-(4-vinylbenzyl)-imidazolium-3-yl)hexyl) imidazolium chloride ([BMMVBImHIm][Cl]), *N*,*N*-dimethyl-*N*-(4-vinylbenzyl) butylaminium chloride ([DMVBDA][Cl]), *N*-butyl-*N*′-(4-vinylbenzyl)-*N*, *N*, *N*′, N′-tetramethylhexane-1, 6 diaminium chloride ([BVBTMHDA][Cl]), 1-butyl-1-(4-vinylbenzyl)pyrrolidinium chloride ([BVBPy][Cl]), 1-butyl-1-(6-(1-(4-vinylbenzyl)pyrrolidinium-1-yl)hexyl) pyrrolidinium chloride ([BVBPyHPy][Cl]) were grafted on to the polymer by one-step general photo-initiated radical polymerization method. Experimental data revealed that, the di-cations easily form hydrated hydroxide ion cluster than mono-cations, which promoted the water uptake, swelling ratio, and ionic conductivity properties of the AEMs. The more uniform electron distribution in di-cation reduced the probability of collision between the cation and hydroxyl ions and imparted excellent alkaline stability to the system [64]. Ahmed et al. compared ethylene–tetrafuluoroethylene copolymers having three vinylimidazolium type (*N*-vinyl imidazolium (NVIm), 2-methyl-NV imidazolium (2MVIm), 1,2,3-trimethyl-4-vinyl imidazolium (4VIm)) and a styrylimidazolium-type (2-styryl-*N*, *N*-dipropyl imidazolium (StIm)) hydroxyl ion conducting units with the standard vinylbenzyltrimethylammonium (BTMA)-based AEM. Their study concluded that the sterically hindered imidazolium (StIm-98%) AEM exhibited a maximum power density of 710 mW cm^{-2} at a current density of 1389 mA cm^{-2}, which is one of the highest values for IM-based AEM so far. Moreover, 56% initial cell voltage with ADR of 0.42 m Vh^{-1} maintained up to 670 h showed its best long-term *in situ* durability. FTIR after post durability showed stable IM group without any change, revealing the excellent stability of StIm [65].

Using cargo-type molecules with a pendant functional group, such as fluorine and tetraphenyl phthalazine, as polymer side chains is a very promising way to develop ion-conducting membrane materials with excellent physical and chemical stabilities. A series of phenolphthalein-based cardo PES block copolymers containing pendant imidazolium group (PI-PESs) with hydroxide conductivity of 0.1 S cm^{-1} at 80 °C, together with good physical and chemical stability under basic conditions, are reported as promising AEM material for direct methanol alkaline fuel cells [45]. Regional aggregation of functional groups without affecting dimensional stability through the surface functionalization of nanocomposites with a large number of functional groups is also an effective way to achieve high conductivity in the AEMs. Imidazolium-functionalized polysulfone/silica (Psf-/Im/SiO_2-Im) and polysulfone/ImGO (graphene oxide) composites are examples of AEMs with enhanced hydroxide ion conductivity without the challenge of excessive swelling. The regional aggregation of functional groups in composites promoted the formation of continuous OH^- ion transport channels [11, 66].

The cold start failure of AEMFCs due to the AEM fracture at sub-zero temperature is a subject to be considered when vehicles are exposed to a colder climate for a long time. According to Ning et al., the stability and conductivity of quaternized poly(phenylene) oxide (QPPO) in sub-zero temperature can be improved using imidazolium-modified metal organic frameworks (ImMOFs). *N*-methylimidazole-modified MIL-101(Cr) MOF (ImMOF)/QPPO (7% filler) composite exhibited hydroxide conductivities of 0.73×10^{-3} S cm^{-1} at −25 °C and 42.1×10^{-3} S cm^{-1} at 80 °C. Furthermore, a stable hydroxide ion conductivity even after exposure in 2 M KOH solution at RT for 768 h and at 60 °C for 433 h was obtained. After five cycles in the temperature range of −25–60 °C with long-term exposure in 2 M KOH, a conductivity of 0.87×10^{-3} S cm^{-1} at −25 °C and 14.9×10^{-3} S cm^{-1} at 30 °C achieved revealed its fine alkaline stability [67].

9.3 AEM Based on Triazolium Cations

Discovering new cationic groups to boost alkaline stability and conductivity led researchers to investigate the potential of triazolium compounds. Triazolium is a heterocyclic cation having a structure similar to that of IM but consisting of two carbon atoms and three nitrogen atoms in a five-membered ring. Depending on the arrangement of the three nitrogen atoms, triazoles have four isomers, which are conventionally divided into two structural isomers – 1,2,3-triazoles and 1,2,4-triazoles [68]. 1,2,3-Triazoles are typically synthesized by Cu(I)-catalyzed azide-alkyne cycloaddition (CuAAC), also known as "click chemistry." 1,2,3-Triazoles can either lose a proton (act as a weak acid) or accept a proton via the nitrogen lone pair of electrons (act as a weak Brønsted base) and promote ion transport through inter- and intramolecular ionic transfer. Alkylation of 1,2,3-triazoles results in 1,2,3-triazolium cations with high thermal stability, low flammability, as well as tunable solubility. Triazole moieties have the advantage of facilitating the formation of continuous hydrogen bond networks. Triazoles represent a promising functional group for AEMs but have not been thoroughly studied so far [32, 49]. Based on the versatility of the click reaction, various 1,2,3-triazolium-based polymers have been prepared via direct polymerization of 1,2,3-triazolium-containing monomers or post quaternization of 1,2,3-triazolium units in the polymers. Moreover, a variety of polymers with 1,2,3-triazolium as side chains have also been developed for AEM applications [32, 49, 69, 70]. Triazole-containing AEMs with "side chain-type" architecture prepared using PPOs by Cu-catalyzed "click chemistry" exhibited excellent alkaline stability and retained over 65% of hydroxide conductivity in 5 M or 10 M NaOH at 80 °C for 400 h. The 1,2,3-triazole helped to induce the continuous hydrogen bonded pathways to sustain efficient ionic conduction [50] (Table 9.1). Li et al. also reported similar kind of AEMs with hydroxide ion conductivity of $27.8–62 \times 10^{-3}$ S cm^{-1} at 20 °C in water, which is several times greater than that of a typical PPO-based AEM derived from trimethylamine (5×10^{-3} S cm^{-1}). The peak power density achieved for H_2/O_2 fuel cell at 50 °C was about 188.7 mW cm^{-2} [71].

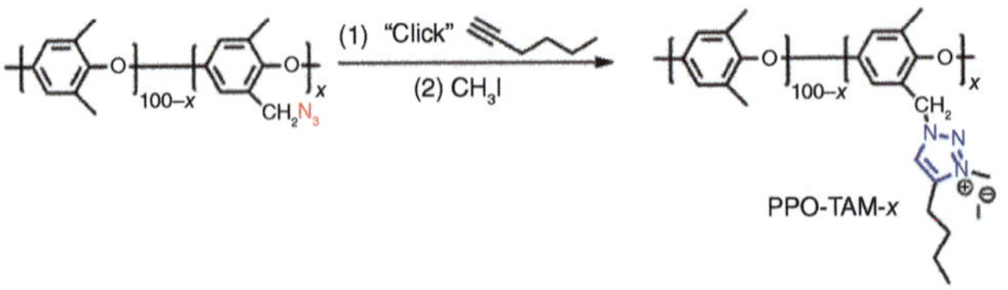

Figure 9.5 Schematic representation of anion-conductive copolymer based on 1,2,3-triazolium and poly(2,6-dimethyl-1,4-phenylene oxide) (PPO-TAM-*x*) [49]. Source: Reproduced with permission from Elsevier.

Nakabayashi et al. proposed a reversible addition-fragmentation chain transfer (RAFT) polymerization method to produce pyrrolidone/1,2,4-triazolium-based block copolymers with phase-separated morphology and good conductivity [33]. AEM based on copolymer of 1,2,3-triazolium and PPO (Figure 9.5) prepared by click chemistry has shown considerably better alkaline stability than that of the corresponding IM-based AEM in 1 M NaOH at 80 °C, but still requires further improvement in long-term stability for alkaline fuel cell application. Their study on degradation mechanism also suggested that the alkaline stability of 1,2,3-triazolium cation could be improved by the introduction of substituents at the C4 and C5 positions, as well as benzylic methylene [49].

1,2,3-triazolium-based polyionic liquid (PIL) obtained by CuAAC step growth polymerization has been actively applied as the new ion-conductive membrane material in the field of fuel cells. Recently, Jang et al. introduced triazolium and QA cations via CuAAC click reaction into the PAES polymer backbones and used them as AEMs. 1,6-diiodohexane used as the cross-linking agent also helped to increase the number of triazolium cation and resulted in improved hydroxide ion conductivity as well as alkaline stability to the membrane owing to the dense and compact structure. Furthermore, the formation of additional triazolium cations from cross-linking also led to balance the trade-off between conductivity and dimensional stability of AEMs [13].

9.4 Summary and Future Perspectives

Currently, AEMFCs have received ever-increasing attention as a clean and highly efficient energy converter, and the focus on exploring AEM, the key component of the fuel cell with high ionic conductivity and excellent alkaline stability, is a great challenge for the practical use and commercialization of AEMFCs. The micromorphology and chemical structure of polymer backbone and cations have a great impact on the properties of AEMs. Among the various approaches attempted in the past few decades, the development of AEMs based on the functional groups having resonance structures such as IM and triazolium has shown promising results in ionic conductivity as well as chemical stability. The ease of synthesis via "click chemistry" and the benefit of facilitating continuous hydrogen bonding networks formation promote ion transport through inter- and intramolecular mechanisms make triazolium cations a promising functional group for AEM, although this moiety has not been studied much thoroughly so far. This chapter provides an up-to-date summary of the latest developments on the IM, and triazolium-based AEMs. The electron-delocalized π-conjugated structure and steric hindrance make IM-based cations relatively stable, but the stability at high pH and temperature remains an open topic. Various polymer architectures including block, side chain type, graft/comb-shaped, clustered with tethered IM and triazolium cationic groups have been synthesized for the preparation of AEMs with increased stability as well as conductivity. Various substituents at different substitution locations of IM cations have a significant role in alkaline stability and other performance properties. The formation of hydrophilic/hydrophobic phase-separated morphology in AEMs has benefited the formation of ion-conducting channels as well as restricted dimensional swelling and improved mechanical and alkaline stability of the membranes.

We believe that the past investigations have laid a solid foundation for a basic understanding of how the various strategies affected the performances of IM and triazolium-based AEMs. Future research should focus on the development of highly alkaline-stable AEMs to improve long-term cell performance without any deterioration. The minimum commercial demand for long-term stability is more than 5000 h at 80 °C, which has not been realized yet. The AEMs are not the only components of AEMFCs that are susceptible to degradation; the catalysts and the supporting layers can also deteriorate. Therefore, optimization of all the components and in-depth studies on the properties of AEMs under practical working conditions are essential for the further development of high-performance AEMFCs.

Acknowledgments

The authors acknowledge the financial assistance received from the National Research Foundation of Korea (NRF), Ministry of Science and ICT (NRF – 2021 M1A2A2038115). This research was also supported by basic science research program funded by the Ministry of Education (NRF – 2020R1A6A1A03038697).

References

1 Sun, Z., Lin, B., and Yan, F. (2018). Anion-exchange membranes for alkaline fuel-cell applications: the effects of cations. *ChemSusChem* 11 (1): 58–70.

2 Fang, J., Lyu, M., Wang, X. et al. (2015). Synthesis and performance of novel anion exchange membranes based on imidazolium ionic liquids for alkaline fuel cell applications. *J Power Sources* 284: 517–523.

3 Yoshimura, K., Zhao, Y., Hasegawa, S. et al. (2017). Imidazolium-based anion exchange membranes for alkaline anion fuel cells: (2) elucidation of the ionic structure and its impact on conducting properties. *Soft Matter* 13 (45): 8463–8473.

4 Nguyen, M.D.T., Dang, H.S., and Kim, D. (2015). Proton exchange membranes based on sulfonated poly(arylene ether ketone) containing triazole group for enhanced proton conductivity. *J Membr Sci* 496: 13–20.
5 Li, Z., Jiang, Z., Tian, H. et al. (2015). Preparing alkaline anion exchange membrane with enhanced hydroxide conductivity via blending imidazolium-functionalized and sulfonated poly(ether ether ketone). *J Power Sources* 288: 384–392.
6 Yang, S., Chu, X., Xue, B. et al. (2021). Polyether ether ketone-based anion exchange membranes with bis-imidazolium cations for all-vanadium redox flow batteries. *ACS Appl Energy Mater* 4: 6787–6796.
7 Hao, J., Li, X., Yu, S. et al. (2015). Development of proton-conducting membrane based on incorporating a proton conductor 1,2,4-triazolium methanesulfonate into the Nafion membrane. *J Energy Chem* 24 (2): 199–206.
8 Song, Y., Liu, C., Zhao, J., and Luo, J. (2016). Imidazolium-functionalized anion exchange polymer containing fluorine group for fuel cell application. *Int J Hydrog Energy* 41 (24): 10446–10457.
9 Dekel, D.R. (2016). Alkaline anion exchange membrane fuel cells (AEM-FC) status. Alkaline Membr. Fuel Cell Work, 3–31.
10 Hu, Q., Shang, Y., Wang, Y. et al. (2012). Preparation and characterization of fluorinated poly(aryl ether oxadiazole)s anion exchange membranes based on imidazolium salts. *Int J Hydrog Energy* 37 (17): 12659–12665.
11 Li, J., Yan, X., Zhang, Y. et al. (2016). Enhanced hydroxide conductivity of imidazolium functionalized polysulfone anion exchange membrane by doping imidazolium surface-functionalized nanocomposites. *RSC Adv* 6 (63): 58380–58386.
12 Guo, M., Fang, J., Xu, H. et al. (2010). Synthesis and characterization of novel anion exchange membranes based on imidazolium-type ionic liquid for alkaline fuel cells. *J Membr Sci* 362 (1–2): 97–104.
13 Jang, J., Ahn, M.K., Lee, S.B. et al. (2021). Conductive and stable crosslinked anion exchange membranes based on poly(arylene ether sulfone). *Macromol Res* 29 (2): 157–163.
14 Cui, Y., Du, J., Shang, Y. et al. (2012). Novel anion exchange membranes from poly(aryl ether)s with quaternary guanidinium groups. *Adv Mater Res* 560–561: 864–868.
15 Sajjad, S.D., Liu, D., Wei, Z. et al. (2015). Guanidinium based blend anion exchange membranes for direct methanol alkaline fuel cells (DMAFCs). *J Power Sources* 300: 95–103.
16 Chen, Y., Tao, Y., Wang, J. et al. (2017). Comb-shaped guanidinium functionalized poly(ether sulfone)s for anion exchange membranes: effects of the spacer types and lengths. *Polym Chem* 55 (8): 1313–1321.
17 Boxin, X., Qian, W., and JIfu Z., Shenghai L., Suobo Z. (2020). Bi-guanidinium-based crosslinked anion exchange membranes: synthesis, characterization and properties. *J Membr Sci* 601: 117923.
18 Ningxin, Z., Ping, L., Xuang, L., and Shaokun, T. (2022). Anion exchange composite membrane based on ionic liquid-grafted covalent organic framework for fuel cells. *Int J Hydrog Energy* 47: 29481–29494.

19 Rao, A.H.N., Nam, S., and Kim, T.H. (2015). Comb-shaped alkyl imidazolium-functionalized poly(arylene ether sulfone)s as high performance anion-exchange membranes. *J Mater Chem A* 3 (16): 8571–8580.

20 Dong, J., Li, H., Ren, X. et al. (2019). Anion exchange membranes of bis-imidazolium cation crosslinked poly(2,6-dimethyl-1,4-phenylene oxide) with enhanced alkaline stability. *Int J Hydrog Energy* 44 (39): 22137–22145.

21 Morandi, C.G., Peach, R., Krieg, H.M., and Kerres, J. (2015). Novel imidazolium-functionalized anion-exchange polymer PBI blend membranes. *J Membr Sci* 476: 256–263.

22 Zeng, L., Zhao, T.S., Wei, L. et al. (2016). Highly stable pyridinium-functionalized cross-linked anion exchange membranes for all vanadium redox flow batteries. *J Power Sources* 331: 452–461.

23 Kang, M.S., Choi, Y.J., and Moon, S.H. (2003). Characterization of anion-exchange membranes containing pyridinium groups. *AIChE J* 49 (12): 3213–3220.

24 Voge, A., Deimede, V., and Kallitsis, J.K. (2014). Synthesis and properties of alkaline stable pyridinium containing anion exchange membranes. *RSC Adv* 4 (85): 45040–45049.

25 Zhang, B., Gu, S., Wang, J. et al. (2012). Tertiary sulfonium as a cationic functional group for hydroxide exchange membranes. *RSC Adv* 2 (33): 12683–12685.

26 Wang, Z., Parrondo, J., Sankarasubramanian, S. et al. (2020). Alkaline stability of pure aliphatic-based anion exchange membranes containing cycloaliphatic quaternary ammonium cations. *J Electrochem Soc* 167 (12): 124504.

27 Zhang, Y., Chen, W., Yan, X. et al. (2020). Ether spaced N-spirocyclic quaternary ammonium functionalized crosslinked polysulfone for high alkaline stable anion exchange membranes. *J Membr Sci* 598 (November): 117650.

28 Pham, T.H., Olsson, J.S., and Jannasch, P. (2017). N-spirocyclic quaternary ammonium ionenes for anion-exchange membranes. *J Am Chem Soc* 139 (8): 2888–2891.

29 Pham, T.H., Olsson, J.S., and Jannasch, P. (2018). Poly(arylene alkylene)s with pendant N-spirocyclic quaternary ammonium cations for anion exchange membranes. *J Mater Chem A* 6 (34): 16537–16547.

30 Jiang, L., Lin, X., Ran, J. et al. (2012). Synthesis and properties of quaternary phosphonium-based anion exchange membrane for fuel cells. *Chin J Chem* 30 (9): 2241–2246.

31 Barnes, A.M., Du, Y., Zhang, W. et al. (2019). Phosphonium-containing block copolymer anion exchange membranes: effect of quaternization level on bulk and surface morphologies at hydrated and dehydrated states. *Macromolecules* 52 (16): 6097–6106.

32 Obadia, M.M., Mudraboyina, B.P., Serghei, A. et al. (2014). Enhancing properties of anionic poly(ionic liquid)s with 1,2,3-triazolium counter cations. *ACS Macro Lett* 3 (7): 658–662.

33 Nakabayashi, K., Umeda, A., Sato, Y., and Mori, H. (2016). Synthesis of 1,2,4-triazolium salt-based polymers and block copolymers by RAFT polymerization: ion conductivity and assembled structures. *Polymer* 96: 81–93.

34 Mudraboyina, B.P., Obadia, M.M., Allaoua, I. et al. (2014). 1,2,3-Triazolium-based poly(ionic liquid)s obtained through click chemistry polyaddition. *Macromol Chem Phys* 215 (22): 2229–2236.

35 Hugar, K.M., Kostalik, H.A., and Coates, G.W. (2015). Imidazolium cations with exceptional alkaline stability: a systematic study of structure-stability relationships. *J Am Chem Soc* 137 (27): 8730–8737.

36 Si, Z., Sun, Z., Gu, F. et al. (2014). Alkaline stable imidazolium-based ionomers containing poly(arylene ether sulfone) side chains for alkaline anion exchange membranes. *J Mater Chem A* 2 (12): 4413–4421.

37 Ran, J., Wu, L., Varcoe, J.R. et al. (2012). Development of imidazolium-type alkaline anion exchange membranes for fuel cell application. *J Membr Sci* 415–416: 242–249.

38 Zhu, X., Wang, B., and Wang, H. (2010). Effects of [Bmim]OH on structure and conductive properties of alkaline PVA/[Bmim]OH membranes. *Polym Bull* 65 (7): 719–730.

39 Wang, D., Wang, Y., Wang, J., and Wang, L. (2019). Synthesized geminal-imidazolium-type ionic liquids applying for PVA-FP/[DimL][OH] anion exchange membranes for fuel cells. *Polymer* 170: 31–42.

40 Qiu, B., Lin, B., Qiu, L., and Yan, F. (2012). Alkaline imidazolium- and quaternary ammonium-functionalized anion exchange membranes for alkaline fuel cell applications. *J Mater Chem* 22 (3): 1040–1045.

41 Yan, X., He, G., Gu, S. et al. (2012). Imidazolium-functionalized polysulfone hydroxide exchange membranes for potential applications in alkaline membrane direct alcohol fuel cells. *Int J Hydrog Energy* 37 (6): 5216–5224.

42 Lu, W., Shao, Z.G., Zhang, G. et al. (2013). Preparation and characterization of imidazolium-functionalized poly (ether sulfone) as anion exchange membrane and ionomer for fuel cell application. *Int J Hydrog Energy* 38 (22): 9285–9296.

43 Rao, A.H.N., Thankamony, R.L., Kim, H.J. et al. (2013). Imidazolium-functionalized poly(arylene ether sulfone) block copolymer as an anion exchange membrane for alkaline fuel cell. *Polymer* 54 (1): 111–119.

44 Jiantao, F., Andrew, G.W., Benjamin, B. et al. (2017). Cationic polyelectrolytes, stable in 10 M KOHaq at 100 °C. *ACS Macro Lett* 6: 1089–1093.

45 Rao, A.H.N., Kim, H.J., Nam, S., and Kim, T.H. (2013). Cardo poly(arylene ether sulfone) block copolymers with pendant imidazolium side chains as novel anion exchange membranes for direct methanol alkaline fuel cell. *Polymer* 54 (26): 6918–6928.

46 Wang, C., Xu, C., Shen, B. et al. (2016). Stable poly(arylene ether sulfone)s anion exchange membranes containing imidazolium cations on pendant phenyl rings. *Electrochim Acta* 190: 1057–1065.

47 Wang, C., Zhou, Y., Xu, C. et al. (2018). Synthesis and properties of new side-chain-type poly(arylene ether sulfone)s containing tri-imidazole cations as anion-exchange membranes. *Int J Hydrog Energy* 43 (45): 20739–20749.

48 Lin, B., Xu, F., Chu, F. et al. (2019). Bis-imidazolium based poly(phenylene oxide) anion exchange membranes for fuel cells: the effect of cross-linking. *J Mater Chem A* 7 (21): 13275–13283.

49 Liu, L., He, S., Zhang, S. et al. (2016). 1,2,3-Triazolium-based poly(2,6-dimethyl phenylene oxide) copolymers as anion exchange membranes. *ACS Appl Mater Interfaces* 8 (7): 4651–4660.

50 Shuqing, H., Lei, L., Xuchao, W. et al. (2016). Azide-assisted self-crosslinking of highly ion conductive anion exchange membranes. *J Membr Sci* 509: 48–56.

51 Si, Z., Qiu, L., Dong, H. et al. (2014). Effects of substituents and substitution positions on alkaline stability of imidazolium cations and their corresponding anion-exchange membranes. *ACS Appl Mater Interfaces* 6 (6): 4346–4355.

52 Lin, B., Dong, H., Li, Y. et al. (2013). Alkaline stable C2-substituted imidazolium-based anion-exchange membranes. *Chem Mater* 25 (9): 1858–1867.

53 Andrew, G.W., Jiantao, F., Benjamin, B. et al. (2016). Hexamethyl-*p*-terphenyl poly(benzimidazolium): a universal hydroxide-conducting polymer for energy conversion devices. *Energy Environ Sci* 9: 2130–2142.

54 Yang, J., Liu, C., Hao, Y. et al. (2016). Preparation and investigation of various imidazolium-functionalized poly(2,6-dimethyl-1,4-phenylene oxide) anion exchange membranes. *Electrochim Acta* 207: 112–119.

55 Pan, J., Sun, Z., Zhu, H. et al. (2020). Synthesis and characterization of main-chain type polyimidazolium-based alkaline anion exchange membranes. *J Membr Sci* 610: 118283.

56 Xiaofeng, L., Zimo, W., Yaohan, C. et al. (2023). Imidazolium-based AEMs with high dimensional and alkaline-resistance stabilities for extended temperature range of alkaline fuel cells. *J Membr Sci* 670: 121352.

57 Boxin, X., Weidong, C., Shengyang, Z. et al. (2021). Facile preparation of highly alkaline stable poly(arylene-imidazolium) anion exchange membranes through an ionized monomer strategy. *Macromolecules* 54: 2202–2212.

58 He, X., Cheng, C., Huang, S. et al. (2020). Alkaline anion exchange membranes with imidazolium-terminated flexible side-chain cross-linked topological structure based on ROMP-type norbornene copolymers. *Polymer* 195: 122412.

59 Lin, B., Chu, F., Ren, Y. et al. (2014). Alkaline stable C2-substituted imidazolium-based cross-linked anion exchange membranes for alkaline fuel cell applications. *J Power Sources* 266: 186–192.

60 Du, X., Wang, Z., Liu, W. et al. (2018). Imidazolium-functionalized poly (arylene ether ketone) cross-linked anion exchange membranes. *J Membr Sci* 566: 205–212.

61 Firouz Tadavani, K., Abdolmaleki, A., Molavian, M.R., and Zhiani, M. (2021). New strategy based on click reaction for preparation of cross-linked poly(benzimidazolium-imide) as an anion-exchange membrane with improved alkaline stability. *Ind Eng Chem Res* 60 (19): 7097–7110.

62 Wei, Y., Kristina, M.H., and Geoffrey, W.C. (2018). Synthesis of alkaline anion exchange membranes with chemically stable imidazolium cations: unexpected cross-linked macrocycles from ring-fused ROMP monomers. *Macromolecules* 51: 3212–3218.

63 Tao, J., Chenxi, W., Tao, W. et al. (2022). Imidazolium structural isomer pyrazolium: a better alkali-stable anion conductor for anion exchange membranes. *J Membr Sci* 660: 120843.

64 Ji, P., Hairong, Z., Huixing, C. et al. (2021). Flexible cationic side chains for enhancing the hydroxide ion conductivity of olefinic-type copolymer-based anion exchange membranes: an experimental and theoretical study. *J Membr Sci* 620: 118794.

65 Ahmed, M.A.M., Kimio, Y., and Yasumari, M. (2021). Alkaline fuel cells consisting of imidazolium-based graft-type anion exchange membranes: optimization of fuel cell conditions to achieve high performance and durability. *J Membr Sci* 620: 118844.

66 Mao, X., Li, Z., He, G. et al. (2019). Enhancing hydroxide conductivity of anion exchange membrane via incorporating densely imidazolium functionalized graphene oxide. *Solid State Ionics* 333: 83–92.

67 Ning, W., Tingting, Z., Ke, L. et al. (2023). Enhancing hydroxide conductivity at subzero temperature of anion exchange membranes based on imidazolium modified metal organic frameworks. *J Mol Liq* 370: 120943.

68 Nolan, S.P. (2006). *N-Heterocyclic Carbenes in Synthesis*. Wiley-VCH Handbook.

69 Dimitrov-Raytchev, P., Beghdadi, S., Serghei, A., and Drockenmuller, E. (2013). Main-chain 1,2,3-triazolium-based poly(ionic liquid)s issued from AB + AB click chemistry polyaddition. *J Polym Sci A* 51 (1): 34–38.

70 Flachard, D., Serghei, A., Fumagalli, M., and Drockenmuller, E. (2019). Main-chain poly(1,2,3-triazolium hydroxide)s obtained through AA+BB click polyaddition as anion exchange membranes. *Polym Int* 68 (9): 1591–1598.

71 Li, N., Guiver, M.D., and Binder, W.H. (2013). Towards high conductivity in anion-exchange membranes for alkaline fuel cells. *ChemSusChem* 6: 1376–1383.

10

Radiation-Grafted and Cross-linked Polymers-Based Anion Exchange Membranes

Ana Laura G. Biancolli, Bianca P. Silva Santos, and Elisabete I. Santiago

Nuclear and Energy Research Institute, IPEN/CNEN, Fuel Cell and Hydrogen Centre, Avenida Lineu Prestes, 2242 - Cidade Universitária, São Paulo, 05508-000, Brazil

10.1 Historic Overview

Radiation-induced grafting (RIG) has been used for polymer functionalization since the 1950s [1–3]. RIG involves modifying the polymer matrix through the addition of polar or non-polar monomers that covalently bind to the polymer backbone, affecting its physicochemical properties, Figure 10.1. The widespread use of this technique rests on its simplicity and the great cost-effectiveness involved. The correct choice of monomers with appropriate functional groups results in chemically modified polymers that can be used for various applications, such as biomedical products, hydrogels, environmentally friendly products, and the automotive industry.

Until the 1960s, RIG polymerization was adopted to improve the properties of polymers used in industries [5] and to modify cellulose [6]. In the 1970s/1980s, works describing polymers capable of transporting ions synthesized via RIG began to gain space in the literature [7, 8]. As an example, Chapiro et al. have performed several pioneering studies [7, 9, 10] in the production of different ionic membranes through RIG, mainly for application in reverse osmosis.

Despite advances in the RIG technique, radiation-grafted anion-exchange membranes (RG-AEMs) only gained visibility in the 1990s. Momose et al. [11] were one of the first groups to present pre-irradiated grafted AEMs. The authors irradiated poly(ethylene-tetrafluoroethylene) (ETFE) and grafted the monomer α,β,β-trifluorostyrene (TFS) onto the polymer film, followed by chloromethylation and quaternization of the grafted film. The ion-exchange capacity (IEC) of their AEMs

Alkaline Anion Exchange Membranes for Fuel Cells: From Tailored Materials to Novel Applications, First Edition. Edited by Jince Thomas, Alex Schechter, Flavio Grynszpan, Bejoy Francis, and Sabu Thomas.

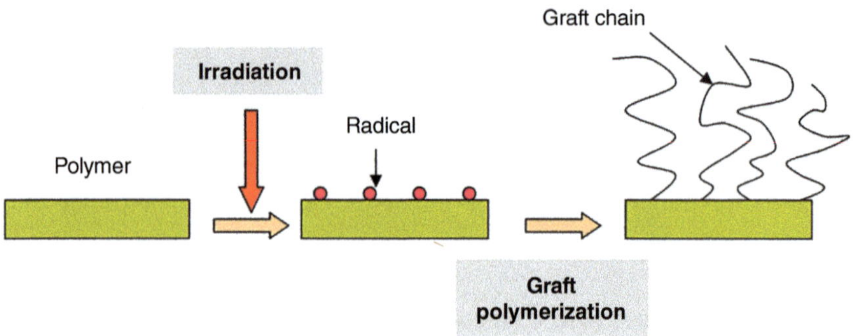

Figure 10.1 Illustration of radiation-induced graft polymerization. Source: Fujiwara [4]/ Elsevier.

reached up to 1.22 mmol g^{-1}. In addition, Lee et al. [12], also in the early 1990s, prepared urea-permeable RG-AEMs for peritoneal dialysis.

Quaternary ammonium (QA) salts were introduced onto nonporous polyethylene (PE) films by grafting vinyl monomers such as glycidyl methacrylate (GMA), diethylaminoethyl methacrylate (DEAEMA), and vinyl benzyl trimethyl ammonium chloride (VBTAC), followed by subsequent chemical modifications, and by cografting hydroxyethyl methacrylate (HEMA). Kobayashi et al. [13] developed anion-exchange PE hollow-fiber membranes for microfiltration using the RIG technique, having DEAEMA, vinyl pyridine (VP), and GMA as monomers. In addition to these applications, RG-AEMs were used for desalination by electrodialysis [14], protein purification [15], etc. The first reports of applications of RG-AEMs in fuel cells were published only in the early 2000s.

The alkaline fuel cell technology, specifically, was stagnant for about three decades due to operational difficulties, especially because of CO_2 intolerance and the consequent formation of carbonates typically found in alkaline fuel cells based on liquid electrolytes. Also, there was a relatively greater interest in proton-exchange membrane fuel cells (PEMFCs) [16]. In the early 2000s, interest in alkaline fuel cells was restored thanks to the possibility of using solid polymeric anion-exchange membranes (AEMs) [17], which would avoid the problem of carbonation during operation. The high cost of PEMFCs due to the need for noble metals in the electrodes is another reason behind the reborn research on the alkaline medium [18]. Varcoe and coworkers [19, 20] were the pioneers to present works with RG-AEMs applied to fuel cells. The group is responsible for most of the scientific articles involving RG-AEMs since the early 2000s. The first studies involved poly(vinylidene fluoride) PVDF- and poly(tetrafluoroethylene-co-hexafluoropropylene) FEP-based RG-AEMs for use in direct methanol fuel cells (DMFCs) [19, 20]. Later, works using RG-AEMs in fuel cells fed by H_2/O_2 proved that this type of membrane was extremely promising in this application, showing a maximum power density of 130 mW cm^{-2} at that time [21, 22].

Over the years, RIG techniques for the synthesis of AEMs were explored by several other research groups and improved, enabling RG-AEMs to reach very high ionic conductivities, over 200 mS cm^{-1} at 80 °C [23, 24]. In parallel to AEM, the development of novel anion-exchange ionomers (AEIs) contributed significantly to boost anion-exchange membrane fuel cell (AEMFC) performances [25, 26]. Nowadays, it is possible to find maximum power densities with values superior to 2000 mW cm^{-2} using these devices [27, 28].

10.2 Sources of Radiation

The ionizing radiations used in the RIG method can be classified into three categories: charged particle radiation (electrons, positrons, and heavy ions), photon radiation (gamma, X-ray, and light), and neutron radiation.

The typical ionization potential for organic or polymeric materials is ~10 eV [29]. The equivalent photon energy in mercury vapor ultraviolet (UV) curing lamps (240 and 270 nm and between 350 and 380 nm wavelengths) is between 5.2 and 3.3 eV [29]. Therefore, UV radiation is not effective in producing ionization without a radical initiator. In this way, light and UV radiations are limited to effects on the surface of materials. To be used for RIG, this type of radiation source needs expensive photosensitive compounds that decompose upon exposure and thus initiate chemical reactions, making this source not commonly used for RIG applications. On the other hand, accelerated electrons and photons, such as X-rays (4.1×10^{-3} nm) and low-wavelength gamma-rays (1.0×10^{-3} nm), interact with matter at the atomic level. For example, for X-rays, the most probable photon energy is ~300 keV, which is far above the ionization potential of any material [30]. The choice of radiation source depends on the desired degree of penetration and availability. There are eight to nine times more commercial electron-beam (EB) accelerator units in operation worldwide than commercial gamma-ray irradiators [29]. In the case of EB, there are two fundamental properties to be considered: the electron energy and the beam current. As electrons have mass and electrical charge, their penetration into materials is limited by their kinetic energy as well as the mass and density of the target material. The amount of exposure to electrons is called the absorbed dose. The measurement of absorbed dose is called "Gray" by the International System of Units (SI), more commonly known as kilograys (kGy), where 1 kGy = 1 J g^{-1} absorbed energy per mass. The EB energies can range from 75 keV to 10 MeV. Low energies, between 80 and 300 keV, have the power to penetrate up to 0.4 mm. On the other hand, energies ranging from 3 to 10 MeV have a penetrating power 100 times higher, reaching up to 40 mm in thickness [30]. Both X-rays and gamma rays (γ-rays) have a comparable depth of penetration, which is much greater than the highest energy industrial EB systems, 10 MeV [29]. Currently, the most used gamma radiation source in industrial radiation processing and research facilities is cobalt-60 (^{60}Co). Cobalt-60 decays into a stable (non-radioactive) nickel isotope mainly emitting one negative beta particle (of maximum energy 0.313 MeV) with a half-life of about 5.27 years [31]. Nickel-60 produced is in an excited state, and it immediately emits two photons of energy, 1.17 and 1.33 MeV, in succession to reach its stable state. These two gamma-ray photons are responsible for radiation processing in the

^{60}Co gamma irradiators [31]. In addition to the difference in penetrating power, the three sources of ionizing radiation also differ in dose rate. EB sources can reach up to 100 kGy s^{-1}, X-rays can reach up to 0.027 kGy s^{-1}, and ^{60}Co gamma sources reach up to 0.0028 kGy s^{-1} (fresh source) [29]. The increasing availability of EB accelerators compared to γ-rays irradiators makes EB more accessible for large-scale AEM production [32].

The formation of free radicals, which will be used for grafting the monomer into the polymer structure, depends on the interaction of ionizing radiation with the material. Primary interactions include ionization, electron stabilization through the generation of hot electrons, ion neutralization, and free radical formation [33]. Free radicals are created through chain scission or by dissociation of side chains. In addition to the formation of free radicals, other reactions are involved during irradiation such as hydrogen abstraction, rearrangements, and formation of new bonds, in addition to the double bonds, chain scission, oxidation, cross-linking, and branching. All these reactions depend on the irradiation conditions and the polymer being irradiated.

10.3 Types of Radiation-Induced Grafting

The preparation of functional materials and ion-exchange membranes by irradiation-induced graft copolymerization can be performed by grafting non-functionalized or functionalized monomers with functional groups that facilitate the quaternization process. Two methods of RIG are most employed: simultaneous (SM), also called direct or mutual, or pre-irradiation method (PIM), which can be performed in the presence of air (peroxidation), vacuum, or inert atmosphere (N_2 or Ar). The atmosphere used during the irradiation process is related to the type of radical species that will be formed. Various types of high-energy radiation can be used to initiate the polymerization reaction, but gamma-ray sources from ^{60}Co and EB from accelerators are the most used [34].

In the SM, the base polymer is irradiated in the presence of the monomer, which can be in the form of vapor, liquid, or solution. In this case, active free radicals are formed in the base polymer and monomer. Despite being, in principle, the most efficient, copolymerization by SM has serious limitations due to the high level of homopolymer formation [34, 35].

The PIM involves a combination of two steps: (i) irradiation of the base polymer to form active free radicals; and (ii) grafting of the pre-irradiated polymer with the monomer by immersion. If irradiation occurs in air, the generated free radicals react with oxygen to form peroxides and hydroperoxides, which, by thermal decomposition, initiate the grafting process after being placed in contact with the monomer. On the other hand, if the irradiation is carried out in a vacuum or in an inert atmosphere, the generated free radicals remain trapped in the base polymer and will initiate grafting in the presence of the monomers [34, 35].

Due to the concomitant exposure of monomers and base polymers to ionizing radiation in the SM, it is necessary to use lower absorbed doses than in PIM to obtain a similar degree of grafting (DoG) in the polymer matrix [23]. Despite the high possibility of homopolymer formation [34, 36] during SM, the low absorbed doses employed in this methodology can result in higher integrity of the polymer backbone due to reduced radiation exposure. On the other hand, PIM allows better control of the post-irradiation grafting process by varying the synthesis parameters.

During irradiation, the formation of free radicals in the polymeric backbone leads to cross-links, end-links, and disproportionation reactions [37–39], which result in modifications in the physicochemical properties of the base polymers [40]. The processes occurring during irradiation depend on the type of polymer being irradiated. In general, the reactions have two main impacts on the polymer: mechanical and chemical reinforcement, and degradation; the first is normally a result of cross-linking and the second of scission processes [41, 42]. Examples of these reactions for a PE film are shown in Figure 10.2 [23]. Figure 10.2 presents possible reactions happening during irradiation either by e-beam (EB, in red) in the PIM or gamma-rays (γ, in blue) in the SM. Reactions 1–8 happen in both PIM and SM to different extents for each technique. Reaction 1 shows the formation of macroradicals upon irradiation, with subsequent formation of the unsaturated structure due to C—C scission. Reaction 2 is the abstraction of hydrogen and formation of alkyl macroradicals, with subsequent formation of cross-linking structure by reaction between two adjacent radicals (reaction 3), or formation of unsaturated structure by releasing H_2 (reaction 4). Reaction 5 occurs only if the irradiation takes place in presence of oxygen, which results in the formation of peroxyl-type radicals after contact with O_2. Moreover, reaction 5 can lead to reaction 6 with the formation of hydroperoxides, and/or to reaction 7 with peroxyl radicals reacting to form alkylperoxides [43], and/or to reaction 8 with the formation of carbonyl group [44], which results in chain scission [37, 45]. Reactions 9, 10, and 11 occur only in the SM, where the solvent (S), monomer (M), and polymer matrix (PE) are simultaneously exposed to ionizing radiation. Reaction 9 demonstrates the grafting of M into the polymer matrix through radical reaction, with a subsequent chain growing of M. Reaction 10 is related to the homopolymerization due to reaction between monomer radicals. Reaction 11 is the possible radical formation in S with subsequent transfer of radical to the polymer matrix.

Figure 10.3 shows an example of reactions occurring during irradiation of a partially fluorinated polymer – ethylene tetrafluoroethylene (ETFE) [46]. Reactions 1–3 present the C—F, C—H, and C—C bonds cleavage in the backbone, forming macroradicals. Reaction 4 demonstrates a macroradical reacting with atmospheric oxygen leading to peroxidation and the subsequent reactions, which are similar to reactions 5, 6, and 7 observed in Figure 10.2. Reaction 7, Figure 10.3, shows the formation of unsaturated structure by C—C scission with subsequent dehydrofluorination, which characterizes degradation. Reaction 8 is the dehydrofluorination with the consecutive formation of cross-linking structure by reaction between two adjacent radicals [34, 42].

The extent of the reactions shown in the examples of Figures 10.2 and 10.3 depends on the absorbed dose, dose rate, atmosphere, and temperature during irradiation. In the SM, there are also variables such as the type and concentration of monomer and reaction time. It is important to mention that the modifications on

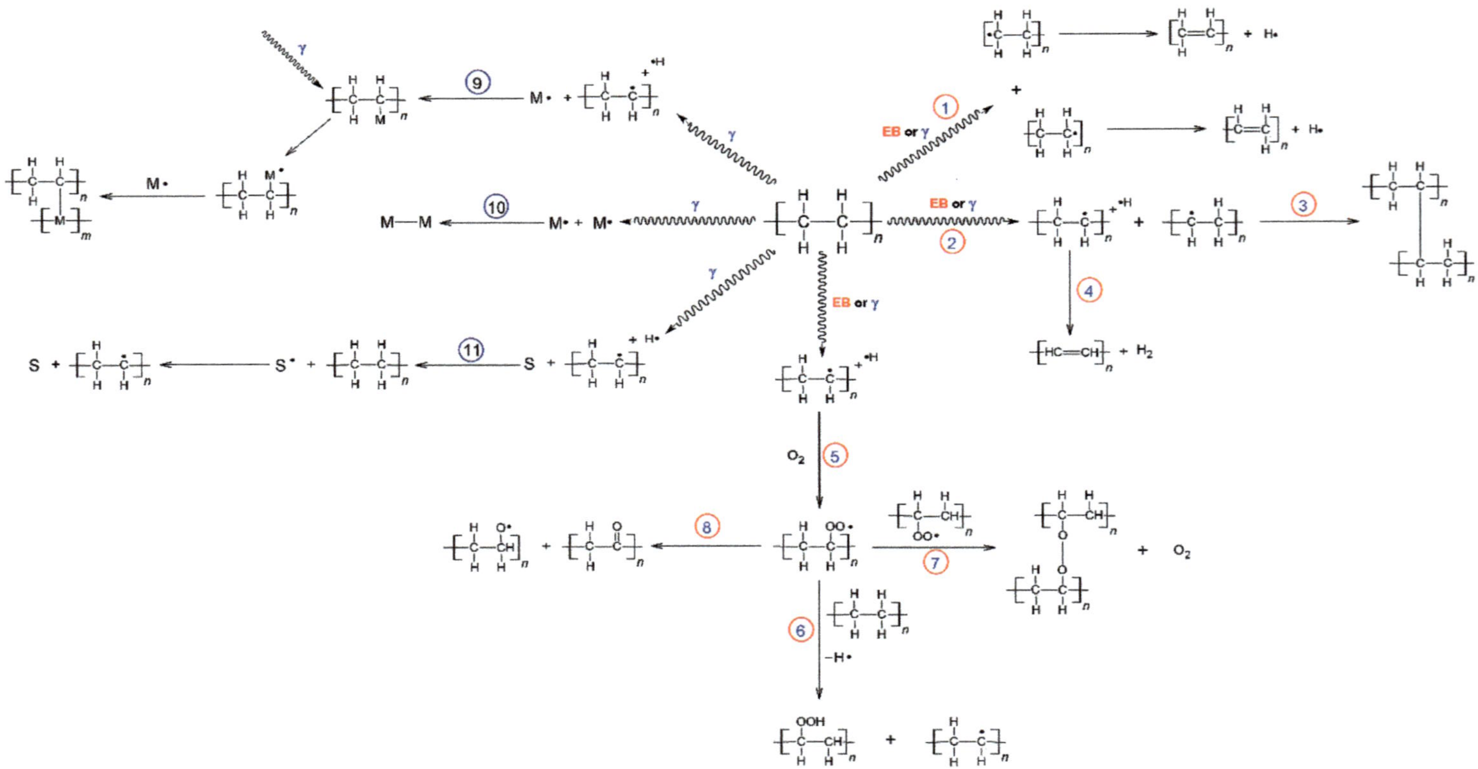

Figure 10.2 Possible reactions happening during irradiation of polyethylene either by e-beam (EB, in red), pre-irradiation method, or gamma-rays (γ, in blue), simultaneous method. Source: Extracted with permission from [23]. Copyright holder: Elsevier.

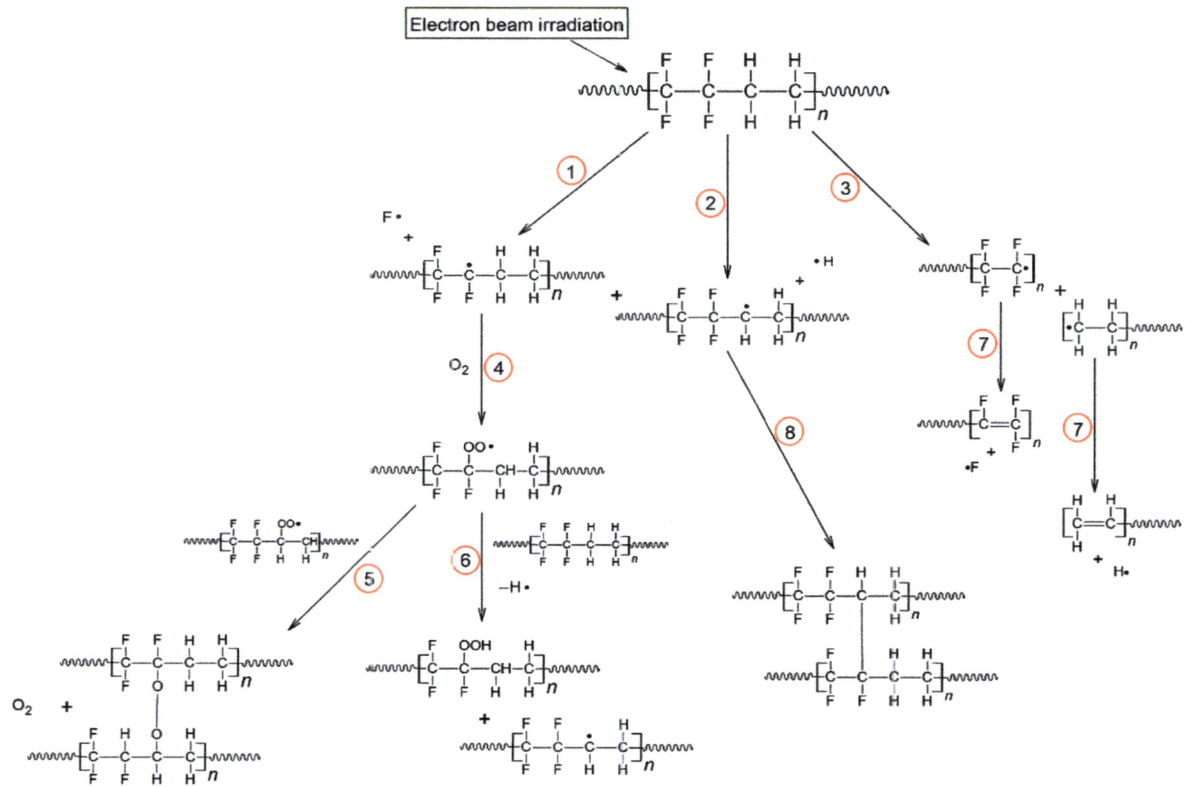

Figure 10.3 Possible reactions occurring during irradiation of ethylene tetrafluoroethylene (ETFE) by the pre-irradiation method. Source: Extracted with permission from [46]. Copyright holder: Elsevier.

the polymer backbone during irradiation affect directly the ion conduction in AEMs as well as degradation mechanisms [23, 47].

10.3.1 Absorbed Dose

The absorbed dose is directly related to the DoG and the mechanical properties of the AEM. The higher the absorbed dose, the more radicals will be formed in the polymer matrix during irradiation, which, in turn, will be available for grafting – regardless of the atmosphere or temperature [36]. An increase in the formation of free radicals is likely associated with higher DoG and IEC in AEMs. It is important to highlight that during the irradiation using the PIM, especially promoted by an EB source, high absorbed doses result in an increase of local temperature in the polymer film. This can raise the rate of recombination between free radicals, resulting in more cross-linking and/or chain scissions [48]. Too high absorbed doses can lead to polymer degradation and loss of mechanical properties due to excessive chain scissions [49]. In SM, the increase of the absorbed dose implies longer reaction times and possibly more homopolymer formation.

10.3.2 Dose Rate

The dose rate is defined as the dose delivered in a specific unit of time, and it is related to the source of radiation. As previously mentioned, EB sources can reach up to 100 $kGy\,s^{-1}$ and a fresh source of γ-rays (^{60}Co) up to 0.0028 $kGy\,s^{-1}$ [29]. Therefore, the irradiation time using a gamma source is much longer (hours or days) than when using a high-energy EB (minutes or seconds). That is the reason why the use of γ-rays is usually associated with SM and the use of an EB accelerator is more suitable for PIM [50]. The dose rate affects the concentration and lifetime of radicals and the extent of oxidative degradation [44]. In SM, for the same absorbed dose, an increase in the dose rate results in lower efficiency of grafting [50]. In this case, the high concentration of radicals increases the recombination rate and favors rapid polymer termination reactions, leading to more homopolymerization and decreasing the chance of reactions between backbone and monomer molecules [34, 51]. Another outcome of using high dose rates in the SM is that the grafting reaction is not diffusion-controlled, taking place only on the surface of the polymer film [36]. Low dose rates, using γ-rays, can be used for PIM [39, 50], but the obtained DoG is usually lower than in polymer films irradiated with the same dose provided by EB. The concentration of radicals per period of time is low, and the long time required to reach a certain dose allows the available free radicals to recombine with each other in the backbone itself or for the polymer to undergo further oxidative degradation (if irradiated in air atmosphere) [33]. Under these circumstances, the polymer will exhibit a reduced number of available sites for subsequent grafting, decreasing the DoG.

10.3.3 Atmosphere During Irradiation

The atmosphere during irradiation is related to the nature of free radicals being formed. Irradiation under oxidative atmosphere favors the formation of peroxyl radicals and peroxide/hydroperoxide species [37, 42, 43]. Therefore, the free radicals formed on the polymer backbone are scavenged by O_2, which decreases their availability for other reactions, such as cross-linking or the formation of unsaturated bonds. In addition, the oxygenated species can lead to oxidative chain degradation during the irradiation process [45, 52]. Graft polymerization is initiated by the decomposition of these peroxide/hydroperoxide species, which are sensitive to temperature. Despite being a very simple and convenient way of irradiation, normally high doses are needed to achieve sufficient available grafting sites (number of peroxide/hydroperoxide species) to obtain a reasonable DoG [36]. The irradiation can also be conducted in an inert atmosphere (N_2, Ar) or vacuum. In these cases, free radicals do not suffer inactivation by oxygen, increasing their availability for both grafting [37] and cross-linking, resulting from reactions between adjacent free radicals. Irradiation of polymers in an inert atmosphere or vacuum can be an alternative to prevent oxidative chain degradation and to obtain higher DoGs [46].

10.3.4 Temperature During Irradiation

The temperature during irradiation is a very important factor to be considered, and it is related to the mobility of the formed radicals. At low temperatures, especially below 0 °C (273 K), the polymer chains have lower mobility, thus the recombination between free radicals is hampered [53, 54]. As a consequence, high DoGs can be obtained due to the stabilization of free radicals [46]. On the other hand, high temperatures during irradiation can favor cross-linking and chain scission reactions, leading to a decrease in DoG due to the reduced availability of free radicals [42, 55].

Oshima et al. [55] have demonstrated the effects of absorbed dose and temperature on the mechanical properties of ETFE films. Figure 10.4a shows the elongation at break of 0.2 mm thick ETFE films after irradiation at 270 °C, room temperature

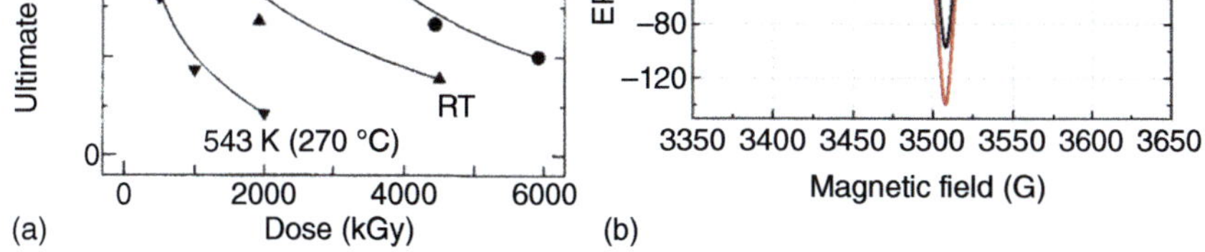

Figure 10.4 (a) Elongation at break of 0.2 mm thick ETFE films after irradiation at 270°C, RT, and −196 °C and with different absorbed doses (oxygen-free atmosphere). Source: Extracted with permission from [55]. Copyright holder: Elsevier. (b) EPR spectra of ETFE film samples (25 μm) irradiated with 100 kGy, by an EB source, in air atmosphere, at room (RT ~ 28 °C) and low temperatures (LT < −10 °C). Source: Extracted with permission from [46]. Copyright holder: Elsevier.

(RT), and −196 °C and with different absorbed doses (oxygen-free atmosphere). This mechanical property of the films is less affected as the irradiation temperature decreases and when the film is exposed to lower doses of irradiation. Figure 10.4b was adapted from Biancolli et al. [46] and shows the electron paramagnetic resonance (EPR) spectra of ETFE film samples (25 μm) irradiated with 100 kGy, by an EB source, in air atmosphere at room (RT ~ 28 °C) and low temperatures (LT < −10 °C). The peak-to-peak height (pph) is proportional to the number of free radicals in the sample, and it is higher for the film irradiated at low temperature. Increasing pph values indicates an increase in the formation of free radicals, which in turn results in higher DoGs [46].

10.4 Base Polymer

The use of a solid polymer as electrolyte in a fuel cell is a good choice because polymer films can be thin, flexible, and mechanically robust. The different responses to radiation for the polymers are related to the chemical structures of the backbone. The requirements for the base polymer to become an AEM using RIG are [53, 56]:

(i) not degrade upon irradiation and yield active sites during irradiation that can initiate the graft polymerization reaction;
(ii) be hydrophobic to promote hydrophilic–hydrophobic phase separation in the final ion-exchange membrane;
(iii) allow the diffusion of the grafting monomer into the polymer;
(iv) be thermally stable and mechanically robust;
(v) be chemically stable against the reactants used during the synthesis of the membrane and the high alkaline medium encountered during AEMFC operation.

AEMs are formed from a variety of base polymers, including non-fluorinated, partially fluorinated, or fully fluorinated polymers. Some polymeric films are particularly interesting for this type of application due to their morphological, mechanical, and chemical characteristics. The polymers can be divided into three categories: cross-linking types, degradation types, and radiation-resistant ones. All of the highly aromatic polymers are radiation-resistant, compared to the non-aromatic polymers, such as polyimides (PIs), poly(aryl ether ether ketone) (PEEK), poly(aryl ether sulphone) (PES), etc. The polymers of the degradation type undergo deterioration of their mechanical properties even at low absorbed doses. Among the fluoropolymers, Figure 10.5, higher hydrogen content favors cross-linking, and the trend can be expressed as follows [37]:

For cross-linking: PVF > PVDF > ETFE > FEP > PFA > PTFE
For degradation: PTFE > PFA ≈ FEP > ETFE > PVDF > PVF

Polystyrene (PS) is an example of a cross-linking-type polymer, despite being relatively resistant to radiation due to the “protective” effect of the aromatic groups. In the same way, PE also mainly cross-links upon irradiation, which generally enhances the mechanical properties due to the high content of cross-linking [37].

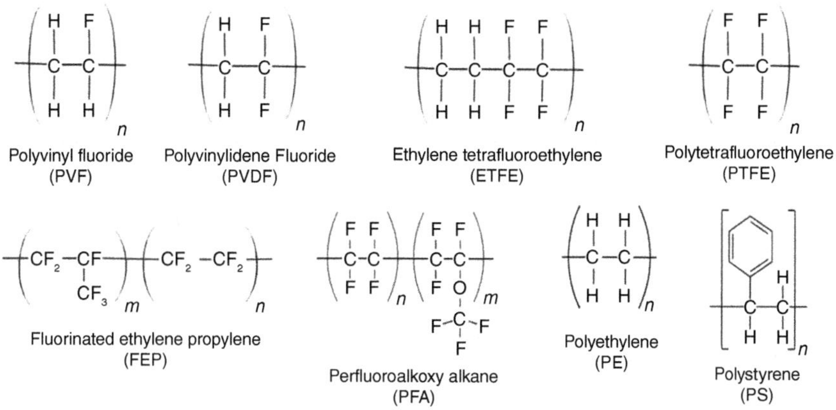

Figure 10.5 Chemical structures of few non-fluorinated, partially fluorinated, or fully fluorinated base polymers.

By considering the fact that highly fluorinated polymers tend to undergo degradation upon irradiation, partially fluorinated (such as ETFE) and non-fluorinated polymers (such as PE) have been presented as the most promising alternatives for application as base polymers for AEMs. In the case of ETFE, its copolymer structure consists of alternating components of PE and PTFE, providing unique properties that correspond to the physicochemical properties of both polymers (fluorocarbons and hydrocarbons). This type of polymer also offers excellent thermal and chemical stabilities and it is a promising choice for use in AEMFCs [21, 57]. In the case of PEs, their properties are even more attractive for this type of application [28], since they have high impact and traction resistances, high flexibility, good workability, and thermal and chemical stability. Basically, PEs of interest for use in AEM are low-density polyethylene (LDPE), high-density polyethylene (HDPE), and ultrahigh molecular weight polyethylene (UHMWPE). The difference among them is associated with the molecular weight and degree of branching of the carbon chains in their molecular structure [58].

10.5 Grafting Solution

Among the parameters that affect the synthesis of RG-AEMS, those directly related to the radiation source, explained in Section 10.3, the nature of the base polymer, explained in Section 10.4, and others related to the graft mixture and its components have an important role in the final AEM properties. In addition, the nature of the monomer, concentration of monomers, addition of diluents, addition of cross-linking agents, reaction temperature, and reaction solvent are variables that can directly affect the DoG and grafting homogeneity [36]. Figure 10.6 shows common monomers used for the production of AEMs via RIG.

Generally, grafting reaction occurs by the front mechanism, in which grafting starts on the surface of the film and proceeds toward the bulk [59, 60]. As the

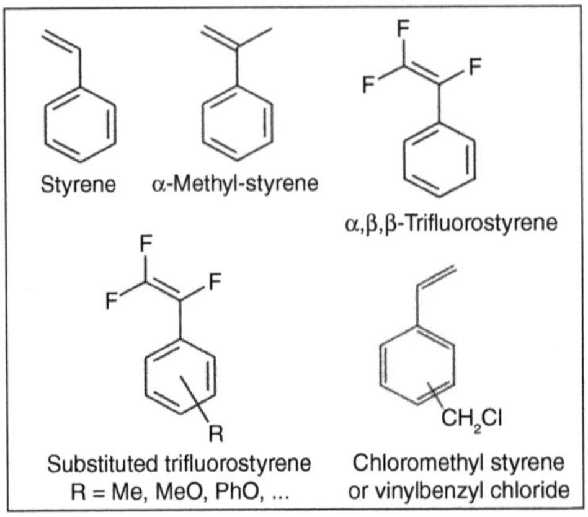

Figure 10.6 Examples of monomers used for the preparation of radiation-induced grafted AEMs.

reaction takes place, the swelling of the film occurs by a gradient of monomer diffusion. Such mechanism and the final grafting homogeneity depend on several parameters, such as time of the reaction, solvent, and temperature [61]. Monomer concentration also significantly influences the grafting process, either in SM or PIM. The grafting tends to proceed smoothly if the monomer has access to the propagating sites. For this reason, an increase in the monomer concentration normally leads to an increase of the DoG [36]. On the other hand, excessive concentration of monomers in SM can produce large amounts of homopolymers. As the reaction proceeds, the viscosity of the polymer matrix increases as a result of the amount of grafted monomer into its layers. Depending on the grafting solution, this increase in viscosity can hinder the monomer from diffusing through the membrane to react with the remaining free radicals [34].

In the case of AEMs prepared through SM, a more homogenous grafting is expected than in the ones prepared via PIM (for the same DoG) due to the low absorbed dose rate and consequently longer reaction time [23]. The diffusion of the monomer controls chain growth and chain termination in the internal structures of the polymer [58]. The solvent is another important parameter that helps to disrupt intermolecular cohesive forces between polymer chains, allowing monomer diffusion. For example, toluene as solvent can enhance the accessibility of chloromethylstyrene in PE films through base polymer swelling, favoring bulk grafting [36]. The use of a very polar solvent, such as water, despite enhancing the reaction kinetics, requires an appropriate surfactant and may favor surface grafting [61, 62].

The reaction temperature also has a significant influence on the DoG. The initial rate of grafting increases with grafting temperature, probably due to an increase in the monomer transport and the initiation and propagation reaction rates. However, after a certain reaction time at high temperatures, there is also an increase in the rate of termination due to higher macroradical mobility. This results in a higher rate of radical combination reactions, decreasing the final graft yield [63]. Thus, each RIG has an optimal reaction temperature depending on the base polymer, monomer, and solvent, which normally varies from 40 to 80 °C in the PIM.

10.6 Physicochemical Properties of RG-AEMs

Despite RIG being a well-explored grafting technique, several factors related to the influence of radiation on the physicochemical and electrochemical properties of RG-AEMs are still unclear. Some important studies using RIG to produce ion-exchange membranes, especially ETFE films, aiming at PEMFC applications can be found in the literature [47, 49, 64–67]. However, until recently, there were no works directly relating the effects of irradiation to the physicochemical and electrochemical properties of the AEMs for AEMFC applications. Recent works [23, 46, 68] have shown that variations in irradiation parameters, such as temperature, atmosphere, absorbed dose, dose rate, as well as the choice of the type of RIG methods – SM or PIM– affect not only the grafting step but also the microstructure of the AEMs. As a result, the ions and water transportation is affected, primarily influencing the performance and stability of membranes. A property normally introduced by the irradiation of polymeric films is the mechanical reinforcement of the polymer-matrix backbone through cross-linking. Cross-linking is desirable to increase the membrane's lifetime, but it can also affect some other critical parameters such as AEM's water uptake, ion conductivity, and thermal properties. In addition, an uncontrolled degree of cross-linking can make the AEM too brittle.

The type of amine used for functionalization after the grafting step also plays a fundamental role in the physicochemical properties of AEMs. The stability of the cationic groups attached to the base polymer, in highly alkaline environments, is extremely important for the AEM durability. Therefore, many research groups [18, 69–71] have focused on the search for alternative cationic groups (anion-exchange) that are less susceptible to degradation aiming at the production of more stable AEMs. Several cationic groups, including phosphonium, guanidinium, imidazolium, and others (Figure 10.7), have been receiving strong attention as potential substitutes for the traditional QA groups due to their improved stability under alkaline conditions [71–74].

For application in electrochemical devices, both the conductivity and the chemical and dimensional stabilities are considered key parameters for obtaining AEMFC with satisfactory performance and durability. In order to achieve high ion conduction, high IEC is required, which means that a large number of functional groups must be incorporated in the grafted polymer. On the other hand, high IECs can lead to an appreciable increase in the water absorption capacity (or water uptake) by the AEM, which can result in severe losses in dimensional/mechanical properties. In this context, cross-linking can be considered an effective method to

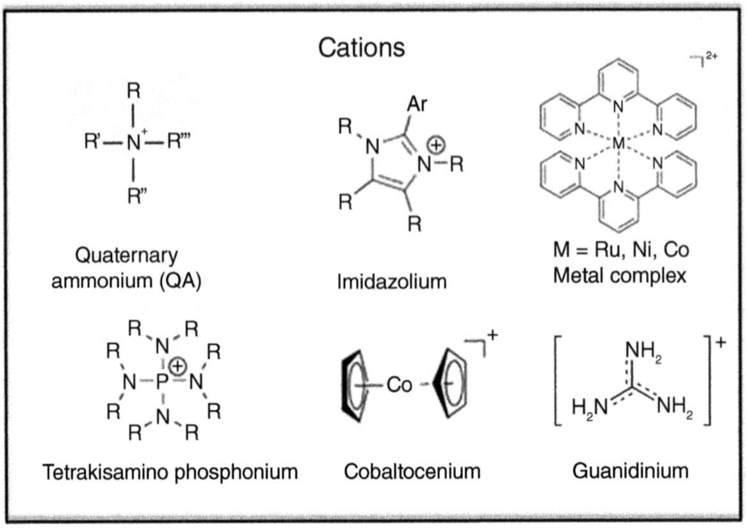

Figure 10.7 Cations structures that can be used in AEMs.

enhance mechanical properties, promoting more stable AEMs and, consequently, more durable AEMFCs [75].

10.7 Cross-linking in AEMs

As discussed in the previous sections, cross-linking processes in polymers can be promoted by using high-energy radiation, such as X-rays, gamma-rays, and EB accelerators. However, physical and chemical cross-linking can also be achieved by using cross-linking agents [76].

One of the strategies to maintain or enhance ionic conductivity and dimensional stability of AEMs is to build cross-linking structures. Cross-linking can reduce the swelling ratio and control the water uptake. It can also contribute to the formation of hydrophilic/hydrophobic phase-separation morphology and ion-conducting channels in AEMs. Another advantage of a cross-linked AEM is the decreased fuel crossover, the enhanced mechanical, thermal and chemical stabilities as well as the improved solvent resistance [77]. Although, cross-linked membranes tune many properties in the AEM, they are likely insoluble in many types of solvents, making it difficult to characterize them. In addition, depending on the cross-linking chemistry, it can be economically unfeasible [78].

Clemens et al. [78] recently summarized the effect of cross-linking in membranes. The addition of cross-link agents can affect the AEM mechanical properties by preventing or causing rearrangement in the polymer chains. Another property affected by the cross-linking is the ion conductivity, which was found to be decreased in AEMs with similar IECs. The cross-linking reduces the hydrophilic phases, which, in turn, restricts ionic mobility. On the other hand, highly cross-linked

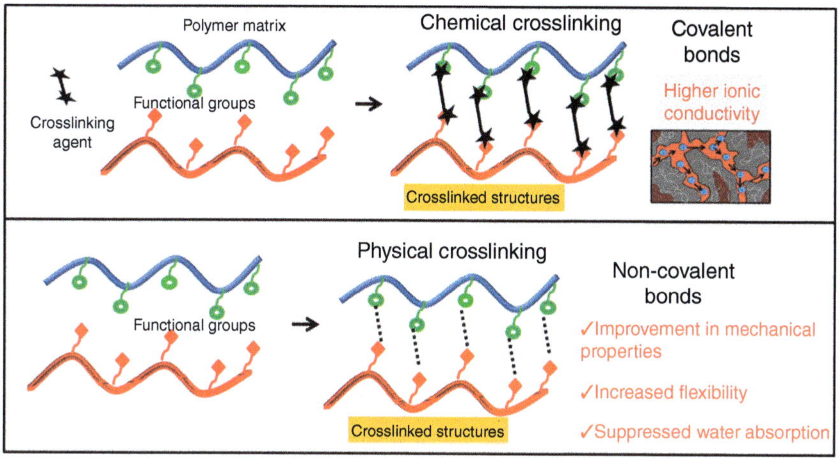

Figure 10.8 Scheme of chemical and physical cross-linking methods. Source: Adapted from [79].

AEMs often show superior alkaline stability in comparison with non-cross-linked membranes as a result of improved water management, and reinforcement in mechanical and chemical properties.

Cross-linking can be classified as chemical or physical (Figure 10.8). Chemical cross-linking involves covalent, electrovalent, or coordinate bonds. Physical cross-linking binds molecules reversibly through hydrogen bonds, ionic, and van der Waals interactions [76, 80, 81].

10.7.1 Physical Cross-linking

Studies focusing on the physical cross-linking technique that employs physical interaction between molecules can be used to outline problems in the chemical approach, such as reduced solubility and flexibility of the polymer, which can cause severe damage to the mechanical properties. Thus, it is necessary to control cross-linking degrees aiming to avoid the packing stage and membrane embrittlement [82, 83].

Physical techniques, such as the addition of ion-conducting additives, pore-filling, reinforcement, and block copolymer architectures have been applied to increase fuel cell performance [82]. For physical cross-linking, cross-linking agents, such as *N,N′*-didodecyl hexyl-bis(quaternary ammonium iodide) (DHBQA) containing both a long alkyl chain (C12) and a QA group have been widely employed in AEMs as depicted in Figure 10.9. In this case, the diammonium groups act as ion conductors, while the physical cross-linkers as well as the long alkyl chains bind to the alkyl chains of the C_{12}-PPO-QA (poly(2,6-dimethyl-1,4-phenylene oxide) (PPO)) polymer through van der Waals interactions [84]. The AEMs containing additives have improved water uptake when compared to non-modified AEMs, which can be related to the presence of the diammonium group. In addition, the use of a physical

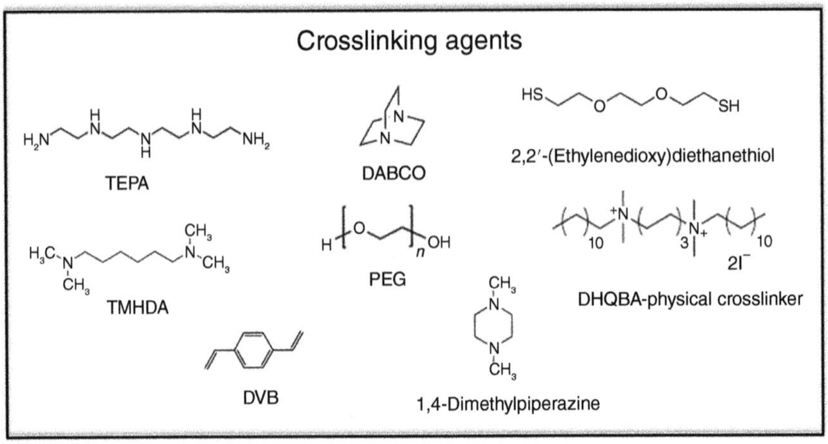

Figure 10.9 Cross-linking agents employed in AEM synthesis.

cross-linking through van der Waals interactions increases by almost 1.7 times the maximum power density when compared to non-cross-linked AEM [84].

Cross-linked AEMs can also be produced by using the non-covalent hydrogen-bonding approach. For that, poly(2,6-dimethylphenylene oxide) (PPO) matrix functionalized with an amide functional group as a hydrogen-bonding cross-linker in the side chain of the backbone was synthesized by Wang et al. [85]. As a result, AEM with amide-containing side chains demonstrated an improvement in many properties such as ion conductivity, stretchability, flexibility, and thermal stability.

10.7.2 Chemical Cross-linking

Chemical cross-linking is more widely employed than physical cross-linking due to the strong covalent forces among polymer chains that permanently interconnect the polymer backbone, which lead to enhanced ionic conductivity [86].

In general, the cross-linking method consists of the addition of cross-linkable molecules, followed by thermal (casting) or UV irradiation treatments, which can increase the complexity of reactions and reduce the stability of cationic groups. During membrane fabrication, cross-linking occurs simultaneously with membrane casting, making it difficult to produce thin and durable ion exchange membranes. In this context, many efforts have been employed to outline the chemical cross-linking problems, promoting difficult control over membrane fabrication, membrane embrittlement due to excessive addition of cross-linking agents, an increase in the manufacturing cost, and degradation of the functional groups during the cross-linking step [87].

10.7.2.1 Cross-linking with Diamine Agents

Covalent cross-linked AEMs are still in development, and a few studies employing cross-linking via quaternization by diamine cross-linking agents have been reported [88–92].

The cross-linking of covalent molecules is mainly produced by Friedel–Crafts electrophilic substitutions and olefin metathesis techniques, direct cross-linking through epoxides, and thiol-ene click chemistry [93–98].

The use of mutual cross-linking and amination agents is widely reported in the literature. In this context, cross-linking agents such as (1,4-diazabicyclo[2.2.2]octane) (DABCO), diethylamine (DEA), tertiary amines (4,4′-trimethylene-bis (1-methylpiperidine), and 1,4-bis (2-methylimidazol-1-yl)butane) are employed the formation of cross-linking structure and amination concomitantly [89–91, 99]. The modified membrane using this cross-linking method can cause an additional steric hindrance for the attack of hydroxyl ions promoting reasonable stability [89]. Besides, *N,N,N,N*-tetramethyl-1,6-hexanediamine (THMDA) has been highlighted as a promising cross-linking agent providing enhanced dimensional stability and high conductivity for AEMs [100–104]. Such feature can be attributed to low swelling promoted by diamine characteristics while providing ion exchange function sites.

Cross-linking in perfluorinated-based AEM is a challenge due to the low reactivity of C—F bonds. In this context, fluorinated cross-linked AEMs using poly(aryl ether oxadiazole)s (FPAEO), *N*-methyl imidazole, and TMHDA, as a polymer matrix, quaternization, and cross-linking agents, respectively, have been proposed [100]. The tensile strength values for cross-linked membranes have been slightly improved when compared with non-cross-linked AEM. Moreover, an original chloromethylation of poly(ether ether ketone) (PEEK), poly(ether sulfone) (PES), and poly(phthalazinone ether sulfone ketone) (PPESK) by using 1,4-bis(chloromethoxy)butane (BCBM) and TMHDA as chloromethylation and quaternization/cross-linking agents, respectively, was reported in the literature [101]. The advantages of using BCBM consist in its low cost and non-toxicity, while TMHDA can improve the IEC and hydroxide transport in comparison with conventional methods that employ trimethylamine (TMA) as QA.

Another strategy to produce AEMs with remarkable enhancements in dimensional stability and conductivity properties involves the selective grafting of ion groups in phenolphthalein units, which are cross-linked with TMHDA to produce hydrophilic ionic regions, forming ion clusters [102]. The water absorption is reduced with an increase of TMHDA amount due to cross-linking structure, which can restrict the mobility of polymer chains. The control of water absorption is essential to avoid problems with AEM swelling and consequent loss of mechanical stability. In this work, mechanical properties (tensile strength and elongation at break), ion conductivity, and maximum power density of cross-linked AEM showed better results compared to non-cross-linked samples.

A solvent-free cross-linked AEM is an interesting alternative to decrease the use of toxic organic solvents, extensive preparation time, difficult processes, and disposal of large quantities of residue during AEM synthesis [105]. This type of AEM has been prepared by using cardo polyether ketone (PEK-C), benzoyl peroxide (BPO), a mixture of vinylbenzyl chloride (VBC) with divinylbenzene (DVB), and tetraethylenepentamine (TEPA), which correspond to polymer intensifier, initiator, monomers, and cross-linking agent, respectively. According to the authors [105], the increase in the amount of cross-linking agent brings a reduction of free volume in the matrix for water absorption. However, poor performance of such AEM in AEMFC was evidenced, making it necessary to optimize the synthesis parameters.

10.7.2.2 Chemical Cross-linking Reaction with Other Agents

Cross-linking agents, such as carboxylic acid (carbodiimide or aziridine), acetoacetyl groups, activated amines with carbonyl functions, and pH-sensitive reactants, have been considered as cross-linking agents for several polymer backbones [76]. Specifically for AEMs, other cross-linking agents, such as DVB [106], brominated poly(phenylene oxide) (Br-PPO) [83, 107], and polybenzimidazole (PBI) [103], have been widely employed.

Luo et al. [106] produced a poly(methyl methacrylate-*co*-vinyl benzyl chloride) (PMV)-based AEM, which was cross-linked into a semi-interpenetrating network (s-IPN) using DVB to decrease the water absorption. In this work, the authors found that the membranes modified with cross-linking agents presented a slight decrease in ion exchange capacity, while Young's modulus was practically unaltered in comparison to the unmodified membrane. In addition, swelling and water absorption decreased with the increase in cross-linking ratio. Satisfactory stability during 146 h at 70 °C was obtained in a modified membrane with 10% of DVB, while the non-cross-linked sample could last only 62 h in the operating fuel cell, demonstrating the beneficial effect of the cross-linking agent on the AEM properties.

A method to synthesize AEMs associating grafted copolymers and cross-linking strategy was proposed by Ran et al. [83]. In this work, the authors synthesized a copolymer of brominated poly(phenylene oxide) (Br-PPO) grafted with 4-vinylbenzil chloride (QVBC) via atom transfer radical polymerization (ATRP), followed by cross-linking reaction using a Br-PPO macromolecular agent. This cross-linking agent is extremely efficient and allows the development of ion channels derived from its unique structure. According to the authors, membranes without cross-linking agents became brittle, and the size of hydrophilic channels in the cross-linked AEM was smaller, suggesting a dense cross-linking network in the hydrophobic domains.

Recent studies have reported some alternative stable cations aiming at increasing the AEM stability in alkaline environment, such as guanidinium, imidazolium, and phosphonium, as well as metal cations, which can slow down the degradation of AEMs. Nowadays, cross-linked AEMs containing these cations have been reported in the literature [72, 88, 108–111]. For example, poly(vinylbenzyl chloride)/polybenzimidazole (PVBC/PBI) cross-linked membranes functionalized with 1-butyl-4-aza-1-azaniabicyclo [2.2.2] octane bromide (BDABCO) groups were used to produce AEMs by Hao et al. [103]. PBI is a cross-linking agent with high alkaline resistance and mechanical strength, while BDABCO is used to increase the compatibility between polymer and QA groups. In terms of performance in AEMFC, the cross-linked AEMs reveal appreciable performance (340 mW cm^{-2}), and the stability data indicate good resistance to oxidative environments.

Polynorborene and its derivatives have received special attention as polymer backbones for reconciling chemical stability in alkaline environments with improved thermal and mechanical properties [112]. As an example, ring-opening metathesis polymerization (ROMP)-type norbornene copolymer membranes using monomers 5-norbornene-2-methylene glycidyl ether (NB-MGE) and 1-(6-(5-norbornene-2-carboxylate)-hexyl)-3-methyl-imidazolium bromide (NB-ImBr-) were reported by He et al. [112]. ROMP is one of the most promising ways to synthesize norbornene and copolymers using a moderate temperature and normal pressure. Additionally, the epoxy groups are partially opened to form a rotatable ether-contained cross-linked structure under alkaline conditions. Such AEM has shown OH^- conductivity of 90.2 mS cm^{-1} at 80 °C and a maximum power density of 118.4 mW cm^{-2}. According to the authors, an optimization of synthesis parameters is required to overcome the instability of ester linkages at high pH.

Triazolium cations are considered a promising functional group to be applied to cross-linked AEMs. Cross-linked poly(arylene ether sulfone) (PAES) containing QA and triazolium cations using “click chemistry” was reported in the literature [113]. Click chemistry is the name given to the synthesis of 1,2,3-triazoles via Cu(I)-catalyzed azide–alkyne cycloaddition (CuAAC) [114]. Moreover, 1,6-diiodohexane (DIH) is used as a cross-linking agent during the synthesis step to produce membranes with sufficient toughness and flexibility. In this case, the modified AEM presents IEC that is appreciably higher than that observed for the non-cross-linked one due to new triazolium cations added during the reaction. No significant variations in conductivity values were identified for cross-linked and non-cross-linked (PAES-TA-OH) AEMs. On the other hand, the alkaline stability tests for both membranes in NaOH solution at 80 °C indicate that cross-linking leads to an improvement of chemical stability in severe alkaline environments.

Guanidinium cation-based AEMs have also been employed to solve the stability limitation of AEMs. For that, brominated poly(2,6-dimethyl-phenylene oxide) (BPPO) and bi-guanidine cross-linker were employed [108]. All modified membranes showed a gel fraction above 80%, which confirms that the covalently cross-linked structure was formed. In addition, the alkaline stability of membranes was measured by immersion of resulting AEMs in NaOH solution and testing modifications in the conductivity values. In the first five days, a reduction in conductivity of modified membranes was identified due to the possible degradation of guanidinium groups on the surface of the sample. However, after the five days period, the degradation immediately stops. The best result showed a loss of 5% of hydroxide conductivity after 30 days in the modified AEM with a higher number of ionic groups. The work supports the excellent chemical stability of the guanidinium cations due to the large steric hindrance.

Organic metal cations have been used in order to prevent degradations (Hoffmann elimination and nucleophilic substitution) inherent to QA groups. Then, metal-cation-based AEMs based on bis(terpyridine)Ru(II) complexes cross-linked with dicyclopentadiene (DCPD) by using ROMP of a bis-(terpyridine)ruthenium(II) complex-functionalized norbornene have been reported [109]. Ruthenium complex can increase the ion-exchange capacity of the membrane due to the presence of two cation–anion pairs. Furthermore, these modified AEMs possess high conductivity and mechanical strength comparable to those of conventional QA-based AEMs [109].

Cross-linked AEMs can also be synthesized with a nitrogen-free approach using phosphonium cations that could be used in place of traditional QA cations. Cross-linked membrane structures formed by quaternary phosphonium-functionalized PEEK (QPPEEK) and poly(ethylene glycol) (PEG) cross-linking agents were reported by Kumari et al. [115]. The synthesis involves chloromethylation of PEEK, functionalization with quaternary phosphonium group, and cross-linking with PEG. In addition, a series of cross-linked membranes using different QPPEEK:PEG molar ratios were synthesized. The membrane with higher PEG content showed an increase in water uptake due to the enhancement in hydrophilic properties. Such a feature reflects in the conductivity since ion conduction is a water-dependent process. The modified membrane (80 : 20) has shown an increase of 10% in conductivity after 400 h when compared with the non-cross-linked one. The performance of the fuel cell with QPPEEK-PEG (80 : 20) was higher than PEEK-based AEMs reported in the literature.

10.7.2.3 Other Methods of Producing Cross-linked Membranes

Another approach for producing cross-linking AEMs is the Grubbs-catalyzed olefin metathesis. The desired cross-linking degree can be achieved by the addition of the Grubbs II catalyst in different amounts. This reaction occurs at RT and does not necessarily require the addition of a cross-linker, since the cross-linking network is formed during membrane casting and drying [116, 117]. In general, the cross-linked membrane is prepared by solution-casting the cross-linkable quaternized poly(2,6-dimethyl-1,4-phenylene oxide) PPO in bromide form with Grubb's second-generation catalyst via olefin metathesis reaction [118]. In this work, the alkaline stability of the cross-linked PPO-based membranes was investigated by monitoring the change in ionic conductivity as a function of time. After 960 h, almost 100% of ion conductivity remained. The authors associated the good alkaline stability with the hindered attack of hydroxide ions and the dense cross-linking network structure.

An approach called self-cross-linking, in which it is not necessary to use a cross-linking agent or a catalyst, was proposed by Huang et al. [119]. The method involves ROMP of an epoxy group, followed by condensation of the hydroxyl group in the drying step. The main advantages result from the creation of a rigid structure, as commonly occurs with the conventional cross-linking reaction. In this work, both cross-linked rP(NB-MGE-*b*-NB-O-Im+Cl-)-*x* and non-cross-linked rPNB-O-Im-*x* membranes were compared. The cross-linked AEM showed an enhancement in tensile strength, while elongation at break was above 10%. Also, the thermal property indicated that the modified membrane decomposes slower than the non-cross-linked AEM, while water uptake and swelling properties were highly reduced, demonstrating an excellent mechanical properties.

Another synthetic route to produce cross-linked AEMs is the employment of a multi-cationic oligomer and flexible long-chain cross-linking to increase the ionic group's amount and free volume, aiming to improve the mobility of polymer chains [120]. Poly(1,4-diazabiciclo[2,2,2]octane (PDABCO) oligomer, multi-cationic oligomer modified with graphene oxide (QBGO), and cross-linked membrane poly(arylene ether sulfone) (QPAE) copolymers were prepared by Lu et al. [120]. The mechanical properties of cross-linked AEMs showed an increase in Young's modulus and reduction of elongation at break with the increase of cross-linking degree. According to the authors, the addition of graphene-based inorganic materials (cross-linking agents) increases the mechanical strength and the interfacial adhesion between the polymer backbone and particles. Furthermore, flexible long-chain cross-linkers contribute to the entanglement of polymer chains, increasing the stiffness property of the membrane. The use of QBGO cross-linker up to 2.0% by weight increases AEM water absorption, but after that, the AEM experiences a decrease in water uptake due to a high degree of cross-linking, which hinders the polymer mobility and makes the structure more rigid.

UV irradiation can also be used to produce cross-linking. The main advantages of this method are easier control of side reactions, faster reactions, and high control of membrane thickness. A comparative study involving thermal and photo-cross-linking using polyisoprene-ran-poly(vinylbenzyl chloride) (PI-ran-PVBCl) copolymers as polymer matrices was reported by Ertem et al. [87]. For photo-cross-linking, membranes with different dithiol cross-linkers, such as 10-decanedithiol (DT) and 2,2′-(ethylenedioxy)diethanethiol (EDDT), were employed. The first has an intrinsic hydrophobic character due to its aliphatic hydrocarbon structure, while EDDT presents a slight hydrophilic character offered by the oxygen atoms. Both photo-cross-linked and thermal-cross-linked AEMs presented similar IEC, while the water absorption was lower for AEM produced by the photo method. These results suggest that cross-linking density is more controlled using the photo approach. In another work [121], the AEM synthesis was conducted by using photo-cross-linking method coupled to UV irradiation, using styrene, acrylonitrile, 1-methyl-3-(4-vinylphenyl) imidazolium chloride ([MVBIm][Cl]), and *N,N,N*-trimethyl-1-1-(4-vinylphenyl) methanaminium chloride([TMVPMA][Cl]). The values of IEC, water (or methanol) uptake, and swelling ratio were slightly higher using TMVPMA due to the ease of the copolymerization reaction with styrene and acrylonitrile. On the other hand, the AEM containing the imidazolium group showed better chemical stability under the same experimental conditions. This result was associated with the resonance effect of the conjugated imidazole rings.

10.8 Conclusions

AEMs, in general, have great potential to support the advanced development of electrochemical devices and solve current energy challenges. RG-AEMs can provide a significant reduction in manufacturing, raw material, and overall costs of fuel cells or other electrochemical devices compared to those currently available. Cross-linking is one important strategy to overcome the dimensional stability and durability problems of this type of membranes. Despite the advantages of using this approach for polymer modification, finding the optimal cross-linking degree in AEMs and issues related to the degree of complexity and process costs when using cross-linking agents are still major challenges. RG-AEMs are still considered new materials and require considerable scientific research and development to enable the production of large-scale sustainable technology from them.

References

1 Burlant, W.J. and Green, D.H. (1958). Radiation induced craft copolymers several. *J Polym Sci* 28: 252–255. https://doi.org/10.1038/157475a0.

2 Chen, W.K.W., Mesrobian, R.B., Ballantine, D.S. et al. (1957). Studies on graft copolymers derived by ionizing radiation. *J Polym Sci* 23: 903–913. https://doi.org/10.1002/pol.1957.1202310434.

3 Cooper, W., Sewell, P.R., and Vaughan, G. (1959). Radiation graft copolymerization in aqueous dispersion. *J Polym Sci* 41: 167–176. https://doi.org/10.1002/pol.1959.1204113813.

4 Fujiwara, K. (2007). Separation functional fibers by radiation induced graft polymerization and application. *Nucl Instrum Methods Phys Res Sect B Beam Interact Mater Atoms* 265: 150–155. https://doi.org/10.1016/j.nimb.2007.08.042.

5 Kawase, K. and Hayakawa, K. (1967). Radiation-induced graft copolymerization of alkyl methacrylates onto polypropylene. *Radiat Res* 30: 116–128.

6 Huang, R.Y.M. (1966). Structure and properties of cellulose graft copolymers. I. Radiation-induced rayon-styrene graft copolymers. *J Appl Polym Sci* 10: 325–341. https://doi.org/10.1002/9783527807437.ch2.

7 Chapiro, A. and Jendrychowska-Bonamour, A.-M. (1980). Synthesis of permselective membranes by radiation induced grafting of hydrophilic monomers into poly(tetrafluoroethylene) films. *Polym Eng Sci* 20: 202–205. https://doi.org/10.1002/pen.760200306.

8 Ishigaki, I., Kamiya, N., Sugo, T., and Machi, S. (1978). Synthesis of an ion exchange membrane by radiation-induced grafting of acrylic acid onto poly(tetrafluoroethylene). *Polym J* 10: 513–519.

9 Gouloubandi, R. and Chapiro, A. (1980). Membranes permselectives obtenues par greffage radiochimique des acides acrylique et methacrylique sur des films de poly(chlorure de vinyle). *Eur Polym J* 16: 957–964.

10 Chapiro, A., Jendrychowska-Bonamour, A.M., Morel, G., and Oppelt, R. (1973). Membranes hydrophiles non-ionisables obtenues par le greffage radiochimique de la n-vinylpyrrolidone dans des films de polytetrafluoroethylene. *Eur Polym J* 9: 847–862. https://doi.org/10.1016/0014-3057(73)90049-9.

11 Momose, T., Kitazumi, T., Ishigaki, I., and Okamoto, J. (1990). Radiation grafting of α,β,β-trifluorostyrene onto poly(ethylene–tetrafluoroethylene) film by preirradiation method. III. Properties of anion-exchange membrane obtained by chloromethylation and quaternization of the grafted film. *J Appl Polym Sci* 39: 1221–1230. https://doi.org/10.1002/app.1990.070390601.

12 Lee, W., Saito, K., Furusaki, S. et al. (1993). Design of urea-permeable anion-exchange membrane by radiation-induced graft polymerization. *J Membr Sci* 81: 295–305. https://doi.org/10.1016/0376-7388(93)85181-U.

13 Kobayashi, K., Tsuneda, S., Saito, K. et al. (1993). Preparation of microfiltration membranes containing anion-exchange groups. *J Membr Sci* 76: 209–218. https://doi.org/10.1016/0376-7388(93)85218-L.

14 Choi, S., Jeong, Y.H., Ryoo, J.J., and Lee, K. (2001). Desalination by electrodialysis with the ion-exchange membrane prepared by radiation-induced graft polymerization. *Radiat Phys Chem* 60: 503–511.

15 Saito, K., Tsuneda, S., Kim, M. et al. (1999). Radiation-induced graft polymerization is the key to develop high-performance functional materials for protein purification. *Radiat Phys Chem* 54: 517–525. https://doi.org/10.1016/S0969-806X(98)00256-4.

16 Mclean, G.F., Niet, T., and Djilali, N. (2002). An assessment of alkaline fuel cell technology. *Int J Hydrog Energy* 27: 507–526. https://doi.org/10.1016/S0360-3199(01)00181-1.

17 Agel, E., Bouet, J., and Fauvarque, J.F. (2001). Characterization and use of anionic membranes for alkaline fuel cells. *J Power Sources* 101: 267–274.

18 Varcoe, J.R., Atanassov, P., Dekel, D.R. et al. (2014). Anion-exchange membranes in electrochemical energy systems. *Energy Environ Sci* 7: 3135–3191. https://doi.org/10.1039/c4ee01303d.

19 Danks, T.N., Slade, R.C.T., and Varcoe, J.R. (2003). Alkaline anion-exchange radiation-grafted membranes for possible electrochemical application in fuel cells. *J Mater Chem* 13: 712–721. https://doi.org/10.1039/b212164f.

20 Danks, T.N., Slade, R.C.T., and Varcoe, J.R. (2002). Comparison of PVDF- and FEP-based radiation-grafted alkaline anion-exchange membranes for use in low temperature portable DMFCs. *J Mater Chem* 12: 3371–3373. https://doi.org/10.1039/b208627a.

21 Varcoe, J.R. and Slade, R.C.T. (2006). An electron-beam-grafted ETFE alkaline anion-exchange membrane in metal-cation-free solid-state alkaline fuel cells. *Electrochem Commun* 8: 839–843. https://doi.org/10.1016/j.elecom.2006.03.027.

22 Varcoe, J.R., Slade, R.C.T., Wright, G.L., and Chen, Y. (2006). Steady-state dc and impedance investigations of H_2/O_2 alkaline membrane fuel cells with commercial Pt/C, Ag/C, and Au/C cathodes. *J Phys Chem B* 110: 21041–21049. https://doi.org/10.1021/jp064898b.

23 Biancolli, A.L.G., Barbosa, A.S., Kodama, Y. et al. (2021). Unveiling the influence of radiation-induced grafting methods on the properties of polyethylene-based anion-exchange membranes for alkaline fuel cells. *J Power Sources* 512: https://doi.org/10.1016/j.jpowsour.2021.230484.

24 Zhegur-Khais, A., Kubannek, F., Krewer, U., and Dekel, D.R. (2020). Measuring the true hydroxide conductivity of anion exchange membranes. *J Membr Sci* 612: 118461. https://doi.org/10.1016/j.memsci.2020.118461.

25 Poynton, S.D., Slade, R.C.T.T., Omasta, T.J. et al. (2014). Preparation of radiation-grafted powders for use as anion exchange ionomers in alkaline polymer electrolyte fuel cells. *J Mater Chem A* 2: 5124–5130. https://doi.org/10.1039/c4ta00558a.

26 Biancolli, A.L.G., Konovalova, A., Santiago, E.I., and Holdcroft, S. (2023). Measuring the ionic conductivity of solid polymer electrolyte powders. *Int J Electrochem Sci* 18 (10): 100288. https://doi.org/10.1016/j.ijoes.2023.100288.

27 Peng, X., Kulkarni, D., Huang, Y. et al. (2020). Using operando techniques to understand and design high performance and stable alkaline membrane fuel cells. *Nat Commun* 11: https://doi.org/10.1038/s41467-020-17370-7.

28 Wang, L., Peng, X., Mustain, W.E., and Varcoe, J.R. (2019). Radiation-grafted anion-exchange membranes: the switch from low- to high-density polyethylene leads to remarkably enhanced fuel cell performance. *Energy Environ Sci* 12: 1575–1579. https://doi.org/10.1039/c9ee00331b.

29 Auslender, V. L., Berejka, A. J., Bol, J. L. et al. (2011). Industrial radiation with electron beams and X-rays, Revision 6, 2011, International Atomic Energy Agency.

30 Gueven, O. (2004). An overview of current developments in applied radiation chemistry of polymers. *Adv Radiat Chem Polys* IAEA-TECDOC-1420. 33–39.

31 Międzynarodowa Agencja Energii Atomowej (2005). Gamma irradiators for radiation processing. Institute of Nuclear Chemistry and Technology. https://inis.iaea.org/search/search.aspx?orig_q=RN:37081743 (accessed 1 March 2022).

32 Clough, R.L. (2001). High-energy radiation and polymers: a review of commercial processes and emerging applications. *Nucl Instrum Methods Phys Res Sect B Beam Interact Mater Atoms* 185: 8–33. https://doi.org/10.1016/S0168-583X(01)00966-1.

33 Sun, Y. and Chmielewski, A.G. (2017). *Applications of Ionizing Radiation in Materials Processing*, 1e. Warszawa: Institute of Nuclear Chemistry and Technology http://www.ichtj.waw.pl/ichtj/publ/monogr/m2017_1.htm.

34 Nasef, M.M. and Hegazy, E.-S.A.S.A. (2004). Preparation and applications of ion exchange membranes by radiation-induced graft copolymerization of polar monomers onto non-polar films. *Prog Polym Sci* 29: 499–561. https://doi.org/10.1016/j.progpolymsci.2004.01.003.

35 Couture, G., Alaaeddine, A., Boschet, F., and Ameduri, B. (2011). Polymeric materials as anion-exchange membranes for alkaline fuel cells. *Prog Polym Sci* 36: 1521–1557. https://doi.org/10.1016/j.progpolymsci.2011.04.004.

36 Alkan Gürsel, S., Gubler, L., Gupta, B., and Scherer, G.G. (2008). Radiation grafted membranes. *Adv Polym Sci* 215: 157–217. https://doi.org/10.1007/12_2008_153.

37 Dawes, K., Glover, L.C., and Vroom, D.A. (2007). The effects of electron beam and g-irradiation on polymeric materials. In: *Physical Properties of Polymers Handbook* (ed. J.E. Mark), 867–887. New York: Springer New York https://doi.org/10.1007/978-0-387-69002-5_52.

38 Bhattacharya, A. and Misra, B.N. (2004). Grafting: a versatile means to modify polymers: techniques, factors and applications. *Prog Polym Sci* 29: 767–814. https://doi.org/10.1016/j.progpolymsci.2004.05.002.

39 Sherazi, T.A., Yong Sohn, J., Moo Lee, Y., and Guiver, M.D. (2013). Polyethylene-based radiation grafted anion-exchange membranes for alkaline fuel cells. *J Membr Sci* 441: 148–157. https://doi.org/10.1016/j.memsci.2013.03.053.

40 Ashfaq, A., Clochard, M.-C., Coqueret, X. et al. (2020). Polymerization reactions and modifications of polymers by ionizing radiation. *Polymers* 12: 2877. https://doi.org/10.3390/polym12122877.

41 Lyons, B.J. (1995). Radiation crosslinking of fluoropolymers-a review. *Radiat Phys Chem* 45: 159–174. https://doi.org/10.1016/0969-806X(94)E0002-Z.

42 Nasef, M.M., Saidi, H., and Dahlan, K.Z.M. (2003). Electron beam irradiation effects on ethylene-tetrafluoroethylene copolymer films. *Radiat Phys Chem* 68: 875–883. https://doi.org/10.1016/S0969-806X(03)00209-3.

43 Guilmeau, I., Esnouf, S., Betz, N., and Le Moël, A. (1997). Kinetics and characterization of radiation-induced grafting of styrene on fluoropolymers. *Nucl Instrum Methods Phys Res Sect B Beam Interact Mater Atoms* 131: 270–275. https://doi.org/10.1016/s0168-583x(97)00332-7.

44 Arakawa, K., Seguchi, T., Watanabe, Y., and Hayakawa, N. (1982). Radiation-induced oxidation of polyethylene, ethylene–butene copolymer, and ethylene–propylene copolymer. *J Polym Sci* 20: 2681–2692. https://doi.org/10.1002/pol.1982.170200926.

45 Chapiro, A. (1964). Radiation chemistry of polymers. *Radiat Res Suppl* 4: 179–191. https://doi.org/10.2307/3583578.

46 Biancolli, A.L.G., Bsoul-Haj, S., Douglin, J.C. et al. (2022). High-performance radiation grafted anion-exchange membranes for fuel cell applications: effects of irradiation conditions on ETFE-based membranes properties. *J Membr Sci* 641: https://doi.org/10.1016/j.memsci.2021.119879.

47 Nasef, M.M. (2014). Radiation-grafted membranes for polymer electrolyte fuel cells: current trends and future directions. *Chem Rev* 114: 12278–12329. https://doi.org/10.1021/cr4005499.

48 Svetlin Mitov, E.R., Hubner, G., Brack, H.-P., and Scherer, G.G. (2006). In situ electron spin resonance study of styrene grafting of electron irradiated fluoropolymer films for fuel cell membranes. *J Polym Sci Part B Polym Phys* 44: 3323–3336. https://doi.org/10.1002/polb.

49 Ben Youcef, H., Gürsel, S.A., Buisson, A. et al. (2010). Influence of radiation-induced grafting process on mechanical properties of ETFE-based membranes for fuel cells. *Fuel Cells* 10: 401–410. https://doi.org/10.1002/fuce.200900200.

50 Espiritu, R., Golding, B.T., Scott, K., and Mamlouk, M. (2017). Degradation of radiation grafted hydroxide anion exchange membrane immersed in neutral pH: removal of vinylbenzyl trimethylammonium hydroxide due to oxidation. *J Mater Chem A* 5: 1248–1267. https://doi.org/10.1039/c6ta08232g.

51 Zhou, T., Shao, R., Chen, S. et al. (2015). A review of radiation-grafted polymer electrolyte membranes for alkaline polymer electrolyte membrane fuel cells. *J Power Sources* 293: 946–975. https://doi.org/10.1016/j.jpowsour.2015.06.026.

52 Vohlídal, J. (2020). Polymer degradation: a short review. *Chem Teach Int* 3: 1–8. https://doi.org/10.1515/cti-2020-0015.

53 Gubler, L. (2014). Polymer design strategies for radiation-grafted fuel cell membranes. *Adv Energy Mater* 4: 1300827. https://doi.org/10.1002/aenm.201300827.

54 Mélot, M., Ngono-Ravache, Y., and Balanzat, E. (2003). Very low temperature irradiation of aliphatic polymers: role of radical migration on the creation of stable groups (O-127). *Nucl Instrum Methods Phys Res Sect B Beam Interact Mater Atoms* 208: 345–352. https://doi.org/10.1016/S0168-583X(03)00892-9.

55 Oshima, A., Ikeda, S., Seguchi, T., and Tabata, Y. (1997). Temperature effect on radiation induced reactions in ethylene and tetrafluoroethylene copolymer (ETFE). *Radiat Phys Chem* 50: 519–522. https://doi.org/10.1016/S0969-806X(97)00080-7.

56 Chand, K. and Paladino, O. (2023). Recent developments of membranes and electrocatalysts for the hydrogen production by anion exchange membrane water electrolysers: a review. *Arab J Chem* 16: 104451. https://doi.org/10.1016/j.arabjc.2022.104451.

57 Fang, J., Yang, Y., Lu, X. et al. (2012). Cross-linked, ETFE-derived and radiation grafted membranes for anion exchange membrane fuel cell applications. *Int J Hydrog Energy* 37: 594–602. https://doi.org/10.1016/j.ijhydene.2011.09.112.

58 Drobny, J.G. (2013). Electron beam processing of commercial polymers, monomers, and oligomers. In: *Ionizing Radiation Polymers*, 101–147. Elsevier https://doi.org/10.1016/b978-1-4557-7881-2.00005-5.

59 Brack, H.P. and Scherer, G.G. (1998). Modification and characterization of thin polymer films for electrochemical applications. *Macromol Symp* 126: 25–49. https://doi.org/10.1002/masy.19981260105.

60 Lee, W.H., Crean, C., Varcoe, J.R., and Bance-Soualhi, R. (2017). A Raman spectro-microscopic investigation of ETFE-based radiation-grafted anion-exchange membranes. *RSC Adv* 7: 47726–47737. https://doi.org/10.1039/c7ra09650j.

61 Gubler, L., Gürsel, S.A., and Scherer, G.G. (2005). Radiation grafted membranes for polymer electrolyte fuel cells. *Fuel Cells* 5: 317–335. https://doi.org/10.1002/fuce.200400078.

62 Poynton, S.D. and Varcoe, J.R. (2015). Reduction of the monomer quantities required for the preparation of radiation-grafted alkaline anion-exchange membranes. *Solid State Ionics* 277: 38–43. https://doi.org/10.1016/j.ssi.2015.04.013.

63 Brack, H.P., Buhrer, H.G., Bonorand, L., and Scherer, G.G. (2000). Grafting of pre-irradiated poly(ethylene-alt-tetrafluoroethylene) films with styrene: influence of base polymer film properties and processing parameters. *J Mater Chem* 10: 1795–1803. https://doi.org/10.1039/b001851l.

64 Nasef, M.M., Gürsel, S.A., Karabelli, D., and Güven, O. (2016). Radiation-grafted materials for energy conversion and energy storage applications. *Prog Polym Sci* 63: 1–41. https://doi.org/10.1016/j.progpolymsci.2016.05.002.

65 Gürsel, S.A., Youcef, H.B., Wokaun, A., and Scherer, G.G. (2007). Influence of reaction parameters on grafting of styrene into poly(ethylene-alt-tetrafluoroethylene) films. *Nucl Instrum Methods Phys Res Sect B Beam Interact Mater Atoms* 265: 198–203. https://doi.org/10.1016/j.nimb.2007.08.050.

66 Lappan, U., Geißler, U., Gohs, U., and Uhlmann, S. (2010). Grafting of styrene into pre-irradiated fluoropolymer films: influence of base material and irradiation temperature. *Radiat Phys Chem* 79: 1067–1072. https://doi.org/10.1016/j.radphyschem.2010.05.001.

67 Gubler, L., Prost, N., Gürsel, S.A., and Scherer, G.G. (2005). Proton exchange membranes prepared by radiation grafting of styrene/divinylbenzene onto poly(ethylene-alt-tetrafluoroethylene) for low temperature fuel cells. *Solid State Ionics* 176: 2849–2860. https://doi.org/10.1016/j.ssi.2005.09.045.

68 Barbosa, A.S., Biancolli, A.L.G., Lanfredi, A.J.C. et al. (2022). Enhancing the durability and performance of radiation-induced grafted low-density polyethylene-based anion-exchange membranes by controlling irradiation conditions. *J Membr Sci* 659: 120804. https://doi.org/10.1016/j.memsci.2022.120804.

69 Ponce-Gonzalez, J., Whelligan, D.K., Wang, L. et al. (2016). High performance aliphatic-heterocyclic benzyl-quaternary ammonium radiation-grafted anion-exchange membranes. *Energy Environ Sci* 9: 3724–3735. https://doi.org/10.1039/c6ee01958g.

70 Gonçalves Biancolli, A.L., Herranz, D., Wang, L. et al. (2018). ETFE-based anion-exchange membrane ionomer powders for alkaline membrane fuel cells: a first performance comparison of head-group chemistry. *J Mater Chem A* 6: 24330–24341. https://doi.org/10.1039/C8TA08309F.

71 Kuppusamy, H.G., Dhanasekaran, P., Nagaraju, N. et al. (2022). Anion exchange membranes for alkaline polymer electrolyte fuel cells—a concise review. *Materials* 15: https://doi.org/10.3390/ma15165601.

72 Sun, Z., Lin, B., and Yan, F. (2018). Anion-exchange membranes for alkaline fuel-cell applications: the effects of cations. *ChemSusChem* 11: 58–70. https://doi.org/10.1002/cssc.201701600.

73 Salma, U. and Shalahin, N. (2023). A mini-review on alkaline stability of imidazolium cations and imidazolium-based anion exchange membranes. *Results Mater* 17: 100366. https://doi.org/10.1016/j.rinma.2023.100366.

74 Wang, X. and Lammertink, R.G.H. (2022). Dimensionally stable multication-crosslinked poly(arylene piperidinium) membranes for water electrolysis. *J Mater Chem A* 10: 8401–8412. https://doi.org/10.1039/d1ta10820d.

75 Zhu, L., Zimudzi, T.J., Li, N. et al. (2016). Crosslinking of comb-shaped polymer anion exchange membranes via thiol-ene click chemistry. *Polym Chem* 7: 2464–2475. https://doi.org/10.1039/c5py01911g.

76 Tillet, G., Boutevin, B., and Ameduri, B. (2011). Chemical reactions of polymer crosslinking and post-crosslinking at room and medium temperature. *Prog Polym Sci* 36: 191–217. https://doi.org/10.1016/j.progpolymsci.2010.08.003.

77 Lee, K.H., Cho, D.H., Kim, Y.M. et al. (2017). Highly conductive and durable poly(arylene ether sulfone) anion exchange membrane with end-group cross-linking. *Energy Environ Sci* 10: 275–285. https://doi.org/10.1039/c6ee03079c.

78 Clemens, A.L., Jayathilake, B.S., Karnes, J.J. et al. (2023). Tuning material properties of alkaline anion exchange membranes through crosslinking: a review of synthetic strategies and property relationships. *Polymers* 15: 1534.

79 Ramachandran, R., Jung, D., and Spokoyny, A.M. (2019). Cross-linking dots on metal oxides. *NPG Asia Mater* 11: 9–12. https://doi.org/10.1038/s41427-019-0119-9.

80 Shangguan, Y., Yang, J., and Zheng, Q. (2017). Rheology of nitrile rubber with hybrid crosslinked network composed of covalent bonding and hydrogen bonding. *RSC Adv* 7: 15978–15985. https://doi.org/10.1039/c7ra01106g.

81 Salvatore, D.A., Gabardo, C.M., Reyes, A. et al. (2021). Designing anion exchange membranes for CO_2 electrolysers. *Nat Energy* 6: 339–348. https://doi.org/10.1038/s41560-020-00761-x.

82 Gu, S., Skovgard, J., and Yan, Y.S. (2012). Engineering the van der Waals interaction in cross-linking-free hydroxide exchange membranes for low swelling and high conductivity. *ChemSusChem* 5: 843–848. https://doi.org/10.1002/cssc.201200057.

83 Ran, J., Wu, L., Ge, Q. et al. (2014). High performance anion exchange membranes obtained through graft architecture and rational cross-linking. *J Membr Sci* 470: 229–236. https://doi.org/10.1016/j.memsci.2014.07.036.

84 Lee, B., Lim, H., Chae, J.E. et al. (2019). Physically-crosslinked anion exchange membranes by blending ionic additive into alkyl-substituted quaternized PPO. *J Membr Sci* 574: 33–43. https://doi.org/10.1016/j.memsci.2018.12.053.

85 Wang, X., Wang, P., Sun, Y. et al. (2017). A mechanically strong and tough anion exchange membrane engineered with non-covalent modalities. *Chem Commun* 53: 12369–12372. https://doi.org/10.1039/c7cc07284h.

86 Sung, S., Mayadevi, T.S., Min, K. et al. (2021). Crosslinked PPO-based anion exchange membranes: the effect of crystallinity versus hydrophilicity by oxygen-containing crosslinker chain length. *J Membr Sci* 619: 118774. https://doi.org/10.1016/j.memsci.2020.118774.

87 Ertem, S.P., Tsai, T.H., Donahue, M.M. et al. (2016). Photo-cross-linked anion exchange membranes with improved water management and conductivity. *Macromolecules* 49: 153–161. https://doi.org/10.1021/acs.macromol.5b01784.

88 Lin, C.X., Zhuo, Y.Z., Hu, E.N. et al. (2017). Crosslinked side-chain-type anion exchange membranes with enhanced conductivity and dimensional stability. *J Membr Sci* 539: 24–33. https://doi.org/10.1016/j.memsci.2017.05.063.

89 Katzfuß, A., Gogel, V., Jörissen, L., and Kerres, J. (2013). The application of covalently cross-linked BrPPO as AEM in alkaline DMFC. *J Membr Sci* 425–426: 131–140. https://doi.org/10.1016/j.memsci.2012.09.022.

90 Tongwen, X. and Zha, F.F. (2002). Fundamental studies on a new series of anion exchange membranes: effect of simultaneous amination-crosslinking processes on membranes ion-exchange capacity and dimensional stability. *J Membr Sci* 199: 203–210. https://doi.org/10.1016/S0376-7388(01)00698-6.

91 Zhao, Y., Yu, H., Yang, D. et al. (2013). High-performance alkaline fuel cells using crosslinked composite anion exchange membrane. *J Power Sources* 221: 247–251. https://doi.org/10.1016/j.jpowsour.2012.08.053.

92 Li, H., Yu, N., Gellrich, F. et al. (2021). Diamine crosslinked anion exchange membranes based on poly(vinyl benzyl methylpyrrolidinium) for alkaline water electrolysis. *J Membr Sci* 633: https://doi.org/10.1016/j.memsci.2021.119418.

93 Wu, L., Pan, Q., Varcoe, J.R. et al. (2015). Thermal crosslinking of an alkaline anion exchange membrane bearing unsaturated side chains. *J Membr Sci* 490: 1–8. https://doi.org/10.1016/j.memsci.2015.04.046.

94 Hossain, M.M., Wu, L., Liang, X. et al. (2018). Anion exchange membrane crosslinked in the easiest way stands out for fuel cells. *J Power Sources* 390: 234–241. https://doi.org/10.1016/j.jpowsour.2018.04.064.

95 Weissbach, T., Wright, A.G., Peckham, T.J. et al. (2016). Simultaneous, synergistic control of ion exchange capacity and cross-linking of sterically-protected poly(benzimidazolium)s. *Chem Mater* 28: 8060–8070. https://doi.org/10.1021/acs.chemmater.6b03902.

96 You, W., Noonan, K.J.T., and Coates, G.W. (2020). Alkaline-stable anion exchange membranes: a review of synthetic approaches. *Prog Polym Sci* 100: 101177. https://doi.org/10.1016/j.progpolymsci.2019.101177.

97 Dong, D., Xiao, Y., Zhang, M. et al. (2022). Crosslinked anion exchange membranes with regional intensive ion clusters prepared from quaternized branched polyethyleneimine/quaternized polysulfone. *Int J Hydrog Energy* 47: 24991–25006. https://doi.org/10.1016/j.ijhydene.2022.05.237.

98 Yang, W., Chen, J., Yan, J. et al. (2022). Advance of click chemistry in anion exchange membranes for energy application. *J Polym Sci* 60: 627–649. https://doi.org/10.1002/pol.20210819.

99 Li, X., Ji, D., Yu, B. et al. (2021). Boosting piezoelectric and triboelectric effects of PVDF nanofiber through carbon-coated piezoelectric nanoparticles for highly sensitive wearable sensors. *Chem Eng J* 426: 130345. https://doi.org/10.1016/j.cej.2021.130345.

100 Wang, W., Wang, S., Li, W. et al. (2013). Synthesis and characterization of a fluorinated cross-linked anion exchange membrane. *Int J Hydrog Energy* 38: 11045–11052. https://doi.org/10.1016/j.ijhydene.2013.03.166.

101 Lu, W., Shao, Z.G., Zhang, G. et al. (2013). Preparation of anion exchange membranes by an efficient chloromethylation method and homogeneous quaternization/crosslinking strategy. *Solid State Ionics* 245–246: 8–18. https://doi.org/10.1016/j.ssi.2013.05.005.

102 Lai, A.N., Guo, D., Lin, C.X. et al. (2016). Enhanced performance of anion exchange membranes via crosslinking of ion cluster regions for fuel cells. *J Power Sources* 327: 56–66. https://doi.org/10.1016/j.jpowsour.2016.07.043.

103 Hao, J., Gao, X., Jiang, Y. et al. (2018). Crosslinked high-performance anion exchange membranes based on poly(styrene-*b*-(ethylene-*co*-butylene)-*b*-styrene). *J Membr Sci* 551: 66–75. https://doi.org/10.1016/j.memsci.2018.01.033.

104 Long, C., Wang, Z., and Zhu, H. (2021). High chemical stability anion exchange membrane based on poly(aryl piperidinium): effect of monomer configuration on membrane properties. *Int J Hydrog Energy* 46: 18524–18533. https://doi.org/10.1016/j.ijhydene.2021.02.209.

105 Lin, X., Gong, M., Liu, Y. et al. (2013). A convenient, efficient and green route for preparing anion exchange membranes for potential application in alkaline fuel cells. *J Membr Sci* 425–426: 190–199.

106 Luo, Y., Guo, J., Wang, C., and Chu, D. (2012). Fuel cell durability enhancement by crosslinking alkaline anion exchange membrane electrolyte. *Electrochem Commun* 16: 65–68. https://doi.org/10.1016/j.elecom.2012.01.005.

107 Rao, A.H.N., Nam, S., and Kim, T.H. (2014). Crosslinked poly(arylene ether sulfone) block copolymers containing pendant imidazolium groups as both crosslinkage sites and hydroxide conductors for highly selective and stable membranes. *Int J Hydrog Energy* 39: 5919–5930. https://doi.org/10.1016/j.ijhydene.2014.01.191.

108 Xue, B., Wang, Q., Zheng, J. et al. (2020). Bi-guanidinium-based crosslinked anion exchange membranes: synthesis, characterization, and properties. *J Membr Sci* 601: 117923. https://doi.org/10.1016/j.memsci.2020.117923.

109 Zha, Y., Disabb-miller, M.L., Johnson, Z.D. et al. (2012). Metal-cation-based anion exchange membranes. *J Am Chem Soc* 134: 4493–4496.

110 Zhu, T., Xu, S., Rahman, A. et al. (2018). Cationic metallo-polyelectrolytes for robust alkaline anion-exchange membranes. *Angew Chem Int Ed* 57: 2388–2392. https://doi.org/10.1002/anie.201712387.

111 Yang, W., Yan, J., Liu, S. et al. (2021). Macromolecular crosslink of imidazole functionalized poly(vinyl alcohol) and brominated poly(phenylene oxide) for anion exchange membrane with enhanced alkaline stability and ionic conductivity. *Int J Hydrog Energy* 46: 37007–37016. https://doi.org/10.1016/j.ijhydene.2021.08.184.

112 He, X., Cheng, C., Huang, S. et al. (2020). Alkaline anion exchange membranes with imidazolium-terminated flexible side-chain cross-linked topological structure based on ROMP-type norbornene copolymers. *Polymer* 195: 122412. https://doi.org/10.1016/j.polymer.2020.122412.

113 Jang, J., Ahn, M.K., Bin Lee, S. et al. (2021). Conductive and stable crosslinked anion exchange membranes based on poly(arylene ether sulfone). *Macromol Res* 29: 157–163. https://doi.org/10.1007/s13233-021-9023-6.

114 Marrocchi, A., Facchetti, A., Lanari, D. et al. (2016). Click-chemistry approaches to π-conjugated polymers for organic electronics applications. *Chem Sci* 7: 6298–6308. https://doi.org/10.1039/c6sc01832g.

115 Kumari, M., Douglin, J.C., and Dekel, D.R. (2021). Crosslinked quaternary phosphonium-functionalized poly (ether ketone) polymer-based anion-exchange membranes. *J Membr Sci* 626: 119167.

116 Li, N., Wang, L., and Hickner, M. (2014). Cross-linked comb-shaped anion exchange membranes with high base stability. *Chem Commun* 50: 4092–4095. https://doi.org/10.1039/c3cc49027k.

117 Huang, J., Yu, Z., Tang, J. et al. (2022). A review on anion exchange membranes for fuel cells: anion-exchange polyelectrolytes and synthesis strategies. *Int J Hydrog Energy* 47: 27800–27820. https://doi.org/10.1016/j.ijhydene.2022.06.140.

118 Zhang, X., Cao, Y., Zhang, M. et al. (2020). Olefin metathesis-crosslinked, bulky imidazolium-based anion exchange membranes with excellent base stability and mechanical properties. *J Membr Sci* 598: 117793. https://doi.org/10.1016/j.memsci.2019.117793.

119 Huang, S., He, X., Cheng, C. et al. (2020). Facile self-crosslinking to improve mechanical and durability of polynorbornene for alkaline anion exchange membranes. *Int J Hydrog Energy* 45: 13068–13079. https://doi.org/10.1016/j.ijhydene.2020.03.013.

120 Lu, Y., Liu, L., Pu, Y. et al. (2021). Towards performance improved anion exchange membrane: cross-linking with multi-cations oligomer modified graphene oxide. *Int J Hydrog Energy* 46: 23855–23867. https://doi.org/10.1016/j.ijhydene.2021.04.167.

121 Qiu, B., Lin, B., Qiu, L., and Yan, F. (2012). Alkaline imidazolium- and quaternary ammonium-functionalized anion exchange membranes for alkaline fuel cell applications. *J Mater Chem* 22: 1040–1045. https://doi.org/10.1039/c1jm14331j.

11

Degradation Mechanisms of Anion Exchange Membranes due to Alkali Hydrolysis and Radical Oxidative Species

Jeet Sharma and Vaibhav Kulshrestha

CSIR-Central Salt and Marine Chemicals Research Institute, Membrane Science and Separation Technology (MS & ST) Division, Bhavnagar 364002, Gujarat, India
Academy of Scientific and Innovative Research (AcSIR), Ghaziabad 201002, Uttar Pradesh, India

11.1 Introduction

Significant progress, innovative research, and rational insights in designing stable cationic head groups have placed anion exchange membranes (AEMs) at the heart of promising low-cost electrochemical energy conversion and storage devices [1–3]. Unlike proton exchange membrane fuel cells (PEMFCs), the AEM integrated devices can operate using affordable and abundant transition metal-based electrocatalysts, making AEMs a promising constituent in economically viable fuel cell technology, water electrolyzers, and alkaline redox flow batteries. Moreover, AEM integrated electrochemical energy appliances offer dual advantages. (i) They can deliver power comparable to that with proton exchange membranes (PEMs) attributed to low fuel crossover, rapid O_2 reduction kinetics, and higher fuel oxidation rates, and (ii) they can operate at higher current densities. However, the chemical degradation of AEM and carbonation hampers the deployment of these energy conversion devices significantly. The former results in mechanical failure, while the latter affects the ion conduction in AEM [4]. In addition, the intrinsic degradation aspects like adsorptive fouling, morphological relaxation/interfacial delamination with catalyst layers, and cell efficiency issues are directly or indirectly associated to performance losses in the electrochemical device [2]. In AEMs, the most common alkaline exchanger is quaternary ammonium groups (QAGs, [P]-N^+R_3) because of its better hydroxide (^-OH) dissociation in hydrophilic phase, and hence high ^-OH conductivity ($\kappa^m{}_{^-OH}$) and ^-OH exchange capacities can be achieved in comparison

Alkaline Anion Exchange Membranes for Fuel Cells: From Tailored Materials to Novel Applications, First Edition. Edited by Jince Thomas, Alex Schechter, Flavio Grynszpan, Bejoy Francis, and Sabu Thomas.

to those of sulfonium ([P]-S^+R_2) and phosphonium ([P]-P^+R_3)-based functional groups. Interestingly, in stark contrast to QAGs, the docking and functional grafting of inorganic multivalent cationic sites are gaining sustained surveillance in AEM fabrication. Thus, strategic designing of polymer backbone and ion conducting groups becomes important to fabricate stable and efficient AEMs. In addition, understanding the cause of failure is even more important. The context of the chapter includes the following:

(i) Necessity to study degradation mechanism in AEMs under an alkaline and oxidative environment.
(ii) Structure and degradation mechanism of tailored hydroxide conducting groups, this includes various cationic head-groups of organic, inorganic (metallo-polyelectrolytes), and hybrid origin, which shows nucleophilic substitution reaction (S_N), β-H elimination reaction, ylide formation, and heterocyclic cation ring-opening reactions.
(iii) Evidences of alkaline degradation and oxidative radical specie formation through various spectroscopic and *in situ* techniques.
(iv) Recent advances in AEM designs with respect to hydrophilic/hydrophobic phase separation, bulky group masking type functionalization (the shielding through σ-π hyperconjugation effect, rat-trap effect, decoy effect, backbone-spacer alkyl tuning, etc.), hydration environment management, the cation group effect, and long-side chain puckered architectures with novel chemical designs of AEMs.
(v) Benchmarking polymeric materials for electrochemical energy applications with summary on half-life of various $^-$OH conducting functionalities.
(vi) Prospect and outlook on the economic commercialization of AEMs.

11.2 Necessity to Investigate the Degradation Mechanism in AEMs

Conscientious designing and development of the anion exchange membrane fuel cells (AEMFCs) has achieved the performance equivalent to the state-of-the-art PEMFC with remarkably high hydroxide conductivity ($\kappa^m{}_{\text{-OH}} \geq 120\,\text{mS cm}^{-1}$) in poly(olefin)- and poly(arylene)-based AEMs. The AEMFC devices have sustainably been able to deliver the peak power density ≥2000 and ≥1500 mW cm^{-2} at 60–80°C, respectively. Since these devices with AEM operate using affordable and abundant rare-earth metal-based electrocatalysts, the designing of such advanced materials can bring economic implementation for sustainable energy goals. In the beginning of AEMFC developments, the $\kappa^m{}_{\text{-OH}}$ trait gained significant attention. However, in another prospect, various reports claim that the most decisive parameter to the terrestrial utility lies in the durability of AEM in fuel cell devices. Due to this instability issue, the operational lifetime of AEMFCs is less than that of the PEMFC devices [2]. In brief, the low hydration state in the conducting domains of the AEM perturbates the activation energy (or nucleophilicity) of hydroxide ions to breach cationic head groups and cause faster degradation in fuel cell devices [5,6]. Yet another issue that underlies performance degradation is interference from atm CO_2. It is evident that both hydration state of AEM and the CO_2 uptake dictate the degradation and cause severe reduction of electrochemical/^{-}OH transport properties and increases overall cell resistance significantly. Thus, studies concerned to nanophase morphology of membranes, which causes pore opening and swelling for different CO_2 regime during AEM synthesis, are desired [7,8]. Interestingly, the stability of functionally modified polymer backbone also plays a crucial role to navigate the mechanical strength of AEMs. With different chemical structures, the rate of alkaline hydrolysis depends on the ease of reacting with main polymer chains, and this in turn causes rupture of AEMs [2]. Various factors, viz. the stereoelectronic effects, steric/bulky group effects, and water dynamics affect the degradation mechanisms in AEM [3–5]. The “triggered polymer scissoring” is a conventional polymer degradation pathway [9–11]. Other issues that affect the durability of AEMFC/alkaline membrane water electrolyzer (AEMWE) include water-flooding management, delamination of interfaces, and enhanced carbonation at elevated temperatures influencing the performance and longevity of electrochemical energy devices. Evidently, the dual nature of ^{-}OH pertaining both the nucleophilicity and base behavior (i.e. amphiphilic) is a real challenge which offers a novel platform to explore, optimize, and establish the most stable chemistry between a free predator (^{-}OH) and confined prey (i.e. cationic functionalities) which seems impossible without understanding the crux of degradation routes. Thus, hydroxide conductivity is now most likely not a hurdle in the success of AEMFC/AEMWE. It is the durability of AEM we are trying to breach for and to design state-of-the-art electrochemical energy devices.

11.3 Structure and Degradation Mechanism of Tailored Anion Exchange Groups and Polymers

The benzylic ammonium groups are studied much comprehensively in AEM synthesis due to their broad working pH stability, high ion-dissociation traits, and ease of modification in polymer matrix. Similarly, the benzylic phosphonium and sulfonium functional group offers hydroxide conduction; however, due to low ion-dissociation affinity, high sensitivity toward alkaline pH, and severe environmental impact, the latter are not preferred in AEM synthesis. Depending on the type of the amine precursor, the ammonium groups include aliphatic, cycloaliphatic, and heterocyclic aromatic ammonium ion systems. Other novel groups include guanidinium cation, Tröger’s base groups, metallocenium cationic groups, and crown ethers. Moreover, for sulfonium- and phosphonium-based AEMs, the functional groups are generally dimethyl sulfide, trimethyl phosphine, and triphenyl phosphine (shown in Figure 11.1) [1,12]. However, recent advances have shown novel designs to screen stable aromatic and cycloaliphatic groups in phosphonium- and sulfonium-based AEMs.

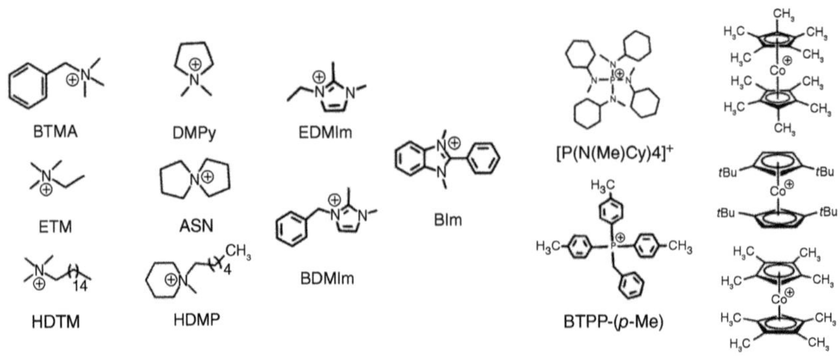

Figure 11.1 Structures of various anion exchange head groups studied in AEM synthesis [1]. Source: Reproduced with permission of Elsevier.

11.3.1 Alkaline Hydrolysis of Cationic Head Groups

In AEM degradation, the bimolecular nucleophilic substitution (S_N2) and E_2 elimination (also, Hoffmann elimination) reactions are the most prominent reaction mechanisms. Moreover, the rearrangement reactions like Steven's rearrangement and Sommelet–Hauser rearrangement are reported in literatures. These degradation pathways are called alkaline hydrolysis. E_1 eliminations are rare occurring mostly with bulky cationic head groups [13]. The degradation through rearrangements proceeds via *N*-ylide intermediates. In addition, depending upon the structures and net electron density around these groups, the degradation rate may vary in alkaline conditions because hydration environment around these head groups influence the vulnerability for ^-OH attack. Also, the nucleophilicity of ^-OH varies depending upon the counterions. For instance, the potassium hydroxide (KOH) is more nucleophilic than sodium hydroxide (NaOH) (according to Fajan's rule) and cause much faster degradation of cationic functional motifs. Other factors like operational temperature and the pressure of fuel gas affect the membrane performance in fuel cell devices. Studies by Marino and Kreuer confer that, with electron-withdrawing functionalities in the proximity, the S_N2 rates also increase, which reduces the $t_{1/2}$ of ion-conducting ammonium groups. For example, the methoxy (–OMe)-substituted BTMA (MBTMA) has ~4 times longer ($t_{1/2}$ ~ 16.6 h) $t_{1/2}$ life than benzyl trimethylammonium (BTMA) ($t_{1/2}$ ~ 4.18 h) and ~25 times longer $t_{1/2}$ than nitro ($^-NO_2$) substituted BTMA (NBTMA, $t_{1/2}$ ~ 0.66 h). Similarly, tetra methyl ammonium (Me_4N^+) has ~15 times longer ($t_{1/2}$ ~ 61.9 h) life than benzyl-substituted analogs (BTMA) in 6 M NaOH at 160 °C [14]. The reports by Chen et al. proposed a novel method to enhance the durability of AEM by incorporating a much more electrophilic substituent seems appropriate strategy (Figure 11.2a,c) [15]. In this work, they synthesized poly(sulfone) (PSf)-based AEM with bifunctional groups bearing the carbonyl group at the γ-position can provide reversible protection to the ammonium group. In addition, the H-atoms at β-position were replaced by bulky methyl groups, eliminating any possibility of Hoffmann elimination. As confirmed by density functional theory (DFT) method and van der Waals surfaces of the molecules, these groups evidently perturbate the local electron density in aromatic rings (reduces electro-static potential [ESP] on aromatic ring) and hence reduce alkaline hydrolysis via reversible tetrahedral ketal formation (Figure 11.2b). The membranes also exhibit an improved hydroxide conductivity of 35.3 $mS\,cm^{-1}$ at 20 °C (Figure 11.2d) and only 5% loss when kept in 1 M NaOH for 10 days. This phenomenon was termed as "The Rat-trap effect." In further studies, the effects of multi-cationic long-side chain architectures and alkyl-oxy extenders were explored and validated by Zhu et al. and Li et al., respectively [16,17]. The fine-tuning between the spacer chain length and ionic conductivity was studied by Zhu et al. The trade-offs at lower spacer length and higher fixed charge densities were explained with the notion that such morphologies bring strong interactions, causing agglomeration of charge and hence lowering the water uptake. Thus, asymmetric spacer length with triple cation morphology (T25NC6NC5N membrane) was proposed best in

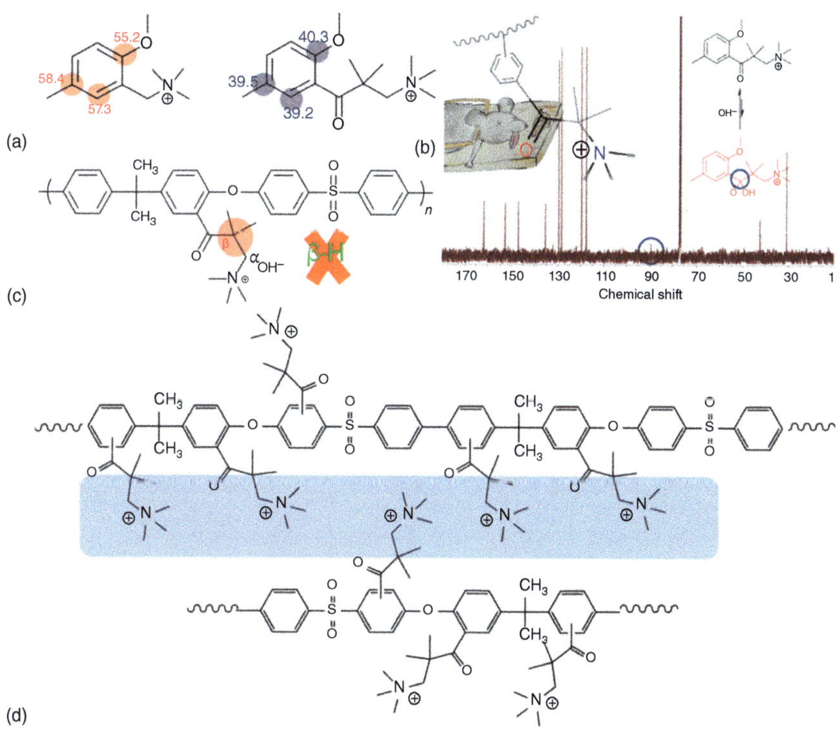

Figure 11.2 (a) The electro-static potential (ESP) values on poly(sulfone) polymer chains with the benzyl quaternary ammonium type hydrophilic functional group (numerical values highlighted), (b) carbonyl groups can provide reversible protection to QA groups. The QA group attracts ^{-}OH ions like cheese attract mice. The carbonyl group can capture ^{-}OH to form gem-diol anion, just as a mouse is captured by a rat trap. When gem-diol anion is formed, the other ^{-}OH ions will not attack QA anymore, like the deterrent effect by the rat trap, (c) eliminating the possibility of β-H elimination with the two methyl groups in 3-chloro-2,2-dimethylpropionyl chloride occupying the position of quaternary ammonium group, and (d) the continuous water aggregation zone in designed AEM [15]. Source: Reproduced with permission of Royal Society of Chemistry.

terms of hydroxide conductivity and ion-exchange capacity (IEC) ($\kappa^{m}{}_{\text{-OH}} = 99\,\text{mS}\,\text{cm}^{-1}$ at room temperature with an IEC value of $2.87\,\text{mmol}\,\text{g}^{-1}$). In contrast, the T25NC6NC8N and T25NC6NC3N with 6, 8 and 6, 3 carbon chained spacer exhibit the hydroxide conductivity of 57 and $56\,\text{mS}\,\text{cm}^{-1}$ showing an IEC of 2.76 and $2.94\,\text{mmol}\,\text{g}^{-1}$ under the same experimental conditions, respectively. At 80 °C, the T25NC6NC5N AEM exhibit the highest hydroxide conductivity of $176\,\text{mS}\,\text{cm}^{-1}$. Moreover, the alkaline hydrolysis was studied for different membranes and the T25NC6NC5N exhibits 90% conductivity retention in 1 M NaOH at 80 °C after 500 h (Figure 11.3a). Li et al. developed structurally similar yet modified morphologies bearing alkoxy extenders in the side chain of AEM. The mutual benefits from the hydrophobic alkyl chain and alkoxy extenders impart better durability in AEMs. The alkaline stability studies suggest more than 70% of electrochemical property retention, whereas the conventional quaternized AEMs were able to retain only 24% of their initial values. This strategic design of AEM to endure alkaline hydrolysis was termed as "the Decoy effect." In this, the $^{-}$OH ion gets confined in the densely populated alkoxy extender-based co-network through H-bonding and hence reducing the $^{-}$OH to breach the barrier to reach the ammonium group (Figure 11.4b,c).

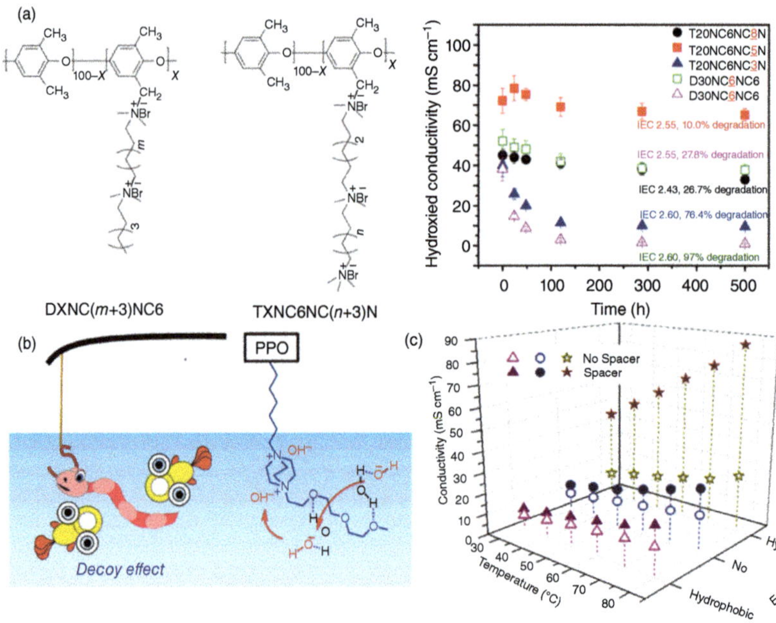

Figure 11.3 (a) Multi-cation side chain type AEMs with different spacer lengths with a stability profile of different membranes (in the graph) studied by Zhu et al. using 1 M NaOH solution at 80 °C ($^{-}$OH conductivity as a function of duration time measured at 20 °C), (b) schematic illustration of "The Decoy Effect"; ether groups in the long-side chain (akin decoy) can influence the hydrated $^{-}$OH ions through hydrogen bonding and hence reduce alkaline hydrolysis of ammonium ion-conducting groups, and (c) variation in the hydroxide conductivity of poly(phenylene oxide)-based AEMs as a function of temperature (ion exchange capacity [IEC] = 1.10–1.19 mmol g^{-1}) [16, 17]. Source: Reproduced with permissions of Elsevier.

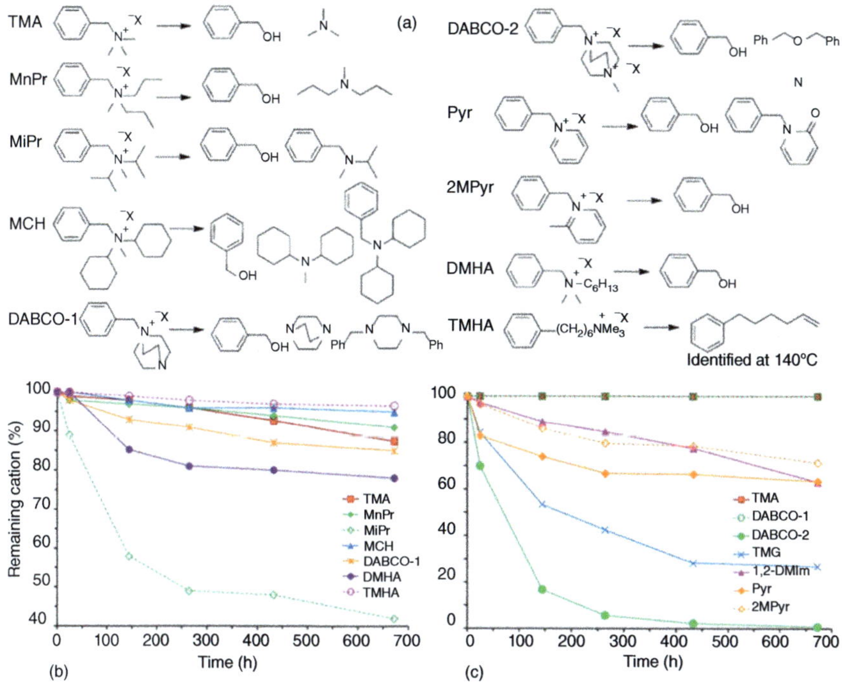

Figure 11.4 (a) Alkaline hydrolysis by-products of variable cationic head groups identified from $CDCl_3$ extraction by Mohanty and Bae (where X = OH or OD), (b) percentage retention of cationic head groups in AEMs over time at 60 °C, and (c) percentage retention of cationic head groups in AEMs over time at 120 °C. Source: Mohanty and Bae [18]/Royal Society of Chemistry.

In addition to the optimization of local electronic density, the bulky architecture of amines plays a significant role and dictates alkaline hydrolysis in AEMs. For instance, the influence on the alkaline hydrolysis of sterically crowded quaternary amines was explored by Mohanty and Bae using mass spectroscopy analysis and retention in cationic groups in model compounds. Nearly 10 different amines of various origin and bulk orientation around the N-atom (Figure 11.4) were explored. With >95% cationic retention in bulky methyl cyclohexyl (MCH), methyl *n*-propyl (MnPr) and long-side chain type trimethyl hexyl amine (TMHA) at 120 °C for 700 h, the results strongly illustrate the influence of bulk orientation in AEMs.

In the same work, she corroborated that even with resonance stabilization in the heterocyclic amines, viz., imidazolium and pyridinium, the stability was diagnosed to be lower than that of the tri-methyl amine derivative. However, the substitution of the methyl group next to cationic head of imidazolium and pyridinium might improve the alkaline stability through steric crowding. But, in conclusion, a note was mentioned for the conductivity–stability trade-offs in bulky amine-based anion exchange models [18]. Nuñez et al. comprehensively studied the influence of alkaline hydrolysis with *N*-alkyl interstitial spacers and terminal pendent-type ammonium functionalities on the percentage degradation via S_N2 and E_2 reactions, respectively [19] (Figure 11.5).

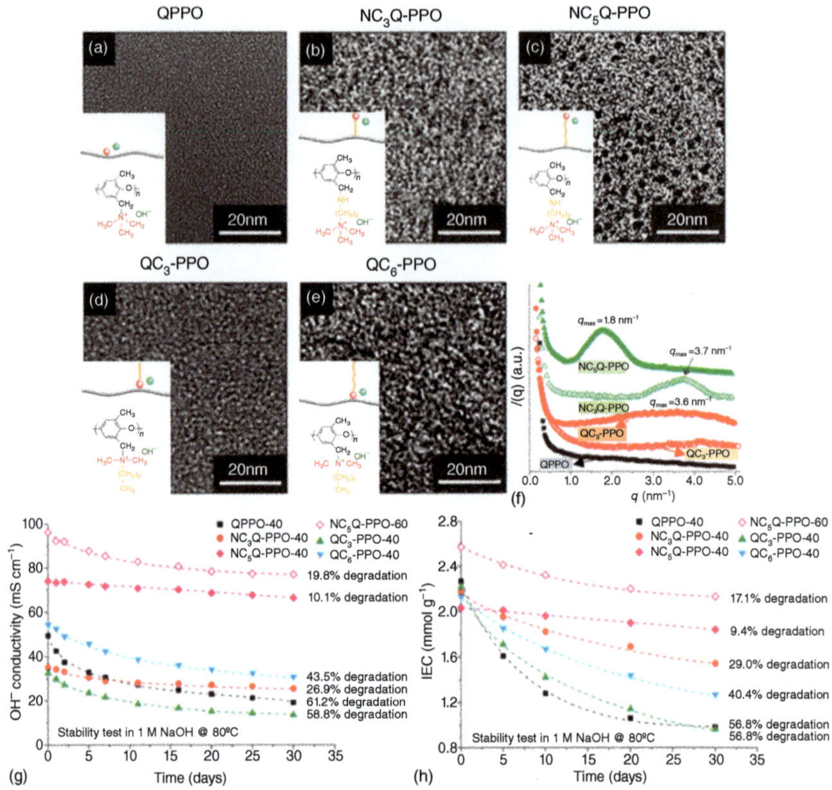

Figure 11.5 Influence of microstructure perturbation on alkaline hydrolysis of AEMs: (a–e) transmission electron microscopy (TEM) showing the phase morphology with variable cationic architecture (the dark contrast in the TEM image corresponds to the hydrophilic domains with bromide anions reside), (f) small angle X-ray scattering (SAXS) profile of different AEMs in dry state, (g, h) determination of quaternary ammonium cation stability in terms of variation in hydroxide conductivity (at 80 °C in 1 M NaOH) and IEC (function of time) of AEMs. Source: Reproduced with permissions from Pan et al. [20]/American Chemical Society.

The results suggest that trialkyl ammonium with *N*-alkyl interstitial spacers are more stable toward alkaline hydrolysis in comparison to conventional BTMA and benzyl dimethyl ammonium cation. Moreover, a small improvement is observed in the alkaline durability of AEMs when the synergy of *N*-alkyl interstitial spacer and *N*-alkyl pendent-type ammonium groups is incorporated in a single cation. These results strongly affirm to the observations reported by Zhu et al. [17]. Moreover, results confer that at higher temperatures (~120 °C), the poly(styrene)-based AEMs are more prone to degradation than poly(phenylene oxide) (PPO)-based AEMs due to the triggering of anionic-1, 4-induced elimination reaction.

In contrast, the report by Pan et al. examine that the length of the side chain and its grafting behavior in solid polyelectrolytes have some limitations. With only three C-atom long chain (i.e. dimethyl propyl amine), the stability and conductivity both were compromised because of facile β-H elimination. However, on extending the chain to six C-atoms, the enhancement in alkaline stability and hydroxide conductivity was more profound attributed to the increment in the affinity barrier for ^{-}OH after introducing long and flexible spacer groups. It is stated that such morphologies induce distinct hydrophobic/hydrophilic microphase separation causing higher $\kappa^{m}{}_{-OH}$ and reducing swelling behavior of AEMs [20].

Interestingly, the guanidinium cationic groups were entertained as an alternative amine model and studied for its influence on chemo-mechanical stability of AEM. Attributed to bulky size, optimum basicity, higher achievable hydration level, and delocalized charge distribution around three nitrogen atoms and one carbon center, it gained a curious interest in AEM synthesis [21]. Moreover, some studies have stated that the kinetics of oxidation and oxygen reduction reactions (ORRs) were greatly improved by using guanidinium-based materials. Sherazi et al. designed AEM using 1,1,3,3-tetramethyl-2-butylguanidine and γ-induced vinyl benzyl chloride grafted poly(ethylene). These membranes exhibit highest ^{-}OH conductivity of 27.7 mS cm^{-1} at 90 °C with only 12.5% degree of grafting. The degradation mechanism was visualized by Fourier transform infrared (FT-IR) spectroscopy analysis showing peculiar increment in peak for the carbonyl group at 1740 cm^{-1} due to alkaline hydrolysis. The degradation mechanism of the guanidinium group is shown in Figure 11.6 (a-c) [22].

Xue et al. studied the substituent effect on the alkaline hydrolysis of guanidinium groups. Attributed to high electron delocalization and absence of β-H atom, these functionalities can impart high conductivity (due to high basicity) and can protect from conventional S_N and E_2 degradation reactions, respectively. The degradation studies suggest hydrolysis of guanidinium to form small amine molecules and urea of high molecular weights. Moreover, the influence of the substituent show that the phenyl substitution offers improved susceptibility for alkaline hydrolysis over dodecyl- and hexyl-substituted guanidinium, respectively. This effect can be termed as the "shielding effect", because the bulky group shields the ionic group from degradation [23,24]. In his subsequent study, he designed a PPO-based di-cationic cross-linked AEM with optimized swelling attributes and high hydroxide conductivity. These membranes show good alkaline durability in 1 M NaOH for 30 days (retaining 90% of its initial electrochemical property) [24]. Liu et al. explored guanidinium functionalized graphene oxide (GGO) incorporation in hydroxyl modified poly(ether sulfone) (PESf) AEM for fuel cell applications. The AEM with 25 wt% GGO exhibited higher alkaline stability showing only 25% degradation in 1 M NaOH at 60 °C for 120 h. As evidenced, this strategy to design GGO composites to hydroxyl modified PESf lacking the ammonium group in the main polymer chain could acquire better durability in alkaline fuel cell application [25]. Despite bulky scaffold and unique charge delocalization in guanidinium ion, the lifetime assessment studies by Marino and Kreuer summarized that the half-life of the guanidinium ion was too short to measure in contrast to trimethylamine (TMA) groups [14]. The plausible reason to this anomaly could be the counter-effect of charge delocalization at the sp^2 carbon center, which makes it more susceptible to ^{-}OH attack via the electronic effects. However, the precise chemical modification of guanidinium-based AEMs is gaining sustained surveillance to improvise

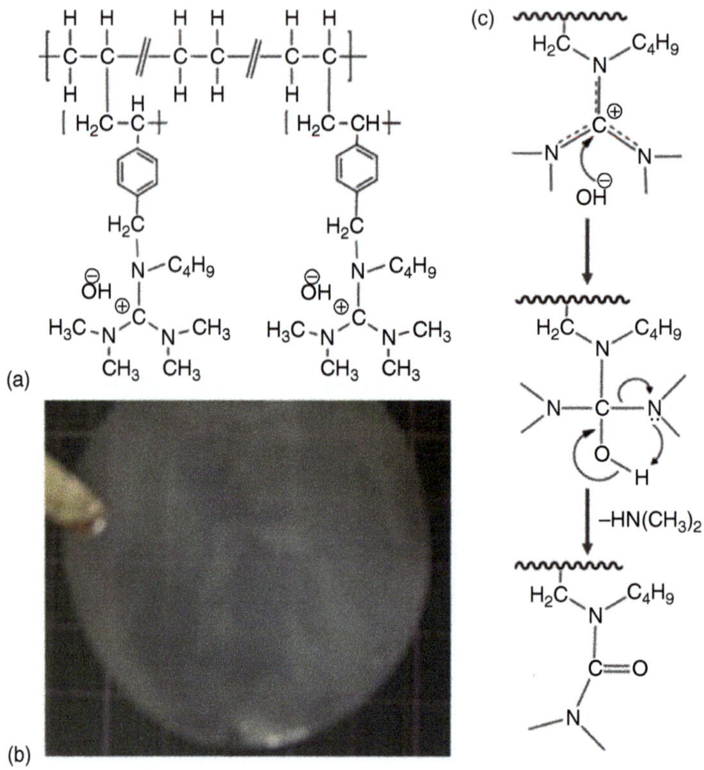

Figure 11.6 (a) Chemical structure of 1,1,3,3-tetramethyl-2-butylguanidinium functionalized on to the radiation-grafted poly(ethylene)-*g*-VBC AEM, (b) optical image of AEM with 12.5% degree of grafting, and (c) degradation mechanism of 1,1,3,3-tetramethyl-2-butylguanidinium functional group via alkaline hydrolysis. Source: Reproduced with permissions from Sherazi et al. [22], Elsevier.

durability and performance for fuel cell applications. Due to high alkaline stability and availability of cycloaliphatic ammonium groups, various AEMs have been reported. Attributed to strained and chair conformation, the free attacking sites are vulnerable to breach. In many reports, the stability of cycloaliphatic ammonium groups is higher than that of tetra methyl ammonium. For example, the $t_{1/2}$ of tetra methyl ammonium is 61.9 h at 160 °C in 6 M NaOH, while dimethyl piperidinium sustains for 87.3 h. Similar to those for benzyl trimethyl ammonium and benzyl *N*-methyl piperidinium, the degradation times were 4.18 and 7.3 h, respectively. It is evident that cyclic amines might improve the alkaline stability of AEMs. Another prospect is that the ring opening of piperidinium groups in AEMs is more complex than that in the alkyl amines [1, 14, 26, 27]. It is likely that the conformation (geometric factor) and electronic factors dictate the alkaline hydrolysis rate in AEMs. Moreover, the stability of six-membered amine (viz. *N*-methyl piperidine [NMPip], $t_{1/2} = 87$ h) over the rigid five-membered system (viz. *N*-methyl pyrrolidine [NMPy], $t_{1/2} = 37$ h) can be justified on carefully evaluating the conformation of their structures [26, 27]. During the β-H elimination, the anti-periplanar conformation affords rapid eliminations over conformationally less reactive *syn*-periplanar orientation (Figure 11.7b). Figure 11.7c illustrates the degradation mechanism via trigonal bipyramidal transition state (TBP#) in dimethyl piperidine type cations, and it prefers in-plane and out-of-plane bond angles of 120° and 90°, respectively.

Furthermore, Dang and Jannasch diagnosed the degradation trend of AEM with various cyclic ammonium ions functionalized on the PPO polymer (Table 11.1). The methyl (–CH_3) substitution at 2- and 6- positions of the *N*-methyl piperidinium group does not possess significant alkaline durability and degrades preferably via

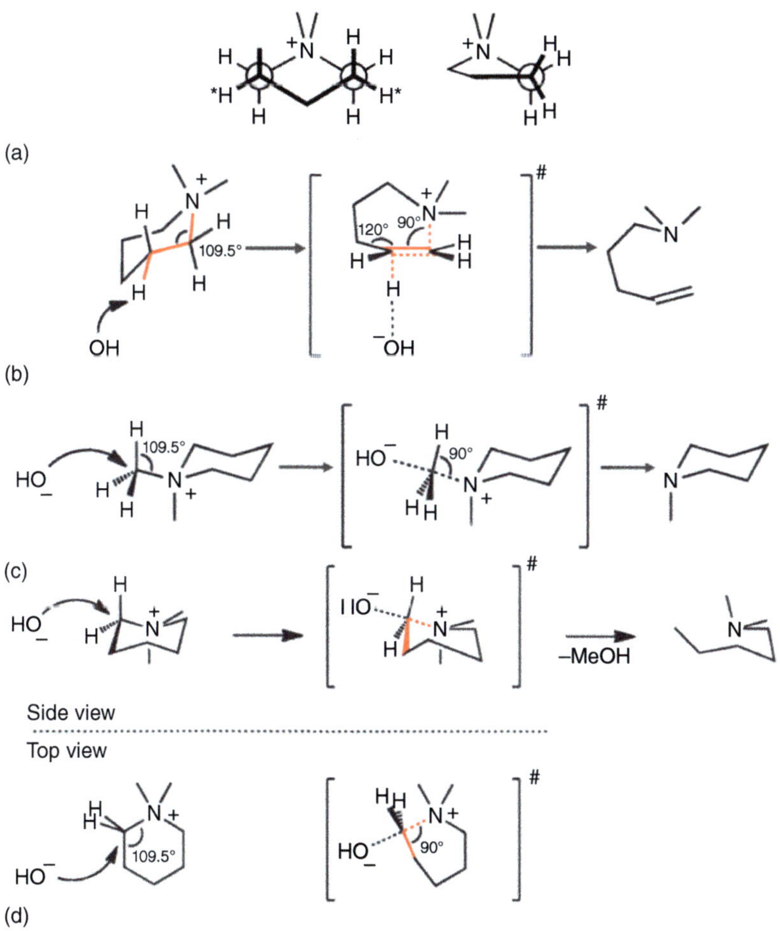

Figure 11.7 Alkaline hydrolysis mechanism and stereochemistry in *N*-methyl piperidinium and *N*-methyl pyrrolidinium ion (a) Newman projection of *N*-methyl piperidinium (DMP) and *N*-methyl pyrrolidinium (DMPy) ion, showing rotatable protons (star marked) in DMP. In contrast, DMPy possesses unfavorable conformation due to the ring strain; (b) transition state formed during the elimination of β-proton in DMP or ASU demands energetically expensive bond angles and lengths; and (c, d) chemical scheme for the S_N type reaction in DMP [14]. Source: Reproduced with permission of Wiley.

Table 11.1 The properties of AEMs with variable cycloaliphatic and hetero-cycloaliphatic functionalized quaternary amines with corresponding scattering vector, hydroxide conductivity, and mode of alkaline hydrolysis [28].

Anion exchange membrane	$^{1}q_{max}$ (nm^{-1})	$^{2}\kappa^{m}{}_{-OH}$ (mS cm^{-1})	^{3} Alkaline hydrolysis pathway
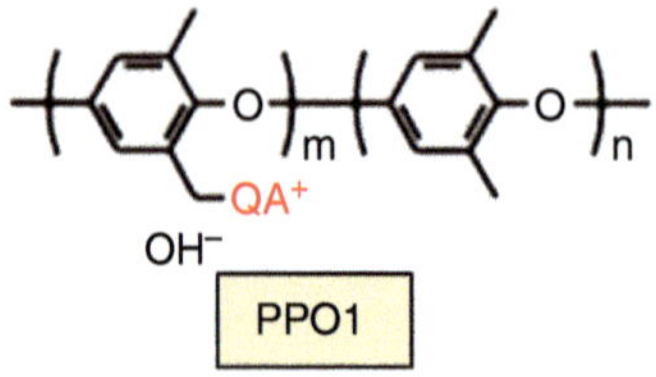		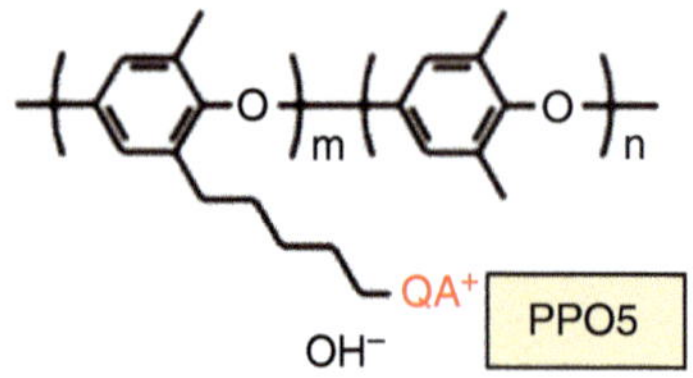	
 PPO5-Pip	2.1	117	Not determined
 PPO5-Pyr	2.1	89	Not determined
 PPO5-Aze	2.2	78	Hoffmann elimination
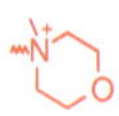 PPO5-Mor	2.1	101	Hoffmann elimination/ nucleophilic substitution
 PPO5-Qui	2.1	151	Not determined
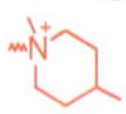 PPO5-4MPip	Not measured	94	Not determined
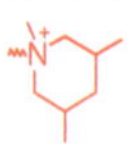 PPO5-3,5DMPip	Not measured	52	Not determined
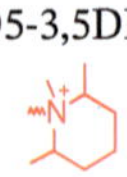 PPO5-2,6DMPip	2.0	86	Hoffmann elimination
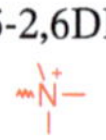 PPO5-TMA	2.1	139	Not determined
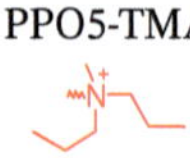 PPO5-DPMA	2.2	64	Not determined
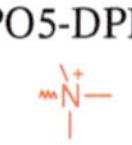 PPO1-TMA	2.1	14	Highly degraded
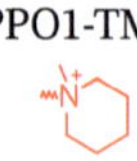 PPO1-Pip	2.0	11	Highly degraded

Conditions: 1 – measured using SAXS profiling on dehydrated AEM; 2 – fully hydrated state at 80 °C; and 3 – in 1 M NaOH at 90 °C for 384 h (by ^{1}H-NMR analysis).Source: Dang and Jannasch [28]. Reproduced with permissions of Royal Society of Chemistry.

the elimination route. On substituting $-CH_3$ at 3-, 5-, and 4-positions, the AEMs exhibit higher alkaline durability and do not show loss in property even after 16 days. Moreover, the quinuclidinium-based AEM exhibit improved ion-clustering attributes and high ^{-}OH conductivity. With morpholium derivatives, the stability was less and can be explained based on the inductive effect (−I effect) of heteroatoms in the cyclic system [28]. A similar study showing the influence of the pK_a value of amines on alkaline stability and ^{-}OH conductivity of PPO-based AEM was studied by Khan et al. [29]. The conductivity attributes were explained based on the pK_a values. For instance, the AEM with NMPy ($pK_a = 10.32$) exhibits higher conductivity than TMA, *N*-methyl imidazolium (NMIm), and NMPip in order;

$$\mathrm{PPO-NMPy > PPO-NMPip > PPO-TMA > PPO-NMIm}$$

Similarly, the complimentary results for alkaline stability of PPO-TMA and PPO-NMPip in 1 M NaOH for 30 days at 60 °C evidently illustrate the similar trend. Modification of cycloaliphatic systems to more stable ^{-}OH conducting groups was reported by Pham and Jannasch, a newer quaternary ammonium model, viz, the *N*-spirocyclic cation-based AEMs were developed [30]. Attributed to high transition state energies for alkaline hydrolysis, the introduction of these *spiro*-pyrrolidinium, *spiro*-piperidinium, and *spiro*-azepinium in poly(arylene ether sulfone) could be a preferred choice. Studies suggest that the influence of ring size on alkaline hydrolysis was not significant; however, the alkaline stability is better than that of 6-azonia-spiro[5.5]undecane (ASU)-type AEMs. In addition, the degradation in spirocyclic motifs preferably occurs at benzylic positions. This is likely because, in contrast to conventional freely rotatable systems, the possibility of β-H elimination gets reduced due to the puckered configuration. Thus, only strained rings undergo ring opening in alkaline conditions. This also ramifies that the introduction of alkyl systems would plausibly ensure better durability than aryl systems.

Olsson et al. approached a *de novo* synthesis of poly(*N*,*N*-diallylazacycloalkane)s tethered with variable spirocyclic cations, viz, the highly cationic poly(diallyl dimethylammonium chloride), poly(diallyl pyrrolidinium chloride), poly(diallyl piperidinium chloride), poly(diallyl morpholium chloride), and poly(diallyl azepinium chloride) were synthesized. These AEMs show no fragmentation of rings; and exhibit profound stability with minor degradations at 120 °C for >330 h (Figure 11.8) [31].

The alkaline stability for poly(*N*,*N*-diallylazacycloalkane)-based spirocyclic cations was in order;

$$Spiro\left[5,6\right]\mathrm{alkane} > Spiro\left[5,5\right]\mathrm{alkene} > Spiro\left[5,7\right]\mathrm{alkane} >$$

$$Spiro\left[5,6\right]\mathrm{alkyl\ morpholium}$$

Application of the spiro[5,6]alkane-based poly(ionene) was explored with PPO-based cross-linked AEM. The membrane exhibited 101 mS cm^{-1} ^{-}OH conductivity at 80 °C. Moreover, in subsequent work by the same group, a cyclo-polycondensation methodology was established to design highly alkaline stable poly(ionene) ($t_{1/2}$ = 1800 h in 1 M KOD/D_2O at 80 °C). These ionenes were blended with the

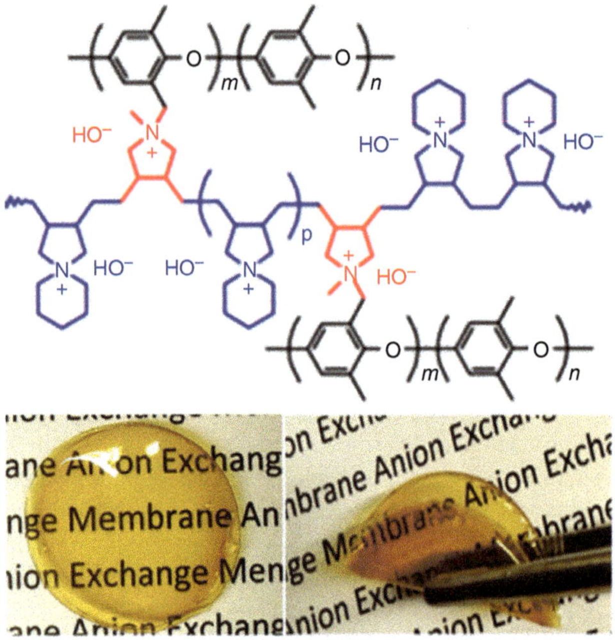

Figure 11.8 The photograph of PPOPip1.7 AEM showing transparent yellow color and flexible character attributes [31]. Source: Reproduced with permissions from Olsson et al. [31], American Chemical Society.

poly(benzimidazole) (PBI) matrix and possess an $^{-}$OH conductivity of 120 mS cm^{-1} at 90 °C. Such high conductivity is attributed to the well-established electrostatic interaction via donor–acceptor motifs, eventually these spiro-ionene-based AEM showed excellent stability at 120 °C and constitutes a new class of alkaline stable polyelectrolyte for fuel cell applications [31, 32]. Wang et al. studied the influence of different *N*-cyclic cations in a twisted poly(terphenylene)-based AEMs. The results were corroborated using proton nuclear magnetic resonance (^{1}H-NMR) studies and through lowest unoccupied molecular orbital (LUMO) iso-surface energy calculations. With better nanophase morphology, the piperidinium-type cation was found more stable than the spiro-cyclic cation. However, the LUMO energy values depict that the spiro[5,6]alkane exhibits higher LUMO energy and hence is more stable. This discrepancy in result was explained because the high phase miscibility and segregation between hydrophobic/hydrophilic might favor the alkaline durability. The rich hydration environment can solvate in the $^{-}$OH ions significantly in piperidinium systems and hence reducing the rate of alkaline hydrolysis of AEMs [33]. Various reports confer that the long-side chain morphologies impart better alkaline stability and high $^{-}$OH conductivity. Liu et al. proposed the ether spacer side chain type AEM with 8-hydroxyl-5-azonia-spiro[4,5]decane and 3-hydroxyl-6-azonia-spiro[5,5]undecane type cations on to the PESf polymer. The membranes exhibit a high conductivity of 85.7 mS cm^{-1} at 80 °C and delivered up to 110 mW cm^{-2} at 300 mA cm^{-2} of peak power density in the H_2–O_2 fuel cell. Moreover, the alkaline stability results suggest 86.2% and 84.3% retention of $^{-}$OH exchange capacities and conductivity after 864 h of test in 2 M NaOH at 80 °C, respectively [34]. A fluorinated AEM-bearing pendent-type spirocyclic cation was synthesized by Lin et al. via simple Williamson reaction. The membranes exhibit excellent nanophase morphology as confirmed by atomic force microscopy. On varying the functional concentration, improvements in electrochemical properties were observed. Moreover, these membranes exhibit high alkaline stability with >94% property retention at 80 °C in 1 M KOH for 480 h and delivered a peak power density of 138.6 mW cm^{-2} in H_2–O_2 fuel cell at 60 °C [35]. Subsequently, Zhang et al. reported ether tethered on PSf-based AEM, and it was anticipated that the ether functionality will defame the possibility of benzylic substitution and can provide a much better hydration environment, which can significantly increase the activation energy barrier for $^{-}$OH attack. As a result, these AEMs exhibit >95% property retention after 720 h in 1 M KOH at 80 °C [36]. To overcome the stability issue of aryl ether-based backbones, Olsson et al. fabricated poly(styrene)-tethered spirocyclic quaternary ammonium-based AEM via facile Lewis acid catalyzed Friedel–Craft alkylation reaction [37]. With all hydrocarbon backbone and methyl-substituted benzylic positions, the degradation by S_N at the benzylic position is ruled out, and hence better alkaline durability can be achieved. These membranes exhibit almost 99% of ionic functionalities for 30 days at 90 °C in 2 M NaOH. However, at higher temperature (120 °C), losses of >10% were observed. The mechanism was proposed through β-H elimination and ring opening route. Zhang et al. proposed a rod-coil grafting strategy to improvise nanophase morphology in side-chain type ASU grafted PSf for stable and high power density in H_2–O_2 fuel cell application (Figure 11.9). In proposed architectures, the cationic

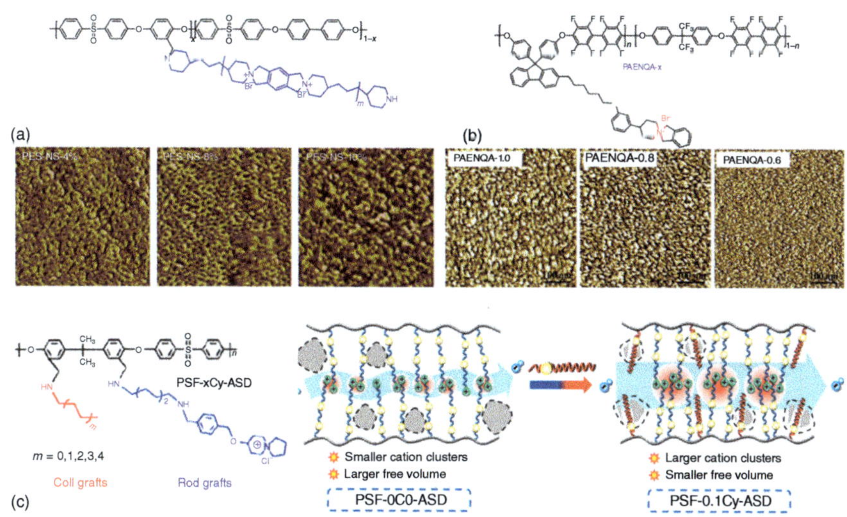

Figure 11.9 (a, b) Influence of concentration of spirocyclic cations on the nanophase morphology of AEMs and (c) the strategic designing of rod-coil morphology in AEM for high performance in fuel cell (peak power density deliverance of 850.1 mW cm^{-2} at 80 °C in H_2–O_2 fuel cell, 1 l min^{-1} fuel and oxidant flow with 100.0% humidity) [34, 35, 38].
Source: Reproduced with permissions of Elsevier.

aggregation compensates the free volume in contrast to the conventional side chain-based AEMs. These membranes possess the desired benchmark performance of $850\,mW\,cm^{-2}$ of peak power density and $\kappa^{m}{}_{-OH}$ ~ $136\,mS\,cm^{-1}$ at 80 °C [38]. Moreover, with the increase in coil grafting, the AEM (PSF-0.1C8-ASD) retains 80% of hydroxide conductivity and 78% of mechanical stability in extreme conditions (8 M hot KOH). As discussed earlier, the alkaline stability depends on various parameters like hydration number, operation temperature, nature of cationic head, and polymer backbone. For this, Chen et al. performed a systematic study on 24 cationic heads (i.e. *N*-heterocyclic ammonium groups) using combined DFT calculations (computational) and experimental studies having different hydration number at 80 °C [27]. The results suggest that the degradation mechanism is governed by the nature of cationic head and the hydration number around these groups. For instance, the *N*-heterocyclic ammonium degrades via S_N reaction, while the *N*-heterocyclic ammonium-based AEMs predominately degrade via Hoffmann elimination reaction. In addition, the models with low hydration numbers, viz, $\lambda = 4.8$ and 7, the membrane with ASU cationic heads exhibit high alkaline durability than *N, N*-dimethyl piperidinium (DMP). However, at $\lambda = 10$, the DMP-type cationic groups exhibit comparable alkaline stability to that of ASUs. Symmetry of *N*-heterocyclic ammonium is also found to influence the durability. For example, the ASU, DMP, Bis-ASU, and Bis-DMP exhibit higher stability than B-DMP and B-ASU. Substitution effects were also corroborated and found that the γ-substituted amines possess higher alkaline stability than ones with α- and β-substituted amine. And with various observation in this study, a detailed insight was presented for strategic designing of AEMs for high-performance fuel cells. Summarizing these results, the cyclo-aliphatic amines are the preferred choice in AEM fabrication over conventional TMA-based groups. The effect of symmetry on the stability of 2-azabicyclo[2.2.2]octane (ABCO or quinuclidine) was studied by Klein et al. using the (CAM-B3LYP/aug-*cc*-pVDZ) level of DFT and equation-of-motion coupled cluster (EOM-CCSD/aug-*cc*-pVDZ) approaches (Figure 11.10a–c).

Although this study was not directly linked to the alkaline stability, however, the information on stabilization of ABCO-based motifs by lowering of energy via Jahn-Teller distortion confers that the ABCO attains the most stable state which eliminates out the favorable geometry for facile β-H elimination via the Hoffmann ring opening reaction route. Attributed to the locked mono-conformeric nature in this 3p-state, one of the N—C bonds gets shorter than other two N—C bonds. Thus, bicyclic amine systems can show a unique non-adiabatic dynamic through restricted cage-type structure in ABCO. Similarly, the 1,4-diazabicyclo[2.2.2]octane

Figure 11.10 (a) The degradation time via alkaline hydrolysis of variable quaternary ammonium groups at λ = 4.8 in $NaOD/D_2O/CD_3OH$ at 80 °C [27]. Source: Reproduced with permission of Wiley. (b) The degradation time via alkaline hydrolysis of variable quaternary ammonium groups at λ = 7.0 in $NaOD/D_2O/CD_3OH$ at 80°C [27]. Source: Reproduced with permission of Wiley. (c) The degradation time via alkaline hydrolysis of variable quaternary ammonium groups at λ = 10.0 in $NaOD/D_2O/CD_3OH$ at 80 °C [27]. Source: Reproduced with permission of Wiley.

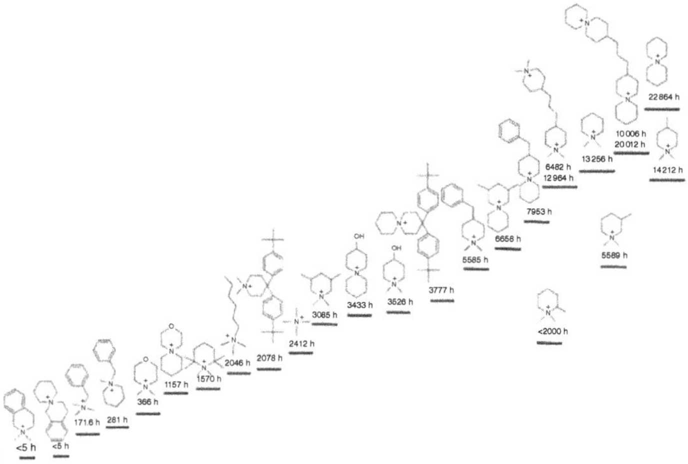
<5 h
<5 h
171.6 h
281 h
366 h
1157 h
1570 h
2046 h
2078 h
2412 h
3085 h
3433 h
3526 h
3777 h
5585 h
6658 h
7953 h
6482 h
12964 h
13256 h
10006 h
20012 h
22864 h
14212 h
5589 h
<2000 h

(a)

<5 h
<5 h
282 h
440 h
549 h
684 h
2144 h
4170 h
5328 h
6127 h
6201 h
7788 h
8256 h
17907 h
21681 h
35170 h
38329 h
>40000 h
>80000 h
>80000 h

(b)

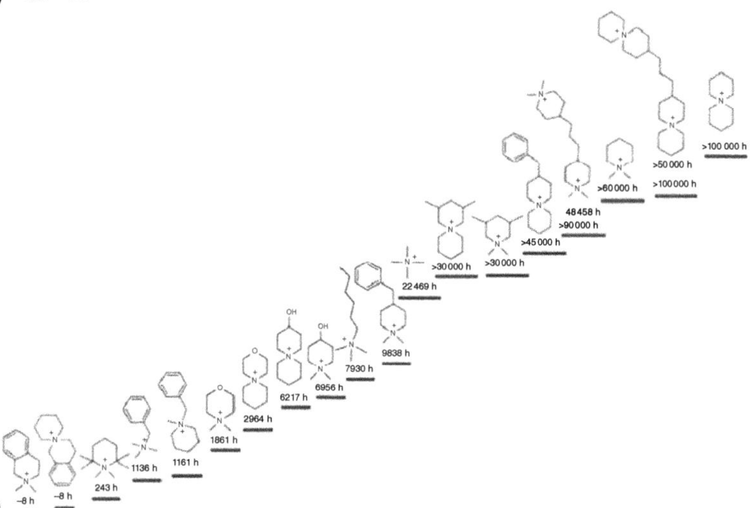
−8 h
−8 h
243 h
1136 h
1161 h
1861 h
2964 h
6217 h
6956 h
7930 h
9838 h
22469 h
>30000 h
>30000 h
>45000 h
48458 h
>90000 h
>60000 h
>50000 h
>100000 h
>100000 h

(c)

(DABCO) could be desired alternative amines. The D_{3h}-6/m2 point group model in DABCO confers the lone pair-through space interaction which provides extra stability under alkaline hydrolysis condition [39]. Moreover, in the di-quaternized state of DABCO, the cationic groups are in proximity; this unfolds the facile geometry to trigger faster degradations, causing rapid ring opening via the elimination route (E_2 elimination). Thus, a much optimized ionic structure and morphology must be explored with a point-of-care to prevent AEMs from alkaline hydrolysis. Limited to the alkaline stability of aromatic heterocyclic ionic groups, more studies are confined to improvise structures of heterocyclic amines to resist alkali degradation. Interestingly, a prominent scope on the C-2 substitution of imidazolium ions has been observed to dramatically improve stability in fuel cell conditions. The electron delocalization in the π-system improves the positive charge distribution helping in higher water retention through high phase separated percolating matrix. In additions, the imidazole-based AEMs exhibit better thermal stability to dehydration and impart oxidative resilience [40,41]. In stark contrast, attributed to the high susceptibility to $^{-}$OH attack at the α-position of pyridinium groups, the literature is limited to vinyl pyridine (VP), dimethyl amino pyridine (DMAP), and simple pyridine moieties in PBI-based AEMs. In a report by Xue et al., the group comprehensively reviewed the designs, stability aspects, and degradation mechanism of AEMs based on cycloaliphatic cationic groups for electrochemical energy applications. Various hydrolysis mechanisms and structures for the ASU cations, Tröger base, and other AEMs were compiled, and the fuel cell performance of recent AEMs was summarized. Figure 11.11 illustrates the synthesis and alkaline hydrolysis of Tröger base-based highly ion conducting membranes with intrinsic microporosity [43]. These membranes exhibit high alkaline stability with no significant degradation in 2 M NaOH at 60 °C for 240 h. These results suggest that Tröger base-based polymers with sophisticated ion conducting morphology are emerging materials for fuel cells. Mahmoud et al. studied the influence on the stability of various alkyl-protected imidazolium AEMs prepared via γ-radiation-induced grafting of monomers on ethylene tetrafluoro ethylene (ETFE)-based films. Membrane with styrene modified substituted imidazole illustrated highest stability and performance in H_2–O_2 fuel cell.

The *in situ* degradation analysis illustrated that the hydration structures dictate the stability. The membranes deliver the peak power density up to 710 mW cm^{-2} at 1389 mA cm^{-2} current density. Moreover, the 56% retention of initial voltage after 670 h was observed inside the cell and surpassed the standard BTMA-based AEMs [44] (Figure 11.12). Ryu et al. designed a chitosan and 4-formyl pyridine-based AEM via the reductive amination strategy for alkaline direct methanol fuel cell (DMFC). The higher positive charge was created by functionalizing the free hydroxyl ends by poly(vinyl imidazolium) groups. In alkaline DMFCs, the membrane exhibits a peak power density of 10.42 mW cm^{-2} at 28.76 mA cm^{-2} current density for the 3 M methanol and 2 M KOH solutions. The design was interesting due to the non-toxic polymer; however, the performance of these AEMs required to improvise, and no degradation mechanism was provided [45]. Evidently, in AEMFC devices, various fuels can be employed and the influence of fuel on AEM degradation remained elusive.

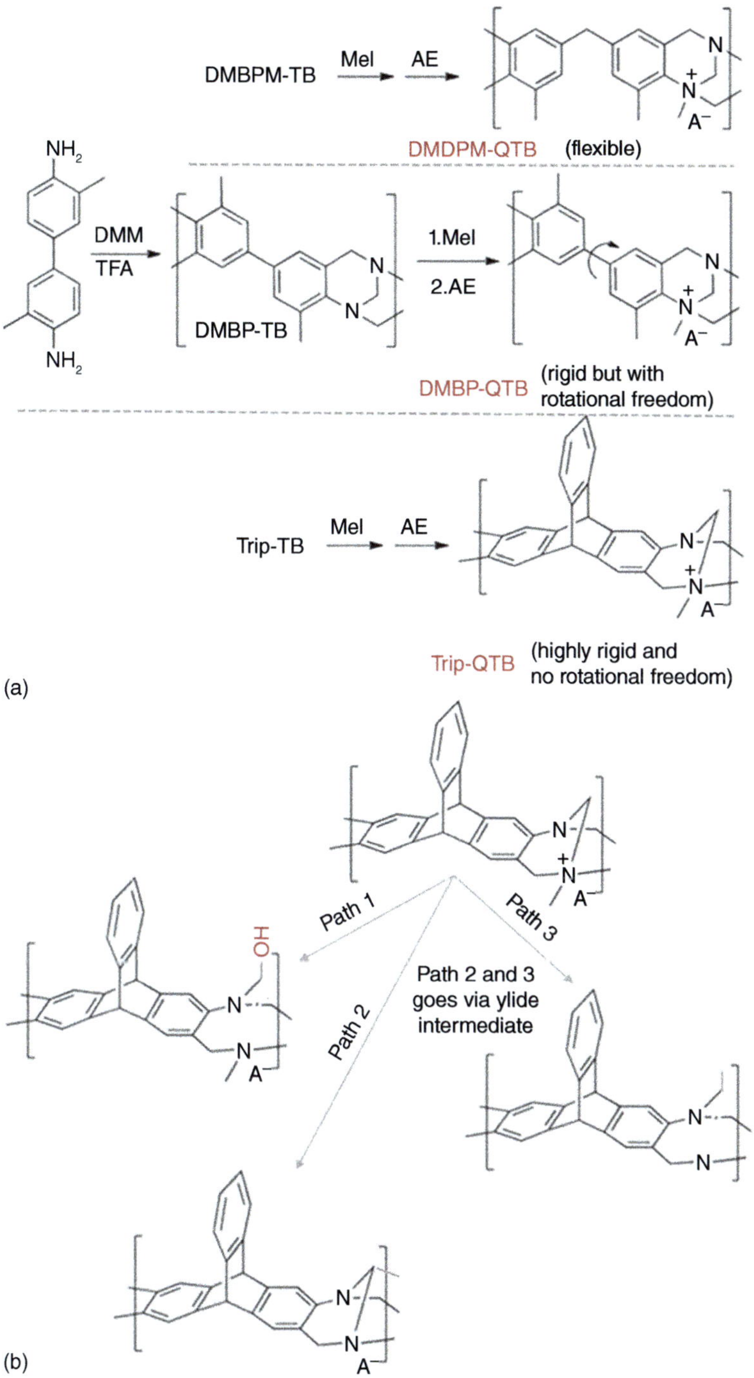

Figure 11.11 (a) Synthesis of various Tröger base-based polymeric materials and (b) degradation mechanism in alkaline conditions [42]. Source: Reproduced with permission of Wiley.

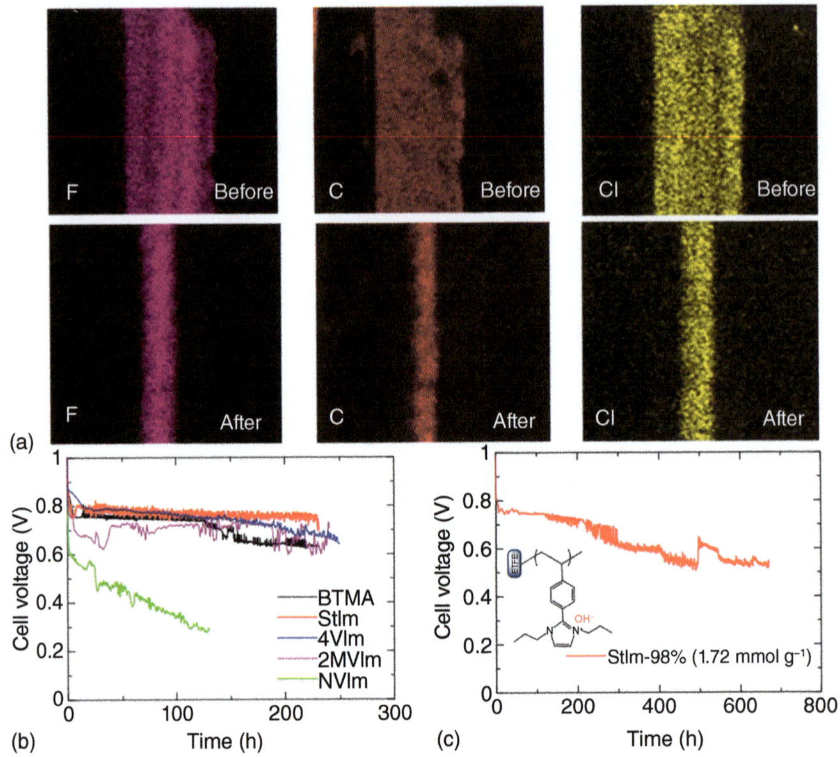

Figure 11.12 (a) *In situ* degradation analysis: EDX-SEM elemental mapping of fluorine (F), carbon (C), and chloride (Cl) atoms along the cross-sectional direction before and after 670 h of operations in StIm AEM. Accelerated durability test showing cell voltage as a function of operation time (b) at 50 mA cm^{-2} for 230 h and (c) for 670 h in a single H_2/O_2 fuel cell at 60 °C. Source: Reproduced with permissions from Mahmoud et al. [44], Elsevier.

Thus, another prospect on the alkaline degradation was studied by Sun et al. [46]. The study suggests that the alkaline durability is a cumulative influence of the nature of cationic head and alkaline fuel employed in fuel cell devices. The addition of co-solvents like dimethyl sulfoxide (DMSO), methanol, or ethanol accelerates the degradation of cations in AEM. Moreover, the alkaline stability order for cationic polymers was in order (Table 11.2);

$$[\mathrm{PVBMPL}]^+ > [\mathrm{PVBTMA}]^+ > [\mathrm{PDMVBIm}]^+ > \left([\mathrm{PMVBIm}]^+, [\mathrm{PVBPy}]^+, \text{and} [\mathrm{PVBTMP}]^+\right)$$

Subsequently, complementary to the research like Chen et al. [27], Fan et al. report the systematic study on the alkaline stability of the novel arylimidazoliums and bis-arylimidazolium compounds. The initial library of 14 imidazolium molecules with bulky substitution at C-2, C-4, and C-5 and at N-1 and N-3 was strategically screened for their hydrolyzing stability in 3 M $NaOD/D_2O/CD_3OD$ at 80 °C. Figure 11.13 illustrates the half-life of various aryl- and bi-aryl-imidazolium

Table 11.2 The lowest unoccupied molecular orbital energy (LUMO) values of various organic cations in different alkaline media.

Cations	LUMO energy (eV)			
	Water (ε: 78.54)	**DMSO (ε: 46.70)**	**Methanol (ε: 32.63)**	**Ethanol (ε: 24.30)**
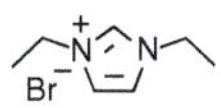 $[DeIm]^+$	−1.761	−1.814	−1.870	−1.934
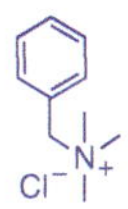 $[BTMA]^+$	−1.698	−1.743	−1.783	−1.833
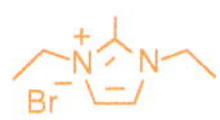 $[DemIm]^+$	−1.503	−1.553	−1.606	−1.666
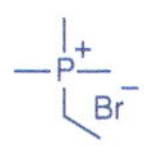 $[ETMP]^+$	0.222	0.168	0.111	0.046
^-OH (HOMO)	−3.083	−2.979	−2.869	−2.745

Source: Sun et al. [46]. Reproduced with permissions of Wiley.

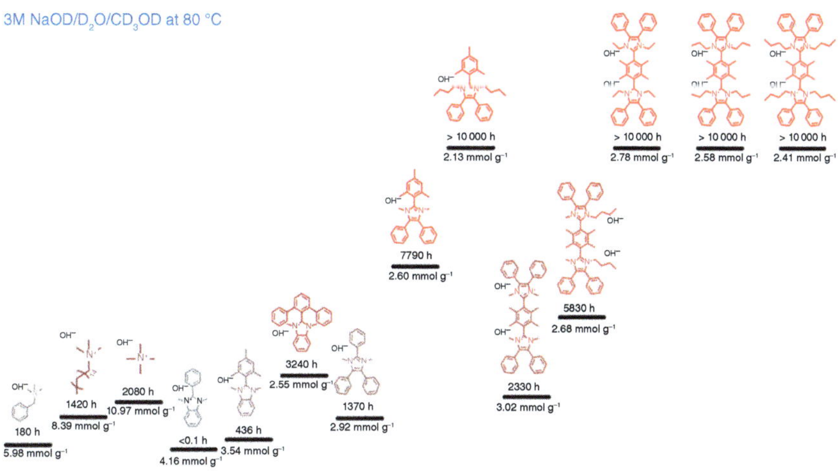

Figure 11.13 The degradation time via alkaline hydrolysis of variable imidazolium groups in 3 $NaOD/D_2O/CD_3OH$ at 80 °C [47]. Source: Reproduced with permissions of Nature Communications, Springer.

groups presented in the study. Subsequently, the stability limitations of imidazolium-based groups were improved and achieved to desired benchmarks [47]. Moreover, to design stable imidazolium-based AEMs, it is highly desirable to understand the substitution effect of alkyl chains on variable available N-atoms in the five-membered rings. Studies by Tham and Kim suggest that the resistance to alkaline hydrolysis can be significantly reinforced by the C-2 substitution in imidazole ring through synergistic hyper-conjugative and steric hindrance effect of alkyl substituents. The modification of C-2 methyl and N-3 butyl alkyl chains in polyarylene ether ketone (PAEK)-APMBI-based membranes exhibits superior chemical stability with no change in its IEC attributes for the immersion in 10 wt% NaOH solution at 60 °C. In addition, the influence of substitution is position-specific. For instance, the increasing steric functionalities at the C-2 position impart higher resistance to S_N ring opening degradation. However, the substitution of longer alkyl chains at C-4 and C-5 position does no good; and it was observed that the rate of hydrolysis tends to increase (Figure 11.14) [49].

Similarly, Hugar et al. investigated the prominent effect and better choices of the substituent at C-4 and C-5 position of imidazolium rings. Despite substituting the

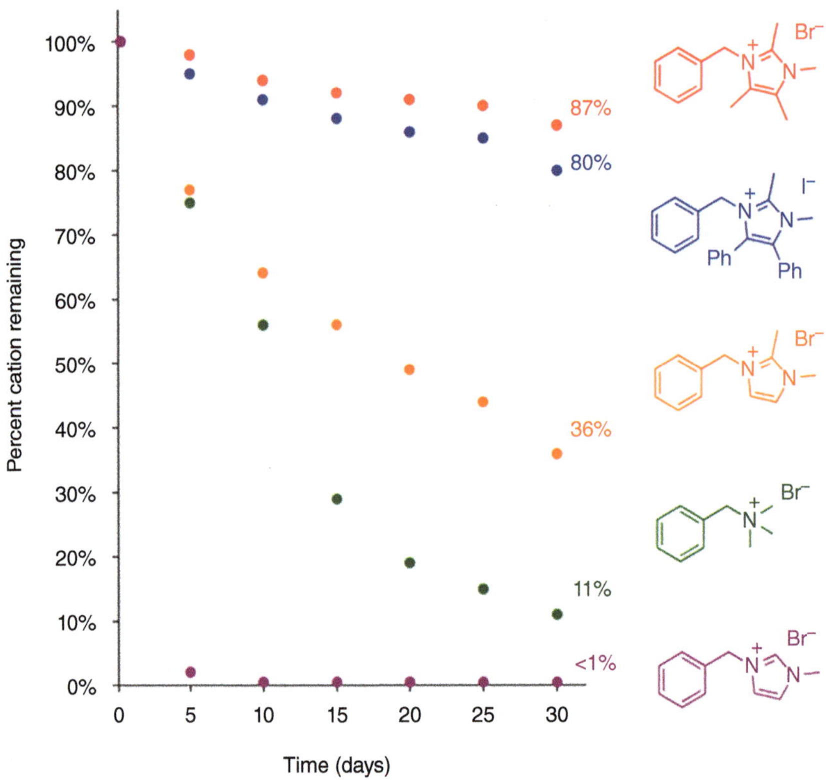

Figure 11.14 Stability trend of C-4 and C-5-substituted imidazolium cations (0.05 M) in 1 M KOH/CD_3OH at 80 °C [48]. Source: Reproduced with permissions of American Chemical Society.

bulky phenyl groups at the C-4 and C-5 position, the methyl substituents afford higher alkaline stability and improved ion-exchange capacities in AEMs [48]. Furthermore, You et al. designed a novel AEM via ring opening metathesis polymerization (ROMP) reaction. The membrane does not exhibit alkali-susceptible polymer backbone; these membranes exhibit remarkable durability at 80 °C for 30 days in 2 M KOH with no detectable changes in ^{1}H-NMR [50].

11.3.2 Alkaline Hydrolysis of Novel Metallocenium Based AEMs

In contrast to conventional N- and P-based functional groups, the metallocenium-based AEMs (often metallo-polyelectrolytes) offer alternative novel platforms to design AEMs with higher positive charge density for $^{-}$OH interaction. With the metallic ion as a ion conducting motif, these ionic centers improve the thermochemical stability of AEMs in real operations. The metal ions in metallo-polyelectrolytes are often transition metals (viz, iron and cobalt) or precious metals (viz, ruthenium) in sandwich structures (within π-rich arene moieties) and as chelates (coordinated ring complexes). Thus, this section included some of the recent advances on metallocenium-based AEMs and their degradation studies [51]. Prima facie, the first metallocenium-based AEM was reported by Zha et al. Attributed to high valence Ru(II) ion and having high temperature/pH stabilities, these AEMs can acquire higher water retention capacities and successively high $^{-}$OH conductivities (Figure 11.15). Interestingly, the redox potential of Ru(II) is more negative and hence does not overlap with the fuel cell potential window. This will ensure that the Ru(II) ions are unlikely to get destabilized in the polyelectrolyte matrix during fuel cell reduction–oxidation process. To study the degradation behavior, ultraviolet–visible (UV–Vis) spectra were obtained for the metallocenium polyelectrolyte membrane in 1 M NaOH at room temperatures, and no significant changes were

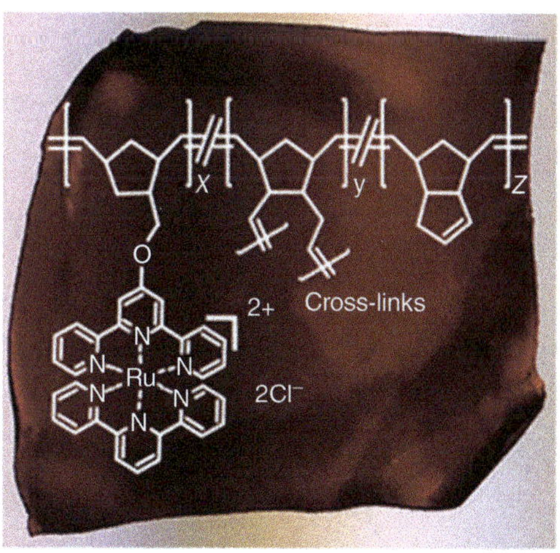

Figure 11.15 A photograph of Ru(II) functionalized AEM prepared by Zha et al. via ring opening metathesis polymerization (ROMP) using second generation Grubb's catalyst [52]. Source: Reproduced with permissions of American Chemical Society.

found after 6 months. Moreover, the mechanical degradation was not significant at 80 °C either and shows ~3% weight loss [52]. Chen et al. developed the cobaltocenium ion-based AEMs. Initially, they studied the degradation mechanisms with different substituents like –H, (Cp_2Co^+ PF_6^-), $-CH_3$, ($DMCp_2Co^+$ PF_6^-), and –COOH ($DCCp_2Co^+$ PF_6^-) to understand the influence of different groups on alkaline hydrolysis. The ion conducting groups with electron-donating substituents, viz, $DMCp_2Co^+$ PF_6^-, exhibit higher durability to ^-OH attack than $DCCp_2Co^+$ PF_6^- when present in the aryl ring of cobaltocene. Furthermore, the $DMCp_2Co^+$ PF_6^- retains 93% of cations after 30 days in 1.0 M KOH at 80 °C, while the $DCCp_2Co^+$ PF_6^- retains only 54% after 10 days in 1 M KOH at 80 °C. This shows strong dependency of the ancillary groups on alkaline stability of metallocenium-based AEMs. Moreover, a series of cobaltocenium containing PBI membranes were designed and systematically studied for the alkaline hydrolysis reaction. Based on the studies, the degradation mechanism involves direct attack of ^-OH on the central metal ion and the inversion of denticity from η^5- to η^1-complex via ^-OH attack on ring forming corresponding metal hydroxide precipitates in the final product (Figure 11.16) [53].

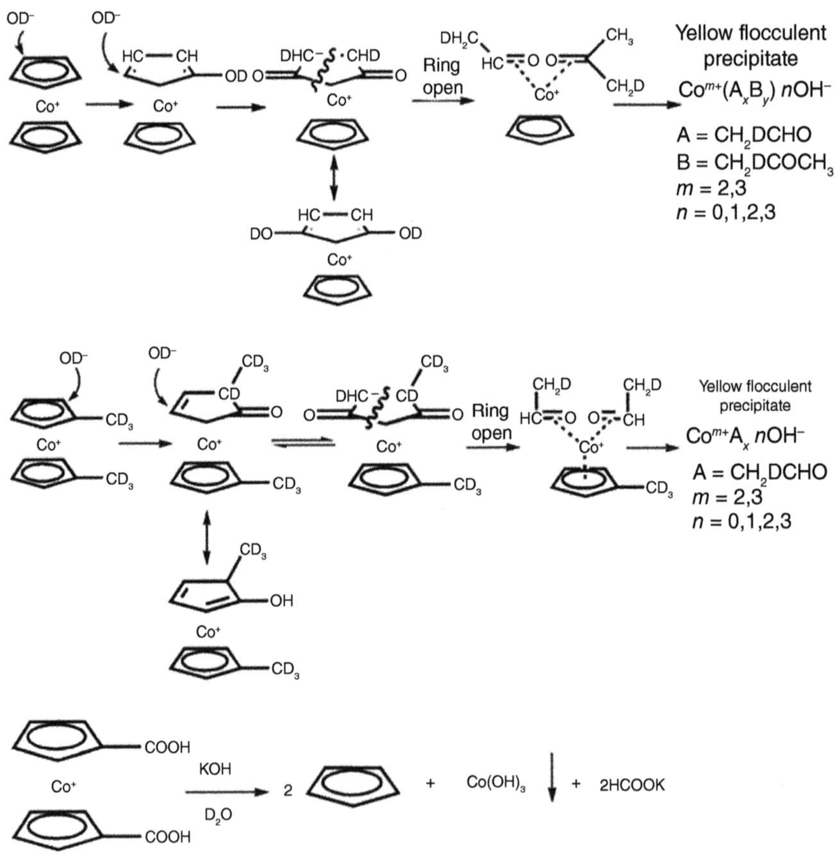

Figure 11.16 Scheme for the alkaline hydrolysis of cobaltocenium-based AEMs [53]. Source: Reproduced with permissions of Royal Society of Chemistry.

Similarly, Gu et al. developed pentamethyl cyclopentadiene cobaltocenium ($Cp^*_2Co^+$)-based AEMs. Confirming the studies from semi-empirical quantum chemistry software MOPAC2012, the penta-methyl substituted ring strength metal–ligand bond (from 499 kJ mol^{-1} in Cp_2Co^+ to 775 kJ mol^{-1} in $Cp^*_2Co^+$) and hence imparts better stability. Furthermore, the membrane was prepared via the diamine bridging strategy on PSf exhibit high alkaline stability and retains 82% and 63% of its initial IEC in 1 M KOH at 80 °C for 1000 h and 2000 h, respectively. At 100 °C, the AEMs retain 50% of IEC after 2000 h. These results suggest dramatic improvement in the stability of cobaltocenium-based AEMs [54]. The onset decomposition temperature of 305 °C in these AEMs shows that this new class of polyelectrolytes could be better than conventional organic cationic groups for intermediate-temperature energy applications, hence suggesting the potential utility of metallo-polyelectrolyte membranes in fuel cell devices. Zhu et al. designed cyclooctene dangled cobaltocenium-based AEM via ROMP. The precursor ionic motif was designed by simple Cu(II) catalyzed click reaction and further allowed to polymerize using Generation II Grubb's catalyst with cyclooctene. The membranes possess excellent morphology and sustained 1 M NaOH at 80 °C for 30 days with 95% $^-$OH conductivity retention. Moreover, attributed to high ion clustering and imparting better mechanical stability, they also designed a cross-linked AEM via thiol-ene click ROMP. In time-dependent $^-$OH conductivity analysis, these membranes retain 14.5 mS cm^{-1} after treating the AEM at 60 °C for 15 days in 3 M KOH. In contrast, the initial $^-$OH conductivity was 18.6 and 53.3 mS cm^{-1} at 20 and 80 °C, respectively. And, such decrease was explained based on the dissolution of free and weakly cross-linked groups in the polyelectrolyte matrix [55, 56].

It is evident that the cationic heads should be alkaline stable as well as redox stable. Thus, Zhu et al. designed variably substituted cobaltocenium ions and corroborated the properties and effects on alkaline hydrolysis and redox stability. A highly substituted cobaltocenium (octa-methyl)-based AEM was designed via ROMP, these AEMs delivered ~350 mW cm^{-2} of peak power density in the H_2–O_2 fuel cell device and retained 91% of electrochemical properties after evaluating in 3 M KOH at 60 °C for 30 days (open air environment) [57]. Tröger base-based crown ether AEMs are gaining sustained surveillance due to high intrinsic porosity and ion selectivity. However, the alkaline stability issue needs critical assessment [58–60]. In addition, You et al. performed a comprehensive study on the degradation of various organic cations, and degradation mechanisms for various cationic heads were corroborated in the literature. Interestingly, the study also highlights that the bulky phosphonium-based groups can also attain high durability in alkaline conditions [61]. To conclude this section, the paradigm shift in AEM synthesis has acquired high alkaline stability materials for alkaline energy applications. However, the *in situ* evaluation conditions are finding their way to reach desired performance strata and lifetime outputs. Moreover, the discussion ramifies that the stability of hydroxide conducting groups depends on various factors, viz, the steric effect, the substituent effect, the stereo-electronic effect, temperature [62], nature of base, the effect of atmospheric CO_2 [63], and degree of solvation (hydration number, λ) [5, 6, 62, 64]. Thus, it is critical to access the stability aspects more accurately to reach state-of-the-art performance. In addition, the studies suggest that the alkaline hydrolysis is inevitable; however, our efforts must emphasize to propose the most durable cationic head group showing high performance and alkaline durability for commercial applications.

11.3.3 Alkaline Hydrolysis of Polymers

Polymers play a pivotal role as a self-standing solid-state support in AEM fabrication. In AEMs, the mechanical instability and performance loss during electrochemical operations are attributed due to the alkaline hydrolysis of main chain polymer backbone having the hydroxide exchange group. During the degradation, the polymer backbone hydrolyzes to much smaller fragments, affecting the polymers' characteristic weight. The degradation mechanism varies based on the nature of backbone, and hence several mechanisms and causes have been studied in literatures. In many works, meticulous efforts were made to stabilize hydroxide conducting groups; however, the polymer resilience is overlooked. Depending upon the AEM designing strategy and application aspects, the polymer backbones can be modified, synthesized, or can be homogenously mixed as an interpenetrating type network (IPN) to form novel AEMs. As discussed earlier, despite the resurgence in the development of various cationic head groups which has surpassed the stability of >10 000 h in alkaline conditions, the AEMs do not possess long life during real-time application. This section of the chapter describes alkaline hydrolysis of various polymer backbones in alkaline conditions. Very less reports comprehensively discuss the polymer degradation, and the most relevant literatures are discussed in this section. Apparently, the polymer backbone chemistry and its functional architecture also play a crucial role in dictating the alkaline hydrolysis durability of AEMs (Figure 11.17) [53–55].

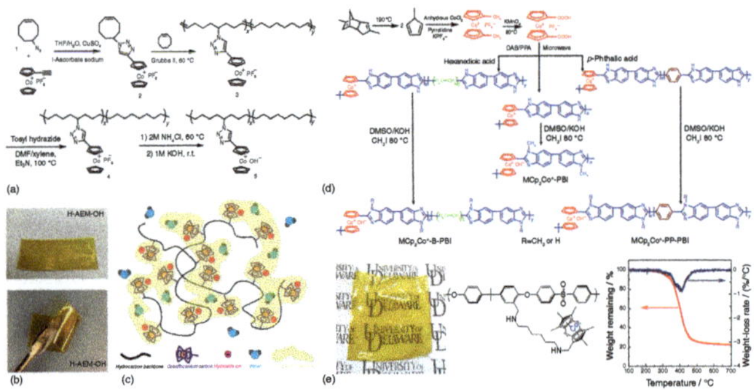

Figure 11.17 (a, b) Scheme for the synthesis of cobaltocenium-based stable AEMs via the ROMP method, (c) matrix structure of cobaltocenium-containing AEMs in a hydroxide solution, (d) synthetic route for the cobaltocenium-based PBI membranes, and (e) membrane image, chemical structures of $Cp^{*}{}_2Co^{+}$ AEM with thermal stability profile (TGA/DTG thermogram) [53–55]. Source: Reproduced with permissions of Royal Society of Chemistry, John Wiley & Sons, Inc., and Springer Nature.

11.3.3.1 Degradation Mechanism in Poly(arylene ethers) (PAEs)

Poly(arylene ethers) (PAEs) are high-performance engineered thermoplastic polymers used to design AEMs. Attributed to high stability and facile sites for chemical modification, these polymers are the most studied polymers in literatures. Moreover, these low-cost poly-hydrocarbons can provide additional cost benefits in device fabrication [65]. These polymers are available in market with trade/commercial names like Victrex®, Udel®, KetaSpire®, Kadel®, and Radel®. Depending upon the functional motifs present in these speciality polymers, they can be poly(ether ketones) (PEEK), PESf, poly(ketones), PSf, PPO, etc. In a report by Gottesfeld et al., the influence of polymers' molecular weight on the mechanical stability was illustrated and discussed as to how it influenced the device performance. Polymers with high molecular weight tend to possess higher alkaline stability due to the chain entanglement; however, the physicochemical and electrochemical properties like water retention and $^{-}$OH conductivity remain independent. Moreover, the synthesis of high molecular weight PAEs is difficult because they are conventionally prepared by step-growth polymerization and if synthesized, then high purity monomers are required, which will increase the processing cost significantly. Thus, a careful strategy to overcome polymer scissoring is required to explore to fabricate durable AEMs [65]. A detailed mechanistic insight on the alkaline hydrolysis of PESf was corroborated by Arges and Ramani The degradation analysis was evaluated employing 2D NMR spectroscopy, viz, the heteronuclear multiple quantum correlation spectroscopy (HMQC) and correlation spectroscopy (COSY). His study concludes that the polymer scissoring is the triggered dynamics of QAGs on PESf backbones. For example, the PESf and chloromethylated PESf are alkaline stable in 1 M and 6 M NaOH at 60 °C. However, the quaternized membranes become reactive to the $^{-}$OH ions due to electronic effects and impart a high positive dipole moment on the ether linkage of the PESf backbone. They also suggested that the polymer backbone stability can be improved through adding alkyl spacers between PESf and conducting groups in AEM [9]. Studies by Miyanishi and Yamaguchi proposed a comprehensively revised degradation mechanism of hydroxide conducting groups and PAE backbones. In the conventional hydrolysis mechanism, direct hydrolysis of the polymer and groups was reported; however, the new mechanism was corroborated as the "ether cleavage-triggered cation degradation." In this, the ether cleavage catalyzes the degradation of cationic groups in AEMs. Also, the results suggest that the polymer stability is critically important and emphasis must be given to improve endurance under alkaline conditions [8]. Figure 11.18 illustrates the mechanism of polymer degradation of PESf-based AEM via alkaline hydrolysis. Interestingly, the alkaline hydrolysis mechanism of PAEs in AEM does not differ to a significant extent. However, the secondary functional groups can dramatically influence the aryl-oxy bond cleavage in alkaline conditions.

Studies by Mohanty et al. illustrate that the substitution of electron withdrawing groups which stabilize the Meisenheimer intermediates catalyze the polymer degradation. In contrast, the polymers without aryl-oxy bonds show no sign of degradation [66]. The degradation mechanisms for all PAEs like PPO and PEEK also follow the same mechanism (Figure 11.19).

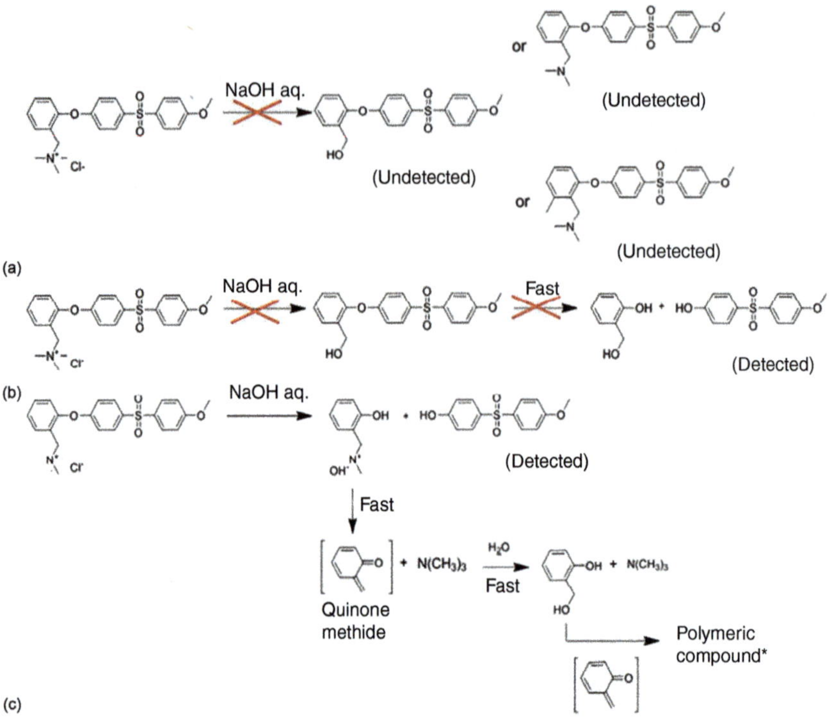

Figure 11.18 Scheme showing the (a, b) conventional and (c) proposed degradation mechanisms of PESf-based AEM in aq. NaOH [8]. Source: Reproduced with permissions of Royal Society of Chemistry.

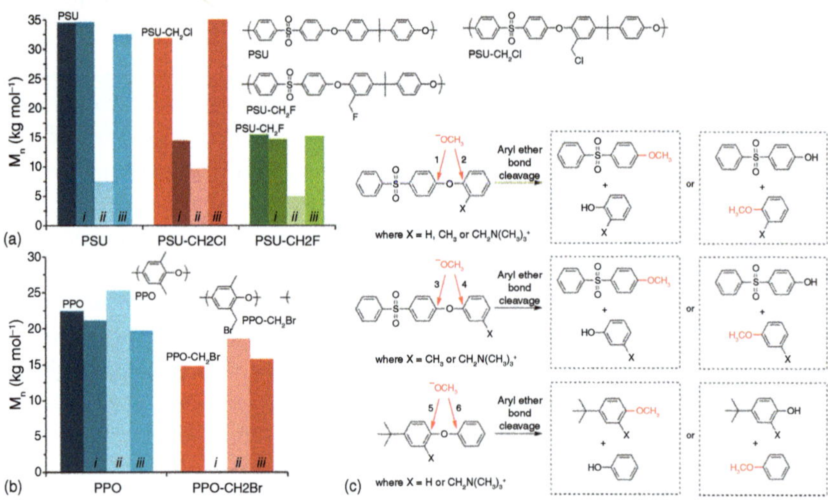

Figure 11.19 (a) Changes in number-average molecular weight (M_n) of (a) PSU, (b) PPO in alkaline conditions: (i) 3 eq. of KOH, 18-crown-6, THF/H_2O (4 : 1, v/v), 60 °C for 2 h and then rt for 12 h; (ii) 10 eq. of $NaOCH_3$, THF/CH_3OH, 60 °C for 14 h; (iii) polymer film (20–40 μm thick) immersed in 2 M $NaOCH_3$/CH_3OH, 60 °C for seven days and (c) potential routes of CH_3O^- attack and their plausible by-products for PESf-based compound models [66]. Source: Reproduced with permissions of American Chemical Society.

11.3.3.2 Degradation Mechanism in Fluorinated Polymer

Fluorinated polymers are attractive polymers in membrane synthesis due to excellent physical properties. Attributed to strong C—F (~110 kcal mol^{-1}) and C—C bonds (97 kcal mol^{-1}), these are the materials with excellent chemical stability and oxidative stability. However, due to the dehydrofluorination reaction in alkaline conditions, the scope of fluorinated polymers in alkaline electrochemical energy devices is not much explored [67, 68]. In recent reports, McHugh et al. designed vinylbenzyl (trimethylammonium) and *N*-vinyl imidazole co-polymerized poly(vinylidene fluoride-*co*-hexafluoropropylene)-based membranes for lignin electrolyzers. The electrolysis performance was parallel to commercial benchmarks; however, the long-term mechanical and chemical stability was not sufficient. They concluded that an alternate approach is required to optimize the durability of these materials in alkaline conditions [69]. Ahmed et al. revisited and studied the dehydrofluorination mechanism of poly(vinylidene fluoride) in propan-2-ol and proposed the dehydrofluorination insights (Figure 11.20).

Mechanism of alkaline degradation in vinylidene fluoride-based polymers: the reaction of poly(vinylidene fluoride) degradation initiates through a hydrogenation step after the elimination reaction. The feasibility of this reaction underlies the partial instability of the unsaturated fragment due to the repulsion between fluorine lone pairs and induced π-system of double bond. In such reactions, the propan-2-ol in NaOH acts as an H-donor, forming sodium isopropyl alcoholate, which in turn is dehydrofluorinated via the six-membered type intermediate. And finally, the decomposition of the intermediate in the dehydrofluorination step induces the double bond (-CH=CF-), causing degradation of poly(vinylidene fluoride) [68]. Recently, Sharma et al. investigated the site selectivity of dehydrofluorination in PVDF and also, highlighted on the phase transformation over alkali treatment and detailed dehydrofluorination mechanism was studied [70].

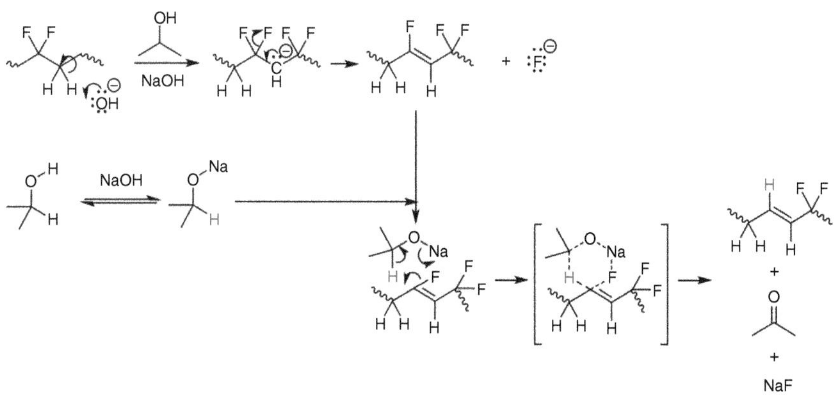

Figure 11.20 Dehydrofluorination mechanism of poly(vinylidene fluoride) in alkaline propan-2-ol [68]. Source: Reproduced with permissions of Elsevier.

11.3.3.3 Degradation Mechanism in Poly(benzimidazole) Based Polymers

Excellent ^{-}OH conductivity attributes, high decomposition temperature, and gas-tightness of the PBI-based AEMs have gained tremendous interest as a separator for high-temperature alkaline exchange membrane fuel cell (HT-AEMFC) applications (mostly $T_{operations} \geq 120\,°C$). Thus, PBI and its modified architectures have been explored tremendously for alkaline energy applications [71] (Figure 11.21). Aili et al. studied the alkaline degradation reaction of poly(2,2′-(*m*-phenylene)-5,5′-bibenzimidazole) (*m*-PBI) and poly(2,2′-(*m*-mesitylene)-5,5′-bibenzimidazole) (*mes*-PBI), respectively. Under accelerated condition of 50 wt% KOH solution at 88 °C, the *m*-PBI spontaneously degrades after 100 days, whereas the *mes*-PBI shows visually intact mechanical stability even after 207 days. The degradation was probed using size exclusion chromatography (SEC) traces after the aging of *m*-PBI and *mes*-PBI for 200–207 days in 10 wt% alkaline solution.

Results suggest relatively higher M_n value of *mes*-PBI over *m*-PBI and hence lower scission rate of *mes*-PBI-based AEM. The *in situ* analysis in the electrolyzer was

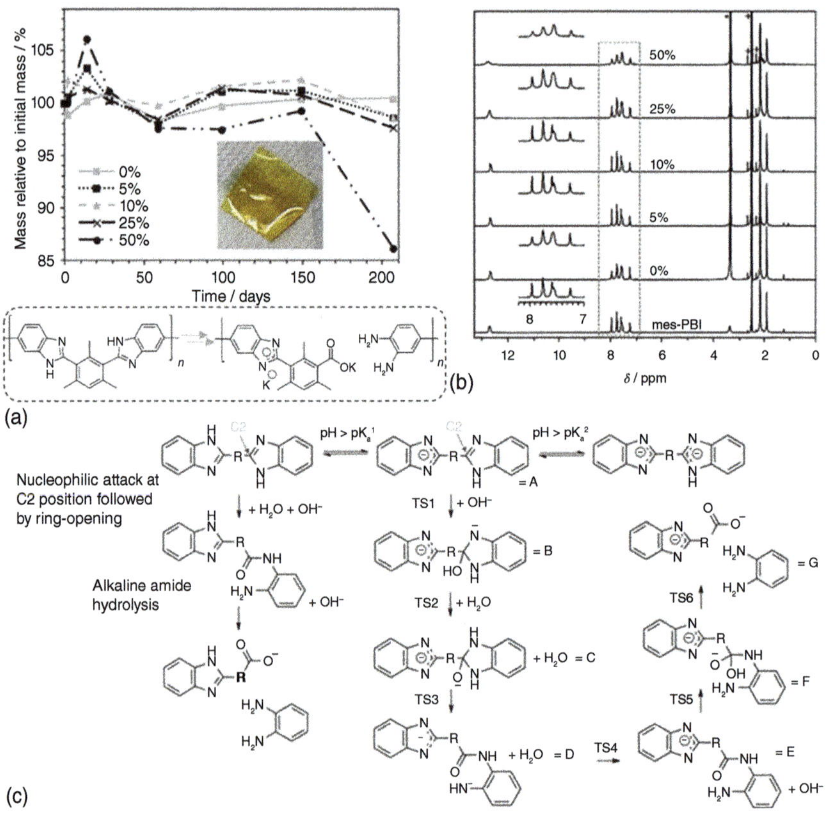

Figure 11.21 Degradation mechanism of poly(benzimidazole)-based AEM (a) *mes*-PBI degradation in the KOH environment, (b) influence of KOH composition on the alkaline hydrolysis rate using ^{1}H-NMR, and (c) detailed mechanism of alkaline hydrolysis of PBI membranes [71, 72]. Source: Reproduced with permissions of American Chemical Society and Royal Society of Chemistry.

carried out for three days with 25 wt% KOH electrolyte at 88 °C, the diagnosis suggests no sign of degradation in *mes*-PBI for three days, while the oxidative degradation and polymer chain scission were evident after two days of experiment in *m*-PBI membranes. Thus, the influence of bulky groups in the PBI-based backbone imparts better resilience to alkaline hydrolysis [71]. In a review by Wu et al., it is suggested that the NaOH doped PBIs do not significantly degrade up to 570 °C. Interestingly, the alkali-doped PBIs are the niche polymers which successfully removed the limitation of fuel cell operation between 60 and 80 °C. In addition, this review summarizes the important factors which can affect the PBI membrane performance. Mechanical and thermal stability was discussed in detail. Moreover, the fast electrochemical kinetics, high CO tolerance at elevated temperatures, etc. assure the PBIs assembled fuel cell stacks for practical applications [73]. In addition, the KOH-doped PBIs possess higher ionic conductivities (~90 mS cm^{-1} at 25 °C) than H_2SO_4 (~50 mS cm^{-1} at 25 °C) and H_3PO_4 (20 mS cm^{-1} at 25 °C), which is 45% and 75% higher, respectively, and thus make it an excellent membrane material [74]. Hou et al. reported the changes in power density through *in situ* stability examination of KOH doped shows that the peak power density was 30, 15, 14, and 9 mW cm^{-2} at 0, 210, 220, and 240 h, respectively. However, the voltage fluctuations were mainly of concern due to the nature of the electrocatalyst. The *ex situ* durability results show decline in ^{-}OH conductivity from 23 to 10 mS cm^{-1} at room temperature within 100 h at a degrading rate of 1.3×10^{-4} S cm^{-1} h^{-1}, and this might be attributed to the alkaline hydrolysis of the imidazolium ring in PBI [75]. Sana et al. designed stable poly(butylated pyridinium benzimidazolium)-based cross-linked AEM with variable aryl motifs in the polymer backbone. The designed AEMs with the aryl ether moiety show highest ^{-}OH conductivity (up to 39.4 and 111 mS cm^{-1} at 30 and 80 °C, respectively) and exhibit an IEC of 3.97 meq g^{-1}. *Ex situ* durability analysis shows no degradation of membranes when tested for 16 days in 5 M KOH at 80 °C. The study suggests that the excellent alkaline durability of designed AEM is attributed to the cross-linked network and steric protection of the substituent at C-2 imidazolium moieties [76]. Serhiichuk et al. comprehensively studied the alkaline degradation mechanism of functionally arylene-linked bis-benzimidazoles for 3164 h at 80 °C. And, the degradation mechanism was proposed. The degradation initiates from the attack of ^{-}OH at the benzimidazolium C-2 position followed by ring opening from the tetrahedral adduct to produce amides and finally, the hydrolysis of amides causing polymer backbone scission (Figure 11.21) [72]. Similar ring opening degradation was observed by Wang et al. Moreover, in his study, the influence of the butyl spacer in modulating the nanophase morphology for better ionic conductivity of PBI-based AEMs was also discussed [77]. Improving the electrochemical attributes for high-performance devices is critically important. Jheng et al. investigated ethylene-oxy spacer side chain grafting of imidazolium moieties in PBI membranes. The membrane exhibits the highest ever ionic conductivity of 272 mS cm^{-1} at 80 °C and undergoes alkaline hydrolysis at the imidazolium center [78]. Fan et al. designed highly stable AEMs which exhibit $t_{1/2}$ > 5000 h at 100 °C in 10 M aq. KOH [79]. This class of methyl-protected poly(arylene-imidazolium) (HMT-PMPI) seems to be the AEM benchmark of the present time, and the design exhibits excellent combinatorial influence of steric, electronic, and

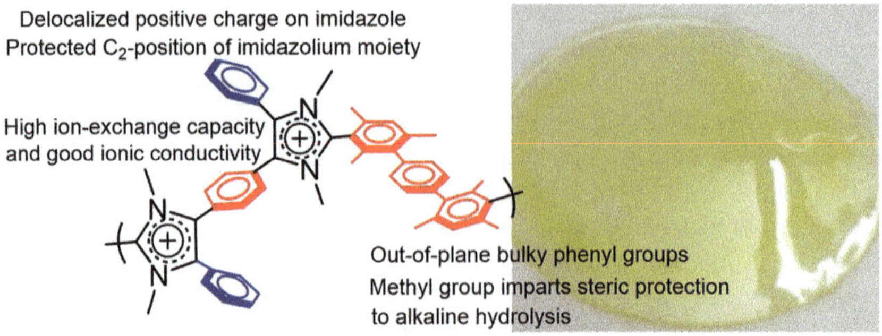

Figure 11.22 Benchmark alkaline stability of the HMT-PMPI membrane in literature: the polymer electrolyte designing and properties. Source: Reproduced with permissions from Fan et al. [79], American Chemical Society.

stereochemical shielding effects. Despite these, the PBI membranes do not possess an aryl-oxy center, but the imine motifs have been known to hydrolyze and degrade in aqueous alkaline conditions. To conclude, the C-2 centers in poly(benzimidazolium)-based AEMs are the most vulnerable site in the alkaline environment (Figure 11.22).

11.3.3.4 Degradation Mechanism in Poly(alkyl) and Poly(arene) Based Polymers

Resurgence in studying the alkaline degradation of various polymers ramifies that the heteroatomic linkage, viz, C–O and C–N in polymer backbone is hydrolyzable and causes mechanical as well as functional failures. To overcome this drawback, an alternate strategy is to explore the durability of poly(alkyl) and poly(arene) with all the hydrocarbon frameworks in AEM synthesis [80]. Unlike conventional polymers, viz, PAE, PBIs, per(fluoro alkanes), and poly(norbornenes), the poly(alkyl) and poly(arene) lack labile groups and show high resilience in wide pH conditions. More often, the reports in such AEMs with such aryl-ether free backbones reveal that the degradation is attributed to the ionic group alkaline hydrolysis discussed in Section 11.2. Moreover, the poly(alkyls) with a flexible hydrocarbon backbone are low-cost elastomeric polymer with appreciable glass transition temperature. Attributed to high stability, the poly(alkyls) and poly(arenes) have gained sustained surveillance to design high-performance and durable AEMs for energy devices. Depending upon the synthetic route adopted, the polymers can be synthesized by free radical polymerization, radiation-induced grafting of the monomer, ring-opening metathesis polymerization, superacid polycondensations [81], organocatalytic polymerization (viz, the coupling reactions), etc. [1, 82]. Zeng et al. designed a poly(vinyl alcohol)-based semi-IPN incorporating the co-polymer of divinyl benzene and NMPip functionalized vinyl benzylchloride. The AEMs achieved a high ^{-}OH conductivity of 257 mS cm^{-1} at 80 °C and show no sign of polymer degradation in ^{13}C solid-state NMR. However, the cationic group shows alkaline hydrolysis [83]. Mohanty et al. systematically studied the alkaline stability of polystyrene-*b*-poly(ethylene-*co*-butylene)-*b*-polystyrene (SEBS) polymer. The results validate that no significant degradation was noticed in the pristine SEBS and chloromethylated SEBS under variable test conditions (Figure 11.23) [66].

Like the observations by Mohanty et al., Shi et al. also affirm the alkaline hydrolysis resilience of SEBS-based AEM [84]. The retention of functional fingerprints was confirmed by FT-IR spectroscopy. Thus, the mechanism of polymer degradation are not well-postulated in studies. However, Wang et al. suggested some discrete degradation pattern in the radiation-grafted low-density polyethylene (LDPE)-based AEM. After 28 days of observation, the brittleness confirms polymer degradation. Further assessment with ^{13}C solid-state NMR confirms that upon alkali treatment, the peak intensity at $\delta = 146$ and 136 ppm found to decrease, and thus some traces of minor degradations were summarized. In addition, the ETFE-grafted AEMs show extreme polymer degradation [85, 86]. Studies by Gao et al. suggested that the elastomeric polymer is an excellent alternative to conventional aryl-ether-bearing polymer backbones [87]. Superacid polycondensation-based AEMs are gaining considerable insight in fabricating durability and show very stable attributes in the alkaline environment with no polymer degradation. However, the concern is synthetic route, which require higher mole ratios of super acids (viz, CF_3SO_3H, CF_3COOH, and CH_3SO_3H) [88]. Various research studies confer that the polymers replacing conventional aryl-ether bonds with all C—C bonds can acquire the AEM designs of desired stability. To conclude, the key to acquire long-term mechanical stability requires an alkali inert polymer of hydrocarbon origin.

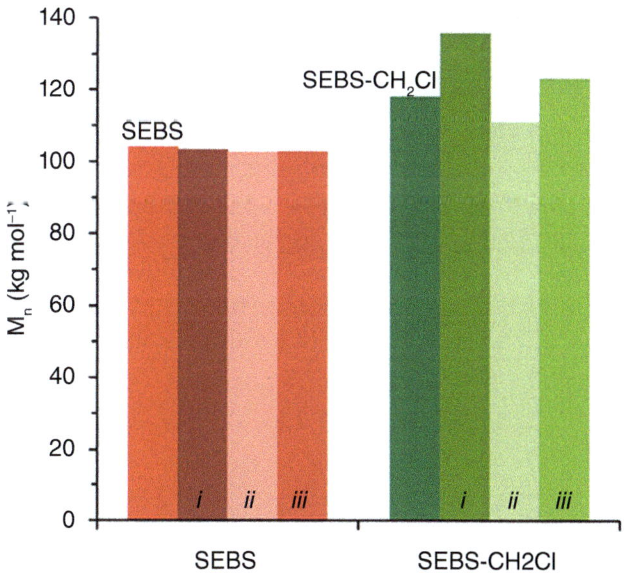

Figure 11.23 Changes in number-average molecular weight (M_n) of SEBS and chloromethylated SEBS polymer in alkaline conditions: (i) 3 eq. of KOH, 18-crown-6, THF/H_2O (4 : 1, v/v), 60 °C for 2 h and then rt for 12 h; (ii) 10 eq. of $NaOCH_3$, THF/CH_3OH, 60 °C for 14 h; (iii) polymer film (20–40 μm thick) immersed in 2 M $NaOCH_3$/CH_3OH, 60 °C for seven days [66]. Source: Reproduced with permissions of American Chemical Society.

11.3.4 Free Radical Oxidative Degradation of AEM

This section of the chapter discusses briefly the degradation mechanism of cationic groups in the presence of radical oxidative species. Oxidative degradation serves the most destructive mode of performance decay in AEMs. Unlike nucleophilic ^{-}OH-based chemical degradations (viz, alkaline hydrolysis), the *in operando* conditions during electrochemical operations are rather more fugitive. The reaction of polymer electrolytes with reactive oxygen species (ROS) generated due to interfacial reactions at both the cathode and anode side imparts the mechanical instability in AEMs [2].

Literatures on the oxidative degradation mechanism of PEMs are well-reported by Mustain et al. and Parrondo et al. [2, 89]. However, the oxidative degradation in AEMs has been documented less in real-time conditions, and hence the testing is performed employing some hydroxide ($^{\cdot}OH$), peroxide ($^{\cdot}OOH$), and superoxide ($^{\cdot}O_2^-$) radicals generated *ex situ* (viz, Fenton's reagent). During the electrochemical operation, $^{\cdot}OH$, $^{\cdot}H$, $^{\cdot}OOH$, and $^{\cdot}O_2^-$ radicals are formed via electron transfer at the membrane–catalyst interface under high alkaline conditions (pH > 10). Gubler et al. performed a systematic study on the radical formation and suggests that the oxidative strength of these ROS is in order; $[^{\cdot}OH] > [^{\cdot}H] > [^{\cdot}OOH] > H_2O_2$ [90]. Studies by Parrondo et al. ramify the influence of dissolved oxygen in generating more hydroxide radicals and hence causing faster oxidative degradation of AEM, leading to mechanical failure [89]. Interestingly, the AEM degradation occurs at the cationic group and polymer backbone. The mechanism of cationic head group degradation is simple and proceeds via *N*-ylide intermediate and can be referred from Figure 11.24. Similarly, the literatures on the mechanism of oxidative degradation of various polymer backbones remained silent. However, few reports on the mechanism of polymer degradation suggest that the mechanism is governed by the chemical nature of backbone. For instance, Espiritu et al. studied the low-density polyethylene grafted vinyl benzylchloride (LDPE-g-VBC)-based AEM with all hydrocarbon polymer backbone, and the nature on the oxidative degradation mechanism was proposed. The mechanism of polymer degradation is through triggered radical peroxy-polymer structure formation followed by simultaneous disintegration from the subsequent active benzylic position, leading to oxidative polymer scissoring (Figure 11.25a), and the driving force for such a reaction is attributed to the

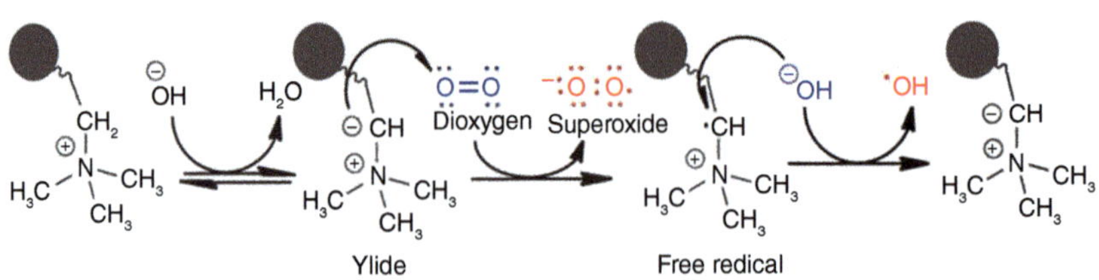

Figure 11.24 The reaction mechanism showing the generation of ROS by one electron reduction of oxygen under alkaline conditions, causing oxidative degradation of the cationic group [89]. Source: Reproduced with permissions of Royal Society of Chemistry.

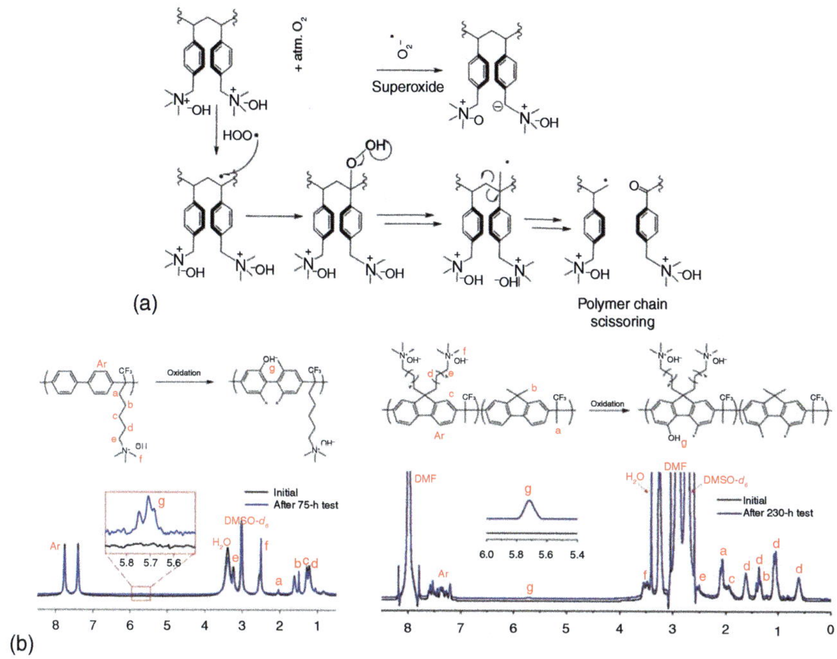

Figure 11.25 Proposed oxidative degradation mechanism of (a) LDPE-*g*-VBC-based AEM at low alkaline pH by Espiritu et al. and (b) the phenyl oxidation to phenol in a cathode BPN ionomer with NMR studies by Maurya et al. (values in *x*-axis in ppm) [91, 92]. Source: Reproduced with permissions of Royal Society of Chemistry and Elsevier.

feasibility in benzylic radical formation [91]. In contrast, the degradation mechanism of novel Friedel–Crafts polymerized aryl-backbones affords a different route. A detailed insight on the pristine AEM and oxidatively treated end-products was studied using NMR spectroscopy (Figure 11.25b). This study by Maurya et al. corroborated that the loss in the AEM performance (cathode ionomer) is due to the phenyl oxidation. The energetically favorable adsorption of aryl-motifs was correlated. It was suggested that the lesser number of phenyl groups can favor better durability, and simultaneous choice of less adsorptive doped electrocatalysts may help in improvisation of the AEM and device life-time [92]. To conclude, it is important to note here that these mechanisms can be more complex *in operando*, and hence this suggests that to understand the combinatorial effect of alkaline hydrolysis and oxidative degradation on AEMs, more real-time *in situ* diagnosis for the degradation mechanism should be explored.

11.4 Prospects and Outlook

Recent developments in AEM fabrication have gained meticulous insights for various novel synthetic strategies. It is evident that high hydroxide conductivity, excellent *ex situ* durability, low fuel cell device cost using non-precious metal catalyst, and acquiring power outputs ≥3 W cm^{-2} have paved its way to future implementations. Also, the half-lives of almost every class of quaternizable amines have been corroborated in past years and allow us to shortlist the most elite and stable scaffolds. Apart from polyelectrolyte degradations, bicarbonate–carbonate generation, membrane electrode assembly (MEA) delamination, flooding issues, etc. is important to be thought off. Summarizing the fuel cell technology, the increasing number of studies and overcoming the technological limitations seem to be challenging for next few years. Moreover, this sustainable technology will eventually earn its credit in the rapidly growing hydrogen economy. Based on the literatures, our conscience to develop polyelectrolytes with following features marked in brief details

(i) The stability of cationic head groups depends upon several factors, viz, the steric/substitution effects, electronic effects, and the hydration environment. The long-side chain morphologies with cycloaliphatic (viz, ASU) type cationic heads acquire high nanophase separation imparting to achieve high hydroxide conductivities. In addition, the stability in the cationic head has achieved stabilities of more than one year under alkaline conditions [93].
(ii) The polymer degradation triggers dynamic phenomena, and hence the cationic motifs should be placed away bearing an optimum spacer length of 3 C to 6 C. In contrast to conventional aryl-ether-based polymers, all hydrocarbon polymers can be a viable option to design alkaline durable AEM [94–96]. Moreover, the PBI-based polyelectrolyte can be a viable alternative to design highly stable AEM for high-temperature fuel cell/electrolyzer applications.
(iii) The formation of ROS is governed by electrocatalysts in membrane electrode assemblies; this ROS generation is inevitable in the aqueous alkaline environment. Thus, radical scavenging materials with high surface oxygen vacancies or electron quenching attributes must be studied, provided the intercalations do not alter the physicochemical and electrochemical attributes of AEM.
(iv) Post mortem analysis of AEMs often suffers edge-deflection and bursting. This mechanical failure defames the device performance. It is observed that at the areas of maximum stress, the punctures significantly occur; thus, high molecular weight polymers with tensile strengths ≥50 MPa at 80–100 °C in alkaline environments must be at the stimulus plan during fuel cell installations.

11.5 Conclusion

The alkaline hydrolysis and oxidative degradation of AEM restrict the terrestrial utilization of alkaline membrane fuel cell and alkaline membrane water electrolyzer devices in the energy market. However, recent advances in AEM fabrication are trailing a positive impact on achieving the highest possible performance and stability standards. Moreover, the inaccurate durability examination protocols in comparison to the original state of operations have limited the detailed scientific understanding. Thus, simultaneous strategies and realistic examination methods should be developed to overcome the real-time instability dilemma. Evidently, the

AEM designing and its characteristics will achieve the desired performance in the near future. Along with the academic interest in developing novel advanced materials, it is highly desirable to design alkali stable polyelectrolyte membranes via benign synthetic routes and cost-effective strategies for large-scale implementations. Although various aspects of AEM design are remained uncovered in this chapter due to limited scope, it is well-understood and positively expected that this technology will grab a faster pace toward global sustainable energy goals in coming years of its development.

References

1 Chen, N. and Lee, Y.M. (2021). Anion exchange polyelectrolytes for membranes and ionomers. *Prog Polym Sci* 113: 101345. https://doi.org/10.1016/j.progpolymsci.2020.101345.

2 Mustain, W.E., Chatenet, M., Page, M., and Kim, Y.S. (2020). Durability challenges of anion exchange membrane fuel cells. *Energy Environ Sci* 13: 2805. https://doi.org/10.1039/D0EE01133A.

3 Merle, G., Wessling, M., and Nijmeijer, K. (2011). Anion exchange membranes for alkaline fuel cells: a review. *J Membr Sci* 377: 1. https://doi.org/10.1016/j.memsci.2011.04.043.

4 Yang, Z., Ran, J., Wu, B. et al. (2016). Stability challenge in anion exchange membrane for fuel cells. *Curr Opin Chem Eng* 12: 22. https://doi.org/10.1016/j.coche.2016.01.009.

5 Dekel, D.R., Amar, M., Willdorf, S. et al. (2017). Effect of water on the stability of quaternary ammonium groups for anion exchange membrane fuel cell applications. *Chem Mater* 29: 4425. https://doi.org/10.1021/acs.chemmater.7b00958.

6 Dekel, D.R., Willdorf, S., Ash, U. et al. (2018). The critical relation between chemical stability of cations and water in anion exchange membrane fuel cells environment. *J Power Sources* 375: 351. https://doi.org/10.1016/j.jpowsour.2017.08.026.

7 Pham, T.H., Allushi, A., Olsson, J.S., and Jannasch, P. (2020). Rational molecular design of anion exchange membranes functionalized with alicyclic quaternary ammonium cations. *Polym Chem* 11: 6953. https://doi.org/10.1039/D0PY01291B.

8 Miyanishi, S. and Yamaguchi, T. (2016). Ether cleavage-triggered degradation of benzyl alkylammonium cations for polyethersulfone anion exchange membranes. *Phys Chem Chem Phys* 18: 12009. https://doi.org/10.1039/C6CP00579A.

9 Arges, C.G. and Ramani, V. (2013). Two-dimensional NMR spectroscopy reveals cation-triggered backbone degradation in polysulfone-based anion exchange membranes. *Proc Natl Acad Sci* 110: 2490. https://doi.org/10.1073/pnas.1217215110.

10 Hyun, J., Yang, S.H., Doo, G. et al. (2021). Degradation study for the membrane electrode assembly of anion exchange membrane fuel cells at a single-cell level. *J Mater Chem A* 9: 18546. https://doi.org/10.1039/D1TA05801K.

11 Shin, D., Bae, C., and Kim, Y.S. (2018). Chapter 8, Anion exchange membranes: stability and synthetic approach. In: *The Chemistry of Membranes Used in Fuel Cells*

(ed. S. Schlick), 195–226. Wiley. ISBN: 978-1-119-19608-2. https://www.wiley.com/en-us/The+Chemistry+of+Membranes+Used+in+Fuel+Cells:+Degradation+and+Stabilization-p-9781119196082 (accessed 13 November 2023).

12 Sun, Z., Lin, B., and Yan, F. (2018). Anion-exchange membranes for alkaline fuel-cell applications: the effects of cations. *ChemSusChem* 11: 58. https://doi.org/10.1002/cssc.201701600.

13 Cheng, J., He, G., and Zhang, F. (2015). A mini-review on anion exchange membranes for fuel cell applications: stability issue and addressing strategies. *Int J Hydrog Energy* 40: 7348. https://doi.org/10.1016/j.ijhydene.2015.04.040.

14 Marino, M.G. and Kreuer, K.D. (2015). Alkaline stability of quaternary ammonium cations for alkaline fuel cell membranes and ionic liquids. *ChemSusChem* 8: 513. https://doi.org/10.1002/cssc.201403022.

15 Chen, J., Li, C., Wang, J. et al. (2017). A general strategy to enhance the alkaline stability of anion exchange membranes. *J Mater Chem A* 5: 6318. https://doi.org/10.1039/C7TA00879A.

16 Li, L., Wang, J., Hussain, M. et al. (2021). Side-chain manipulation of poly(phenylene oxide) based anion exchange membrane: alkoxyl extender integrated with flexible spacer. *J Membr Sci* 624: 119088. https://doi.org/10.1016/j.memsci.2021.119088.

17 Zhu, L., Yu, X., and Hickner, M.A. (2018). Exploring backbone-cation alkyl spacers for multi-cation side chain anion exchange membranes. *J Power Sources* 375: 433. https://doi.org/10.1016/j.jpowsour.2017.06.020.

18 Mohanty, A.D. and Bae, C. (2014). Mechanistic analysis of ammonium cation stability for alkaline exchange membrane fuel cells. *J Mater Chem A* 2: 17314. https://doi.org/10.1039/C4TA03300K.

19 Nuñez, S.A., Capparelli, C., and Hickner, M.A. (2016). N-Alkyl interstitial spacers and terminal pendants influence the alkaline stability of tetraalkylammonium cations for anion exchange membrane fuel cells. *Chem Mater* 28: 2589. https://doi.org/10.1021/acs.chemmater.5b04767.

20 Pan, J., Han, J., Zhu, L., and Hickner, M.A. (2017). Cationic side-chain attachment to poly(phenylene oxide) backbones for chemically stable and conductive anion exchange membranes. *Chem Mater* 29: 5321. https://doi.org/10.1021/acs.chemmater.7b01494.

21 Liu, L., Li, Q., Dai, J. et al. (2014). A facile strategy for the synthesis of guanidinium-functionalized polymer as alkaline anion exchange membrane with improved alkaline stability. *J Membr Sci* 453: 52. https://doi.org/10.1016/j.memsci.2013.10.054.

22 Sherazi, T.A., Zahoor, S., Raza, R. et al. (2015). Guanidine functionalized radiation induced grafted anion-exchange membranes for solid alkaline fuel cells. *Int J Hydrog Energy* 40: 786. https://doi.org/10.1016/j.ijhydene.2014.08.086.

23 Xue, B., Dong, X., Li, Y. et al. (2017). Synthesis of novel guanidinium-based anion-exchange membranes with controlled microblock structures. *J Membr Sci* 537: 151. https://doi.org/10.1016/j.memsci.2017.05.030.

24 Xue, B., Wang, Q., Zheng, J. et al. (2020). Bi-guanidinium-based crosslinked anion exchange membranes: synthesis, characterization, and properties. *J Membr Sci* 601: 117923. https://doi.org/10.1016/j.memsci.2020.117923.

25 Liu, Y., Dai, J., Zhang, K. et al. (2017). Hybrid anion exchange membrane of hydroxyl-modified polysulfone incorporating guanidinium-functionalized graphene oxide. *Ionics* 23: 3085. https://doi.org/10.1007/s11581-017-2100-3.

26 Mohanty, A.D., Tignor, S.E., Sturgeon, M.R. et al. (2017). Thermochemical stability study of alkyl-tethered quaternary ammonium cations for anion exchange membrane fuel cells. *J Electrochem Soc* 164: F1279. https://doi.org/10.1149/2.0141713jes.

27 Chen, N., Jin, Y., Liu, H. et al. (2021). Insight into the alkaline stability of n-heterocyclic ammonium groups for anion-exchange polyelectrolytes. *Angew Chem Int Ed* 60: 19272. https://doi.org/10.1002/anie.202105231.

28 Dang, H.-S. and Jannasch, P. (2017). A comparative study of anion-exchange membranes tethered with different hetero-cycloaliphatic quaternary ammonium hydroxides. *J Mater Chem A* 5: 21965. https://doi.org/10.1039/C7TA06029G.

29 Khan, M.I., Li, X., Fernandez-Garcia, J. et al. (2021). Effect of different quaternary ammonium groups on the hydroxide conductivity and stability of anion exchange membranes. *ACS Omega* 6: 7994. https://doi.org/10.1021/acsomega.0c05134.

30 Pham, T.H. and Jannasch, P. (2015). Aromatic polymers incorporating bis-N-spirocyclic quaternary ammonium moieties for anion-exchange membranes. *ACS Macro Lett* 4: 1370. https://doi.org/10.1021/acsmacrolett.5b00690.

31 Olsson, J.S., Pham, T.H., and Jannasch, P. (2017). Poly(N,N-diallylazacycloalkane)s for anion-exchange membranes functionalized with N-spirocyclic quaternary ammonium cations. *Macromolecules* 50: 2784. https://doi.org/10.1021/acs.macromol.7b00168.

32 Pham, T.H., Olsson, J.S., and Jannasch, P. (2017). N-Spirocyclic quaternary ammonium ionenes for anion-exchange membranes. *J Am Chem Soc* 139: 2888. https://doi.org/10.1021/jacs.6b12944.

33 Wang, X., Lin, C., Gao, Y., and Lammertink, R.G.H. (2021). Anion exchange membranes with twisted poly(terphenylene) backbone: effect of the N-cyclic cations. *J Membr Sci* 635: 119525. https://doi.org/10.1016/j.memsci.2021.119525.

34 Liu, F.H., Yang, Q., Gao, X.L. et al. (2020). Anion exchange membranes with dense N-spirocyclic cations as side-chain. *J Membr Sci* 595: 117560. https://doi.org/10.1016/j.memsci.2019.117560.

35 Lin, C., Yu, D., Wang, J. et al. (2019). Facile construction of poly(arylene ether)s-based anion exchange membranes bearing pendent N-spirocyclic quaternary ammonium for fuel cells. *Int J Hydrog Energy* 44: 26565. https://doi.org/10.1016/j.ijhydene.2019.08.092.

36 Zhang, Y., Chen, W., Yan, X. et al. (2020). Ether spaced N-spirocyclic quaternary ammonium functionalized crosslinked polysulfone for high alkaline stable anion exchange membranes. *J Membr Sci* 598: 117650. https://doi.org/10.1016/j.memsci.2019.117650.

37 Olsson, J.S., Pham, T.H., and Jannasch, P. (2020). Functionalizing polystyrene with N-alicyclic piperidine-based cations via friedel–crafts alkylation for highly alkali-stable anion-exchange membranes. *Macromolecules* 53: 4722. https://doi.org/10.1021/acs.macromol.0c00201.

38 Zhang, Y., Chen, W., Li, T. et al. (2021). A rod-coil grafts strategy for N-spirocyclic functionalized anion exchange membranes with high fuel cell power density. *J Power Sources* 490: 229544. https://doi.org/10.1016/j.jpowsour.2021.229544.

39 Klein, L.B., Morsing, T.J., Livingstone, R.A. et al. (2016). The effects of symmetry and rigidity on non-adiabatic dynamics in tertiary amines: a time-resolved photoelectron velocity-map imaging study of the cage-amine ABCO. *Phys Chem Chem Phys* 18: 9715. https://doi.org/10.1039/C5CP07910A.

40 Li, W., Fang, J., Lv, M. et al. (2011). Novel anion exchange membranes based on polymerizable imidazolium salt for alkaline fuel cell applications. *J Mater Chem* 21: 11340. https://doi.org/10.1039/C1JM11093D.

41 Mondal, R., Pal, S., and Chatterjee, U. (2021). Alkylated imidazole moieties in a cross-linked anion exchange membrane facilitate acid recovery with high purity. *ACS Appl Polym Mater* 3: 1544. https://doi.org/10.1021/acsapm.0c01383.

42 Yang, Z., Guo, R., Malpass-Evans, R. et al. (2016). Highly conductive anion-exchange membranes from microporous Tröger's base polymers. *Angew Chem Int Ed* 55: 11499. https://doi.org/10.1002/anie.201605916.

43 Xue, J., Zhang, J., Liu, X. et al. (2022). Toward alkaline-stable anion exchange membranes in fuel cells: cycloaliphatic quaternary ammonium-based anion conductors. *Electrochem Energy Rev* 5: 348. https://doi.org/10.1007/s41918-021-00105-7.

44 Mahmoud, A.M.A., Yoshimura, K., and Maekawa, Y. (2021). Alkaline fuel cells consisting of imidazolium-based graft-type anion exchange membranes: optimization of fuel cell conditions to achieve high performance and durability. *J Membr Sci* 620: 118844. https://doi.org/10.1016/j.memsci.2020.118844.

45 Ryu, J., Seo, J.Y., Choi, B.N. et al. (2019). Quaternized chitosan-based anion exchange membrane for alkaline direct methanol fuel cells. *J Ind Eng Chem* 73: 254. https://doi.org/10.1016/j.jiec.2019.01.033.

46 Sun, Z., Pan, J., Guo, J., and Yan, F. (2018). The alkaline stability of anion exchange membrane for fuel cell applications: the effects of alkaline media. *Adv Sci* 5: 1800065. https://doi.org/10.1002/advs.201800065.

47 Fan, J., Willdorf-Cohen, S., Schibli, E.M. et al. (2019). Poly(bis-arylimidazoliums) possessing high hydroxide ion exchange capacity and high alkaline stability. *Nat Commun* 10: 2306. https://doi.org/10.1038/s41467-019-10292-z.

48 Hugar, K.M., Kostalik, H.A.I.V., and Coates, G.W. (2015). Imidazolium cations with exceptional alkaline stability: a systematic study of structure–stability relationships. *J Am Chem Soc* 137: 8730. https://doi.org/10.1021/jacs.5b02879.

49 Tham, D.D. and Kim, D. (2019). C2 and N3 substituted imidazolium functionalized poly(arylene ether ketone) anion exchange membrane for water electrolysis with improved chemical stability. *J Membr Sci* 581: 139. https://doi.org/10.1016/j.memsci.2019.03.060.

50 You, W., Padgett, E., MacMillan, S.N. et al. (2019). Highly conductive and chemically stable alkaline anion exchange membranes via ROMP of trans-cyclooctene derivatives. *Proc Natl Acad Sci* 116: 9729. https://doi.org/10.1073/pnas.1900988116.

51 Zhu, T., Zhang, J., and Tang, C. (2020). Metallo-polyelectrolytes: correlating macromolecular architectures with properties and applications. *Trends Chem* 2: 227. https://doi.org/10.1016/j.trechm.2019.12.004.

52 Zha, Y., Disabb-Miller, M.L., Johnson, Z.D. et al. (2012). Metal-cation-based anion exchange membranes. *J Am Chem Soc* 134: 4493. https://doi.org/10.1021/ja211365r.

53 Chen, N., Zhu, H., Chu, Y. et al. (2017). Cobaltocenium-containing polybenzimidazole polymers for alkaline anion exchange membrane applications. *Polym Chem* 8: 1381. https://doi.org/10.1039/C6PY01936F.

54 Gu, S., Wang, J., Kaspar, R.B. et al. (2015). Permethyl cobaltocenium (Cp*2Co+) as an ultra-stable cation for polymer hydroxide-exchange membranes. *Sci Rep* 5: 11668. https://doi.org/10.1038/srep11668.

55 Zhu, T., Xu, S., Rahman, A. et al. (2018). Cationic metallo-polyelectrolytes for robust alkaline anion-exchange membranes. *Angew Chem Int Ed* 57: 2388. https://doi.org/10.1002/anie.201712387.

56 Zhu, T. and Tang, C. (2020). Crosslinked metallo-polyelectrolytes with enhanced flexibility and dimensional stability for anion-exchange membranes. *Polym Chem* 11: 4542. https://doi.org/10.1039/D0PY00757A.

57 Zhu, T., Sha, Y., Firouzjaie, H.A. et al. (2020). Rational synthesis of metallo-cations toward redox- and alkaline-stable metallo-polyelectrolytes. *J Am Chem Soc* 142: 1083. https://doi.org/10.1021/jacs.9b12051.

58 Ge, X., He, Y., Guiver, M.D. et al. (2016). Alkaline anion-exchange membranes containing mobile ion shuttles. *Adv Mater* 28: 3467. https://doi.org/10.1002/adma.201506199.

59 Yang, Q., Li, L., Gao, X.L. et al. (2019). Crown ether bridged anion exchange membranes with robust alkaline durability. *J Membr Sci* 578: 230. https://doi.org/10.1016/j.memsci.2019.02.038.

60 Lin, C., Gao, Y., Li, N. et al. (2020). Quaternized Tröger's base polymer with crown ether unit for alkaline stable anion exchange membranes. *Electrochim Acta* 354: 136693. https://doi.org/10.1016/j.electacta.2020.136693.

61 You, W., Hugar, K.M., Selhorst, R.C. et al. (2021). Degradation of organic cations under alkaline conditions. *J Org Chem* 86: 254. https://doi.org/10.1021/acs.joc.0c02051.

62 Yassin, K., Rasin, I.G., Willdorf-Cohen, S. et al. (2021). A surprising relation between operating temperature and stability of anion exchange membrane fuel cells. *J Power Source Adv* 11: 100066. https://doi.org/10.1016/j.powera.2021.100066.

63 Sharma, J., Mishra, S., Rathod, N.H., and Kulshrestha, V. (2022). Investigation on flexible and thermally crosslinked bis-piperidinium-PPO anion exchange membrane (AEM) for electro-kinetic desalination and acid recovery. *J Membr Sci* 664: 121082. https://doi.org/10.1016/j.memsci.2022.121082.

64 Berber, M.R. (2020). Molecular weight impact of poly(2,5-benzimidazole) polymer on film conductivity, ion exchange capacity, acid retention capability, and oxidative stability. *Front Energy Res* 8: https://doi.org/10.3389/fenrg.2020.571651.

65 Gottesfeld, S., Dekel, D.R., Page, M. et al. (2018). Anion exchange membrane fuel cells: current status and remaining challenges. *J Power Sources* 375: 170. https://doi.org/10.1016/j.jpowsour.2017.08.010.

66 Mohanty, A.D., Tignor, S.E., Krause, J.A. et al. (2016). Systematic alkaline stability study of polymer backbones for anion exchange membrane applications. *Macromolecules* 49: 3361. https://doi.org/10.1021/acs.macromol.5b02550.

67 Taguet, A., Ameduri, B., and Boutevin, B. (2005). *Crosslinking in Materials Science*, 127. Springer Berlin Heidelberg: Berlin, Heidelberg.

68 Ahmed, M.M., Hrůza, J., Stuchlík, M. et al. (2021). Revisiting the polyvinylidene fluoride heterogeneous alkaline reaction mechanism in propan-2-ol: an additional hydrogenation step. *Eur Polym J* 156: 110605. https://doi.org/10.1016/j.eurpolymj.2021.110605.

69 McHugh, P.J., Das, A.K., Wallace, A.G. et al. (2021). An investigation of a (vinylbenzyl) trimethylammonium and *N*-vinylimidazole-substituted poly(vinylidene fluoride-*co*-hexafluoropropylene) copolymer as an anion-exchange membrane in a lignin-oxidising electrolyser. In: *Membranes*, vol. 11 (ed. L. Dammak), 425–440, Membranes MDPI. https://doi.org/10.3390/membranes11060425.

70 Sharma, J., Totee, C., Kulshrestha, V., and Ameduri, B. (2023). Spectroscopic evidence and mechanistic insights on dehydrofluorination of PVDF in alkaline medium. *Eur Polym J* 201: 112580. https://doi.org/10.1016/j.eurpolymj.2023.112580.

71 Aili, D., Wright, A.G., Kraglund, M.R. et al. (2017). Towards a stable ion-solvating polymer electrolyte for advanced alkaline water electrolysis. *J Mater Chem A* 5: 5055. https://doi.org/10.1039/C6TA10680C.

72 Serhiichuk, D., Patniboon, T., Xia, Y. et al. (2023). Insight into the alkaline stability of arylene-linked bis-benzimidazoles and polybenzimidazoles. *ACS Appl Polym Mater* 5: 803. https://doi.org/10.1021/acsapm.2c01769.

73 Wu, Q.X., Pan, Z.F., and An, L. (2018). Recent advances in alkali-doped polybenzimidazole membranes for fuel cell applications. *Renew Sust Energ Rev* 89: 168. https://doi.org/10.1016/j.rser.2018.03.024.

74 Xing, B. and Savadogo, O. (2000). Hydrogen/oxygen polymer electrolyte membrane fuel cells (PEMFCs) based on alkaline-doped polybenzimidazole (PBI). *Electrochem Commun* 2: 697. https://doi.org/10.1016/S1388-2481(00)00107-7.

75 Hou, H., Wang, S., Jiang, Q. et al. (2011). Durability study of KOH doped polybenzimidazole membrane for air-breathing alkaline direct ethanol fuel cell. *J Power Sources* 196: 3244. https://doi.org/10.1016/j.jpowsour.2010.11.104.

76 Sana, B., Das, A., and Jana, T. (2019). Polybenzimidazole as alkaline anion exchange membrane with twin hydroxide ion conducting sites. *Polymer* 172: 213. https://doi.org/10.1016/j.polymer.2019.03.078.

77 Wang, Y., Qiao, X., Liu, M. et al. (2021). The effect of $-NH^{-}$ on quaternized polybenzimidazole anion exchange membranes for alkaline fuel cells. *J Membr Sci* 626: 119178. https://doi.org/10.1016/j.memsci.2021.119178.

78 Jheng, L.-C., Hsu, S.L.-C., Lin, B.-Y., and Hsu, Y.-L. (2014). Quaternized polybenzimidazoles with imidazolium cation moieties for anion exchange membrane fuel cells. *J Membr Sci* 460: 160. https://doi.org/10.1016/j.memsci.2014.02.043.

79 Fan, J., Wright, A.G., Britton, B. et al. (2017). Cationic polyelectrolytes, stable in 10 M KOHaq at 100 °C. *ACS Macro Lett* 6: 1089. https://doi.org/10.1021/acsmacrolett.7b00679.

80 Noh, S., Jeon, J.Y., Adhikari, S. et al. (2019). Molecular engineering of hydroxide conducting polymers for anion exchange membranes in electrochemical energy conversion technology. *Acc Chem Res* 52: 2745. https://doi.org/10.1021/acs.accounts.9b00355.

81 Wang, J., Zhao, Y., Setzler, B.P. et al. (2019). Poly(aryl piperidinium) membranes and ionomers for hydroxide exchange membrane fuel cells. *Nat Energy* 4: 392. https://doi.org/10.1038/s41560-019-0372-8.

82 Tao, Z., Wang, C., Zhao, X. et al. (2021). Progress in high-performance anion exchange membranes based on the design of stable cations for alkaline fuel cells. *Adv Mater Technol* 6: 2001220. https://doi.org/10.1002/admt.202001220.

83 Zeng, L., He, Q., Liao, Y. et al. (2020). Anion exchange membrane based on interpenetrating polymer network with ultrahigh ion conductivity and excellent stability for alkaline fuel. *Cell Res* https://doi.org/10.34133/2020/4794706.

84 Shi, Y., Zhao, Z., Liu, W., and Zhang, C. (2020). Physically self-cross-linked sebs anion exchange membranes. *Energy Fuel* 34: 16746. https://doi.org/10.1021/acs.energyfuels.0c03017.

85 Wang, L., Brink, J.J., Liu, Y. et al. (2017). Non-fluorinated pre-irradiation-grafted (peroxidated) LDPE-based anion-exchange membranes with high performance and stability. *Energy Environ Sci* 10: 2154. https://doi.org/10.1039/C7EE02053H.

86 Meek, K.M., Reed, C.M., Pivovar, B. et al. (2020). The alkali degradation of LDPE-based radiation-grafted anion-exchange membranes studied using different ex situ methods. *RSC Adv* 10: 36467. https://doi.org/10.1039/D0RA06484J.

87 Gao, X., Yu, H., Xie, F. et al. (2020). High performance cross-linked anion exchange membrane based on aryl-ether free polymer backbones for anion exchange membrane fuel cell application. *Sustain Energy Fuel* 4: 4057. https://doi.org/10.1039/D0SE00502A.

88 Lcc, W.-H., Park, E.J., Han, J. et al. (2017). Poly(terphenylene) anion exchange membranes: the effect of backbone structure on morphology and membrane property. *ACS Macro Lett* 6: 566. https://doi.org/10.1021/acsmacrolett.7b00148.

89 Parrondo, J., Wang, Z., Jung, M.-S.J., and Ramani, V. (2016). Reactive oxygen species accelerate degradation of anion exchange membranes based on polyphenylene oxide in alkaline environments. *Phys Chem Chem Phys* 18: 19705. https://doi.org/10.1039/C6CP01978A.

90 Gubler, L., Dockheer, S.M., and Koppenol, W.H. (2011). Radical (HO•, H• and HOO•) formation and ionomer degradation in polymer electrolyte fuel cells. *J Electrochem Soc* 158: B755. https://doi.org/10.1149/1.3581040.

91 Espiritu, R., Golding, B.T., Scott, K., and Mamlouk, M. (2017). Degradation of radiation grafted hydroxide anion exchange membrane immersed in neutral pH: removal of vinylbenzyl trimethylammonium hydroxide due to oxidation. *J Mater Chem A* 5: 1248. https://doi.org/10.1039/C6TA08232G.

92 Maurya, S., Lee, A.S., Li, D. et al. (2019). On the origin of permanent performance loss of anion exchange membrane fuel cells: electrochemical oxidation of phenyl group. *J Power Sources* 436: 226866. https://doi.org/10.1016/j.jpowsour.2019.226866.

93 Moreno-González, M., Mardle, P., Zhu, S. et al. (2023). One year operation of an anion exchange membrane water electrolyzer utilizing Aemion+® membrane:

minimal degradation, low H2 crossover and high efficiency. *J Power Source Adv* 19: 100109. https://doi.org/10.1016/j.powera.2023.100109.

94 Pan, J., Chen, C., Zhuang, L., and Lu, J. (2012). Designing advanced alkaline polymer electrolytes for fuel cell applications. *Acc Chem Res* 45: 473. https://doi.org/10.1021/ar200201x.

95 Cao, H., Pan, J., Zhu, H. et al. (2021). Interaction regulation between ionomer binder and catalyst: active triple-phase boundary and high performance catalyst layer for anion exchange membrane fuel cells. *Adv Sci* 8: 2101744. https://doi.org/10.1002/advs.202101744.

96 Xu, L., Wang, H., Min, L. et al. (2022). Anion exchange membranes based on poly(aryl piperidinium) containing both hydrophilic and hydrophobic side chains. *Ind Eng Chem Res* 61: 14232. https://doi.org/10.1021/acs.iecr.2c01722.

12

Computational Approaches to Alkaline Anion Exchange Membranes

Minu Elizabeth Thomas

Mahatma Gandhi University, School of Polymer Science and Technology, Kottayam 686560, Kerala, India

12.1 Introduction

Anion exchange membranes (AEMs) are specialized membranes that facilitate the transport of anions while hindering the movement of cations. These membranes find extensive application in devices that are employed in energy conversion and storage, such as batteries, fuel cells, and electrolyzers [1, 2]. The functional groups present in AEMs exhibit a positive charge, effectively promoting anion transport [2–8]. The polymer's backbone provides the necessary mechanical strength, which enables optimum ion conductivity, mechanical stability, and chemical stability. The performance of AEMs can be optimized by adjusting the degree of ion exchange, polymer morphology, and membrane thickness, which determine their overall efficiency.

Although AEMs show great promise for various applications, their commercial availability is currently limited due to several challenges. The prime challenge is the chemical stability of AEMs. The functional groups in AEMs can degrade over time, limiting their long-term performance and durability. Additionally, they swell in water or at high humidity, which can cause dimensional changes and reduced mechanical integrity. The development and scale-up manufacturing of AEMs with consistent quality and performance remain challenging. However, ongoing research efforts are underway to address these challenges and improve the stability, performance, and cost-effectiveness of AEMs. Based on expert analysis, the phase segregation structure of Nafion, a commercial available proton exchange membrane (PEM), is responsible for its high ionic conductivity. Similarly, AEMs with a stiff hydrophobic backbone and flexible hydrophilic side groups show comparable conductivity due to their polymeric structure resembling that of Nafion. In addition to experimental research, many

Alkaline Anion Exchange Membranes for Fuel Cells: From Tailored Materials to Novel Applications, First Edition. Edited by Jince Thomas, Alex Schechter, Flavio Grynszpan, Bejoy Francis, and Sabu Thomas.

scientific teams are utilizing theoretical and computational calculations to determine the relationship between the structure and property and the alkaline stability of cationic groups attached to the AEM synthesis polymeric backbone [1]. Furthermore, *in silico* methods offer transportation mechanisms and novel AEM designs. With advancements in materials science and engineering, it is expected that AEMs will become more readily available for commercial applications in the future. This chapter provides an overview of important factors that will impact the future design of new AEMs using *in silico* approaches. These factors will create AEMs with better performance characteristics like improved chemical stability and higher conductivity.

12.2 Why Computational Studies Are Important in Anion Exchange Membranes?

Computational analysis is essential in comprehending the fundamental mechanisms, refining the design, and anticipating the efficacy of AEM systems. This process provides researchers with valuable knowledge regarding ion transportation, membrane structure, water management, and material creation, ultimately creating more effective and long-lasting AEMs for diverse electrochemical applications. Understanding the significance of computational analysis for AEM systems is crucial due to the following reasons:

- *The Structure and Stability of AEM*. Computational analysis allows for the precise characterization and prediction of the structure and stability of AEM materials. By thoroughly simulating the interactions between polymer chains, solvent molecules, and ions, researchers can investigate the morphological changes, stability, and degradation mechanisms of AEMs as time progresses. This knowledge is crucial in designing AEMs with unparalleled durability and long-term stability.
- *Understanding Ion Transport Mechanisms*. Through simulating ion behavior at the molecular level, computational models can offer valuable insights into processes like diffusion, migration, and adsorption. This understanding can aid in optimizing the design and performance of AEMs by identifying key factors that impact ion transport, like membrane structure, functional groups, and hydration effects. It also helps comprehend the degradation of the polymer and the strength of the interaction between the cation and polymer. This knowledge ultimately leads to the enhanced design of AEMs for electrochemical applications.
- *Predicting Ion Selectivity and Conductivity*. Researchers can use computational methods to predict and enhance the ion selectivity and conductivity of AEMs. Computational models offer valuable insights into membrane performance by calculating energy barriers for ion transport and analyzing factors such as membrane thickness, charge density, and ion hydration. Such insights can guide the design of AEMs with improved ion selectivity and high conductivity, ultimately resulting in greater device efficiency.
- *Water Management*. Managing membrane swelling and water is a crucial challenge in AEM systems. Computational models are utilized to simulate the behavior of

water molecules, quantify their interactions with the polymer matrix and ions, and optimize the polymer structure. This helps regulate water uptake and prevent excessive swelling, which can adversely affect membrane performance.

- *Design of AEM Materials.* Researchers can efficiently screen and design new AEM materials with desired properties by utilizing computational analysis. They can explore numerous chemical structures, functional groups, and polymer compositions using various computational tools. This virtual screening process allows them to identify the most promising candidates for experimental synthesis, ultimately accelerating the discovery of novel AEM materials with improved performance.

Computational simulations offer numerous advantages to the scientific community. In the field of AEM, computational simulations have proven to be highly effective in accelerating research and development progress. These *in silico* tools provide a quick and cost-efficient approach to investigating diverse material compositions, design parameters, and operating conditions. This expedites discovery of new AEM materials with enhanced properties, such as conductivity, stability, and ion selectivity. Additionally, computational simulations provide a deeper understanding of AEM systems' underlying mechanisms and behavior. By revealing intricate molecular-level interactions, ion transport dynamics, and membrane structure–property relationships, researchers can enhance the design and functionality of AEM materials. Such simulations assist in forecasting and analyzing multiple factors, including membrane structure, charge density, and hydration effects, which guide the design process toward AEMs that offer better efficiency, durability, and selectivity. Using computational simulations also reduces the reliance on expensive and time-consuming experimental trials, providing a cost-effective way to explore numerous material compositions and operating conditions virtually. This saves time, resources, and materials, enabling researchers to explore wider possibilities. Moreover, computational simulations open new avenues for research and innovation in AEM by allowing researchers to examine and analyze various material compositions, thereby discovering new materials with superior properties that may have been missed through traditional trial-and-error methods.

12.3 Tools of *In Silico* Approaches in Anion Exchange Membranes

Although many attempts have been made to create effective AEMs, their application in anion exchange membrane fuel cells (AEMFCs) has been hindered by low ionic conductivity and alkaline stability. Both experimental and theoretical studies have been conducted to address these issues. Computational tools and techniques, known as *in silico* approaches, are now essential in the study and design of AEMs. These methods allow researchers to investigate AEM structure, properties, and performance at various scales, from atomic and molecular levels to system-level behavior. *In silico* approaches such as quantum mechanical calculations, molecular dynamics (MD) simulations, continuum electrostatics models, Monte Carlo simulations, kinetic models, and machine learning (ML) provide valuable insights into ion transport mechanisms, membrane structure, electrostatic behavior, kinetics, and material design. They are crucial in understanding fundamental processes, optimizing AEM performance, and developing advanced membranes for various electrochemical applications. These tools enable researchers to explore and analyze AEM systems computationally, contributing significantly to the advancement and innovation of AEM technology. In some cases, computational studies have been utilized to complement experimental findings and make predictions before conducting experiments. In recent years, the use of computational techniques to model AEM and AEMFC components has increased significantly. These techniques are often described in terms of time and length scales, and various methods are available for modeling these materials. In this discussion, we will explore various important computational techniques that can be utilized to model the characteristics of AEM and AEMFC components.

12.3.1 Electronic Structure Methods in Anion Exchange Membranes

In AEM research, the electronic structure and properties of molecules and materials are calculated using quantum mechanical methods. Density functional theory (DFT) is a commonly used technique to analyze the energetics, electronic properties, and charge distribution within AEM materials. By studying the interaction between ions, functional groups, and the polymer matrix, DFT can provide valuable insights into the ion selectivity, reactivity, and other important properties of AEMs. Many research groups use theoretical and computational calculations along with experimental studies to discover the relationship between the structure and properties of cationic groups in AEM synthesis. They aim to determine the alkaline stability of these groups that are attached to the polymeric backbone. The most commonly used analyses for this purpose are the frontier molecular orbital (FMO) analysis, electrostatic potential (ESP), electronic density, and electronic charges, especially Mulliken charges on atoms and atomic groups. Researchers also make inferences on the energy barrier and free energy changes for forming cation-OH^- complexes to predict alkaline stability. Based on FMO analysis, cations with a higher energy value in their lowest unoccupied molecular orbital (LUMO) and a lower energy value in their highest occupied molecular orbital (HOMO) compared to OH^- HOMO are less likely to be attacked by OH^- ions. However, cations with a significant HOMO–LUMO gap (HLG) energy are more stable. Cations with higher ESP values and electron-withdrawing groups with high negative charges (Mulliken charges) are more vulnerable to OH^- attack. In addition, if the energy barrier (E_{barrier} or ΔE) and the change in free energy (ΔG) associated with the transition state (TS) for forming the cation–OH^- complex are low, it indicates that the cations present in the alkaline solution are destabilizing. Therefore, based on quantum mechanical calculations, cations with high LUMO energy, a low ESP value, a low negative charge or charge density, and a high ΔE and ΔG are inferred to have strong alkaline stability. As a result, experimentalists have used these inferences to screen stable cations for AEM applications.

12.3.1.1 Analysis on HOMO–LUMO Energies and Mulliken Charges

Through the application of electronic structure theory, especially the DFT analysis, one can evaluate various chemical properties such as the stability of cations, their resistance to alkaline environments, how they interact with polymers, the adsorption and diffusion of OH^- ions, their capacity for the OH^- transport, and its susceptibility to nucleophilic attacks. By performing DFT calculations, we optimized the molecules and better understood the interactions between the LUMO and HOMO energies of the cations and OH^- ions. Based on existing research, when the HOMO energies of OH^- ions are lower than the LUMO energies of cations, it can impede nucleophilic attacks from OH^- ions. Qiao et al. observed the LUMO energies of poly(*N*-spirocyclic QA) to be in the following order: P-ASN (−1.881 eV) > BP-ASN (−2.087 eV) > N-ASU (−2.656 eV) > BN-ASU (−2.691 eV). The LUMO energies are higher than the HOMO energies of the OH^- ion, indicating alkaline stability in the same order. ^{1}H NMR (nuclear magnetic resonance) has confirmed this stability (see Case Study 12.1) [9].

Case Study 12.1. Alkaline Stability Prediction of AEM Using LUMO Energies of Cation

Qiao et al. have created a variety of tetrakis(bromomethyl) monomers with varying molecular weights and chemical structures to produce poly(*N*-spirocyclic QA) ionenes with adjustable ion exchange capacity (IEC) values. These ionenes will also include 5-/6- and 6-/6-membered rings *N*-spirocyclic quaternary ammonium (QA) (Figure 12.1). The molecular orbital (MO) energy levels of spirocyclic QA molecules were determined through DFT calculations using the GGA-BLYP method and DNP basis set. A Dmol [3] density functional code was used, as implemented in Materials Studio. These calculations were carried out under neutral conditions and in the presence of water as a solvent.

The order of LUMO energies for poly(*N*-spirocyclic QA) in aqueous solution is as follows: P-ASN (−1.881 eV), BP-ASN (−2.087 eV), N-ASU (−2.656 eV), and BN-ASU (−2.691 eV) (shown in Figure 12.2). The LUMO energy level of *N*-spirocyclic QA cations has been found to be higher than the −3.074 eV energy level of OH^- ion in water. Attacks by OH^- ion from an alkaline medium on cations in AEM will result in decreased stability. Quantum mechanical calculations have explained this as the flow of electrons from the HOMO of OH^- ion to the LUMO of cations. The energy barrier between the HOMO of OH^- and the LUMO of the cation determines the alkaline stability. If this energy barrier is very high, the electron is restricted from flowing to the cation from the OH^- ion, resulting in stable cations. In this case, the stability of the cation is in the same order as the increase in the LUMO energies.

The alkaline stability of cations is tested by treating them with a solution of 1 M $NaOD/D_2O/CD_3OD$ at 120 °C. The ^{1}H NMR is aided to study how the reaction time affected the stability (Figure 12.3). It was found that the benzylic position

(Continued)

Case Study 12.1 (Continued)

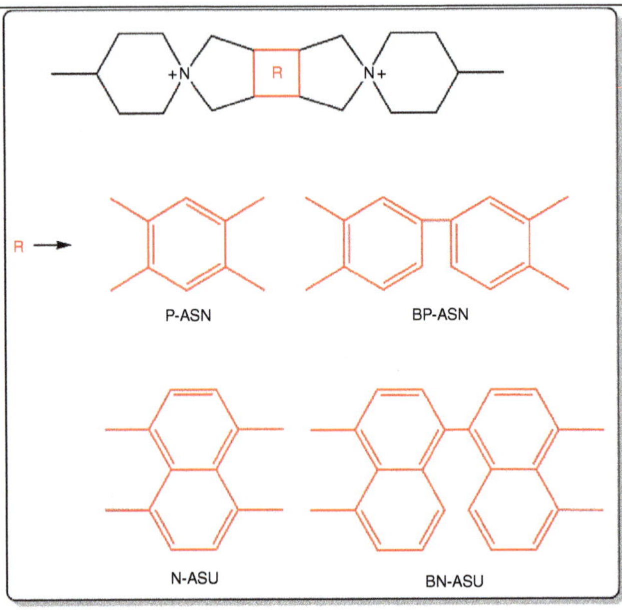

Figure 12.1 Chemical structure of *N*-spirocyclic QA under study.

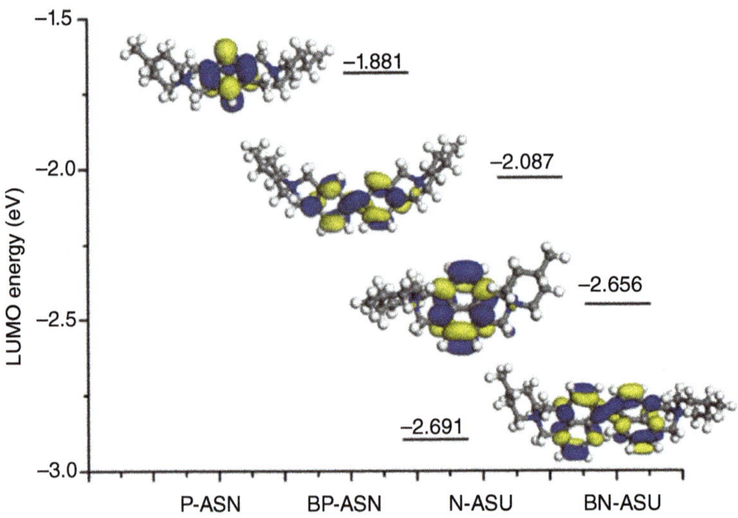

Figure 12.2 LUMO of *N*-spirocyclic QA cations. Source: Reprinted with permission from Qiao et al. [9]. Copyright 2021 Elsevier.

signals disappeared due to H–D exchange, but P-ASN remained stable for 1800 h without any structural degradation under these conditions, indicating its excellent alkaline stability. However, after exposure to 1 M NaOD solution, a new

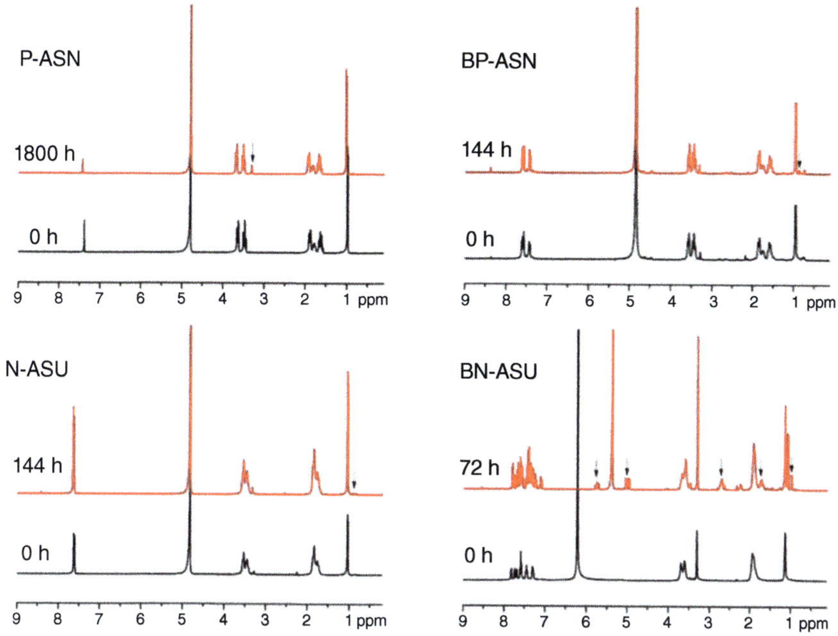

Figure 12.3 ^{1}H NMR of *N*-spirocyclic QA cations after alkali treatment. Source: Reproduce with permission from Qiao et al. [9].

peak around 1 ppm appeared for BP-ASN, which could be attributed to a product resulting from the ring-opening of spirocyclic ionenes. The lower electron-cloud density of spirocyclic QA groups in the biphenyl structure led to lower alkaline stability due to the electron-withdrawing effect of phenyl groups. Therefore, unsurprisingly, the N-ASU and BN-ASU showed poor alkaline strength even after a short treatment time of ~100 h due to the strong electron-withdrawing ability of naphthalene and bi-naphthalene groups. Therefore, the alkaline stability order of the spirocyclic QA cations obtained from DFT analysis aligns with the ^{1}H NMR spectra.

Ma et al. studied the variation in the LUMO energies of OH^- in various positions of *N,N,N,N*-tetramethyl-1,6-hexanediamine functionalized cyclodextrin (TMCD). They found that the LUMO energies of TMCD are much higher than those of single cation-based AEMs like $[GIm]^+$ and $[BMI]^+$ [10]. Thus, the presence of multi or dense cations will enhance the alkaline stability than that of single cations. Based on the FMO studies in ionic liquids (ILs), the LUMO energy level of IL-B (−1.74 eV) is higher than that of IL-M (−1.83 eV) [11]. This indicates that the alkaline stability of IL-B is significantly more robust.

In addition, the stability of side chain AEMs can be predicted by computational studies, considering factors such as substitution, substitution positions, and spacer length of the cations. Experiments have confirmed these predictions. The presence

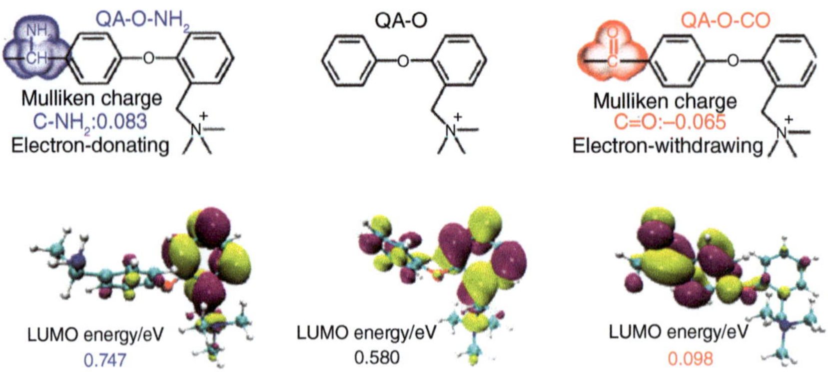

Figure 12.4 Mulliken charges and LUMO energies of QA cations. Source: Reprinted with permission from Zhang et al. [19]. Copyright 2021 ACS.

of resonance effect on cations leads to variations in LUMO energies. Cations with hydroxyl substitutions, bulky groups, longer spacers, and cross-linking tend to have higher LUMO energies than simple cations [12–15]. However, some exceptional cases are observed in the alkaline stability of cations with low LUMO energies [16]. The LUMO energy of cations attached to β-cyclodextrin (β-CD) is higher than that attached to 1-bromohexyl-*N*-methylpiperidinium ionic liquid (bpIL), suggesting that the side chain of cations affects alkaline stability [17]. Furthermore, by LUMO energy analysis, cations grafted on a spacer group are less vulnerable to OH^- attack than cations without spacer groups [18]. Zhang et al. investigated the effect of electron-donating groups on the LUMO energy level. They found that the order of the groups in terms of their impact on the LUMO energy level is electron-donating > non-substituted > electron-withdrawing group. The electronic effect is also a factor in these results, with Mulliken charges of 0.083 a.u for electron-donating C–NH_2 and −0.065 a.u for electron-withdrawing C=O, indicating that cations with lower negative charges are less susceptible to OH^- attack (Figure 12.4) [19].

It is clear from the literature that regardless of the type of the cation, the LUMO energy level is a valuable tool for measuring alkaline stability, and experimental results support these findings. However, Wang et al. found that QA side chains with low LUMO energy have high alkaline stability, despite LUMO energies being used to predict cation alkaline stability [20].

12.3.1.2 Analysis on ESP

The ESP plots are another essential tool for predicting the nucleophilic attack intensity and the attacking sites. Hence, scientists incorporate the ESP results to predict the alkaline stability of cations tethered in the polymer chain. The electron donating linkage group (C–NH_2) attached to the ether-connected carbon reduces the molecule's electrophilicity and results in a lower ESP value than cations with electron-withdrawing groups (C=O) [19]. As a result, C=O link free cations with alkyl spacers have a lower ESP value and are more resistant to alkaline attack [15]. Pan et al. investigated the ESP value of various C_2 substitutions on imidazolium cations, as well as the increasing order of ESP values: $[DHIm]^+$ (0.218 eV) > $[DHMIm]^+$ (0.186 eV) > $[DHTMIm]^+$ (0.179 eV) > $[DHPDMIm]^+$ (0.164 eV) > $[DHTPIm]^+$ (0.059 eV) indicates that $[DHIm]^+$ is very aggressive to OH^- attack, and thus the size of substitution affects alkaline stability (see Case Study 12.2) [21].

Case Study 12.2. Nucleophilic Attack Determination Using ESP Analysis of Cations

In most commercial AEM materials, the presence of C=O groups promotes aryl-ether cleavage and degradation of benzylic QA groups. However, a new approach has been proposed by Zhang et al. that involves converting these C=O groups to electron-donating $C-NH_2$ linkages in traditional poly(arylene ether ketone)s. This is achieved through reductive amination via the Leuckart reaction. The DFT analysis was performed using Gaussian 16 software with the ωB97XD/6-311G (2d, 2p) level of theory and polarizable continuum model (PCM) for solvation. In this work, the quantum mechanical tools like Mulliken charges (Figure 12.4), LUMO energies (Figure 12.4), ESP (Figure 12.5), and free energy analysis of nucleophilic attack (Figure 12.7) are used to predict the alkaline stability. In addition to analyzing alkaline stability, they also investigated the interaction between cations and the polymeric chain (Figure 12.6). In this case study, the analysis of the ESP plot predicting alkaline stability is discussed.

In Figure 12.5, the compound QA-O- (middle) is shown to have an electron density aryl ether carbons are 55.65 and 59.50 $kcal\,mol^{-1}$ and for QA-O-CO (right) 64.98 and 63.02 $kcal\,mol^{-1}$. The benzylic carbon has electron density of 0.041 and 0.064 $kcal\,mol^{-1}$. However, the compound QA-O-NH_2 (left) has a lower electron density on the carbon atom connected to the ether (40.92 and 50.49 $kcal\,mol^{-1}$) and the benzylic carbon (0.023 $kcal\,mol^{-1}$). This indicates that the linkage between C and NH_2 can significantly increase the electron density of the ether-connected carbon atom and decrease the electrophilicity of the benzylic carbon. As a result, the compound is more stable in alkaline conditions.

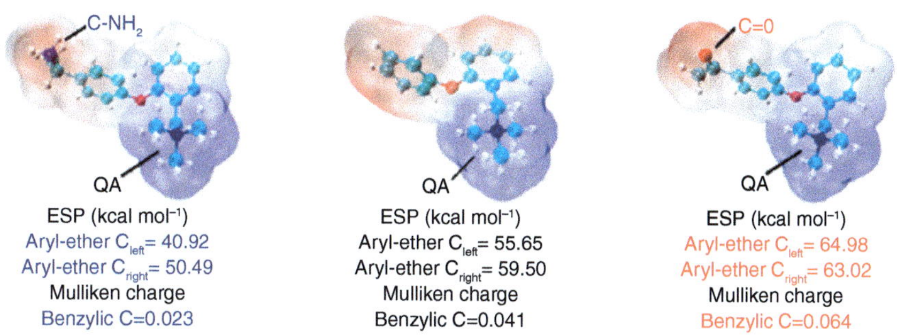

Figure 12.5 ESP plots of QA cations. Source: Reprinted with permission from Zhang et al. [19]. Copyright 2021 ACS.

12.3.1.3 Analysis on Chemical Structure and Bonding Nature

The stability and ionic conductivity of AEMFCs rely heavily on the stability of AEMs. Understanding the bonding interaction between the cation and the polymeric matrix is essential to overcome stability issues. However, experimental studies cannot provide direct evidence for this interaction to an extent. Quantum chemistry has solved this problem by providing evidence to help understand this interaction. In their study, Zuo et al. described the interactions and bonding parameters between the cations and polymer. They suggested that these interactions could be manipulated to enhance the durability of AEMs in highly alkaline environments [22]. Their research showed that the stability of AEMs is influenced by the structure and morphology of the polymer backbone where cations are attached [20]. In addition to the bonding interaction between the polymeric backbone and the cationic group, the bonding analysis also helps predict the distribution of the water molecules in the membrane so as to pin sites for water and hydroxide transport. The DFT analysis was performed using Gaussian 16 software with the ωB97XD/6-311G (2d, 2p) level of theory and PCM for solvation. The non-covalent interactions (independent gradient model, IGM method) were analyzed by Multiwfn software, and the images were created by VMD software. In Figure 12.6, between the repeating units QA-PE-NH_2, there is a strong peak (0.044 a.u.) in the interfragment interactions (δg^{inter}) with a sign(λ_2)ρ of roughly –0.029 a.u.

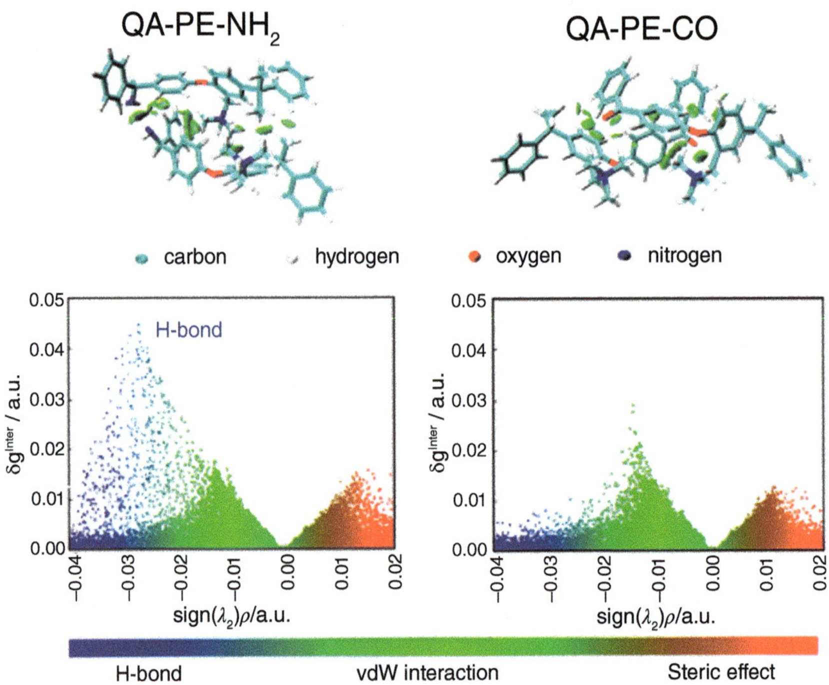

Figure 12.6 Noncovalent interaction isosurface and scatter plot of δg^{inter} versus sign(λ_2)ρ of QA cations. Source: Reproduce with permission from Zhang et al. [19]/American Chemical Society.

However, no δg^{inter} peak can be found between the repeating units QA-PE-CO in the blue region. It is suggested that the C–NH_2 linkages are highly polar and form hydrogen bond networks, leading to increased intermolecular interaction while still maintaining its dynamic nature. This creates sites for long-range water and hydroxide ion transport.

The quantum mechanical calculations analyze the contribution of orbitals in the bond formation of metal-based cations. The $[Cu_6(Pz)_6]$ trimer dimmer has a noteworthy HOMO level that displays significant Cu 3d–Cu 3d antibonding within and between trimers. This occurs within the quadrangle structure of Cu1 and Cu2 atoms in the $[Cu_6(Pz)_6]$ cluster [23]. On the other hand, the Cu 4s and Cu 4p orbitals of the Cu1–Cu2 quadrangle, along with a small amount of N 3s from the coordinated pyrazolate groups, contribute to the LUMO level. Attractive dispersive forces in the FMO populations due to intertrimer cuprophilic interactions suggest that these forces are more powerful than the Pauli repulsion between d^{10} and d^{10} or even $d^{10}s^2$ and $d^{10}s^2$ closed shells. The ΔE for the intertrimer cuprophilic interactions is +4.66 kcal mol^{-1}.

12.3.1.4 Analysis on Degradation Pathways

DFT studies can be used to analyze the degradation pathways of AEMs, in addition to their structural features [12, 24–27]. The DFT method was used by Long et al. to examine how ammonium cations with varying carbon chain lengths degrade [24]. The QA groups are bound to the polymer backbone through a C—C or C—O bond. When the alkyl groups are short ($n = 3$), the C—C bond between the QA groups and polymer backbone is more tolerant to alkalinity than the C—O bond. However, if the flexible spacer is extended beyond $n = 4$, the alkaline stability of AEMs with a C—O bond can be improved significantly and is similar to that of the membrane with the C—C bond. When the alkyl groups connecting the backbone and QA groups are long enough ($n > 4$), the C—C or C—O group far from the QA groups has little effect. It was found that when the carbon chain of alkyl trimethylammonium cations is extended from $n = 2$ to $n = 4$, the ammonium is susceptible to Hofmann elimination. However, it becomes stable when extended to $n = 6$, consistent with Lin et al.'s findings [25]. Therefore, increasing the length of the flexible spacer to more than $n = 4$ between the backbone and QA groups can significantly improve cation stability. Additionally, the ΔG for both the β-elimination and S_N^2 pathways increases as the $-CH_2$-spacer length increases [15]. Under alkaline conditions, the imidazolium cation degrades through a three-step mechanism: a nucleophilic reaction, a ring-opening reaction, and a rearrangement reaction. This degradation process was studied by DFT analysis. Due to the hyper-conjugation of the methyl group at the α-carbon and the imidazole ring, as well as steric effects, substituted imidazolium cations exhibit more excellent alkaline stability compared to unsubstituted imidazolium cations, as demonstrated by their respective LUMO energies. The activation energies (E_a) for ring-opening and nucleophilic attack of TMIM were 107.6 and 57.1 kJ mol^{-1}, respectively, with a methyl substituent. The level of stability is decided by the high activation energy. When the activation energy decreases, the cation becomes more susceptible to nucleophilic attack, which results in its rapid degradation [12]. Chempath et al. discovered that tetramethylammonium (TMA) degrades in alkaline conditions through two distinct mechanisms using DFT analysis. Two different reactions are taking place. The first one involves a hydroxide ion attacking the methyl groups in an S_N^2 mechanism, which results in the formation of methanol. The second reaction consists of a proton taken from a methyl group, forming an ylide called trimethylammonium methylide and a water molecule. It has been observed that the ylide pathway and the S_N^2 pathway share the same TS structure, resulting in the same overall barrier [26]. In an alkaline environment, spirocyclic QA (5-azoniaspiro[4.5]decane ($[ASD]^+$) is more stable than benzyltrimethylammonium $[BTMA]^+$ cations according to Gu et al.'s DFT analysis of degradation kinetics [28]. The Case Study 12.3 will demonstrate the degradation analysis of cations in AEM using DFT analysis.

Case Study 12.3. Degradation Study of QA Cation

Zhang et al. have proposed a new approach that suggests converting the C=O groups to electron-donating C–NH_2 linkages in traditional poly(arylene ether ketone)s. The study also analyzed nucleophilic attack in different positions of the three cations [19]. The thermochemical analysis was performed using Gaussian 16 software with the ωB97XD/6-311G (2d, 2p) level of theory and PCM for solvation. The free energy (ΔG) of the reaction during the nucleophilic attack by OH^- on three positions (Figure 12.7) was calculated.

The cations QA-O-NH_2 have free energy values 143.85, 131.09, and 99.26 kJ mol^{-1} at positions 1, 2, and 3, respectively, during nucleophilic attack. Similarly, the cations QA-O- have free energy values 113.53, 90.23, and 93.73 kJ mol^{-1} at positions 1, 2, and 3, respectively, and QA-O-CO have free energy values 60.61, 82.97, and 68.85 kJ mol^{-1} at positions 1, 2, and 3, respectively, during the nucleophilic attack. From the free energy analysis, the cation with high ΔG is less susceptible to nucleophilic attack; hence, cation QA-O-NH_2 is more stable out of the three cations.

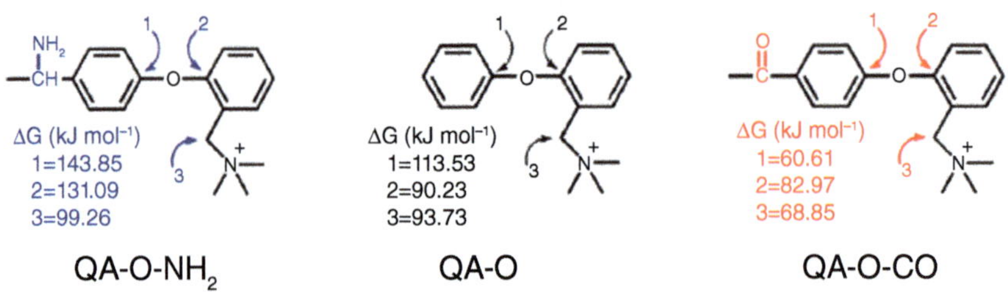

Figure 12.7 Free energy for degradation of QA cations. Source: Reprinted with permission from Zhang et al. [19]. Copyright 2021 ACS.

In addition to the structural and alkaline stability, as well as degradation mechanisms, DFT studies are utilized to predict key AEM performance properties. The DFT method was also used to evaluate the ion transport and water mobility by calculating the binding energy (BE) of fluro-polyolefin with QA-based AEMs. The results showed that the BE of the fluro-polyolefin model to OH^- was 10.6 kJ mol^{-1}, while the BE of the non-fluro-polyolefin model to OH^- was 25.1 kJ mol^{-1} [29]. Similarly, the BE of the fluro-polyolefin model to H_2O was 1.9 kJ mol^{-1}, whereas that of the non-fluro-polyolefin model to H_2O was 5.8 kJ mol^{-1}. The higher BE for the fluro-polyolefin model to both OH^- and H_2O indicates easy ion transfer and increases water mobility for the fluro-polyolefin model, which is essential for AEMFC performance.

Adsorption energies obtained from DFT calculations have been used to explain catalyst deterioration. Pan et al. calculated the adsorption energy between Pt catalyst (Pt(111)) and imidazolium functional cations using the CASTEP program module [21]. The adsorption energy between imidazolium functional cations and Pt(111) follows a specific order, with $[DHPDMIm]^+$ (−7.46 eV) having the lowest adsorption energy and $[DHTMIm]^+$ (−8.59 eV) having the highest adsorption energy. It is concluded that fuel cell performance increases as adsorption energy decreases.

Similarly, the behavior of cation–hydroxide–water and phenyl groups of polymer electrolytes adsorbed on Pd- and Pt-based catalysts was studied by Maurya et al. using DFT analysis [30]. DFT calculations were conducted utilizing the Vienna *ab initio* simulation package (VASP), with the Perdew–Wang (PW91) exchange-correlation functional and projector augmented-wave pseudopotentials. In studying the cation adsorption on different surfaces, they have found that the energy required for adsorption on Pd(111) is −0.16 eV. In comparison, Pd(111) with 0.33 ML hydrogen coverage has a higher energy requirement of +0.1 eV. Pt(111) has a free energy of −0.63 eV.

Interestingly, the addition of hydrogen to Pd causes a shift in the adsorption process from exothermic (negative value) to endothermic (positive value), indicating that hydrogen absorption can prevent cation adsorption on Pd. This is significant for Pd-based catalysts with strong hydrogen absorption properties, as they may inhibit the co-adsorption of cation–hydroxide–water, hindering the hydrogen oxidation process in AEMFCs. Also, the adsorption of the phenyl group exhibits comparable results. Thus, this study demonstrates that Pd-based catalysts are more susceptible to phenyl group adsorption and hydrogenation than Pt. However, incorporating an ionomer with low phenyl group adsorption energy can mitigate these adverse effects.

12.3.2 Molecular Dynamics in Anion Exchange Membranes

Although MD calculations may not be as precise as DFT calculations in determining the specific properties of a system, they can still provide valuable information regarding how a system changes over time, temperature, and pressure. Several research groups utilized a combination of MD simulations and quantum mechanical calculations to elucidate the relationships between the structure and properties of materials [31–36]. In the AEMFCs, MD simulations are highly effective in analyzing factors such as water content [36], ion exchange [35], alkaline stability [32], and OH^- transport [31, 37]. The proposed theoretical models align well with experimental and DFT results. Also, MD calculations are more cost-effective than DFT calculations, rendering them more appropriate for examining larger systems.

Despite of a suggested correlation between water content and ionic conductivity, AEM experimental studies univocally did not provide any insight into water's influence on ion diffusion and OH^- transport through the membrane (see in Case Study 12.4). However, MD simulations offer valuable insights into the transport mechanism of hydroxide ions. In alkaline media, the mobility of both H^+ and OH^- ions is high, with research proposing their similar mobilities via the Grotthuss mechanism, Vehicle mechanism, etc. In hydrated membranes, both *ab initio* molecular dynamics (AIMD) and classical MD reveal the occurrence of two mechanisms, with the Grotthuss mechanism being the primary process. Chen et al. proposed that OH^- conduction can be significantly enhanced by using multi-ion nanochannels assembled with a 2D covalent organic framework sheet embedded into a 1D comb-shaped poly (2,6-dimethyl-1,4-phenylene oxide) (im@PI-2 PPO). This was observed through MD analysis using the COMPASS force field. The ordered pores of imidazolium-based ILs polymerized to form a continuous long-range ionic channel, resulting in a significant improvement in OH^- conductivity as compared to imidazole poly(2,6-dimethyl-1,4-phenylene oxide) (imPPO) [38]. Takaba et al. conducted an MD simulation using Forcite in Materials Studio 6.1, supplied by BIOVIA, Inc., to study OH^- transport in poly(arylene ether sulfone ketone)s containing quaternized ammonio-substituted fluorenyl AEM. The consistent valence force field (CVFF) was used for all calculations. The findings indicate that hydroxide ions move by creating $H_3O_2^-$ through the hydrogen network in water molecules. The $H_3O_2^-$ then splits into H_2O and OH^-, leading to an exchange of protons between a hydroxide ion and a water molecule. This proton transfer mechanism is similar to that observed in PEM. It was observed that increasing the number of QPE units did not significantly affect the conductivity of OH^- ions, which is consistent with experimental observations [39]. Three AEMs made of poly(phenylene oxide) (PPO) were studied by Zhang and Van Duin using ReaxFF reactive MD simulations. The membranes tested were PPO–trimethylamine (PPO–TMA), PPO–dimethylbutylamine (PPO–DMBA), and PPO–dimethyloctylamine (PPO–DMOA). The simulations were performed using the computational chemistry package ADF-2012. The findings showed that as the membranes became more hydrated, they swelled and formed channels, improving OH^- ion diffusion. The study on OH^- diffusion shows that the PPO–TMA hydrated membrane has the highest diffusion rate at a high hydration level. However, when comparing the structural and dynamic properties of the three membranes at the same hydration level, it was found that replacing one methyl group of the QA center with a long alkyl chain group can result in hydrophobic effects, which can block OH^- from approaching nitrogen. This leads to a lower rate of degradation and improved alkaline stability of the PPO–DMOA hydrated membrane [40].

Case Study 12.4. Investigation of the Transport Mechanisms of Hydroxide in Hydrated AEMs

Dong et al. conducted a study investigating hydroxide ion solvation complexes and diffusion mechanisms in model AEMs with low hydration levels [41]. They employed a combination of reactive and non-reactive polarizable MD simulations to explore the transport mechanisms of hydroxide ions within AEMs

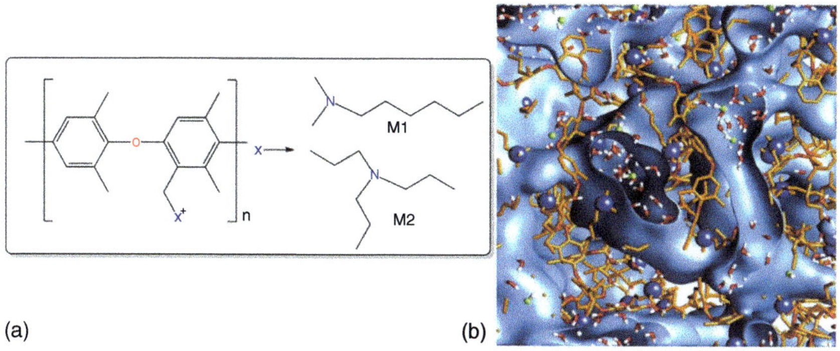

Figure 12.8 (a) Structure of two AEMs with PPO-polymers embedded with two different quaternary ammonium cations. (b) A hydrated membrane snapshot is illustrated by isosurfaces corresponding to 50% bulk water density. Source: Reprinted with permission from Dong et al. [41]. Copyright 2018 ACS.

composed of poly(*p*-phenylene oxide) functionalized with QA cationic groups M1 and M2 (as depicted in Figure 12.8a). The study involved joint MD simulations utilizing both the reactive ReaxFF and the non-reactive atomistic polarizable potential for liquids, electrolytes, and polymers (APPLE&P) force fields for two representative AEMs based on PPO. The research focused on calculating the diffusion coefficients of OH^- (D_{OH-}) and H_2O (D_{H2O}) within these AEMs. In the ReaxFF simulations, the calculated D_{OH-} values for M1 and M2 were 0.236 and 0.162 $Å^2$/ps, respectively, while in the APPLE&P simulations, they were significantly lower at 0.0020 and 0.00079 $Å^2$/ps for M1 and M2, respectively. For D_{H2O}, the ReaxFF simulations yielded values of 0.040 and 0.033 $Å^2$/ps for M1 and M2, respectively, whereas the APPLE&P simulations provided values of 0.0083 and 0.00420 $Å^2$/ps for M1 and M2, respectively.

Remarkably, in the APPLE&P simulations, the D_{OH-} diffusion coefficient was much lower than the D_{H2O} coefficient, with a ratio of 4–5, suggesting a strong dampening effect on hydroxide ion mobility. In contrast, in the ReaxFF simulations, there was an inverse trend, with the diffusivity of H_2O being only approximately 20% of that of OH^-. This indicated a significant enhancement of OH^- dynamics when the Grotthuss mechanism was introduced. Notably, when the Grotthuss mechanism was turned off in the ReaxFF simulations (M1 = 0.031, M2 = 0.068), the diffusion coefficient ratio aligned with the non-reactive simulation predictions. These differences in diffusion coefficient ratios provided clear evidence that the Grotthuss hopping mechanism played a defining role in hydroxide transport within the investigated membranes. To gain further insights into the Grotthuss mechanism's role in hydroxide transport, the study examined changes in the hydroxide ion's environment as it moved through water channels. In simulations using the non-reactive APPLE&P model, it was observed that transporting OH^- through bottlenecks in water channels required a partial

(Continued)

Case Study 12.4 (Continued)

disruption of the anion's immediate hydration shell, constituting the rate-determining step defining hydroxide ion mobility. Therefore, comparing ReaxFF and APPLE&P models for OH^- transport through the same bottlenecks provided straightforward evidence for the Grotthuss mechanism's facilitation of diffusion through AEMs.

Chen et al. used MD simulation to investigate the interaction energy between TQCX and the PSF-ImCl polymer in their study. The total interaction energy (ΔE) consists of two components: a more significant negative electronic energy (ΔE_{elec}) and a smaller negative van der Waals energy (ΔE_{vdw}). The ΔE_{elec} is mainly contributed by the upper rim QA with Im^+ in the polymer matrix via an anion-bridged configuration. In contrast, the lower rim octyl side chain with the polymer backbone contributes to the ΔE_{vdw}. The study found that the interaction between the upper rim QA and Im^+ in the polymer matrix is vital for stability due to the significantly high contribution of ΔE_{elec} with a negative magnitude [42].

Coarse-grained molecular dynamics (CGMD)*:* CGMD modeling is a technique that simplifies complex systems by using a simulated representation. The number of interaction sites is reduced, which results in fewer degrees of freedom in the system. Compared to the fully atomistic model, this reduction leads to a less computationally expensive model. Lu et al. developed a high-resolution CGMD model for PPO-based AEMFC, which considers explicit water models with Martini FF [43]. Their study of the effect of water content and ionomer architecture on the nanostructure and ion conductivity of the AEM implies that the ion conductivity of the AEM is highly dependent on the water content but less sensitive to changes in the architecture of the polymer matrix of the AEM via CGMD [44]. A study was conducted to analyze AEMs based on PPO, poly(ether ketone) (PEEK), and polystyrene-*b*-poly(ethylene-*co*-butylene)-*b*-polystyrene (SEBS) at a coarse-grained scale [45–47].

12.3.3 Continuum Modeling and Simulation in Anion Exchange Membranes

The scientific field of continuum mechanics delves into the behavior of materials treated as a continuous mass, as opposed to individual particles. Although matter comprises atoms and is not truly continuous, representing objects as a continuum is a highly precise approach for length scales greatly exceeding inter-atomic distance. Under the leadership of Machado, the research team conducted an array of studies focused on advancing the understanding of AEMFCs at the continuum level [48–50]. These investigations encompassed several key areas: First, the team delved into the influence of various process parameters, including flow direction, temperature, and relative humidity, on AEMFC performance. Second, they embarked on a comprehensive exploration of an agglomerate model, conducting a parametric study using air at the cathode. Third, the team engaged in entropy generation analysis, utilizing a three-dimensional agglomerate model to gain insights into the system's thermodynamic behavior. Fourth, they conducted an examination of the agglomeration model for 3D AEMFCs, employing a finite-volume modeling approach to enhance understanding. These rigorous investigations yielded several noteworthy conclusions that the team observed that enhancing AEMFC performance could be achieved by reducing the relative humidity on the cathode side while simultaneously increasing the membrane's water content, platinum loading, and ionomer volume fraction. Their analysis pinpointed reversible and irreversible heat as the primary sources of entropy production across all the parameters examined, shedding light on the thermodynamic aspects of AEMFC operation. Importantly, the team noted that the macro-homogeneous model tended to overestimate cell performance when compared to the agglomerate model. This discrepancy could be attributed to the resistances associated with species and ionic transports within the catalyst layer (CL). In essence, Machado's research team conducted a comprehensive suite of studies to deepen our understanding of AEMFCs, leading to valuable insights that can inform the optimization of these promising energy conversion devices using continuum modeling and simulation.

12.3.4 Monte Carlo Simulations in Anion Exchange Membranes

Kim et al. conducted a comprehensive investigation that combined 2D lattice Monte Carlo simulations (2D-LMC) with DFT analysis [51]. Their study focused on designing cross-linked polymer membranes by examining a series of pyrazolium cross-linked poly(triptycene ether sulfone)s. These membranes were explicitly engineered to feature ionic highways formed by charge-delocalized pyrazolium cations and homoconjugated triptycenes. The 2D-LMC model was utilized to construct model systems of ionic networks. The model simulated ionic conductivity in channels with varying percentages of cross-linking (m). The conducting channels were systematically arranged or randomly aligned in two distinct systems, with m values modified to 75, 40, and 10% in each system (Figure 12.9). The Arrhenius equation was used to determine the probability of hydroxide transport between cross-linked and non-cross-linked sites. The ordered system had higher conductivity with 10% cross-linking than the random system with 75% cross-linking. Also, the 2D-LMC model's predicted conductivities match experimental measurements.

12.3.5 Machine Learning in Anion Exchange Membranes

In recent times, there has been a surge in research efforts to explore the development and deployment of ML models in molecular design and enhance the performance of AEMs. ML is essential in AEM research to speed up material discovery, predict properties, optimize composition, ensure durability, and enhance process control. It helps decrease development time and costs and facilitates data-driven decision-making, ultimately promoting the advancement of AEM-based energy conversion and storage technologies.

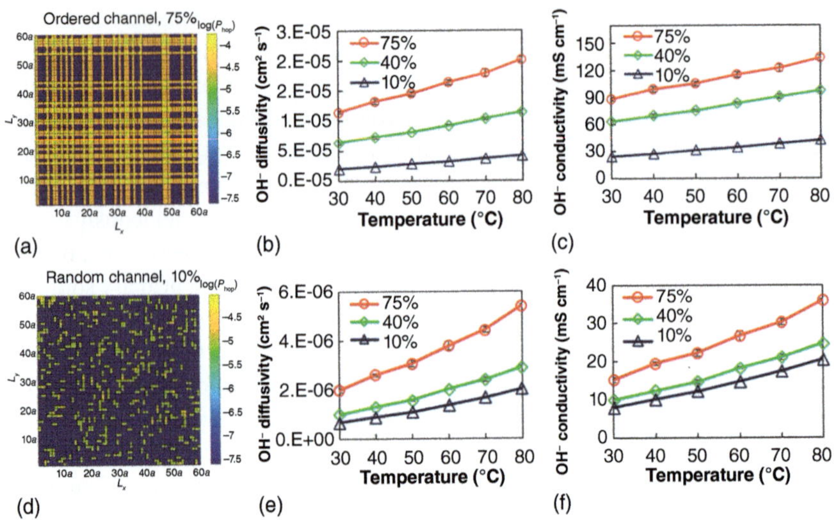

Figure 12.9 2D-LMC results of a study on a polymer network with different cross-linker placement: (a) Anion hopping probability distribution map is shown for a polymer network constructed with orderly placed cross-linkers, with an *m* value of 75%. (b, c) Calculated OH^- diffusivity and conductivity of the polymer network at different *m* values and temperatures of orderly placed cross-linkers, respectively. (d) The anion hopping probability distribution map is shown for a polymer network constructed with randomly placed cross-linkers, with an *m* value of 10%. (e, f) Calculated OH^- diffusivity and conductivity of the polymer network at different *m* values and temperatures of randomly placed cross-linkers, respectively. Source: Reproduce with permission from Kim et al. [51]/American Chemical Society.

Zhai et al. have developed a deep learning model that accurately predicts hydroxide ion (OH^-) conductivity in AEMs made of poly(2,6-dimethyl phenylene oxide) and specific cationic groups [52]. The model has achieved 99.7% accuracy in classifying cationic groups and can predict OH^- conductivity within a reasonable error range of $\pm 0.016\,S\,cm^{-1}$ for QA, $\pm 0.014\,S\,cm^{-1}$ for saturated heterocyclic ammonium, and $\pm 0.07\,S\,cm^{-1}$ for imidazolium cations. This protocol provides a valuable tool for designing AEMs with predictable OH^- conductivity, which could potentially facilitate the development of new and improved AEMs.

In a study by Zou et al., the chemical stability of AEMs was evaluated using Hammett substituent constants and a decision tree classification method based on materials genomics [53]. The team tested various ML algorithms and found that artificial neural networks (ANNs) had the highest accuracy ($R^2 = 0.9978$) in predicting the stability of the AEM. The study highlights the potential of data-driven models in predicting the alkaline stability of the AEM, which can help reduce the need for unnecessary experiments in the development of alkaline membrane fuel cells. The approach was supported by computational analysis and long-term experiments. The same research team has recently introduced a new method known as virtual module compound enumeration screening (V-MCES) that can explore a vast chemical space of over 4.2×10^5 candidates [54]. This eliminates the need for expensive training databases. The V-MCES model's performance was enhanced by supervised learning, which enabled the selection of molecular descriptors. V-MCES generated a list of high-stability AEMs with tremendous potential by connecting AEM molecular structures to predicted chemical stability. This approach helped the research team synthesize highly stable AEMs, which is a significant achievement in AEM science. The structural and performance insights gained from ML will enable advanced architectural design in the future.

12.4 Challenges and Outlook

The field of computational studies in AEM research presents several challenges to researchers. However, despite these obstacles, the prospects for this study area appear promising. After the analysis above, we will further examine the obstacles hindering computational studies' development in AEM research. Furthermore, this section investigates potential paths that may lead to progress in this area.

- *Challenges*

1. Computational models of AEM systems can be difficult to create due to their inherent complexity. These systems entail a combination of various components, including polymers, solvents, ions, and interfaces. Capturing the complete intricacy of these systems can be a challenge.
2. Precise parameterization of force fields, solvation models, and other factors is crucial in the case of computational modeling. However, obtaining accurate and dependable parameters for AEM materials can be difficult due to the limited availability of experimental data.
3. Performing simulations on real-world AEM systems that span long timescales and large-length scales can be challenging regarding computational requirements. There is still a challenge in bridging the gap between atomistic simulations and the behavior of macroscopic systems.
4. Valuing computational studies with experimental data is essential to ensure their accuracy and reliability. However, obtaining experimental data for AEM systems can be challenging, especially for properties that are difficult to measure or transient.

- *Outlook*

1. To advance computational studies in AEM research, it is imperative to develop multiscale modeling approaches. Combining atomistic, mesoscale, and continuum models provides a comprehensive understanding of AEM behavior across various lengths and timescales. Integrating different modeling approaches enables accurate prediction of AEM properties and performance, ultimately leading to significant strides in the field.
2. The combination of computational simulations and advanced experimental techniques, including *in situ* characterization methods and spectroscopic analysis, would be the key to increasing the accuracy and dependability of computational models. When we integrate these approaches, we can achieve a more

comprehensive understanding of AEM systems, which will help us develop next-generation AEM materials with even greater success.

3. Utilizing ML techniques is a powerful way to accelerate AEM research. By analyzing massive amounts of data, recognizing patterns, and predicting outcomes, data-driven approaches can help tackle parameterization, property prediction, and material design challenges. Tailoring ML algorithms specifically for AEM systems can significantly enhance the precision and efficiency of computational studies.

The study of AEMs is a challenging task that involves complex computational research. These challenges include parameterization, limitations in timescale, and validation. Despite these obstacles, there is hope for the future, thanks to the progress made in simulation methods and experiment integration. Furthermore, the ability to create custom materials is expected to contribute to a better understanding of AEM systems, speed up material discovery, and develop more efficient and long-lasting AEMs for various electrochemical applications.

12.5 Conclusion

Computational simulations are necessary for understanding AEM mechanisms, properties, and performance. By studying ion transport mechanisms within AEMs, researchers gain insights into interactions between ions, solvent molecules, and the polymer matrix. Simulations also aid in studying membrane structure and stability, developing AEMs with enhanced durability and longevity, optimizing water regulation, and screening novel materials. This knowledge is crucial for optimizing AEM performance, enhancing durability, and developing more effective and long-lasting membranes for various electrochemical applications. Computational simulations offer a cost-effective approach to exploring new materials and can accelerate research and development, optimize design and performance, and deepen our understanding of AEM systems. However, these simulations face challenges related to the complexity of AEM systems, parameterization of models, limitations in timescales and length scales, and the need for validation against experimental data.

Despite these challenges, the future of computational studies in AEM research is promising, with advancements in multi-scale modeling, ML techniques, simulation methods, integration with experimental techniques, and the design of tailor-made materials expected. These developments will further enhance our understanding of AEM systems, accelerate material discovery, and contribute to designing more efficient and durable AEMs for various electrochemical applications. Ultimately, computational simulations will continue to play a vital role in advancing AEM research and its practical applications.

References

1 Merle, G., Wessling, M., and Nijmeijer, K. (2011). Anion exchange membranes for alkaline fuel cells: a review. *J Membr Sci* 377: 1.

2 Salma, U. and Shalahin, N. (2023). A mini-review on alkaline stability of imidazolium cations and imidazolium-based anion exchange membranes. *Result Mater* 17: 100366.

3 Peltier, C.R., You, W., Fackovic Volcanjk, D. et al. (2023). Quaternary ammonium-functionalized polyethylene-based anion exchange membranes: balancing performance and stability. *ACS Energy Lett* 8: 2365.

4 Li, X., Wang, Z., Chen, Y. et al. (2023). Imidazolium-based AEMs with high dimensional and alkaline-resistance stabilities for extended temperature range of alkaline fuel cells. *J Membr Sci* 670: 121352.

5 Sajjad, S.D., Hong, Y., and Liu, F. (2014). Synthesis of guanidinium-based anion exchange membranes and their stability assessment. *Polym Adv Technol* 25: 108.

6 Chen, Y., Tao, Y., Wang, J. et al. (2017). Comb-shaped guanidinium functionalized poly(ether sulfone)s for anion exchange membranes: effects of the spacer types and lengths. *J Polym Sci A Polym Chem* 55: 1313.

7 Barnes, A.M., Du, Y., Zhang, W. et al. (2019). Phosphonium-containing block copolymer anion exchange membranes: effect of quaternization level on bulk and surface morphologies at hydrated and dehydrated states. *Macromolecules* 52: 6097.

8 Thomas, J., Francis, B., Thomas, S. et al. (2021). Dependable polysulfone based anion exchange membranes incorporating triazatriangulenium cations. *Solid State Ionics* 370: 115731.

9 Qiao, X., Wang, X., Liu, S. et al. (2021). The alkaline stability and fuel cell performance of poly(N-spirocyclic quaternary ammonium) ionenes as anion exchange membrane. *J Membr Sci* 630: 119325.

10 Ma, L., Qaisrani, N.A., Hussain, M. et al. (2020). Cyclodextrin modified, multication cross-linked high performance anion exchange membranes for fuel cell application. *J Membr Sci* 607: 118190.

11 Cha, M.S., Lee, J.Y., Kim, T.H. et al. (2017). Preparation and characterization of crosslinked anion exchange membrane (AEM) materials with poly(phenylene ether)-based short hydrophilic block for use in electrochemical applications. *J Membr Sci* 530: 73.

12 Wang, W., Wang, S., Xie, X. et al. (2014). Density functional theory study of hydroxide-ion induced degradation of imidazolium cations. *Int J Hydrog Energy* 39: 14355.

13 Si, Z., Qiu, L., Dong, H. et al. (2014). Effects of substituents and substitution positions on alkaline stability of imidazolium cations and their corresponding anion-exchange membranes. *ACS Appl Mater Interfaces* 6: 4346.

14 Cheng, J., Yang, G., Zhang, K. et al. (2016). Guanidimidazole-quanternized and cross-linked alkaline polymer electrolyte membrane for fuel cell application. *J Membr Sci* 501: 100.

15 Zhang, F., Li, T., Chen, W. et al. (2021). Highly stable electron-withdrawing CO link-free backbone with branched cationic side chain as anion exchange membrane. *J Membr Sci* 624: 119052.

16 Gu, S., Cai, R., and Yan, Y. (2011). Self-crosslinking for dimensionally stable and solvent-resistant quaternary phosphonium based hydroxide exchange membranes. *Chem Commun* 47: 2856.

17 Ma, L., Hussain, M., Li, L. et al. (2021). Octopus-like side chain grafted poly(arylene piperidinium) membranes for fuel cell application. *J Membr Sci* 636: 119529.

18 Li, L., Wang, J., Hussain, M. et al. (2021). Side-chain manipulation of poly (phenylene oxide) based anion exchange membrane: alkoxyl extender integrated with flexible spacer. *J Membr Sci* 624: 119088.

19 Zhang, F., Li, T., Chen, W. et al. (2021). Electron-donating C-NH2 link backbone for highly alkaline and mechanical stable anion exchange membranes. *ACS Appl Mater Interfaces* 13: 10490.

20 Wang, X., Lin, C., Gao, Y., and Lammertink, R.G.H. (2021). Anion exchange membranes with twisted poly(terphenylene) backbone: effect of the N-cyclic cations. *J Membr Sci* 635: 119525.

21 Pan, J., Sun, Z., Zhu, H. et al. (2020). Synthesis and characterization of main-chain type polyimidazolium-based alkaline anion exchange membranes. *J Membr Sci* 610: 118283.

22 Zuo, P., Su, Y., Li, W. et al. (2016). Comb-like poly(ether-sulfone) membranes derived from planar 6,12-diaryl-5,11-dihydroindolo[3,2-b]carbazole monomer for alkaline fuel cells. *Macromol Rapid Commun* 37: 1748.

23 Li, Z., Zhang, Z., Ye, Y. et al. (2017). Rationally tuning host–guest interactions to free hydroxide ions within intertrimerically cuprophilic metal–organic frameworks for high OH^- conductivity. *J Mater Chem A Mater* 5: 7816.

24 Long, H., Kim, K., and Pivovar, B.S. (2012). Hydroxide degradation pathways for substituted trimethylammonium cations: a DFT study. *J Phys Chem C* 116: 9419.

25 Lin, C.X., Zhuo, Y.Z., Lai, A.N. et al. (2016). Side-chain-type anion exchange membranes bearing pendent imidazolium-functionalized poly(phenylene oxide) for fuel cells. *J Membr Sci* 513: 206.

26 Chempath, S., Einsla, B.R., Pratt, L.R. et al. (2008). Mechanism of tetraalkylammonium headgroup degradation in alkaline fuel cell membranes. *J Phys Chem C* 112: 3179.

27 Chen, N., Jin, Y., Liu, H. et al. (2021). Insight into the alkaline stability of N-heterocyclic ammonium groups for anion-exchange polyelectrolytes. *Angew Chem* 133: 19421.

28 Gu, L., Dong, H., Sun, Z. et al. (2016). Spirocyclic quaternary ammonium cations for alkaline anion exchange membrane applications: an experimental and theoretical study. *RSC Adv* 6: 94387.

29 Zhu, L., Peng, X., Shang, S.L. et al. (2019). High performance anion exchange membrane fuel cells enabled by fluoropoly(olefin) membranes. *Adv Funct Mater* 29: 1902059.

30 Maurya, S., Dumont, J.H., Villarrubia, C.N. et al. (2018). Surface adsorption affects the performance of alkaline anion-exchange membrane fuel cells. *ACS Catal* 8: 9429.

31 Park, C.H., Kim, T.H., Kim, D.J., and Nam, S.Y. (2017). Molecular dynamics simulation of the functional group effect in hydrocarbon anionic exchange membranes. *Int J Hydrog Energy* 42: 20895.

32 Dekel, D.R., Willdorf, S., Ash, U. et al. (2018). The critical relation between chemical stability of cations and water in anion exchange membrane fuel cells environment. *J Power Sources* 375: 351.

33 Dong, D., Wei, X., Hooper, J.B. et al. (2018). Role of cationic groups on structural and dynamical correlations in hydrated quaternary ammonium-functionalized poly(p-phenylene oxide)-based anion exchange membranes. *Phys Chem Chem Phys* 20: 19350.

34 Zhang, N., Huo, J., Yang, B. et al. (2018). Understanding of imidazolium group hydration and polymer structure for hydroxide anion conduction in hydrated imidazolium-g-PPO membrane by molecular dynamics simulations. *Chem Eng Sci* 192: 1167.

35 Luque Di Salvo, J., De Luca, G., Cipollina, A., and Micale, G. (2020). Effect of ion exchange capacity and water uptake on hydroxide transport in PSU-TMA membranes: a DFT and molecular dynamics study. *J Membr Sci* 599: 117837.

36 Zadok, I., Dekel, D.R., and Srebnik, S. (2019). Effect of ammonium cations on the diffusivity and structure of hydroxide ions in low hydration media. *J Phys Chem C* 123: 27355.

37 Wu, L., Zhou, X., Zhang, G. et al. (2021). Tunable OH- transport and alkaline stability by imidazolium-based groups of poly(2,6-dimethyl-1,4-phenylene oxide) anion exchange membranes: a molecular dynamics simulation. *Ind Eng Chem Res* 60: 2481.

38 Chen, J., Li, P., Zhang, N., and Tang, S. (2021). Highly hydroxide-conducting hybrid anion exchange membrane with functional COF-enhanced ion nanochannels. *Electrochim Acta* 391: 138962.

39 Takaba, H., Hisabe, T., Shimizu, T., and Alam, M.K. (2017). Molecular modeling of OH^- transport in poly(arylene ether sulfone ketone)s containing quaternized ammonio-substituted fluorenyl groups as anion exchange membranes. *J Membr Sci* 522: 237.

40 Zhang, W. and Van Duin, A.C.T. (2015). ReaxFF reactive molecular dynamics simulation of functionalized poly(phenylene oxide) anion exchange membrane. *J Phys Chem C* 119: 27727.

41 Dong, D., Zhang, W., Van Duin, A.C.T., and Bedrov, D. (2018). Grotthuss versus vehicular transport of hydroxide in anion-exchange membranes: insight from combined reactive and nonreactive molecular simulations. *J Phys Chem Lett* 9: 825.

42 Chen, W., Wang, X., Li, T. et al. (2021). Amphiphilic cone-shaped cationic calix[4]arene composite anion exchange membranes with continuous ionic channels. *J Membr Sci* 640: 119815.

43 Lu, J., Jacobson, L.C., Perez Sirkin, Y.A., and Molinero, V. (2017). High-resolution coarse-grained model of hydrated anion-exchange membranes that accounts for hydrophobic and ionic interactions through short-ranged potentials. *J Chem Theory Comput* 13: 245.

44 Lu, J., Barnett, A., and Molinero, V. (2019). Effect of polymer architecture on the nanophase segregation, ionic conductivity, and electro-osmotic drag of anion exchange membranes. *J Phys Chem C* 123: 8717.

45 Zhu, Z., Luo, X., and Paddison, S.J. (2019). DPD simulations of anion exchange membranes functionalized with various cationic groups and associated anions. *Solid State Ionics* 340: 115011.

46 Luo, X. and Paddison, S.J. (2019). DPD simulations of anion exchange membrane: the effect of an alkyl spacer on the hydrated morphology. *Solid State Ionics* 339: 115012.

47 Chen, S., Wang, H., Zhang, J. et al. (2020). Effect of side chain on the electrochemical performance of poly (ether ether ketone) based anion-exchange membrane: a molecular dynamics study. *J Membr Sci* 605: 118105.

48 Machado, B.S., Mamlouk, M., and Chakraborty, N. (2019). Three-dimensional agglomerate model of an anion exchange membrane fuel cell using air at the cathode – a parametric study. *J Power Sources* 412: 105.

49 Machado, B.S., Mamlouk, M., and Chakraborty, N. (2020). Entropy generation analysis based on a three-dimensional agglomerate model of an anion exchange membrane fuel cell. *Energy* 193: 116667.

50 Machado, B.S., Chakraborty, N., Mamlouk, M., and Das, P.K. (2018). Three-dimensional agglomerate model of an anion exchange membrane fuel cell. *J Electrochem Energy Convers Storage* 15. https://doi.org/10.1115/1.4037942.

51 Kim, Y., Wang, Y., France-Lanord, A. et al. (2019). Ionic highways from covalent assembly in highly conducting and stable anion exchange membrane fuel cells. *J Am Chem Soc* 141: 18152.

52 Zhai, F.H., Zhan, Q.Q., Yang, Y.F. et al. (2022). A deep learning protocol for analyzing and predicting ionic conductivity of anion exchange membranes. *J Membr Sci* 642: 119983.

53 Zou, X., Pan, J., Sun, Z. et al. (2021). Machine learning analysis and prediction models of alkaline anion exchange membranes for fuel cells. *Energy Environ Sci* 14: 3965.

54 Zou, X., Xu, G., Fang, P. et al. (2023). Unsupervised learning-guided accelerated discovery of alkaline anion exchange membranes for fuel cells. *Angew Chem Int Ed* 62: e202300388.

13

An Overview of Commercial and Non-commercial Anion Exchange Membranes

Yang Zhang†, Fan Zhang†, Liang Wu, Xiaolin Ge, and Tongwen Xu

University of Science and Technology of China, Key Laboratory of Precision and Intelligent Chemistry, Collaborative Innovation Centre of Chemistry for Energy Materials, School of Chemistry and Material Science, Jinzhai Road 96, Hefei, Anhui, 230026, PR China
†These authors contributed equally to this work.

13.1 Introduction

Anion exchange membrane fuel cells (AEMFCs) avoid the drawbacks of carbonate crystallization and the cumbersome equipment in traditional alkaline fuel cells. Meanwhile, compared to proton exchange membrane fuel cells, the electrode reaction rate is faster in alkaline environments, allowing the use of low platinum content or non-platinum catalysts, which can greatly reduce costs. These advantages promote the rapid development of AEMFCs. The anion exchange membrane (AEM) is the core component of AEMFCs, which plays the role of transferring OH^- ions and separating the cathode and anode. Therefore, its performance has a significant impact on the performance and service life of AEMFCs [1, 2]. A desirable AEM is expected to have high conductivity, high ion exchange capacity (IEC), high dimensional stability, as well as high thermal, chemical, and mechanical properties. Tokuyama soda company (Japan) first launched an AEM product in the word (A201); subsequently, some other commercial AEMs such as Fumasep® FAA3 (Fumatech, Germany), AEMION™ (Ionomr, Canada), SUSTAINION® (Dioxide Materials, USA), Xergy® (AEM-Pention, USA), and Alkymer® (China) have been developed in succession. However, due to cumbersome preparation processes, the need for special reagents, and so on, most commercial AEMs are expensive and limited in production, which limits the large-scale commercial deployment of related applications. Therefore, it is crucial to develop AEMs with simpler and cheaper preparation processes.

Alkaline Anion Exchange Membranes for Fuel Cells: From Tailored Materials to Novel Applications, First Edition. Edited by Jince Thomas, Alex Schechter, Flavio Grynszpan, Bejoy Francis, and Sabu Thomas.

Non-commercial AEMs were originally prepared by modifying the aryl ether polymers. Unfortunately, aryl ether cleavage reactions happening in aryl ether polymers (poly(ether–ether ketone) [PEEK], poly(phenylene oxide) [PPO], polysulfone, etc.) and degradation in cationic functional groups through Hoffmann degradation or nucleophilic attack will greatly reduce the usability of the AEMs. In addition, due to the much lower mobility of OH^- compared to H^+, the conductivity of AEMs is generally much lower than that of proton exchange membranes (PEMs). In order to develop AEMs with excellent comprehensive properties, researchers have developed various strategies to improve the properties of AEMs, including (i) the regulation of microphase morphologies, (ii) constructing free volume, (iii) the introduction of cross-linking structures, (iv) other physical methods, such as organic–inorganic hybrid, and electrospinning, and (v) the development of novel cationic functional groups and aryl ether-free main chains with high stability.

We first surveyed the existing commercial alkaline AEMs and summarized their parameters and properties. Then, the non-commercial AEMs were inspected and analyzed, and the development ideas and methods of AEMs were complied with and outlined.

13.1.1 Characteristics and Existing Problems of Commercial Alkaline Anion Exchange Membranes

13.1.1.1 Fumatech: Fumasep

This AEM can be obtained through Fumatech and distributors and is available in multiple thicknesses as non-supported or PP- or PEEK-reinforced AEM. The Fumasep is made of a polyaromatic polymer containing ether bonds as the main chain, attached by quaternary ammonium (QA) groups. Its original generation product AEM (2011) is not cross-linked, while the latest batches of the AEM is mildly cross-linked. FAA-3 as a sort of Fumatech AEM possesses low dimensional swelling, low resistance area, and high selectivity. The research results of several groups indicate that it has an IEC of $2\,\mathrm{mmol\,g^{-1}}$, and it has a 64% water uptake (WU), 15% dimensional expansion, as well as $30\,\mathrm{mS\,cm^{-1}}$ conductivity at 25 °C. However, due to the poor thermal stability of the backbone and poor degradation stability of the QA cation under alkaline conditions, their application in AEMFCs is limited [3, 4].

13.1.1.2 Tokuyama: A201

Tokuyama's AEM can be only obtained through Tokuyama, and the purchase needed signing a confidentiality agreement. The structure of Tokuyama A201 has not been disclosed, except that it involves hydrocarbon-based polymers with QA cations. In the past years, researchers have conducted many *in situ* tests to study the durability, drag coefficient, and CO_2 tolerance of the A201 AEM. For example, water adsorption isotherms of A201 AEM were tested at 50 or 80 °C in equal water vapor. The results exhibited that dimensional expansion of A201 AEM are 9% and 12% for 50 and 80 °C, separately. In addition, the percolation threshold was 0.34 of hydrophilic volume fraction, indicating that A201 AEM has a randomly dispersed cubic

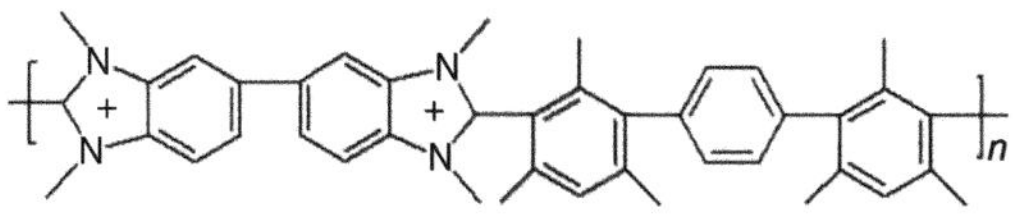

Figure 13.1 Chemical structure of AEMION AEMs [7].

hydrophilic structure domain. The ion conductivity of A201 AEM is only 32.8 mS cm^{-1} at 23 °C, which may result in low fuel cell performance and thus limit its application in fuel cells [5, 6].

13.1.1.3 Ionomr: AEMION

Ionomr currently provides four AEMs of varying thickness (25 or 50 μm) and IEC (1.4–1.7 or 2.1–2.5 meq g^{-1}). AEMION membranes are prepared by the Holdcroft group and consist of methylated polybenzimidazoles (PBIs) (Figure 13.1). In the fully hydrated condition, the Cl^- form HMT-PMBI AEMs exhibit excellent mechanical robustness, with a 33 MPa tensile strength and 225 MPa Young's modulus, respectively, which are obviously higher than those of Nafion 212 (6.5 and 90.4 MPa). The OH^- form AEM shows excellent *ex situ* chemical and mechanical durability; meanwhile, the mechanical strength did not show significant changes after exposure to 1 M NaOH at 80 °C and 6 M NaOH at 25 °C for 168 h. When exposed to 2 M NaOH at 80 °C for 168 h, only 6% chemical decomposition was observed. In AEMION membranes, the attack of OH^- is hindered by the methyl parts in both ortho-positions, and the alkali stability can be further improved by increasing the electron density at C2. However, the low OH^- conductivity of 10 mS cm^{-1} at RT is probably caused by the poor microphase separation, making it not the best candidate for fuel cells [7].

13.1.1.4 Dioxide Materials: Sustainion

The Sustainion AEMs can be obtained from Dioxide Materials. This AEMs were developed by Masel's group, and it consists of the classic poly(4-vinyl benzyl chloride-*co*-styrene) polymers (Figure 13.2). For polystyrene-based materials, the materials are brittle and have cracks under dry conditions. So, Dioxide Materials prepared Sustainion AEMs with a water-soluble plasticizer as a "classic Sustainion." In order to work at higher pressures, Dioxide Materials also provides the AEM containing a polytetrafluoroethylene (PTFE)-reinforced and strengthened with zirconium oxide nanoparticles. Sustainion® 37 AEMs have shown excellent stability in water and carbon dioxide (CO_2) electrolyzers. The Sustainion AEMs can keep 1 A cm^{-2} at 1.9 V and 60 °C for 200 h at 1 M KOH feeding containing less than 5 mV voltage change in water electrolyzer. The CO_2 electrolyzers can function

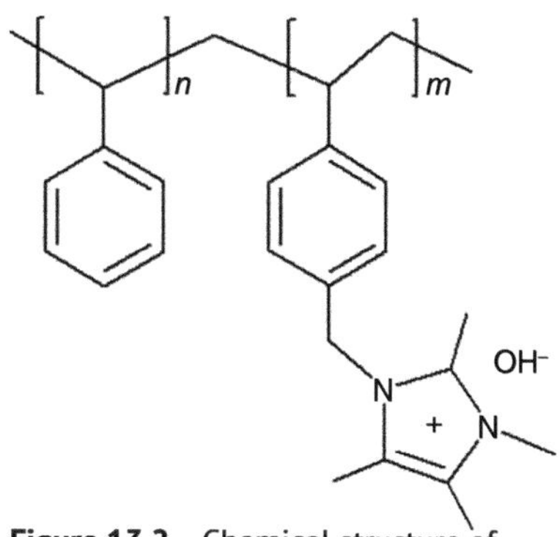

Figure 13.2 Chemical structure of sustainion AEMs [8].

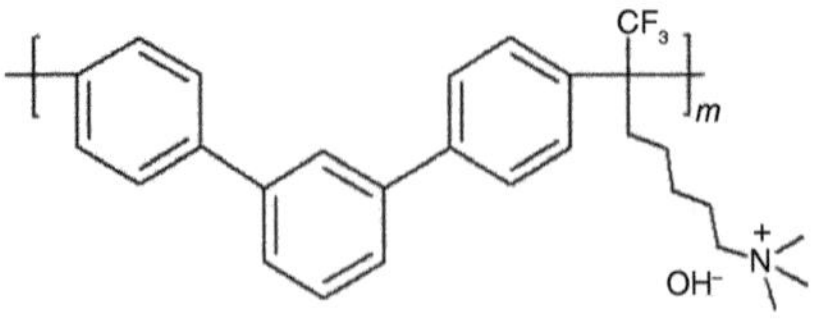

Figure 13.3 The chemical constitution of the Orion TM1 [9].

under 3V at 200 mA cm^{-2} containing a voltage increase between 4 and 6 μV h^{-1} in 3700 h. For AEMFCs, the maximum power density can reach 0.44 W cm^{-2} with a back pressure of 1 atm at 60 °C. However, the low mechanical performance may be a potential factor limiting the application of Sustainion AEMs [8].

13.1.1.5 Orion Polymer: Orion TM1

Orion Polymer's TM1 is a non-reinforced AEMs with an IEC value of 2.19 meq g^{-1}, and thicknesses ranging from 5 to 50 μm (Figure 13.3). TM1 AEMs are developed by Bae's group. Aromatic ether groups which are considered to be the main cause of degradation of other commercial AEMs were removed from the backbone in this membrane. The connection between the rigid aromatic backbones and the substituted methylene parts enhances the flexibility of the polymers, improves its solubility, and enables good treatment of the AEMs. The TM1 AEM showed appropriate water adsorption and swelling (WU is 70% and swelling ratio [SR] is 23% at 80 °C), and it exhibited high conductivity (112 mS cm^{-1} at 80 °C). The membrane electrode assembly (MEA) applying TM1 showed a current density of 400 mA cm^{-2} at 2.4 V in alkaline membrane water electrolysis. Researchers from the National Renewable Energy Laboratory recently measured the alkali stability of over 50 AEMs from at least 10 organizations, with the Orion TM1 AEM exhibiting the best alkali resistance stability, that did not display signs of degradation after being immersed in 1 M KOH at 80 °C for 1000 h. The synthesis of these AEMs seems to be simple and easy to scale up. However, 7-bromo-trifluorheptanone is relatively expensive. In addition, under the conditions of an alkaline membrane water electrolyzer, the TM1 AEM showed low mechanical stability [9, 10].

13.1.1.6 Xergy: Xion-Dappion, Xion-Durion, Xion-Pention

Xergy company has developed three types of AEMs: Xion-Dappion, Xion-Durion, and Xion-Pention. The ionomer structural formula is shown in Figure 13.4. Based on the developed ionomer, the Xergy company not only produces homogeneous membranes, but also adds a reinforcing layer into the AEMs through the composite process. Compared with homogeneous membranes, the mechanical properties of composite membranes are greatly improved due to the existence of the reinforcing layer. In this way, the composite membrane can be made thinner and have higher ionic conductivity without sacrificing mechanical strength. Among them, the Xion-Dappion AEM with trimethylammonium groups on a poly(phenylene) backbone shows Cl^- conductivity of 17.4 mS cm^{-1} with the IEC of 1.74 mmol g^{-1} [12]. Xion-Pention composite AEMs are composed of a resin polynorbornene, the reinforcing

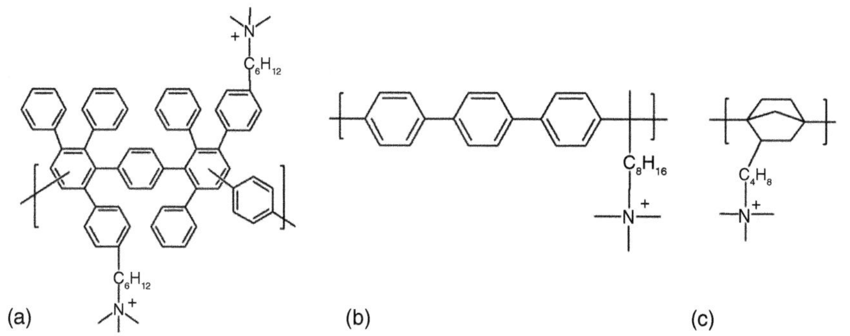

Figure 13.4 The chemical constitution of (a) Xion-Dappion, (b) Xion-Durion, and (c) Xion-Pention [11].

layer integrated with the membrane structure to increase mechanical strength. The enhancement of mechanical properties can make the composite membrane thinner than the independent membrane, provide low area-specific resistance (0.020–0.056 Ω-cm^2), high IEC (3.4–3.6 $mmol\,g^{-1}$), and high strength (66.8–553 MPa). These special AEMs, depending on their thickness and cross-linking degree, have been shown to be up to 9 $A\,cm^{-2}$ in the fuel cell's current density (or greater than 3 $W\,cm^{-2}$ power density), with stability over long periods (>500 h) with no change in resistance [13]. The structure of the Xion-Durion composite AEM contains a polyphenylene main chain with a QA functional group. The Xion-Durion membrane can operate for more than 1000 h soaked in 1 M KOH at temperatures up to 80 °C. After 1000 h of operation at high temperatures, the degradation rate of loss of conductivity and IEC are less than 0.5%. However, the high cost of materials and the cumbersome preparation process have resulted in a high price for this series of membranes, limiting their large-scale application and deployment [11].

13.1.1.7 Versogen: PiperION

Versogen has developed the PiperION AEM. The PiperION AEM consists of a high-molecular-weight, hydrophobic, rigid as well as aryl ether-bond-free backbone and with alkaline-stable piperidinium groups, as shown in Figure 13.5. The partial substitution of piperidone with 2,2,2-trifluoro acetophenone monomer can increase the molecular weight of the polymer while maintaining high alkaline stability. Generally, the mechanical toughness of the polymer allows it to be prepared into ultra-thin AEMs (down to 15 μm). In addition, the resulting AEM simultaneously shows good anti-swelling stability (<20% at 95 °C), high OH^- conductivity (193 $mS\,cm^{-1}$ at 95 °C), and excellent

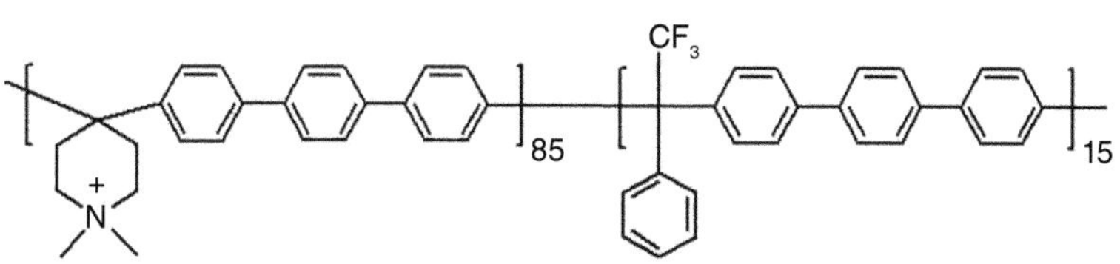

Figure 13.5 The chemical constitution of the PiperION TP-85 AEM [14].

mechanical properties (67 MPa stress and 117% strain at break). A fuel cell based on poly(arylpiperidinium) (PAP) and air (CO_2-free) feeding can achieve a high peak power density of 920 mW cm^{-2}. However, its fuel cell has poor durability, with a voltage decline approximately 11.5% after discharging for 250 h under a 500 mA cm^{-2} density at 95 °C [14, 15].

13.1.1.8 Membranes International Inc.: AMI-7001

The AMI-7001 AEM is a copolymer consisting of vinyl benzyl chloride and styrene which was functionalized with trimethyl amine and then cross-linked with divinylbenzene (Figure 13.6). Heterogeneous AMI-7001 membranes have a backing material to provide greater mechanical strength. A bio-electrochemical reactor based on AMI-7001 AEMs for silver recovery and power generation has achieved good performance, high coulombic efficiency (2.86%), and silver removal (91.28%), although diffusion of Ag^+ ions appeared. Under an initial Ag^+ concentration of 2000 mg l^{-1}, the maximum power density of 3385 mW m^{-2} was obtained in an AMI-based reactor [16, 17]. The poor alkali resistance of the trimethylamine (TMA) functional group results in overall poor alkali resistance of this membrane.

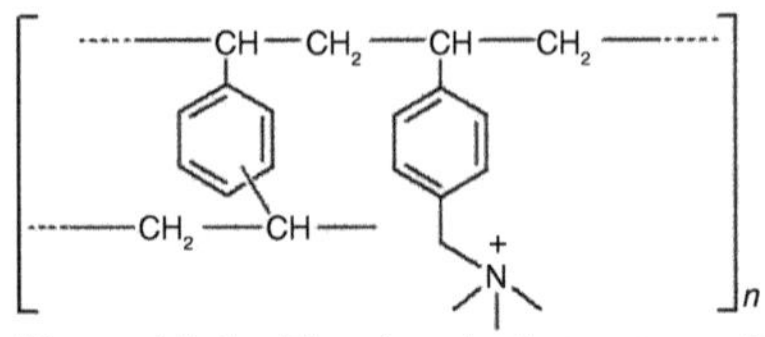

Figure 13.6 The chemical structure of AMI-7001 [16].

13.1.1.9 Asahi Glass: Selemion AMV

AMV is got by the blending method from polyvinyl chloride with a copolymer synthesized from divinylbenzene, styrene, and chloromethyl styrene. AMV is mainly applied for seawater desalination, water purification, separation of different ions, and so on. Using the Selemion AMV AEM resulted in high efficiency of bromate removal (94%). And, the Selemion AMV AEM enabled the bicarbonate flows from 60 to 87% and sulfates (93% of removal) [18]. However, the Selemion AMV AEM exhibited low conductivities of 2 mS cm^{-1} at 25 °C, which makes it unsuitable for application in fuel cells [19].

The comparison of conductivity and mechanical properties of the commercial AEMs is shown in Figure 13.7. It can be seen that commercial AEMs are unable to combine excellent conductivity and mechanical properties at the same time. In addition, to facilitate readers to choose commercial AEMs, we list the applications of these AEMs (Table 13.1). In general, most commercial alkaline membrane preparations require cumbersome synthesis processes, expensive catalysts, and environmentally unfriendly reagents. In addition, the high price further limits their application in fuel cells. Therefore, a better AEM design is needed to get rid of these issues.

13.1.2 Characteristics and Existing Problems of Non-Commercial Alkaline Anion Exchange Membrane

The characteristics of polymer backbones and the concentration of the cationic groups intimately influence the performance of AEMs. Among them, thermal and

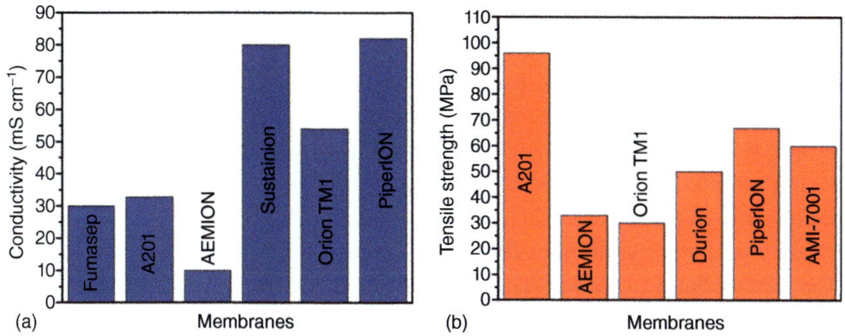

Figure 13.7 The comparison of conductivity and mechanical properties of the commercial AEMs.

mechanical stability is determined by parameters of the polymer backbones, while IEC and ionic conductivity are dominated by parameters related to the cationic groups. Furthermore, the alkali resistance stability of AEMs is related to both the polymer backbones and the cationic groups.

High ionic conductivity is necessary for high current density, and the conductivity of the AEMs is determined by the concentration and mobility of OH^- along with membrane transport properties. The conductivity of the hydroxide ion is 2–8 times lower than that of the proton ion, while the WU of AEMs and PEMs at a specific IEC is similar; this leads to the conductivity of AEMs being much lower than that of PEMs under the same IEC [20]. There are some reasons for the lower OH^- conductivity in AEMs: (i) The lower mobility of OH^- compared to H^+. It is reported that the diffusion coefficient of OH^- (5.3×10^{-9} m^2 s^{-1}) is much lower than that of H^+ (9.3×10^{-9} m^2 s^{-1}). (ii) The lower levels of dissociation of the ammonium hydroxide groups than those of the sulfonic acid groups. This is because the pK_b of the ammonium hydroxide is 4.8 and the pK_a of the sulfonic acid group is −2.8. (iii) The insufficient solvation energy of OH^-. The OH^- in AEMs needs a higher water content

Table 13.1 The applications of commercial AEMs.

AEMs	Applications
Fumasep	Desalination, base/acid recovery
A201	Fuel cell
AEMION	Fuel cell
Sustainion	Water and carbon dioxide electrolyzer, fuel cell
Orion TM1	Water electrolysis, fuel cell
Xion	Fuel cell
PiperION	Water electrolysis, fuel cell
AMI-7001	Silver recovery, fuel cell
Selemion AMV	Desalination, acid recovery

than that of H^+ in PEMs to format a sub-phase of solvated ions, enabling ionic percolation. (iv) The structure morphology of AEMs is worse than that of PEMs. PEMs exhibit more interconnected hydrophilic ionic channels and better-defined phases compared with AEMs at some given water content and IEC. (v) When in contact with air or CO_2, AEMs may undergo partial carbonation, consequently greatly reducing conductivity. This is because the conductivity of carbonate anions is four times lower than that of OH^- [21].

Stability is another vital index for AEMs, which determines long-term operation in the electrochemical device. The degradation of AEMs is divided into cationic group degradation and polymer main chain degradation. For the degradation of cationic groups, there is sufficient evidence that benzyl-linked cations are particularly sensitive to nucleophilic attack [22]. The commonly benzyl trimethylammonium groups are primarily degraded via direct nucleophilic substitution (S_N2). Under this degradation mechanism, hydroxide ions majorly attack the benzylic carbon to form a tertiary amine and an alcohol byproduct or alternatively attack the α-carbon to generate a benzylamine and an alcohol byproduct. When β-hydrogens are present, the QA cations can be cleaved through Hoffmann elimination (E2 elimination) as well as produces both an amine and olefin (Figure 13.8) [23].

In addition to the cationic groups, the alkaline stability of the polymer backbone is important for AEMs. It should be mentioned that polyvinylidene fluoride (PVDF) and VDF-based copolymers degrade easily and rapidly in base solutions via dehydrofluorination reaction, which leads to internal carbon–carbon double bonds [24]. The defluorinated product containing C=C bonds continues to be attacked by OH^-, producing hydroxyl and carbonyl groups in the chains, further resulting in chain

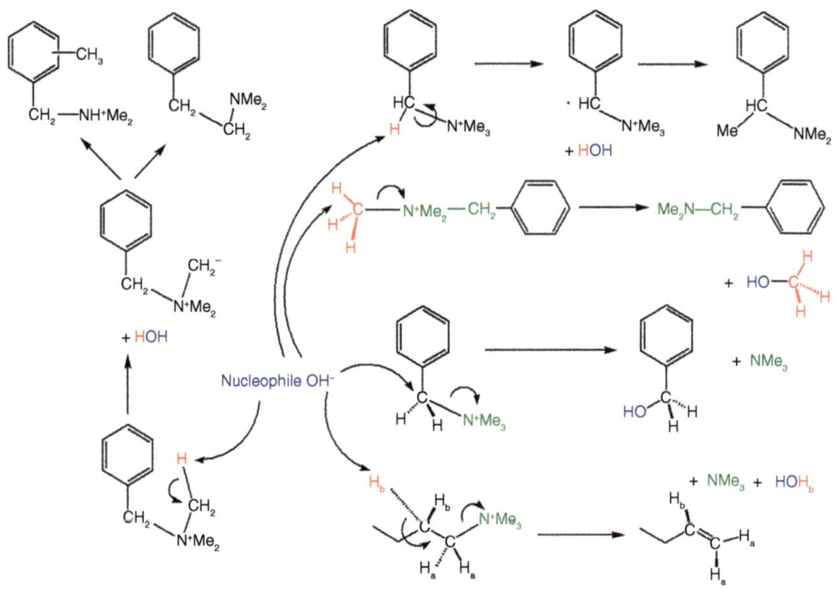

Figure 13.8 Degradation routes of QA groups [22, 23].

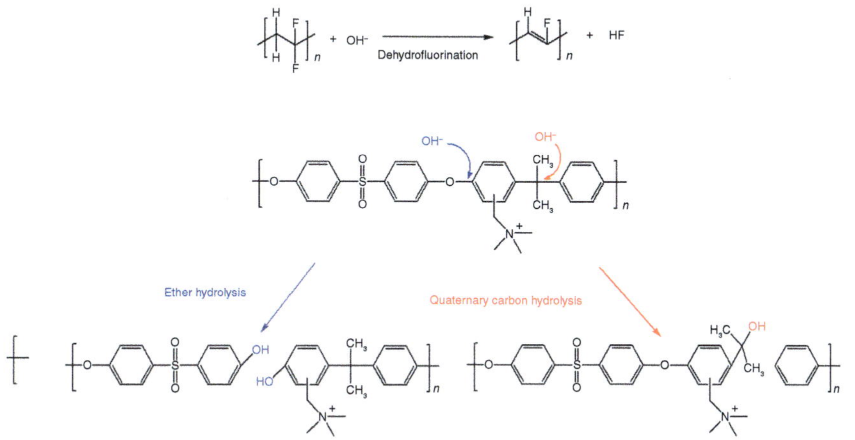

Figure 13.9 Polymer degradation pathways for poly(vinylidene fluoride) and polysulfone [25, 26].

scission. Meanwhile, commonly used poly(arylene ether) backbones (such as PPO, poly(arylene ether ketone)s [PAEKs], partially fluorinated poly(arylene ether)s, and poly(arylene ether sulfone)s [PAEs]) with cations placed at the benzyl positions are prone to hydroxide attack (Figure 13.9) [25]. Further, deuterium exchange experiments combined with density functional theory (DFT) calculations show that (i) aryl-ether cleavage of the polymer backbone occurs faster when the QA group is located closer to the C—O bonds; (ii) the electron-withdrawing parts (such as perfluoro groups, ketone, and sulfone linkages) present in the polymer backbones can accelerate the aryl-ether cleavage; (iii) electron-withdrawing parts connecting to the aromatic rings of polymer backbones exhibit more influence on aryl-ether degradation compared to QA groups [26].

13.1.3 Strategies to Improve the Properties of AEMs

In order to improve the OH^- conductivity and stability of AEMs, researchers have made a lot of efforts, including regulating the microphase morphologies, constructing free volume, introducing the cross-linking structures, and other physical methods as well as applying novel cationic functional groups and aryl ether-free polymer main chains. Herein, we sort out the progress of each strategy.

13.1.3.1 The Regulation of Microphase Morphologies

13.1.3.1.1 Side Chain Type The introduction of flexible side chains creates a distance between the ionic groups and the main chains, which is the cause of the aggregation of hydrophobic main chains to form a hydrophobic phase. When the ionic group is far from the main chain, the ability of OH^- around the ionic group to attack the polymer main chain is weakened, and the alkali resistance can be enhanced to a certain extent. Jannasch and coworker [27] developed side chain-type AEMs with alkyl groups at different positions. As shown in Figure 13.10, the C-series AEM

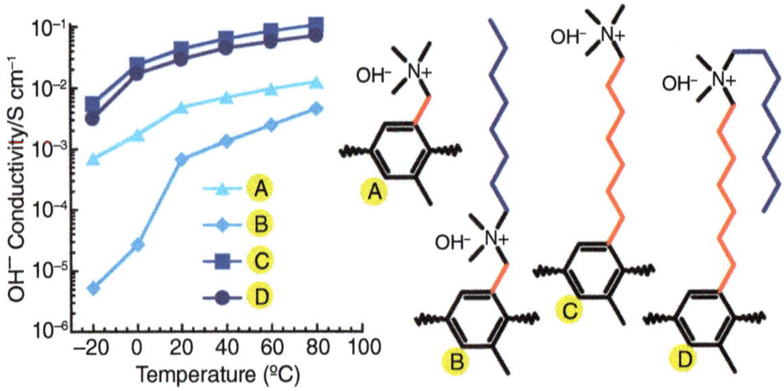

Figure 13.10 Chemical structure and the conductivity of PPO nQm. Source: Reproduced with permission from Dang and Jannasch [27]. Copyright 2015 American Chemical Society.

has the highest conductivity due to the long side chain improving the mobility of functional groups. The steric protection effect of the D-series AEM is enhanced so as to improve the alkali stability of the AEMs. After soaking in 1 M NaOH at 80 °C for 200 h, the AEMs with the spacer chain showed more than 85% conductivity retention, while the conductivity retention of the A-series membrane was only about 15%. To investigate the effects of side-chain polarity and position on AEM performance, He and coworkers [28] designed a series of AEMs containing an alkyl extender/additional side-chain, alkoxy extender/additional side-chain, as well as a side-chain-free structure. The results showed that the additional side-chain along with the alkoxy chain could enhance micro-phase separation, WU, and OH^- conductivity of AEMs. In addition, the alkoxy-containing AEM had excellent fuel cell performance (404 mW cm^{-2} for MImPSf-AO) and longer durability (2.3 h for MImPSf-AO at 100 mA cm^{-2}) due to its excellent hydrogen bond network and the high stability of alkoxyl groups (Figure 13.11).

13.1.3.1.2 Block Polymer

Block AEMs are composed of high-molecular weight polymers with different polymer segment structural units, generally two or three blocks in different combinations.

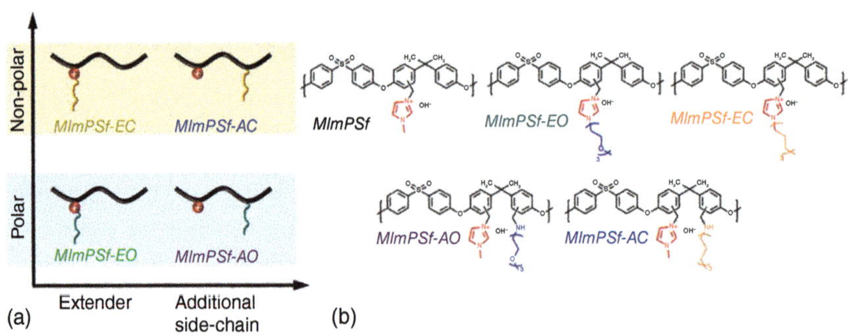

Figure 13.11 (a) Schematic diagram and (b) structures of AEMs containing different side chains. Source: Reproduced with permission from Li et al. [28]. Copyright 2022 Elsevier.

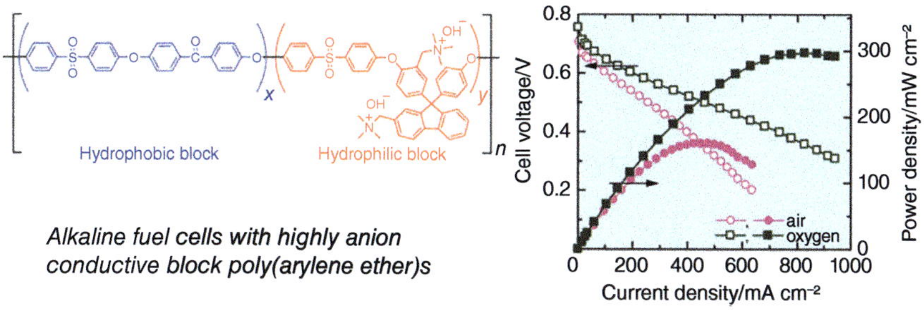

Figure 13.12 Quaternized ammonium-substituted fluorene-based poly(arylene ether)s. Source: Reproduced with permission from Tanaka et al. [29]. Copyright 2011 American Chemical Society.

The hydrophilic segments contain multiple cationic groups to transport OH^-. The hydrophobic segment ensures the mechanical properties of the AEMs. Watanabe and coworkers [29] prepared and studied fluorene-based polyethersulfone containing hydrophilic and hydrophobic segments with different lengths (Figure 13.12). The optimized AEM obtained an excellent conductivity (144 mS cm^{-1} at 80 °C), and the fuel cell power density obtained reached 297 mW cm^{-2}.

13.1.3.1.3 Locally Dense Type

Locally dense type AEMs can help ionic functional groups achieve better self-aggregation behavior in the case of similar or lower IEC. He et al. [30] developed a novel polymer structure, which was characterized by grafting three ion strings onto a benzene ring, as shown in Figure 13.13. The regular ion conduction channel was observed using atomic force microscopy, and a high Br^- conductivity of 50.6 mS cm^{-2} was obtained at 80 °C. Liu and coworkers [31] synthesized ether-free fluoropolymer AEMs containing a long and flexible multi-piperidine side chains, as shown in Figure 13.14. The prepared AEMs showed obvious microphase separation, which resulted in high OH^- conductivity (156.4 mS cm^{-1}) with a low SR (12.9%) at 80 °C. The AEMs also exhibited excellent alkaline stability, with only a 6.8% decrease in OH^- conductivity retained after immersion in 2 M NaOH at 80 °C for 1080 h.

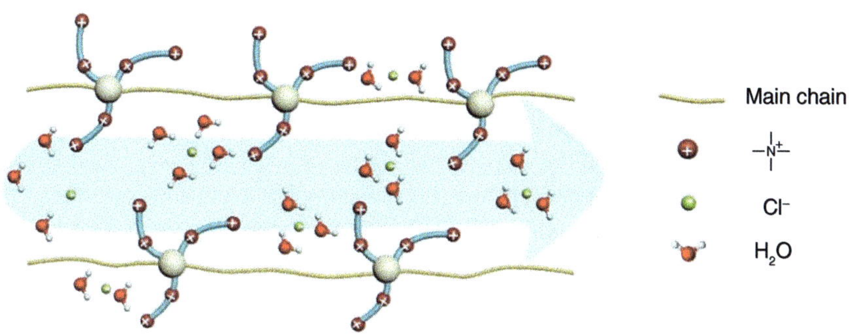

Figure 13.13 Schematic diagram of locally dense type structure. Source: Reproduced with permission from He et al. [30]. Copyright 2018 Elsevier.

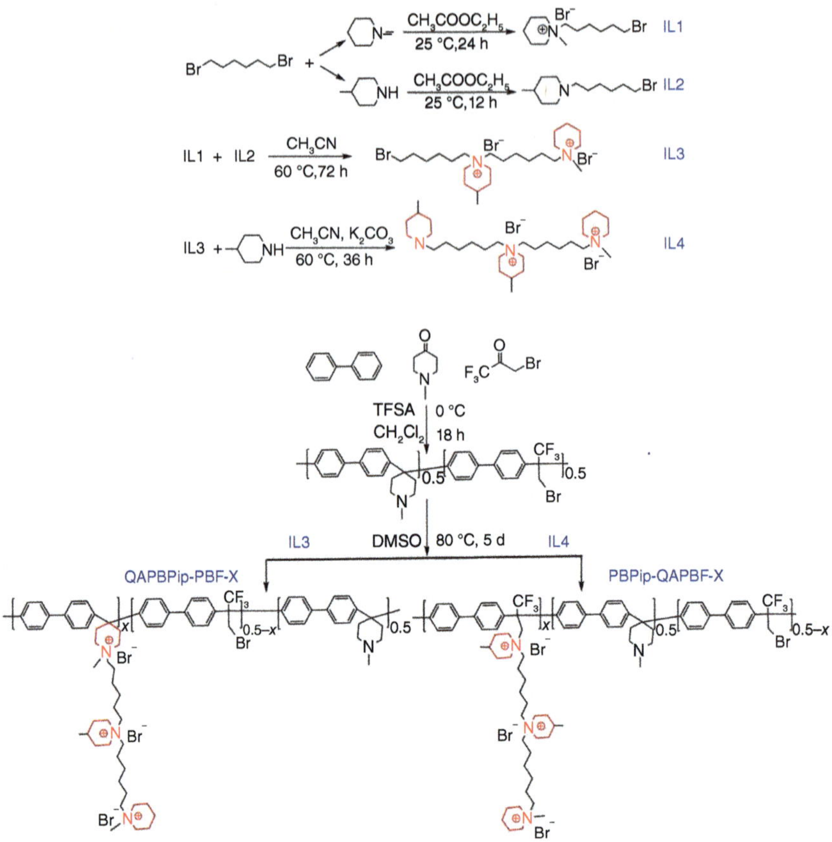

Figure 13.14 Schematic diagram of QAPBPip-PBF-X and PBPip-QAPBF-X. Source: Reproduced with permission from Gou et al. [31]. Copyright 2022 Elsevier.

13.1.3.2 Constructing Free Volumes

Increasing the inherent free volumes in the membranes can also improve the OH^- conductivity of AEMs. The free volumes reduce the transmission resistance of OH^- and promote the conduction of OH^-. More importantly, larger free volumes can provide more water storage space and improve the ability of AEMs to absorb water molecules without excessive expansion of the membranes. Yang et al.[32] first reported an AEM based on a microporous polymer containing V-type rigid Tröger base unit (Figure 13.15). The preparation method of the membrane was simple – it can be prepared in only two steps, and the prepared AEM can achieve a high OH^- conductivity (164.4 mS cm^{-1}, 80 °C) at a relatively low IEC (0.82 mmol g^{-1}). Liu and coworkers [33] introduced the twisted binaphthyl into the backbones to restrain the chain packing meanwhile enlarging the free volume, which can construct a well-developed ion transport channel (Figure 13.16). The poly(binaphthyl piperidinium)-based membrane exhibited a high OH^- conductivity (135.25 mS cm^{-1}) at 80 °C. Furthermore, the conductivity of the AEM only decreased by 10% conductivity retention after being treated in 2 M NaOH at 80 °C for 1080 h. Meanwhile, the fuel cell possessed a high peak power density (1.16 W cm^{-2}).

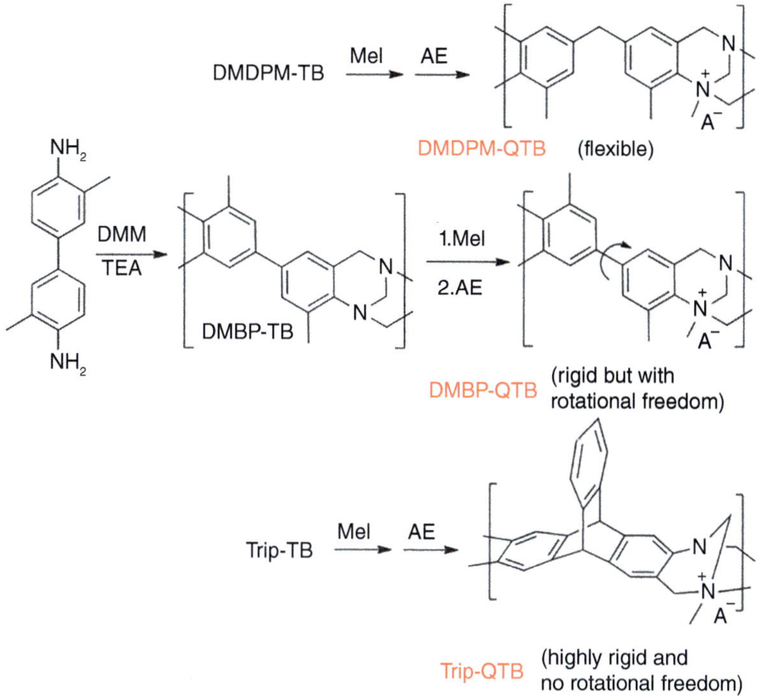

Figure 13.15 Synthesis of the AEM based on the TB structure. Source: Reproduced with permission from Yang et al. [32]. Copyright 2016 Wiley.

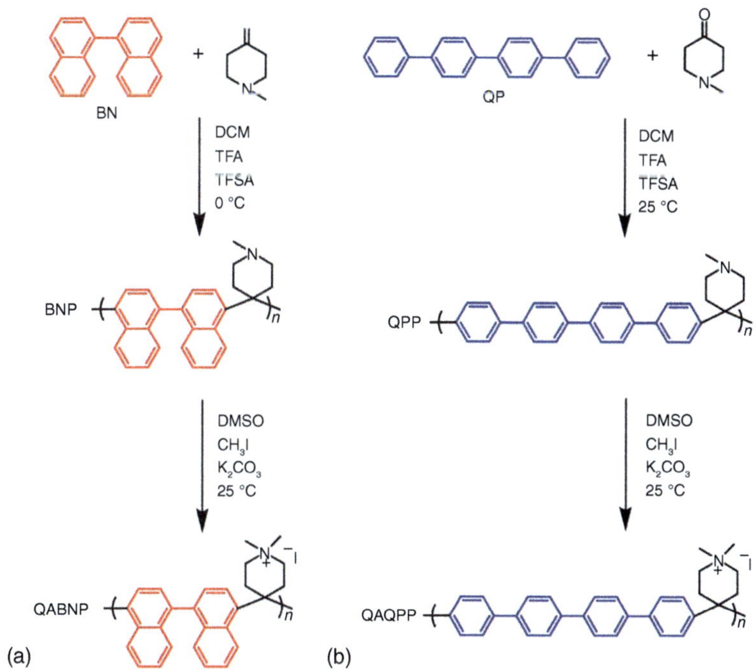

Figure 13.16 Synthesis pathways of QABNP and QAQPP. Source: Reproduced with permission from Gao et al. [33]. Copyright 2022 Elsevier.

13.1.3.3 The Introduction of Cross-linking Structures

Although regulating the microphase morphologies, constructing free volume can significantly improve the conductivity of the AEMs. However, the alkali stability of these AEMs is insufficient to meet the requirements of practical applications. So, it is necessary to develop an improving alkali resistance stability strategy, and the introduction of the cross-linking structure is an effective strategy to improve stability.

The cross-linking strategy can be classified into chemical and physical cross-linking based on their cross-linking mechanism. Chemical cross-linking includes the following: covalent cross-linking [34], ultraviolet (UV) irradiation-induced cross-linking [35], and thermally cross-linking [36]. Physical cross-linking includes ionic cross-linking [37] and metal cation cross-linking [38]. However, the commonly used cross-linking methods usually compromise the conductivity because of the smaller free volume and lower WU in cross-linked AEMs. Therefore, ridding the conductivity–stability dilemma is still a significant challenge. It is crucial to balance conductivity and mechanical properties.

(1) Cross-linking based on two functional components. There are reasons to consider that preparing an AEM with two components (the mechanical-supporting component cross-linked with the hydroxide-conducting function component, respectively) can better conquer the conductivity and stability balance. Lu et al. [39] prepared a novel cross-linked AEMs with PBI as the mechanical-supporting component and 1,4-diazabicyclo[2.2.2]octane (DABCO) functionalized poly(vinylbenzyl chloride) (PVBC) as the conducting component (Figure 13.17a). The resulting PBI-*c*-PVBC/OH AEM exhibits high OH^- conductivity (>25 mS cm^{-1} at RT), and low SR (13%) is obtained. Similarly, Zhuo et al. [40] synthesized a novel composite AEM cross-linked by tertiary amine-based PAE and bromomethyl-containing poly(arylene ethersulfone) (Figure 13.17b). The formed AEMs show 82.4 mS cm^{-1} conductivity at 80 °C and an excellent mechanical property (32.12 MPa).
(2) Cross-linking based on the modified backbone. AEMs with cross-linked ion cluster regions enhance dimensional stability while also ensuring the aggregation of the cation clusters to form microphase separation and ion-conducting channels. Lai et al. successfully synthesized a novel cross-linked ion cluster region AEMs (Figure 13.17c) [41]. In their work, the cross-linking between the ion cluster regions not only enhanced the mechanical properties (39.7 MPa) of the AEMs but also improved hydroxide conductivities (143.4 mS cm^{-1} at 80 °C) and reduced SR (11.3%). Cha et al. prepared a cross-linked AEM based on short hydrophilic block ionomers (Figure 13.17d) [42]. Due to the cross-linked networks and well-developed nanophase-separated morphologies, the membranes exhibited outstanding mechanical properties, low SR, and outstanding conductivity (108 mS cm^{-1} at 80 °C).
(3) Cross-linking based on a functional cross-linking agent. In general, the exploration for an appropriate cross-linking agent without the expense of functional groups is also an effective strategy to conquer the trade-off of conductivity and mechanical stability.

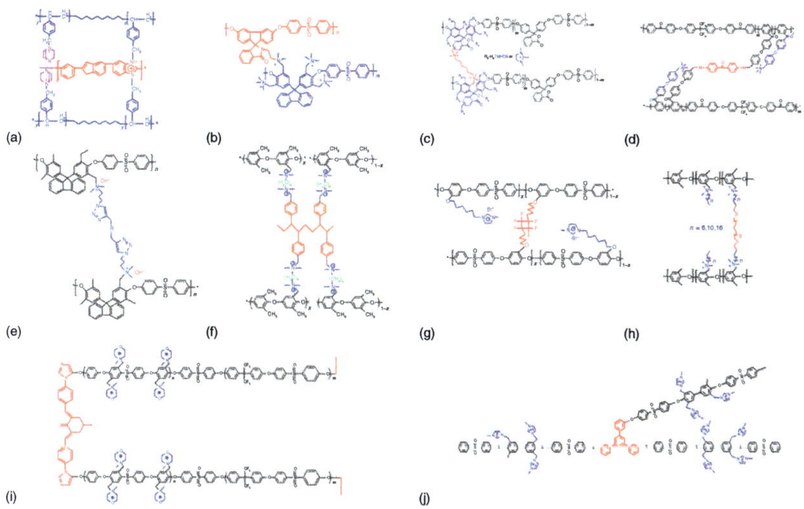

Figure 13.17 Chemical structures of advanced cross-linked AEMs [39–48].

Among these works, a series of novel cross-linked AEMs containing bi-functional cross-linking agents have been synthesized by Hu et al (Figure 13.17e) [43]. In this work, bi-functional cross-linking agents are used for both simultaneous quaternization and cross-linking. The resulting AEM exhibited a high OH^- conductivity (152.7 mS cm^{-1}) at 80 °C; meanwhile, the tightly packed internal construction of membranes after cross-linking also enhanced the mechanical strength (84.8 MPa).

Cross-linked chains with high flexibility and multiple cations have been used in preparing cross-linked AEMs. The cross-linked networks with long, flexible cross-linked chains cannot only improve the mechanical strength of the membranes but also enhance the nanophase separation ability of AEMs. He et al. have designed a particular structure of cross-linked PPO-based AEMs using a long flexible, multi-cation cross-linking chain (Figure 13.17f) [44]. The resulting membranes exhibit a low SR, excellent mechanical properties, and superior high hydroxide conductivity.

(4) Cross-linking based on side-chain type polymer. The side-chain-type AEMs can be prepared by tethering cations to polymer main chains via long spacers. The flexible spacer units can increase the local mobility of the cations and facilitate phase separation in the AEMs, leading to high ionic conductivity. Lin et al. have prepared a novel cross-linked AEMs with a side-chain-type structure (Figure 13.17g) [45]. The induced cross-linked structure makes the resulting AEMs show a low SR (9.6%). Similarly, a novel comb-shaped AEMs with cross-linked networks has been designed by Zhu et al. (Figure 13.17h) [46]. The thiol-ene cross-linking drastically decreases the SR of the AEMs, while the comb-shaped structure effectively improves the conductivity (60 mS cm^{-1} at room temperature [RT]).
(5) Terminal cross-linking and end-group cross-linking. Terminal cross-linked polymer systems are cross-linked only happening at the end of the polymer chain, which shows little loss of conductivity. Meanwhile, the formation of a 3D cross-linked network can lead to exceptionally high proton conductivities. Kwon et al.

have synthesized a novel azide-assisted terminal cross-linking AEMs (Figure 13.17i) [47]. The terminal cross-linking significantly enhances the mechanical properties and conductivity.

Apart from terminal cross-linking, the end-group cross-linking way could also result in morphological transformations that are applied to optimize ion transport. Lee et al. [48] have synthesized a novel end-group cross-linked AEM applying 3-hydroxyphenyl acetylene as the cross-linker (Figure 13.17j). The end-group cross-linking enhanced the tensile strength (>60.2 MPa) and OH^- conductivity (107 mS cm^{-1}) of the resulting AEMs.

As explained above, advanced cross-linking with cross-linker modification and polymer architecture modification is an effective strategy to balancing the conductivity and membrane stability of AEMs.

13.1.3.4 Other Physical Methods

13.1.3.4.1 Physical Blending

As one of the physical modification methods, the blending technique is an extremely promising approach for the development of AEMs, achieving a balance between electrochemical performance and physicochemical stability.

Polymers with a high degree of substitution value are an efficient method to dramatically improve the ion conductivity of AEMs. However, the dimensional stability of these polymers is weakened due to the excessive SR. The blended strategy is a very good option to solve this problem. Blended AEMs based on N1-alkyl-substituted imidazolium-based PVBC homopolymers and aromatic polyethers containing main chain pyridine units were prepared by Deimede and coworkers (Figure 13.18a) [49]. The N1-alkyl-substituted imidazolium-based poly(PVBC-*co*-AA20) copolymers showed excellent chemical stability owing to the hindrance effect of the N1 dodecyl part. Aromatic polyethers containing pyridine units were blended to ensure the high WU, excellent chemical stability, and mechanical toughness of the AEMs. Blended membranes containing hyperbranched/linear polymers also showed enhanced hydroxide conductivity and dimensional stability. Zhang and coworkers have developed a novel blended AEM-containing polymer of intrinsic microporosity (PIM) [50]. The incorporation of PIM increases free volumes in the AEMs, making it easier for cations to cross the backbones to connect with each other, thus enhancing microphase separation (Figure 13.18b). Three-component blended AEMs were synthesized based on tetraarylpolyphosphonium (pTAP)-based copolymers, chitosan, and PVDF (Figure 13.18c) [51]. Among them, chitosan was used as a blend to improve the WU of the base ionomer matrix. And, the goal of blending PVDF was to provide stable mechanical toughness to the composite blend.

The semi-interpenetrating network (SIPN) AEMs can also be prepared by blending. They are composed of cross-linked polymers and one or more linear or grafted polymers, which constitute interpenetrating network (IPN) structures. This type of AEMs combines the advantages of two or more polymers. Pan et al. designed a novel SIPN membranes containing a rigid, conductive functional part and a hydrophilic, flexible network (Figure 13.19a) [52]. The SIPN AEMs kept a good balance between conductivity and SR. The SIPN-60-2 (WQAPPO-60/WPEG-PAGE = 2 : 1,

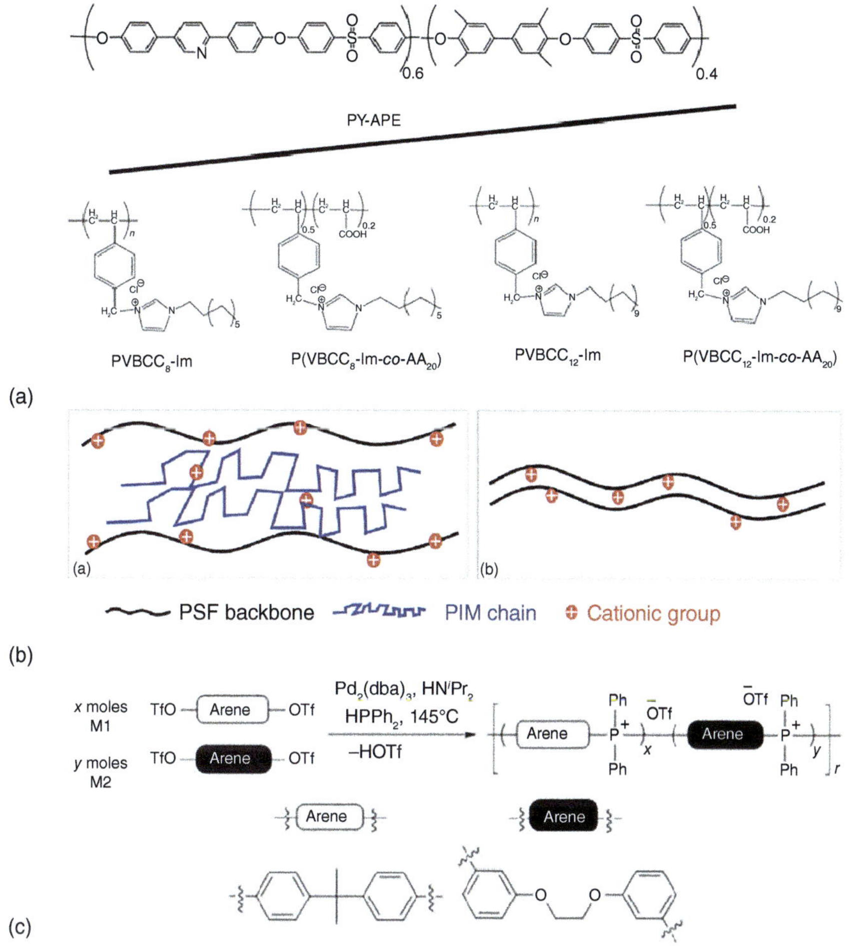

Figure 13.18 Chemical structures of blend-modified AEMs. Source: (a) Reproduced with permission from Tsagdi et al. [49]. Copyright 2021 Springer Nature. (b) Reproduced with permission from Gong et al. [50]. Copyright 2020 Elsevier. (c) Reproduced with permission from Muthumeenal et al. [51]. Copyright 2021 American Chemical Society.

IEC = 1.43 mmol g^{-1}) membrane had a moderate SR (25.5% at 80 °C), and the conductivity reached 67.7 mS cm^{-1} at 80 °C. Similarly, Lan et al. introduced pyrrolidonium-type ionic liquids (IILs), poly(*N*-methyl-*N*-vinyl-pyrrolidonium) (PNVMP) into covalently cross-linked poly(vinyl alcohol) (PVA) network to prepare AEMs (Figure 13.19b) [53]. Due to the high quaternization degree of PNVMP, the SIPN membranes exhibited high ionic conductivity. Moreover, the adding of PVA was very helpful to keep dimensional stability and the cross-linked structure kept membranes possessing sufficient mechanical strength. When the blend ratio of PNVMP/PVA was 0.25/0.45, the PNVMP/PVA membrane exhibited a high conductivity of 2.05×10^{-2} S cm^{-1} and low swelling degree of 5.57%, excellent tensile strength of 18 MPa, and relatively good stability in 6 mol l^{-1} NaOH at 60 °C for 168 h.

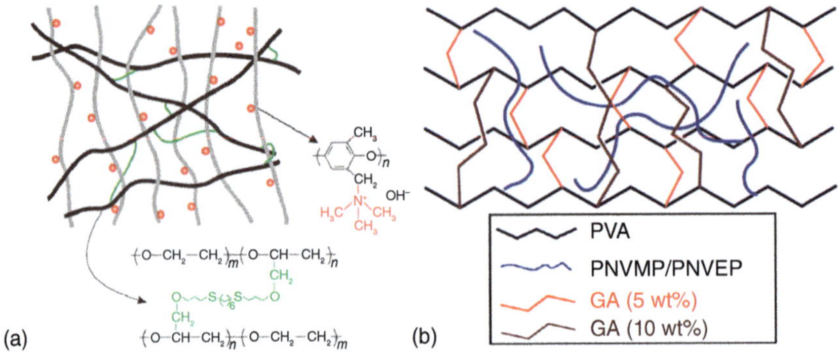

Figure 13.19 Chemical structures of SIPN AEMs. Source: (a) Reproduced with permission from Pan et al. [52]. Copyright 2015 American Chemical Society. (b) Reproduced with permission from Lan et al. [53]. Copyright 2015 Elsevier.

Although polymer blending is a simple and fascinating way to modify the shortcomings of a single part and improve the AEM performances, unfortunately, because the conductive polymers are not covalently bonded to the polymer matrices, they will gradually precipitate with the working of the fuel cell and then lead to a decrease in ion conductivity. Therefore, it still cannot meet the application of AEMFCs.

13.1.3.4.2 Physical Doping

Physical doping methods: The doping methods can make the AEMs combine the superiorities of organic materials and inorganic particles at the same time, thereby improving some of the defects of the pure polymer electrolyte membranes. Inorganic nanoparticles have large specific surface areas and good stability. In earlier studies, inorganic nanomaterials such as carbon nanotubes (CNTs) [54], ZrO_2 [55], and SiO_2 [56] have been added into ion exchange membranes (IEMs) to improve their performance. The works above indicated that the inorganic nanomaterials added to the IEMs may lead to a great enhancement in their mechanical properties, thermal stability, and chemical stability. Due to the intrinsically different surfaces between polymers and inorganic nanomaterials, the prepared IEMs exhibited poor interfacial adhesion, so these fillers tend to agglomerate. Ionic conductivities, mechanical properties, and dimensional stability of the IEMs cannot be apparently improved [57].

Modification of organic groups on the surface of inorganic materials is an effective strategy to solve the compatibility between organic and inorganic materials. Meanwhile, the interaction between the inorganic nanoparticles and the polymers helps construct ion channels at the interface. Lü et al. have prepared novel composite AEMs through incorporating quaternized graphenes into the chloromethylated polysulfone and then carried out quaternization (Figure 13.20a) [58]. The QPSU0.5%-QGs exhibited a fourfold enhancement in the bicarbonate conductivity than that of pristine quaternized polysulfone AEM at 80 °C, and the QPSU0.25%-QGs exhibited a threefold increase in Young's modulus as well as tensile strength. Jiang and coworkers [59] achieved the creation of interconnected ionic channels within the

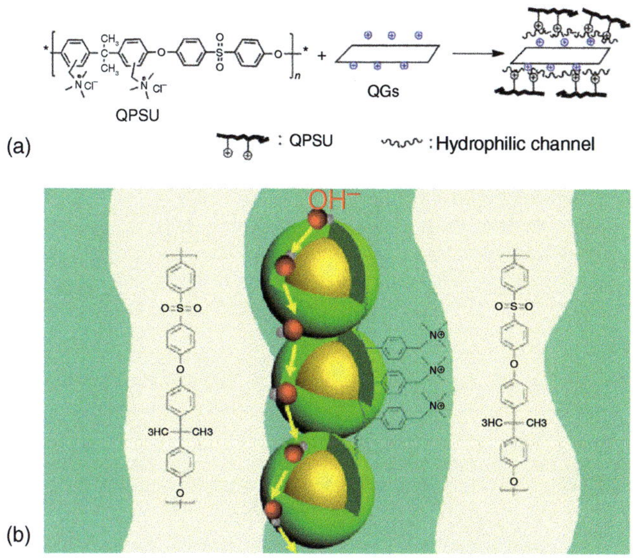

Figure 13.20 Chemical structures of doping modified AEMs. Source: (a) Reproduced with permission from Liu et al. [58]. Copyright 2015 Elsevier. (b) Reproduced with permission from He et al. [59]. Copyright 2017 American Chemical Society.

membrane by embedding the SiO_2-QPSt core–shell structure into a polysulfone (Figure 13.20b). The AEMs with high filler amounts (50–70 wt%) showed 88.4–188.1 mS cm^{-1} OH^- conductivities at 80 °C.

Overall, the nature and quantity of inorganic particles will have different effects on the performance of the hybrid membranes. Choosing the appropriate inorganic particles and the amount is the focus of the polymer modification process. However, despite the functional modification of the surface of inorganic materials, the high content of addition will still cause serious particle aggregation and macroscopic phase separation, generating the degradation of the interface morphologies and the ion conduction channels, reducing the performance of the membranes.

13.1.3.4.3 Physical Electrospinning

Electrospinning is a convenient and simple method using the electrohydrodynamic process to generate continuous thin fibers and can be employed for the preparation of fibrous membranes. A wide range of materials can be electrospun, such as inorganic sols, polymer melts, and polymer solutions. The main components of the typical electrospinning contain a spinneret, a high-voltage power supply, and a grounded collector [60] (Figure 13.21). The spinning solution is pumped through the spinneret by a high-voltage power supply to form a charged solution and a Taylor cone. When the electrostatic force breaks through the surface tension of the solution, a liquid jet is ejected. The jet is drawn to the collector and suffers stretching and bending instabilities when the solvent evaporates, resulting in non-woven nanofibrous mats. The morphologies of polymer fibers can be controlled by adjusting parameters, such as the solution properties (e.g. electrical conductivity and viscosity),

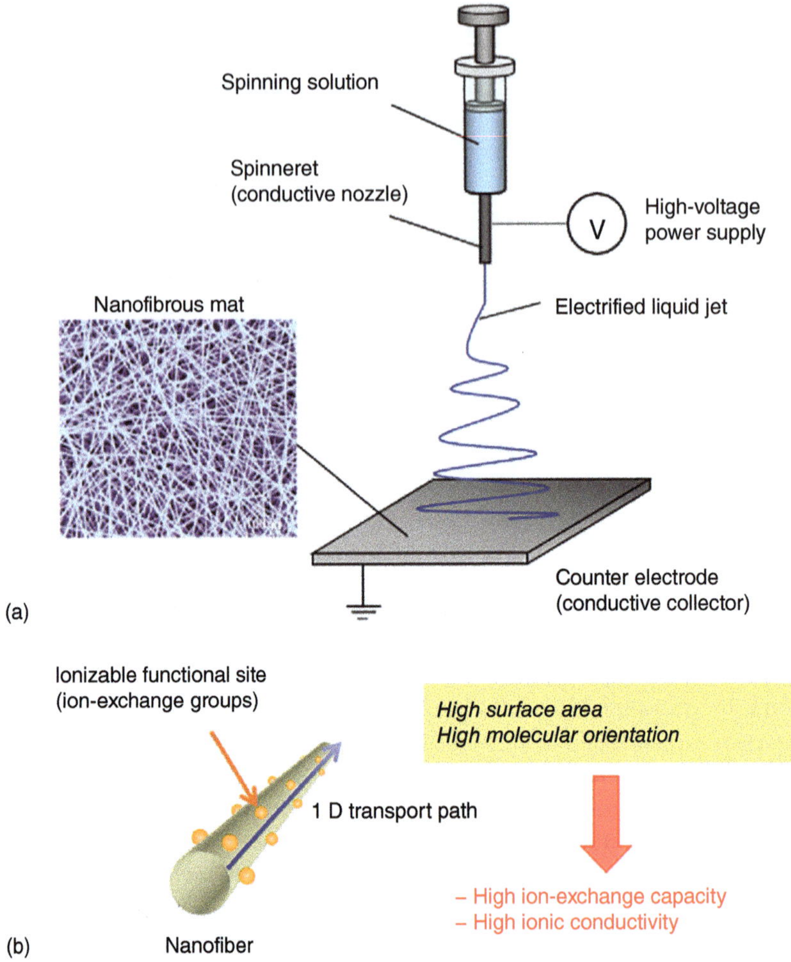

Figure 13.21 Scheme of the electrospinning setup and characteristics of ion-exchange nanofibers. Source: (a) and (b) Reproduced with permission from Zhang et al. [60]. Copyright 2018 Society of Chemical Industry.

operating parameters (e.g. the distance between the spinneret and collector and applied voltage), as well as surrounding conditions.

As electrospinning is gradually used to construct IEMs, anion-conductive nanofibers exhibit potential for improving the membrane properties. In 2011, Pintauro and coworker reported the first electrospun composite AEMs [61]. The heterogeneous composite AEMs were fabricated according to quaternary ammonium polysulfone (QAPS) electrospun nanofibers, where the AEMs were composed of tetramethylammonium-substituted polysulfone fibers embedded in the polyphenyl sulfone matrix. This structure can effectively reduce WU and enhance the mechanical properties of the AEMs. Gong et al. prepared a novel electrospun composite AEM using polysulfone functionalized by imidazolium as the interfiber void fillers and electrospun nanofibrous mats [62]. The resulted composite membranes showed a much higher OH^- conductivity and enhanced mechanical and alkaline stabilities in comparation with the corresponding cast membranes. Wang et al. prepared synthesized novel composite AEMs composed of the PAE containing hexaalkyl guanidinium group (PES-G-Cl) nanofibers and (vinyl benzyl) trimethylammonium chloride (VBTC) ionomers [63]. Compared with the AEMs without nanofibers, the AEMs based on nanofibers exhibited enhanced membrane stability and decreased WU.

Composite AEMs with one or more nanofiber components are one of the most versatile approaches to provide a robust platform to address the balance of conductivity, mechanical properties, and durability. Moreover, the widespread commercialization of nanofiber-based membranes requires simple fabrication lines to incorporate large-scale electro-spinning equipment into traditional roll-to-roll membranes.

13.1.3.5 The Development of Novel Cationic Functional Groups and Aryl Ether-free Main Chains with High Stability

In order to improve the alkali stability while ensuring high conductivity, researchers have gradually developed a variety of novel cationic functional groups and aryl ether-free polymer main chains for the preparation of AEMs.

13.1.3.5.1 The Development of Novel Cationic Functional Groups

QA cations have been widely developed because of simple preparation, high reaction activity, conduction efficiency, and low cost. However, QA salts are prone to nucleophilic substitution and Hoffmann degradation in an alkaline environment, which restricts the development of AEMs [64].

In recent years, nitrogen heterocyclic cations have attracted extensive attention because of their outstanding alkali resistance. Kreuer and coworker[65] systematically explored the alkali resistance of QA cations. Among them, six-membered aza spiro ring cation (ASU) and *N,N*-dimethylpiperidine cation (DMP) showed the highest alkali resistance, and their alkali resistance life was much higher than that of the most stable TMA cations in theory (Table 13.2). This is because the six-membered nitrogen heterocycles have low ring tension and large steric hindrance, which reduces the possibility of Hoffmann degradation and nucleophilic substitution. Since then, Jannasch and coworkers [66] developed *N*-heterospiro cation-loaded violone AEMs (as shown in Figure 13.22). This kind of AEMs hardly degraded after soaked in 1 M KOD/D_2O at 80 °C for 1800 h. After that, they developed a series of different azaspiro polyolefin electrolytes through double-bond polymerization [67]. The alkali resistance of polyolefin electrolytes with 5–6-membered spiro rings was only slightly degraded after soaking at 2 M KOD/D_2O 120 °C for 14 days. However, ionone and polyolefin electrolytes loaded with *N*-heterospiro cation had high water solubility due to high IEC, which limited the application of this kind of polyelectrolytes in AEMs. Therefore, exploring effective grafting methods to use ASU and DMP cations has become a research hotspot.

On the other hand, there are many studies about imidazole cations. However, the alkali resistance of ordinary *N,N′*-dimethylimidazole cation is poor. It is obviously degradation after soaking at 1 M NaOH at 80 °C for 24 h. Yan and coworkers [68] have compared the alkali resistance of imidazole cations with different substituents.

Table 13.2 Half-life of different cations at 160 °C in 6 M NaOH.

Entry	QA	Abbreviation	Half-life (h)
1	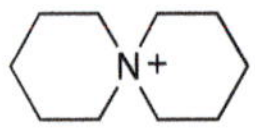	ASU	110
2	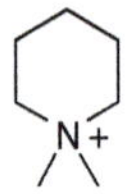	DMP	87.3
3	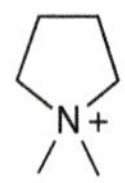	DMPy	37.1
4	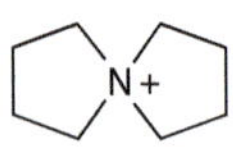	ASN	28.4
5	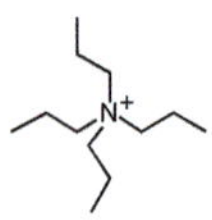	TPA	7.19
6		TMA	61.9
7		MAABCO	13.5

Source: Adapted from Marino and Kreuer [65].

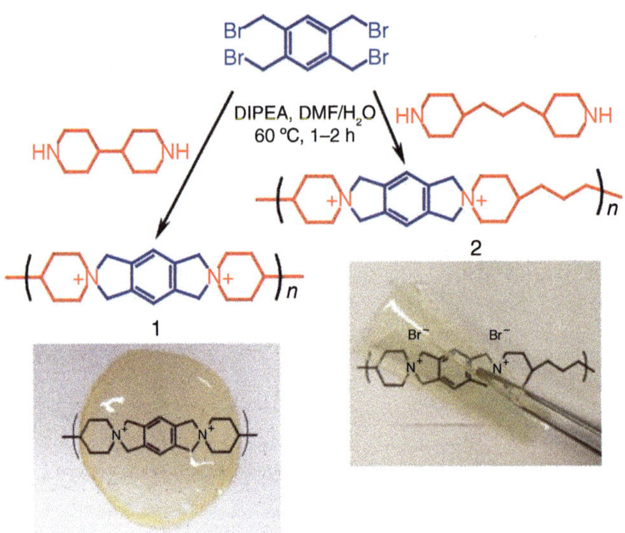

Figure 13.22 Synthetic pathway to spiro-ionene-based AEMs. Source: Reproduced with permission from Pham et al. [66]. Copyright 2017 American Chemical Society.

It was found that the substituents had a great impact on the stability of imidazole cations, especially by the substituents at the C-2 position of imidazole. The alkali resistance of imidazole cations can be effectively improved due to the substituents at the C-2 position of imidazole.

P and N are in the same main group, so quaternary phosphorous cations and QA cations show similar physical and chemical properties. Yan and coworkers [69] first developed benzyl tris(2,4,6-trimethoxy) phenyl-based PSF ionomer (TPQPOH), which reflected the possibility of the application of quaternary phosphorus cations in AEMs and brought new ideas to the researchers and development of AEMs (Figure 13.23). Coates and coworkers [70] developed quaternary phosphorus cations $[P(N(Me)Cy)_4]^+$ and prepared a series of $[P(N(Me)Cy)_4]^+$ supported polyethylene AEMs. $[P(N(Me)Cy)_4]^+$ cation had good alkali resistance, and it hardly degraded after soaked at 1 M $NaOD/CD_3OD$ 80 °C for 20 days. However, due to the difficulty of modification and low grafting activity of such cations, their application in AEMs has not been realized.

Hickner and coworkers [71] proposed using organometallic coordinated rhodium cations as new cationic groups. The design of these structures essentially eliminates the degradation reactions such as Hoffmann elimination, nucleophilic substitution, and phosphorus ylide and provides a new idea for the structural design of highly stable AEMs (Figure 13.24). The conductivity of these membranes at RT was close to that of AEMs supported by ordinary QA cations. However, when such membranes were soaked at 1 M NaOH 80 °C for 2 h, serious degradation reactions such as decoordination occurred, which made them poorly alkali-resistant and also restricted the further development of such membranes.

13.1.3.5.2 The Development of Aryl Ether-free Polymer Main Chains with High Stability

Polymer main chains are the foundation of AEMs, which determine the mechanical properties and stability of AEMs. In previous studies, the main chain structures of

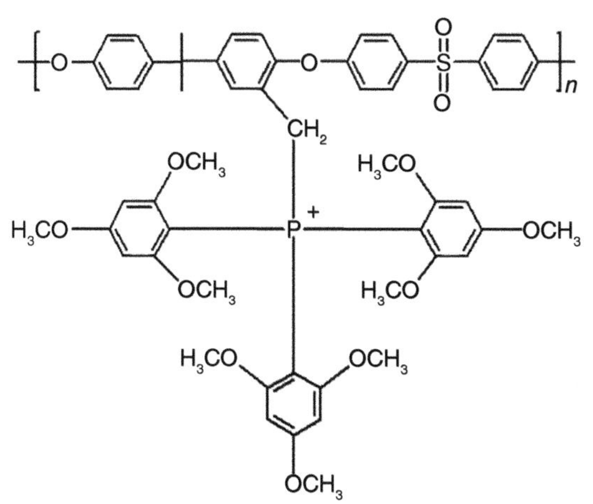

Figure 13.23 Chemical structure of TPQPOH [69].

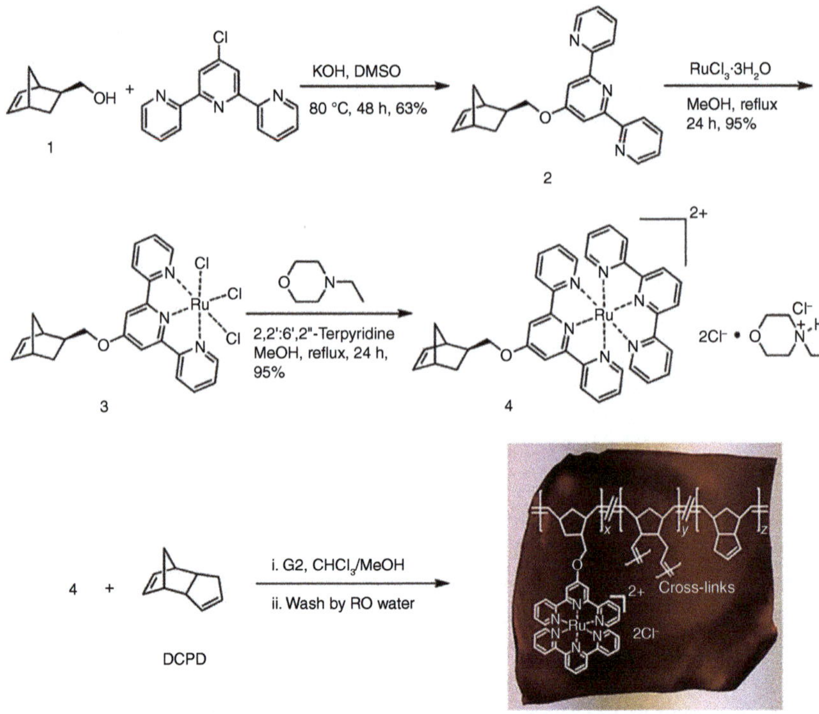

Figure 13.24 Synthesis of the corresponding AEM. Source: Reproduced with permission from Zha et al. [71]. Copyright 2012 American Chemical Society.

polymers were PPO [72], PAEs [73], PAEK [74], PBI [75], polyolefin [76], and polyphenylene [77]. There are significant differences in the physical stability of these polymer chains. Among them, polyphenylene ether, polysulfone, and polyetherketone are engineering plastics with excellent thermal stability and tensile strength, showing good membrane-forming ability and flexibility [78]. However, these polyarylether polymers exhibit poor chemical stability in highly alkaline environments. PBI has advantages such as strong main chain molecule rigidity, hydrogen bonding between molecules, and good mechanical properties. Meanwhile, it has high reaction activity and can be directly converted into AEMs by reacting with halogenated hydrocarbons. However, the stability of ordinary PBI base is not satisfactory [79].

Polyolefin materials are all carbon structures with high stability and excellent membrane toughness. They can be used as ultra-thin (thickness less than 30 μm) and high-power density AEMs. As shown in Figure 13.25, commercial ethylene tetrafluoroethylene has a thickness of 12.7 μm. After electron beam irradiation, the ethylene copolymer was grafted with quaternary amine salt. Under silver as a cathode catalyst, the maximum power density of the fuel cell reached 1110 mW cm^{-2} at 70 °C [80].

Polyphenylene polymers are also polymers without ether bonds, which have high molecular chain rigidity and good mechanical properties. Akiyama et al. prepared a series of quaternized copolymers (QP-QAF) polymers (Figure 13.26) [81]. The

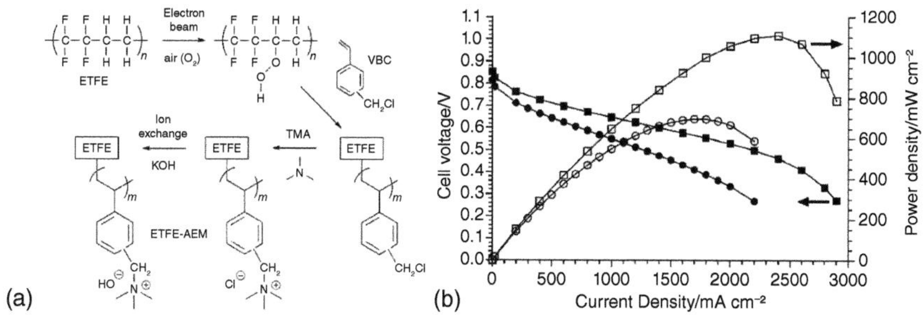

Figure 13.25 (a) Synthetic route of PE-based AEM and (b) corresponding fuel cell. Source: Reproduced from Wang et al. [80] with permission from the Royal Society of Chemistry.

prepared AEMs obtained a high OH^- conductivity (up to 134 mS cm^{-1} at 80 °C) at an IEC of 2.25 mequiv g^{-1}, and it was also stable in 4 M KOH at 80 °C) for 1000 h. However, the synthesis of polyfluorene backbones needs to be catalyzed by expensive palladium metal or a large amount of nickel metal, which seriously limits its large-scale preparation.

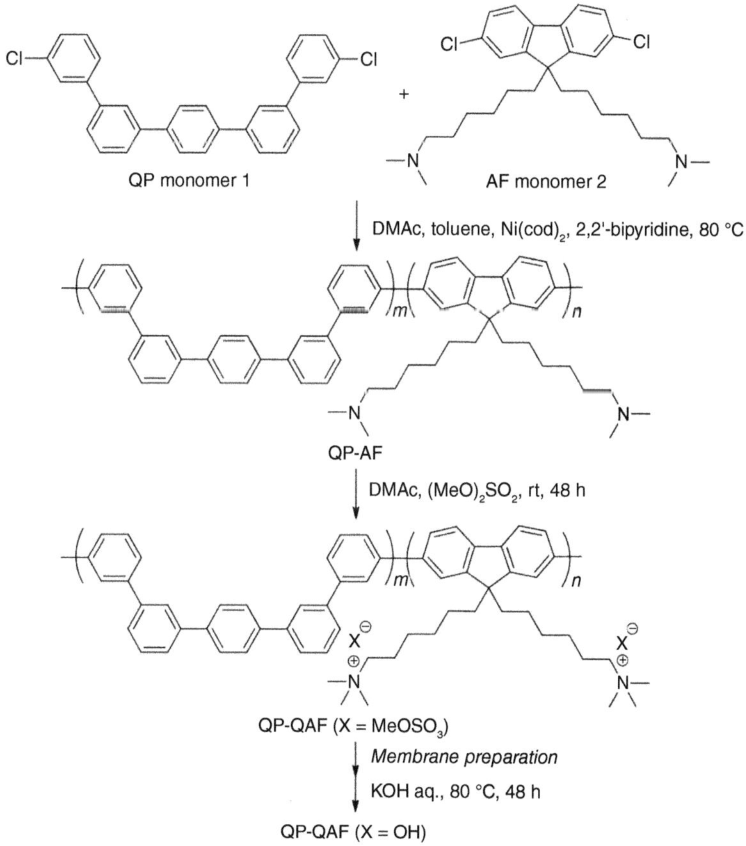

Figure 13.26 The preparation process of QP-AF and QP-QAF. Source: Reproduced with permission from Akiyama et al. [81]. Copyright 2019 American Chemical Society.

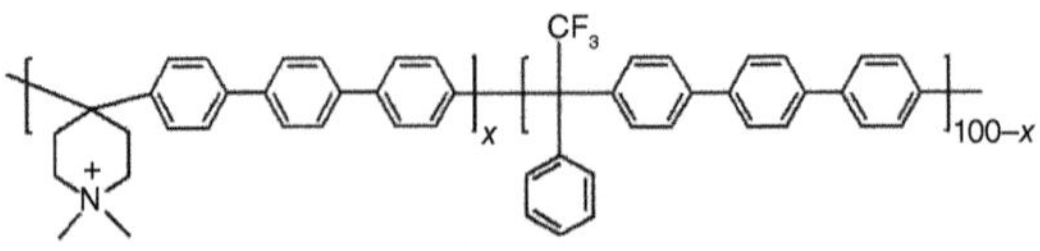

Figure 13.27 General chemical structure of PAP-TP-*x* [15].

Pure aromatic polymer main chains can also be prepared by condensation of ketone or aldehyde with biphenyl or terphenyl through Friedel–Crafts reaction. Yan and coworkers prepared polytriphenyl piperidionium (PAP-TP-85) AEM (Figure 13.27) [15]. The ion conductivity of PAP-TP-85 was as high as 193 mS cm^{-1} at 95 °C. After soaking in 1 M KOH solution at 100 °C for 2000 h, the IEC of the AEM decreased by only 3%, showing excellent alkali stability.

13.2 Summary and Outlooks

Over the past few years, researchers have made numerous efforts to make efficient AEMs for use in AEMFCs. So far, commercial AEMs still have some drawbacks, such as expensive prices and unsatisfactory performance, and are generally still unable to meet the large-scale commercial deployment of fuel cells in the future. Therefore, many different methods have been employed to improve the comprehensive properties and performance of non-commercial AEMs, such as developing novel groups and ether-free main chains, regulating the microphase separation, constructing the free volumes and cross-linking structures, as well as applying physical methods. All these ways seem promising for further enhancement of the performance of AEMs. Although a few AEMs showed high peak power density (~2 W cm^{-2}), the *in situ* chemical durability (~500 h) still cannot research the future requirements (2020 DOE: 2000 h). Moreover, although the *ex situ* stability tests of some AEMs show excellent alkali stability, they are unstable when they *in situ* worked in fuel cells. Hence, it is necessary that *ex situ* and *in situ* stability evaluation methods should be further developed and an effective means of evaluating stability should be exploited in the near future.

Acknowledgment

This research was supported by the National Key R&D Program of China (No. 2020YFB1505601) and the National Natural Science Foundation of China (No. 22278388, 22322811).

References

1 Varcoe, J.R. and Slade, R.C. (2005). Prospects for alkaline anion-exchange membranes in low temperature fuel cells. *Fuel Cells* 5 (2): 187–200.

2 Asalegno, A. and Marchesi, R. (2008). DMFC anode polarization: experimental analysis and model validation. *J Power Sources* 175 (1): 372–382.

3 Marcelo, C., Gustavo, D., Ryan, C.S. et al. (2013). Development and electrochemical studies of membrane electrode assemblies for polymer electrolyte alkaline fuel cells using FAA membrane and ionomer. *J Power Sources* 230: 169–175.

4 Kwon, H.J., Kim, B., Lim, G. et al. (2018). A multiscale-pore ion exchange membrane for better energy efficiency. *J Mater Chem A* 6: 7714–7723.

5 Duan, Q., Ge, S., and Wang, C. (2013). Water uptake, ionic conductivity and swelling properties of anion-exchange membrane. *J Power Sources* 243: 773–778.

6 Marinkas, A., Strużyńska-Piron, I., Lee, Y. et al. (2018). Anion-conductive membranes based on 2-mesityl-benzimidazolium functionalised poly(2, 6-dimethyl-1, 4-phenylene oxide) and their use in alkaline water electrolysis. *Polymer* 145 (6): 242–251.

7 Henkensmeier, D., Cho, H.R., Kim, H.J. et al. (2012). Polybenzimidazolium hydroxides-structure, stability and degradation. *Polym Degrad Stab* 97 (3): 264–272.

8 Yang, H., Kaczur, J.J., Sajjad, S.D. et al. (2017). Electrochemical conversion of CO_2 to formic acid utilizing Sustainion™ membranes. *J CO_2 Util* 20: 208–217.

9 Meek, K.M., Antunes, C.M., Strasser, D. et al. (2019). High-throughput anion exchange membrane characterization at NREL. *ECS Trans* 92 (8): 723–731.

10 Eun, J.P., Christopher, B.C., Katherine, E.A. et al. (2018). Chemically durable polymer electrolytes for solid-state alkaline water electrolysis. *J Power Sources* 375: 367–372.

11 Salvatore, D.A., Gabardo, C.M., Reyes, A. et al. (2011). Designing anion exchange membranes for CO_2 electrolysers. *Nat Energy* 6: 339–348.

12 Michael, R.H. (2013). Alkaline stability of poly(phenylene)-based anion exchange membranes with various cations. *J Polym Sci B Polym Phys* 51: 1736–1742.

13 Garrett, H., Mrinmay, M., Peng, X. et al. (2019). Composite poly(norbornene) anion conducting membranes for achieving durability, water management and high power (3.4 W/cm^2) in hydrogen/oxygen alkaline fuel cells. *J Electrochem Soc* 166 (10): F637–F644.

14 Endrődi, B., Kecsenovity, E., Samu, A. et al. (2020). High carbonate ion conductance of robust PiperION membrane allows industrial current density and conversion in zero-gap carbon dioxide electrolyzer cell. *Energy Environ Sci* 13: 4098–4105.

15 Wang, J., Zhao, Y., Setzler, B.P. et al. (2019). Poly(aryl piperidinium) membranes and ionomers for hydroxide exchange membrane fuel cells. *Nat Energy* 4: 392–398.

16 Aeshala, L.M., Uppaluri, R.G., and Verma, A. (2013). Effect of cationic and anionic solid polymer electrolyte on direct electrochemical reduction of gaseous CO_2 to fuel. *J CO_2 Util* 3–4: 49–55.

17 Hoa, N.A.D., Babela, S., and Kurisu, F. (2017). Bio-electrochemical reactors using AMI-7001S and CMI-7000S membranes as separators for silver recovery and power generation. *Bioresour Technol* 244: 1006–1014.

18 Sylwia, K. and Jacek, A.W. (2011). Removal of bromate and associated anions from water by Donnan dialysis with anion-exchange membrane. *Desalin Water Treat* 35: 158–163.

19 Giffin, G.A., Lavina, S., Pace, G. et al. (2012). Interplay between the structure and relaxations in selemion AMV hydroxide conducting membranes for AEMFC applications. *J Phys Chem C* 116: 23965–23973.

20 Gottesfeld, S., Dekel, D.R., Bae, C. et al. (2018). Anion exchange membrane fuel cells: current status and remaining challenges. *J Power Sources* 375: 170–184.

21 Merle, G., Wessling, M., and Nijmeijer, K. (2011). Anion exchange membranes for alkaline fuel cells: a review. *J Membr Sci* 377: 1–35.

22 Arges, C.G. and Zhang, L. (2018). Anion exchange membranes' evolution toward high hydroxide ion conductivity and alkaline resiliency. *ACS Appl Energy Mater* 1: 2991–3012.

23 Iojoiu, C., Chabert, F., Marechal, M. et al. (2006). From polymer chemistry to membrane elaboration. *J Power Sources* 153: 198–209.

24 Ameduri, B. (2009). From vinylidene fluoride (VDF) to the applications of VDF-containing copolymers: recent developments and future trends. *Chem Rev* 109: 6632–6686.

25 Miyanishi, S. and Yamaguchi, T. (2016). Ether cleavage-triggered degradation of benzyl alkylammonium cations for polyethersulfone anion exchange membranes. *Phys Chem Chem Phys* 18: 12009–12023.

26 Choe, Y.K., Fujimoto, C., Lee, K.S. et al. (2014). Alkaline stability of benzyl trimethyl ammonium functionalized polyaromatics: a computational and experimental study. *Chem Mater* 26: 5675–5682.

27 Dang, H.S. and Jannasch, P. (2015). Exploring different cationic alkyl side chain designs for enhanced alkaline stability and hydroxide ion conductivity of anion-exchange membranes. *Macromolecules* 48 (16): 5742–5751.

28 Li, L., Zhang, N., Wang, J. et al. (2022). Stable alkoxy chain enhanced anion exchange membrane and its fuel cell. *J Membr Sci* 644: 120179–120189.

29 Tanaka, M., Fukasawa, K., Nishino, E. et al. (2011). Anion conductive block poly(arylene ether)s: synthesis, properties, and application in alkaline fuel cells. *J Am Chem Soc* 133 (27): 10646–10654.

30 He, Y., Zhang, J., Liang, X.E. et al. (2018). Achieving high anion conductivity by densely grafting of ionic string. *J Membr Sci* 559: 35–41.

31 Gou, W., Gao, W., Gao, X. et al. (2022). Highly conductive fluorinated poly(biphenyl piperidinium) anion exchange membranes with robust durability. *J Membr Sci* 645: 120200–120211.

32 Yang, Z., Guo, R., Malpass-Evans, R. et al. (2016). Highly conductive anion-exchange membranes from microporous Tröger's base polymers. *Angew Chem Int Ed* 55 (38): 11499–11502.

33 Gao, W., Gao, X., Gou, W. et al. (2022). High-performance tetracyclic aromatic anion exchange membranes containing twisted binaphthyl for fuel cells. *J Membr Sci* 655: 120578–120589.

34 Zhang, C., Guo, X., Fang, J. et al. (2017). A new and facile approach for the preparation of cross-linked sulfonated poly(sulfide sulfone) membranes for fuel cell application. *J Power Sources* 170 (1): 42–45.

35 Bu, F., Zhao, C., Wang, B. et al. (2015). UV irradiation-induced cross-linked bicarbonate anion exchange membranes based on vinylimidazolium-functionalized poly(arylene ether ketone). *RSC Adv* 5 (70): 57067–57075.

36 Tsai, T.H., Ertem, S.P., Maes, A.M. et al. (2015). Thermally cross-linked anion exchange membranes from solvent processable isoprene containing ionomers. *Macromolecules* 48 (3): 655–662.

37 Li, Z., Jiang, Z., Tian, H. et al. (2015). Preparing alkaline anion exchange membrane with enhanced hydroxide conductivity via blending imidazolium-functionalized and sulfonated poly(ether ether ketone). *J Power Sources* 288: 384–392.

38 Gasa, J.V., Weiss, R.A., and Shaw, M.T. (2007). Ionic crosslinking of ionomer polymer electrolyte membranes using barium cations. *J Membr Sci* 304 (1–2): 173–180.

39 Lu, W., Zhang, G., Li, J. et al. (2015). Polybenzimidazole-crosslinked poly(vinylbenzyl chloride) with quaternary 1,4-diazabicyclo (2.2.2) octane groups as high-performance anion exchange membrane for fuel cells. *J Power Sources* 296: 204–214.

40 Zhuo, Y.Z., Nan, L.A., Zhang, Q.G. et al. (2015). Highly ionic-conductive crosslinked cardo poly(arylene ether sulfone)s as anion exchange membranes for alkaline fuel cells. *J Membr Sci* 491: 138–148.

41 Lai, A.N., Guo, D., Lin, C.X. et al. (2016). Enhanced performance of anion exchange membranes via crosslinking of ion cluster regions for fuel cells. *J Power Sources* 327: 56–66.

42 Cha, M.S., Lee, J.Y., Kim, T.H. et al. (2017). Preparation and characterization of crosslinked anion exchange membrane (AEM) materials with poly(phenylene ether)-based short hydrophilic block for use in electrochemical applications. *J Membr Sci* 530: 73–83.

43 Hu, B., Miao, L., Bai, Y. et al. (2017). Facile construction of crosslinked anion exchange membranes based on fluorenyl-containing polysulfone via click chemistry. *Polym Chem* 8 (30): 4403–4413.

44 He, Y., Wu, L., Pan, J. et al. (2016). A mechanically robust anion exchange membrane with high hydroxide conductivity. *J Membr Sci* 504: 47–54.

45 Lin, C.X., Zhuo, Y.Z., Hu, E.N. et al. (2017). Crosslinked side-chain-type anion exchange membranes with enhanced conductivity and dimensional stability. *J Membr Sci* 539: 24–33.

46 Zhu, L., Zimudzi, T.J., Wang, Y. et al. (2017). Mechanically robust anion exchange membranes via long hydrophilic cross-linkers. *Macromolecules* 50 (6): 2329–2337.

47 Kwon, S., Rao, A.H.N., and Kim, T.H. (2017). Anion exchange membranes based on terminally crosslinked methyl morpholinium-functionalized poly(arylene ether sulfone)s. *J Power Sources* 375: 421–432.

48 Lee, K.H., Cho, D.H., Kim, Y.M. et al. (2017). Highly conductive and durable poly(arylene ether sulfone) anion exchange membrane with end-group cross-linking. *Energy Environ Sci* 10 (1): 275–285.

49 Tsagdi, A., Dimitriou, M., Druvari, D. et al. (2021). Blend membranes based on N1-alkyl-substituted imidazolium functionalized polymers and aromatic polyethers: influence of N1-alkyl substituent on properties and alkaline stability. *Polym Bull* 79 (48): 1–22. https://doi.org/10.1007/s00289-021-03581-5.

50 Gong, S., Li, L., Ma, L. et al. (2020). Blend anion exchange membranes containing polymer of intrinsic microporosity for fuel cell application. *J Membr Sci* 595: 117541–117550.

51 Muthumeenal, A., Alessandro, S., Farida, A. et al. (2021). High performance of anion exchange blend membranes based on novel phosphonium cation polymers for all-vanadium redox flow battery applications. *ACS Appl Mater Interfaces* 13: 45935–45943.

52 Pan, J., Zhu, L., Han, J. et al. (2015). Mechanically tough and chemically stable anion exchange membranes from rigid-flexible semi-Interpenetrating networks. *Chem Mater* 27 (19): 6689–6698.

53 Lan, C., Fang, J., Guan, Y. et al. (2015). Cross-linked anion exchange membranes with pendent quaternary pyrrolidonium salts for alkaline polymer electrolyte membrane fuel cells. *J Power Sources* 295: 259–267.

54 Pana, W.H., Lue, S.J., Chang, C.M. et al. (2011). Alkali doped polyvinyl alcohol/ multi-walled carbon nano-tube electrolyte for direct methanol alkaline fuel cell. *J Membr Sci* 376: 225–232.

55 Li, X.H., Yu, Y.F., and Meng, Y.Z. (2013). Novel quaternized poly(arylene ether sulfone)/nanoZrO_2 composite anion exchange membranes for alkaline fuel cells. *ACS Appl Mater Interfaces* 5: 1414–1422.

56 Zuo, X., Yu, S., Xu, X. et al. (2009). New PVDF organic–inorganic membranes: the effect of SiO_2 nanoparticles content on the transport performance of anion-exchange membranes. *J Membr Sci* 340: 206–213.

57 Shalini, S., Amaranadh, J., Mahendra, K. et al. (2010). A green method for the preparation of highly stable organic-inorganic hybrid anion-exchange membranes in aqueous media for electrochemical processes. *Polym Chem* 1: 1302–1312.

58 Liu, L., Tong, C., He, Y. et al. (2015). Enhanced properties of quaternized graphenes reinforced polysulfone based composite anion exchange membranes for alkaline fuel cell. *J Membr Sci* 487: 99–108.

59 He, G., Xu, M., Li, Z. et al. (2017). Highly hydroxide-conductive nanostructured solid electrolyte via predesigned ionic nanoaggregates. *ACS Appl Mater Interfaces* 9: 28346–28354.

60 Zhang, S., Akihiko, T., and Hidetoshi, M. (2018). Nanofibers as novel platform for high-functional ion exchangers. *J Chem Technol Biotechnol* 93: 2791–2803.

61 Park, A.M. and Pintauro, P.N. (2011). Electrospun composite membranes for alkaline fuel cells. *ECS Trans* 41 (1): 1817–1826.

62 Gong, X., He, G., Yan, X. et al. (2015). Electrospun nanofiber enhanced imidazolium functionalized polysulfone composite anion exchange membranes. *RSC Adv* 5: 95118–95125.

63 Wang, L., Dou, L., and Zhang, S. (2013). Nanofiber-based poly(aryl ether sulfone) containing guanidinium groups as novel anion-exchange membranes. *J Polym Res* 20: 232–236.

64 Biancolli, A.L., Go, H.D., Wang, L.Q. et al. (2018). ETFE-based anion-exchange membrane ionomer powders for alkaline membrane fuel cells: a first performance comparison of head-group chemistry. *J Mater Chem A* 6: 24330–24341.

65 Marino, M.G. and Kreuer, K.D. (2015). Alkaline stability of quaternary ammonium cations for alkaline fuel cell membranes and ionic liquids. *ChemSusChem* 8 (3): 513–523.

66 Pham, T.H., Olsson, J.S., and Jannasch, P. (2017). N-spirocyclic quaternary ammonium ionenes for anion-exchange membranes. *J Am Chem Soc* 139 (8): 2888–2891.
67 Olsson, J.S., Pham, T.H., and Jannasch, P. (2017). Poly(N,N-diallylazacycloalkane)s for anion-exchange membranes functionalized with N-spirocyclic quaternary ammonium cations. *Macromolecules* 50 (7): 2784–2793.
68 Si, Z., Qiu, L., Dong, H. et al. (2014). Effects of substituents and substitution positions on alkaline stability of imidazolium cations and their corresponding anion-exchange membranes. *ACS Appl Mater Interfaces* 6 (6): 4346–4355.
69 Gu, S., Cai, R., Luo, T. et al. (2009). A soluble and highly conductive ionomer for high-performance hydroxide exchange membrane fuel cells. *Angew Chem Int Ed* 48 (35): 6499–6502.
70 Noonan, K.J.T., Hugar, K.M., Kostalik, H.A. 4th et al. (2012). Phosphonium-functionalized polyethylene: a new class of base stable alkaline anion exchange membranes. *J Am Chem Soc* 134 (44): 18161–18164.
71 Zha, Y., Disabb-Miller, M.L., Johnson, Z.D. et al. (2012). Metal-cation-based anion exchange membranes. *J Am Chem Soc* 134: 4493–4496.
72 Li, L., Wang, J., Hussain, M. et al. (2021). Side-chain manipulation of poly (phenylene oxide) based anion exchange membrane: alkoxyl extender integrated with flexible spacer. *J Membr Sci* 624: 119088–110998.
73 Shukla, G. and Shahi, V.K. (2019). Amine functionalized graphene oxide containing C16 chain grafted with poly(ether sulfone) by DABCO coupling: anion exchange membrane for vanadium redox flow battery. *J Membr Sci* 575: 109–117.
74 Yang, P.Y., Zhang, B., Wu, H. et al. (2019). Imidazolium-functionalized carbon nanotubes crosslinked with imidazole poly(ether ether ketone) for fabricating anion exchange membranes with high hydroxide conductivity and dimension stability. *Electrochim Acta* 318: 572 580.
75 Lin, B.C., Xu, F., Su, Y. et al. (2019). Ether-free polybenzimidazole bearing pendant imidazolium groups for alkaline anion exchange membrane fuel cells application. *ACS Appl Energy Mater* 3 (1): 1089–1098.
76 Buggy, N.C., Du, Y., Kuo, M.C. et al. (2020). A polyethylene-based triblock copolymer anion exchange membrane with high conductivity and practical mechanical properties. *ACS Appl Polym Mater* 2 (3): 1294–1303.
77 Mayadevi, T.S., Sung, S., Varghese, L. et al. (2020). Poly(meta/para-terphenylene-methyl piperidinium)-based anion exchange membranes: The effect of backbone structure in AEMFC application. *Membranes* 10 (11): 329–345.
78 Liu, D., Lin, L.M., Xie, Y.J. et al. (2021). Anion exchange membrane based on poly(arylene ether ketone) containing long alkyl densely quaternized carbazole derivative pendant. *J Membr Sci* 623: 119079–119087.
79 Ouadah, A., Xu, H., Luo, T. et al. (2017). Synthesis of novel copolymers based on p-methylstyrene, N,N-butylvinylimidazolium and polybenzimidazole as highly conductive anion exchange membranes for fuel cell application. *RSC Adv* 7: 47806–47817.

80 Wang, L., Brink, J.J., and Varcoe, J.R. (2017). The first anion-exchange membrane fuel cell to exceed 1 W cm^{-2} at 70 °C with a non-Pt-group (O_2) cathode. *Chem Commun* 53: 11771–11773.

81 Akiyama, R., Yokota, N., and Miyatake, K. (2019). Chemically stable, highly anion conductive polymers composed of quinquephenylene and pendant ammonium groups. *Macromolecules* 52 (5): 2131–2138.

14

Membrane Electrode Assembly Preparation for Anion Exchange Membrane Fuel Cell (AEMFC): Selection of Ionomers and How to Avoid CO_2 Poisoning

Weihong Yang, Qiuyu Zhang, and Yi Yan

Northwestern Polytechnical University, School of Chemistry and Chemical Engineering, Department of Chemistry, Dongxiang Road, Chang'an District, Xi'an 710129, PR China

14.1 The Preparation of Membrane Electrode Assembly

Membrane electrode assembly (MEA) is the core component of anion exchange membrane fuel cells (AEMFCs) and consists of the anion exchange membrane (AEM) layer, the anode and cathode catalyst layers (CLs), and the gas diffusion layers (GDLs) (Figure 14.1). The GDL acts as a transmission site for reactants and products during electrochemical reactions [1]. The hydroxide ions are first generated by the oxygen reduction reaction (ORR) at the cathode. Afterward, the hydroxide ions pass through the AEM layer and transit to the anode, where the oxidation reaction happens between the fuels (hydrogen, methanol, etc.) and the CL. The CL usually combines the electrochemical catalyst and the ionomers, resulting in a three-phase interface for the electrochemical reactions [2].

According to the different methods of catalyst loading on the AEM, the preparation methods of the MEA can be divided into two categories, gas diffusion electrode (GDE) and catalyst-coated membrane (CCM) (Figure 14.2). The catalyst is first loaded on the surface of the diffusion layer to form a GDE, and then the MEA is formed by hot-pressing between GDE and AEM, which is named the GDE method. According to the specific preparation process, the GDE strategy is composed of the traditional GDE method, dry spray coating [3], electrochemical deposition [4], and electrophoresis [5]. Generally, the traditional GDE method to prepare the MEA is as follows (Figure 14.2A): the catalyst and ionomer are dispersed by mechanical stirring and ultrasonication. The solution composed of the catalyst and ionomer is

Alkaline Anion Exchange Membranes for Fuel Cells: From Tailored Materials to Novel Applications, First Edition. Edited by Jince Thomas, Alex Schechter, Flavio Grynszpan, Bejoy Francis, and Sabu Thomas.

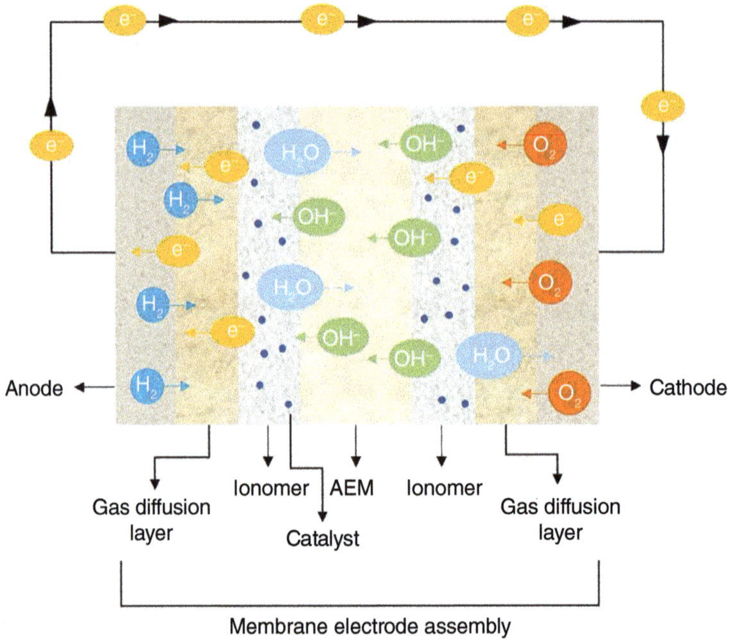

Figure 14.1 The structure of MEA for the AEMFC.

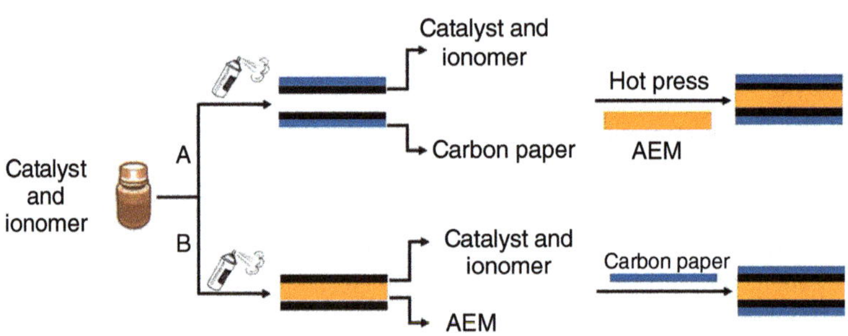

Figure 14.2 Different methods of loading catalysts on the AEM: (A) GDE; (B) CCM.

sprayed uniformly on the carbon paper. The membrane is sandwiched or hot-pressed between two pieces of carbon paper, and then the preparation of the membrane electrode is completed [6]. Since there is no liquid solvent involved, the dry spray coating method is quite simple. Moreover, it is easy to control the catalyst loading by adjusting the spray gas flow, which is suitable for commercial production. The GDE prepared by the electrochemical deposition method shows a higher activity for both oxygen reduction and hydrogen oxidation than the electrodes prepared via the conventional method [7]. A large electrode can be easily produced by electrophoresis, and the process of such a method is simple and cost-effective.

The CCM is formed by directly loading the catalyst on the AEM, which is then laminated with the GDL to fabricate the MEA. The traditional CCM method to prepare the MEA is as follows (Figure 14.2B) [8]: a well-dispersed catalyst ink is prepared by mixing the catalyst and ionomer solution using mechanical stirring and ultrasonication. Then, the as-prepared ink is coated onto the AEM by spraying to achieve the CCMs for the electrodes. Subsequently, the assembly of carbon paper and a CCM are used to afford the MEA. In addition, sputter deposition [9], catalyst ink transfer [10], and chemical-reduction [7] are also used in previous reports. Although the GDE method has been widely used in the current state, it still has certain defects. For example, the thickness of the catalytic layer is relatively large, the mixture of the catalyst and ionomers is nonuniform, and the stability of the interface between the CL and AEM still needs to be improved. The CCM method significantly improves the interface stability between the CL and the AEM, which can also afford a thinner CL. The CCM method to prepare the MEA may be the future direction of the MEA preparation.

The GDE being the most robust method involves coating of the catalyst on wet-proofed GDL using brush coating or spray coating method [11]. MEAs with areas $<50\,cm^2$ are normally fabricated using such a technique. The primary reason for employing the GDE method is its ease of fabrication (especially brush-coated GDEs) compared to other techniques. Such a method is also proven to have the least loss of catalyst, as discussed earlier. CCM, on the other hand, is also a well-established technique commonly used for fabricating MEAs of larger active areas, thereby reducing the catalyst loss per unit area. Triple-phase boundary created by such a technique is also claimed to be better as the CL is brought much closer to the membrane compared to the GDE. The CCM method involves a dedicated equipment for coating the catalyst over the membrane surface and therefore is found exorbitant [11].

Other than the GDE and CCM methods, Li and coworkers [12] proposed a novel MEA fabrication approach. The dispersion was coated on a substrate to form a wet film, and then a GDE was placed on the wet film. After drying, separation from the substrate yields a membrane-coated electrode. The MEA fabrication method not only achieved excellent interfacial contact between the membrane and the CL but also prevented overfilling with the ionomer. Applying this novel MEA method, a high peak power density of $2636\,mW\,cm^{-2}$ was achieved. Mehmood and Ha [13] developed a new roll-press for MEA fabrication and compared the performance of respective MEAs with that of the conventional flat press decal method. The MEA produced by a roll-press (R-MEA) delivers an excellent single-cell performance with power densities more than 30% higher than that fabricated using a flat-press. The new method considerably improved catalyst active sites in both electrodes and renders a high cathode porosity. In addition, an innovative way of reducing the membrane and charge transfer resistance was found by spray coating the ionomer dispersion on the CL followed by spraying the CL [11, 14]. The average coating sequence involved coating of an anode/cathode on the GDL, followed by spray coating the ionomer dispersion for a definite number of turns to develop a uniformly thick membrane over the area. The other electrode was then spray coated by masking the solid membrane. Moreover, the electrode–electrolyte interface is found better than that of the conventional MEAs.

14.2 Selection of Ionomers

As the catalyst binder in the MEA, the ionomer is a key component for electrochemical energy conversion and storage technology [15]. In comparison with AEMs, the study on ionomers is fairly limited in the literature. In the proton exchange membrane fuel cell (PEMFC), ionomers are commonly perfluorinated sulfonic acid polymers, which have the same structure as that of the proton exchange membrane (PEM). When it comes to anion exchange ionomers (AEIs), the requirements for AEM and AEI are not always the same due to their different roles (Table 14.1). For example, AEM requires good gas barrier properties and limited water uptake, as well as lower membrane swelling, while AEI requires high water permeability, minimal interaction with electrocatalysts, and sufficient electro-oxidative stability [16]. Therefore, using the same structure of AEI and membrane is not always a good choice for AEMFCs. Moreover, the water management of AEMFCs is more complicated than that of PEMFCs. The cathode and anode have different requirements for water, and the structure requirements for AEI are also different. The cathode is prone to be dried, while the anode tends to overflow under high current density. Although AEI is a key component in AEMFCs, the related research is not well-developed, and only a few groups have studied AEI. So far, only a few types of AEI have been commercialized. Generally, AEI is synthesized by researchers individually to fabricate the MEA.

There are two approaches that have produced effective CLs in operating AEMFCs [17]. The first approach starts by dispersing catalyst particles with solubilized anionic polymers to create the catalyst ink. This approach is very similar in style to PEMFC inks, but not in concept because this approach relies on the removal of the solvent to set up the physical structure of the ionomer and its incorporation with the catalyst. If the incorrect solvent is selected or solvent removal is not carefully controlled, the resulting CL could be too dense or the structure of the ionomer in the CL might not facilitate facile charge transfer. In either case, the resulting AEMFC would experience severe performance limitations. The second approach is

Table 14.1 A comparison of properties between AEM and AEI.

Properties	AEM	AEI
Ionic conductivity (σ)	High σ	High σ
Water absorption	Suitable water uptake to maintain the durability	High water permeability
Swelling ratio	Low SR	Low–intermediate SR
Mechanical properties	Suitable tensile strength and toughness	No request
Gas permeability	Low gas permeability	High gas permeability
Solubility	No request	Good solubility/dispersion
Alkaline stability	Highly stable	Highly stable

far less common, though the one that has led to the highest reported peak power densities, achievable current densities, and stability. In this approach, large solid ionomer particles were integrated with the catalyst to make the electrode. There has only been one such "solid ionomer" that has been reported in the literature to date: 25 μm ethylene tetrafluoroethylene (ETFE) powder that has been radiation-grafted and quaternized [18]. To make CLs, the ETFE powder was typically exposed to grinding steps with and without the catalyst, and then the solvent (water + isopropyl alcohol) was added in stages and then sonicated to make the catalyst ink that was sprayed onto a low-Teflon GDL.

14.2.1 Commercial Ionomers

After the years of development, Tokuyama Corporation reported two types of cationic ionomers (product codes: A3Ver2 and AS-4), and Fumatech Corporation also reported FAA membranes and Fumion ionomers (Table 14.2) in 2008 [19]. The first study on fuel cell performance with different A3Ver2 ionomer content was investigated by Bunazawa and Yamazaki, and they revealed that the MEA containing 45.4% A3Ver2 ionomer can achieve the peak power density of 58.9 mW cm^{-2} [20]. The peak power density of the AEMFCs rarely exceeded 60 mW cm^{-2} before 2008 due to the unsatisfied performance of AEM and ionomers. In 2009, Gu et al. designed a soluble tri(2, 4, 6-trimethoxyphenyl) polysulfone-methylene quaternary phosphonium hydroxide (TPQPOH) ionomer and used for FAA commercial AEM, which achieved a peak power density of 196 mW cm^{-2} [21]. It is worth noting that Fumion is an aromatic ether-containing polyelectrolyte, which is easily degraded under high pH conditions. The degradation of Fumion reduces the local pH value of the catalyst/ionomer interface and negatively affects the ionic conductivity, which results in poor fuel cell performance and stability [19]. A series of highly conducting cross-linked AEMs were synthesized based on quaternary phosphonium-functionalized poly(ether ketone) (QPPEEK) polymer skeleton and poly(ethylene glycol) (PEG) as the cross-linker [22]. Fumion ionomers were used to fabricate the MEA, and the highest peak power density of 154 mW cm^{-2} was achieved. In addi-

Table 14.2 Technical data sheet of Fumion® FAA-3 solution.

Fumion® solution	FAA-3
Ionomer type	Anion exchange ionomer
Polymer type	Polyaromatic polymer
Appearance/color	Brown, transparent solution
Solvent	*N*-methyl-2-pyrrolidone
Concentration (wt%)	8–12
Solution viscosity at 25 °C (mPa·s)	<20
Functional group	Quaternary ammonium group
Additive (defoamer, wetting, anti-cratering)	Optional

tion, Acta I2 ionomer was also used in quaternary ammonium poly (styrene-ethylene/butylene-styrene) (QASEBS) AEM to fabricate the MEA, and the peak power density can reach $124\,mW\,cm^{-2}$ [23]. However, the output was still lower than the ionomer with the same structure of QASEBS AEM.

AEMFCs based on commercial ionomers (e.g. FAA-3SOLUT-10, AS-4) resulted in fuel cell performance of around $200–500\,mW\,cm^{-2}$, far below the recently reported high AEMFC power density ($>2000\,mW\,cm^{-2}$). More efforts should be focused on the investigation of AEI to achieve a high output of fuel cells.

14.2.2 Custom-made Ionomers

Most of the commercial AEMs and ionomers (e.g. FuMA-Tech, Tokuyama, Gen) exhibit poor lifetime and low power density of AEMFCs ($100–500\,mW\,cm^{-2}$), and thus, a deeper and more thorough understanding of the ionomer properties within the MEA is desired. The key point to constructing the effective CL is that the catalyst ink, especially ionomers, should have a good dispersion and fast solidification during the coating process. Therefore, the solvent to dissolve ionomers should have a high dielectric constant and a low boiling point at the same time [24]. Typically, alcohols with a low boiling point, such as isopropanol or ethanol and deionized water, have been well-proven to be a good solvent system to dissolve ionomers. To enhance the conductivity and alkaline stability of AEMFCs, partially fluorinated poly(arylene ether) ionomers with functionalized claw-type side chains were prepared [25]. Furthermore, the single cell using FPAE-3B-3.0-PD achieved the highest power density of $156\,mW\,cm^{-2}$ under a current density of $350\,mA\,cm^{-2}$ at 80 °C, and the fuel cell performance still needed to be improved. Shao and coworkers synthesized a novel ionomer based on a triblock copolymer with high conductivity and good durability, and the MEA based on this ionomer reached a higher peak power density of $375\,mW\,cm^{-2}$, which was three times higher than that of commercially available Acta I2 ionomer [23]. Hickner and coworkers [26] prepared a series of ionomers with multiple cations per side chain, and the H_2/O_2 fuel cell using T20NC6NC5N (IEC = $2.52\,mmol\,g^{-1}$) AEM and ionomer achieved the highest peak power density of $364\,mW\,cm^{-2}$.

In recent years, the output of fuel cells has greatly increased owing to the optimization of the ionomer, MEA fabrication, and working conditions (Table 14.3). In 2014, Zhuang and coworkers [31] reported the AEMFC cell performance of $1.0\,W\,cm^{-2}$ based on QAPS-S_{14} (IEC = $1.0\,mmol\,g^{-1}$, 20 wt% in the CL) ionomer and Pt–Ru/C anode at 60 °C under fully humidified H_2 and O_2 conditions. Their result was the highest record in the literature and was comparable to the performance of the state-of-the-art PEMFCs. As the commonly used cation of the ionomer, Xu and coworkers synthesized a series of ionomers based on quaternary ammonium cation functionalized poly(2,6-dimethyl phenylene oxide) (QPPO), and a maximum power density of $322\,mW\,cm^{-2}$ at 60 °C was achieved [35]. In addition, the QPPO ionomer was also used in the MEA based on poly(vinyl alcohol) hydrogel AEM, and the fuel cell performance demonstrated a remarkable peak power density of $715\,mW\,cm^{-2}$.

Table 14.3 A series of ionomers with peak power density (P_{max}) higher than 1000 mW cm^{-2} under H_2/O_2 condition are given in the table below.

Ionomer	AEMFC test condition	P_{max} (W cm^{-2})	References
GT78	200 kPa backpressure, H_2/O_2 (1 l min^{-1}) @80 °C	2.7	[27]
PDTP-75	2.0 bar backpressure, H_2/O_2 (1 l min^{-1}) @80 °C	2.58	[28]
QPC-TMA	H_2/O_2 (0.8/1 l min^{-1}) @60 °C	1.61	[29]
QAPPT	0.1 MPa backpressure, H_2/O_2 (120 ml min^{-1}) @80 °C	1.45	[30]
aQAPS-S14	0.1 MPa backpressure, H_2/O_2 (400 ml min^{-1}) @60 °C	1.0	[31]
ETFE powder	H_2/O_2 (1 l min^{-1}) @80 °C	2.0	[32]
ETFE-g-poly (VBTMA$^+$Cl$^-$)	H_2/O_2 (1 l min^{-1}) @60 °C	1.40	[33]
ETFE–BTMA	H_2/O_2 (1 l min^{-1}) @60 °C	1.7	[34]

Elastomeric ionomers were prepared by acid-catalyzed Friedel–Crafts alkylation of the polystyrene block of polystyrene b poly(ethylene-*co*-butylene)-*b*-polystyrene (SEBS) using bromoalkylated tertiary alcohols and triflic acid as the catalyst, followed by quaternization with trimethylamine [36], and a peak power density of 520 mW cm^{-2} was achieved at 60 °C under H_2/O_2 conditions using SEBS-based ionomers. Similarly, Shao and coworkers [37] also prepared the ionomer based on quaternized SEBS (polymer concentration of 8.7 mg ml^{-1}), and the AEMFC performance reached 721.7 mW cm^{-2} at the cell temperature of 60 °C under H_2/O_2 condition. As a kind of excellent engineering plastic, poly(arylene ether sulfone)s with different branching degrees were synthesized as ionomers, and a platinum-catalyzed fuel cell based on hyper branched methoxyl polyaryl ether sulfone (HBMPES)-8 ionomer achieved a peak power density of 160 mW cm^{-2} [38].

Mustain and coworkers [34] created new electrode compositions by systematically manipulating the ionomer and carbon content in the anode CL, which alleviated the mass transport behavior limitations of H_2/O_2 AEMFCs and achieved a peak power density of 1900 mW cm^{-2}. They also synthesized three poly(norbornene) tetrablock copolymer ionomers [27], and the ionomers were employed in the AEMFC anode and cathode electrodes, respectively. The H_2/O_2 fuel cell with GT78 ionomer in the anode and GT32 ionomer in the cathode achieved the highest peak power density of 3.2 W cm^{-2}. This report indicated that the requirements of the ionomer in the cathode and anode are different and therefore greatly affected the output of fuel cells. Aryl ether-free ionomers and AEMs based on 1,2-diphenylethane (DP)-containing poly(aryl piperidinium)s (PAPs) were prepared and employed to fabricate the MEA. Ionomers with low phenyl content and high water permeability displayed an excellent peak power density of 2580 mW cm^{-2} [28].

Besides the soluble ionomers, powder ionomers were also reported. The peak power density of 1.8 W cm^{-2} for an H_2/O_2 AEMFC was achieved based on radiation-grafted ETFE AEM and ionomers [39]. The corresponding high output was mainly attributed to the better water balance achieved through electrode optimization, using the powder form ionomers. Lee and coworkers prepared a novel ionomer based on quaternized poly-carbazole (QPC-TMA) with a rigid ether-free and curved backbone structure comprising carbazole monomers [29]. AEMFCs using QPC-TMA ionomer demonstrated an excellent performance of 1610 mW cm^{-2}.

Figure 14.3 summarizes the structure of current AEIs, and Figure 14.4 is the schematic representation for fabrication of the MEA. Further optimization of ionomer selection, dispersion control, and preparation techniques should be developed to obtain AEMFCs with high performance and stability.

14.3 Effect of CO_2 on AEMFCs

A characteristic of the AEMFC is their susceptibility to poisoning by ambient atmospheric CO_2, which is widely investigated to improve performance and stability [40]. Much efforts have been devoted to exploring the primary mechanisms for the voltage loss caused by CO_2 poisoning [41]: (i) slow mobility of carbonate salts, leading to high area-specific resistance; (ii) carbonates accumulate at the anode, leading to a Nernstian thermodynamic shift and anode potential increase; (iii) a lower OH^- concentration in the anode, leading to an increase in charge transfer resistance. Similarly, the ambient atmospheric CO_2 also has a great impact on the performance of AEMFCs, and more efforts should be devoted to solving this issue.

14.3.1 Effect of CO_2 on *Ex Situ* Measured Conductivity

As reported previously, due to the larger ionic radius of CO_3^{2-} and HCO_3^- compared to that of OH^- anions, CO_3^{2-} and HCO_3^- have a detrimental effect on the ionic conductivity of the AEM, caused by their lower diffusion coefficients and lower mobility [42]. The ionic mobilities of OH^-, CO_3^{2-}, and HCO_3^- were reported to be 20.64, 7.46, and $4.61 \times 10^8\,m^2 s^{-1}\,V^{-1}$, respectively [43]. The ionic conductivity is not only dependent on the ionic radius but also on the number of hydration layers. It is reported that the hydration radii for solvated OH^-, CO_3^{2-}, and HCO_3^- in an aqueous solution are 3.0, 3.94 [44], and 5.6 Å [45], respectively. Due to the highest hydration radius and its $^-1$ charge, HCO_3^- has the biggest influence on the decrease in ionic conductivity of the AEM. In addition, the dissociation equilibrium of anions and fixed cationic groups also has great effects on the accurate conductivity of the AEM. It is calculated that about 32% of the total charge was free in the OH^- form AEM, whereas that number reduced to only 25% in the mixed CO_3^{2-}/HCO_3^- AEM [46]. Therefore, the ionic conductivity of the AEM in OH^- form decreases obviously upon contact with any CO_2-containing gas, including ambient air.

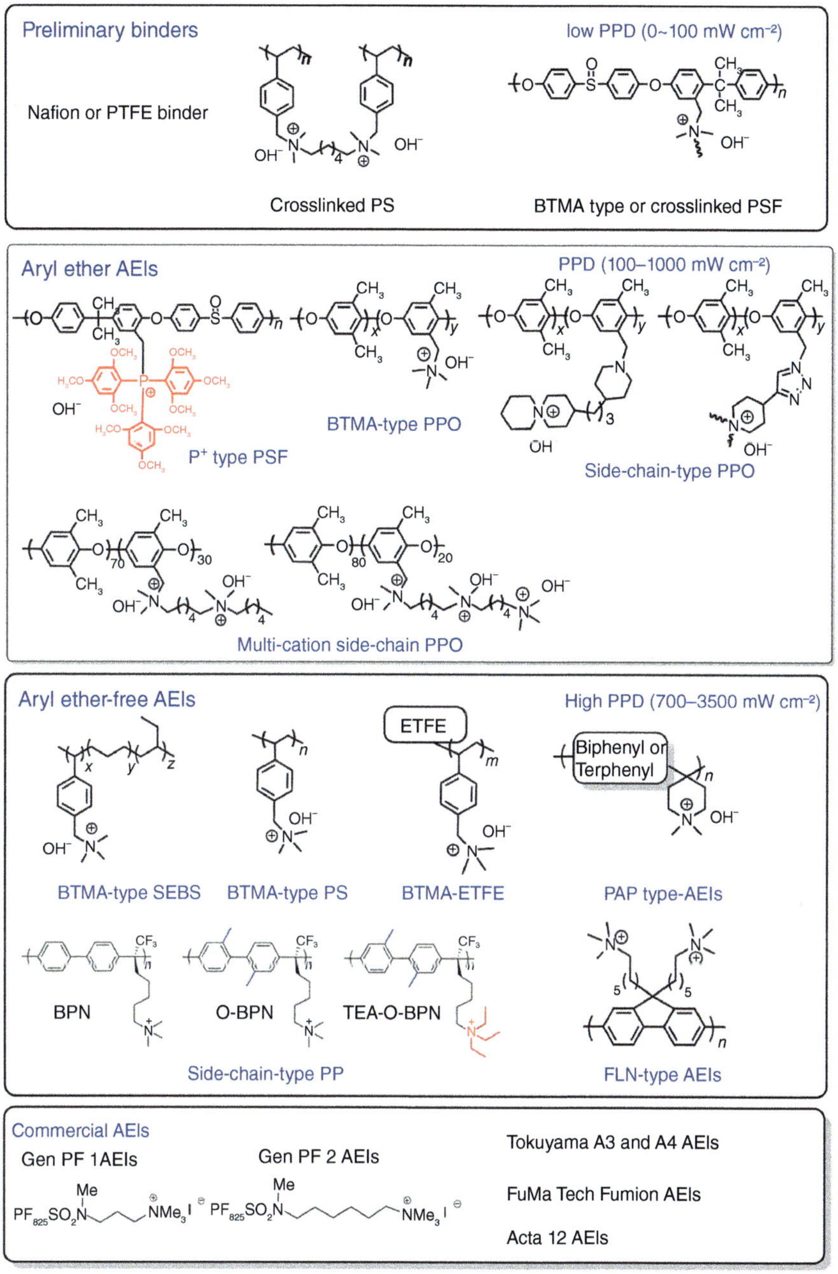

Figure 14.3 Summary of AEIs and binders in current research. Source: Reproduced with permission [16]. Copyright 2021, Elsevier.

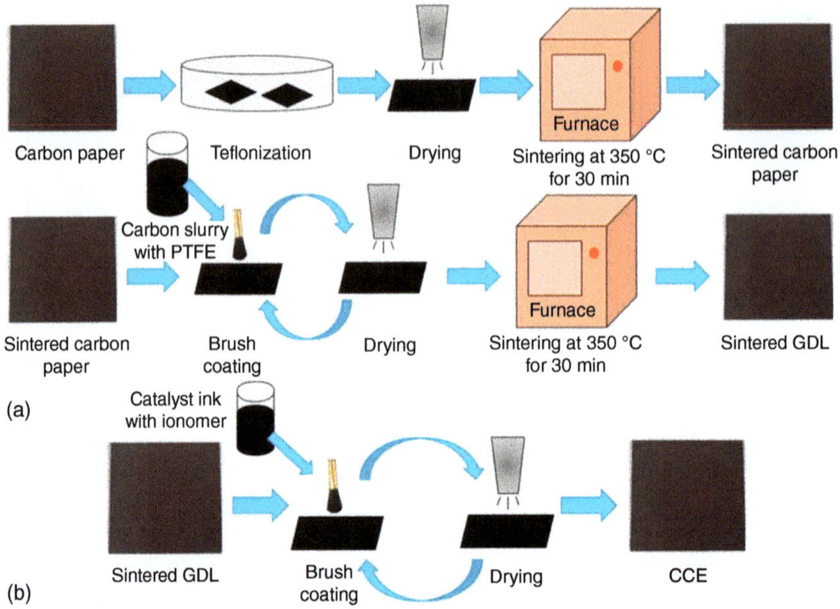

Figure 14.4 Schematic representation for fabrication of the MEA. Source: Reproduced with permission [11]. Copyright 2020, Elsevier.

14.3.2 Effect of CO_2 on Electrochemical Reactions on the Electrodes

In addition to the effect on ionic conductivity, CO_2 may also influence the electrochemical reactions of the membrane electrodes. When CO_2-containing ambient air is present in the electrode, the ORR may occur as follows:

$$O_2 + CO_2 + 4e^- \rightarrow 2CO_3^{2-}$$

Such a result may cause CO_3^{2-} anions to be adsorbed on the electrode, which may reduce the active area of the electrode, as well as decrease the observed ORR rate. In addition, the O_2 diffusion coefficient in AEM with CO_3^{2-} form is relatively lower in comparison with the AEM in OH^- form, which also decreases the electrochemical reaction rate. Inaba et al. found that the anode kinetics and activity may be disrupted by the carbonate of hydroxide, especially at low current densities, which is attributed to the fact that OH^- is a direct reactant in the hydrogen oxidation reaction (HOR) [47]. The accumulation of HCO_3^-/CO_3^{2-} anions in the anode causes the decrease of hydroxide conductivity and activity of OH^-. However, the cathode gas has little effect on the electrode overpotential with the CO_2 content in the range of 100–5000 ppm.

14.3.3 Effect of CO_2 on Fuel Cell Performance

Almost all the fuel cell performance determinations have been performed on pure hydrogen and oxygen or CO_2-free air, and only few reports have investigated the

Table 14.4 A comparison of the effect of CO_2 on AEMFC performance.

AEM	Anion form	Operating temperature (°C)	Operating gas	Peak power density ($mW\,cm^{-2}$)	References
aQAPS-S8	OH^-	60	O_2	610	[48]
			Ambient air	320	
Tokuyama A201	OH^-	50	O_2	290	[49]
			Ambient air	125	
QAPSF	CO_3^{2-}	25	O_2	2.4	[50]
			O_2+CO_2	4.4	
PDTP-25	OH^-	80	O_2	2580	[28]
			Ambient air	1380	

effect of CO_2 on cell performance. As expected, the peak power density of AEMFCs under oxygen or CO_2-free air in the cathode was higher than that of ambient air (Table 14.4). For example, Lee et al. obtained a maximum peak power density of 2.58 $W\,cm^{-2}$ under pure O_2 in the cathode, whereas only 1380 $mW\,cm^{-2}$ was obtained under ambient air in the cathode [28]. In addition, AEMs with CO_3^{2-} counterion indicate much lower output even under pure O_2 condition, demonstrating that the ion species significantly affect the ionic conductivity and corresponding fuel cell performance [50].

Gottesfeld et al. changed the cathode gas from CO_2-free air to ambient air, and the voltage at a current density of 400 $mA\,cm^{-2}$ decreased by almost 50% [51]. Similar results were also reported like the cell voltage decreased by 30% when changing clean/synthetic air to ambient air in the cathode. In addition, the cell resistance also increased with the increase of CO_2 content. The cell resistance increased by 1.7 times and 2.3 times when adding 100 and 5000 ppm CO_2 to pure O_2, respectively [47]. Even if CO_2 exists in a small amount, such as 100 ppm (much lower than the content in ambient air), it can greatly affect the cell performance, and all the reports demonstrate that CO_2 has a clear negative impact on fuel cell performance.

14.4 Strategies to Avoid CO_2 Poisoning

The negative effect of CO_2 on fuel cell performance needs to be addressed. More and more research groups are investigating how to overcome the impact of CO_2, and a series of solutions are proposed, such as self-purging, filtering the feeding air, increasing the operating current density, and optimizing the start-up protocol.

14.4.1 Reducing HCO_3^-/CO_3^{2-} Concentration Through Self-purging

Self-purging refers to the removal of HCO_3^-/CO_3^{2-} anions from the AEM by releasing CO_2 gas in the anode in an operating fuel cell [52]. As shown in Figure 14.5,

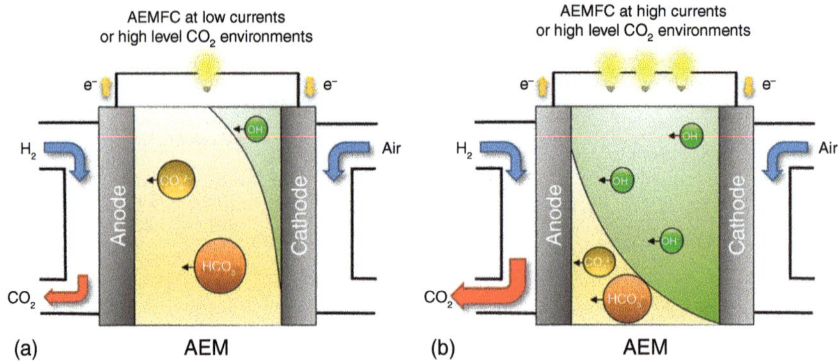

Figure 14.5 The self-purging process of AEM under different current densities. When at low current density, OH^- is slowly formed in the cathode, but quickly converted into HCO_3^-/CO_3^{2-} (a), whereas at high current density, OH^- formation surpasses the carbonation process, and HCO_3^-/CO_3^{2-} is released at the anode as CO_2 (b). Source: Reproduced with permission [42]. Copyright 2018, Wiley.

when the current density is at a relatively high level, the high rate of the ORR in the cathode results in the formation of OH^- anions. The rate of OH^- production in the cathode overcomes the rate of CO_2 absorption into the membrane, and therefore, the overall result is the increase in OH^- concentration at the expense of HCO_3^-/CO_3^{2-} concentration [42]. However, the self-purging process requires several hours after contamination, which was attributed to the fact that the decarbonation of the AEMFC does not occur through a direct electrochemical reaction [53]. Adams et al. found that the CO_3^{2-} content of the AEM decreased after the AEMs in CO_3^{2-} form operated in an H_2/air AEMFC by Raman analysis [54]. Yanagi and Fukuta measured the different anion concentrations in AEM with HCO_3^- form before and after AEMFC operation (H_2/clean air). Interestingly, the anion concentration after the fuel cell operation only included OH^- and CO_3^{2-} (the ratio between these two anions is ca. 1 : 1) [52]. These studies indicated that HCO_3^-/CO_3^{2-} can be removed from AEM to achieve the self-purging process. In addition, the ohmic resistance of AEM in HCO_3^-/CO_3^{2-} form decreased at high current density during the fuel cell test. The start-up of the fuel cell can be operated for a short time under high current density/low voltage to remove HCO_3^-/CO_3^{2-} anions [33], and the MEA should be assembled using AEMs in OH^- form to achieve high performance.

14.4.2 Increasing the Current Density to Improve the Outlet of CO_2

The current densities have a great effect on the performance of fuel cells: as the current densities increase, the CO_2 flow rate at the anode outlet increases [49, 55], which indicates that less CO_2 accumulated in the AEMFC. At low current density, the anode overpotential (operated with pure H_2) of the fuel cell was significantly affected by CO_2 fed to the cathode. In addition, the rate of the ORR in the cathode increases with the increase of current densities, and the self-purge mechanism is the process of removing HCO_3^-/CO_3^{2-} from the AEMFC under high current conditions. Moreover, the ohmic resistance of AEM fuel cells decreased at high current density, according to previous reports.

14.4.3 Filtered the Feeding Air

Owing to the effect of CO_2 on fuel cell performance, the most effective way is to filter the feeding air. All the experiment results indicated that the performance of the AEMFC operated under pure H_2/O_2 is largely higher than that of ambient air, and there is an urgent need for alternative technologies and novel materials for energy- and cost-effective oxygen separation from air. Zeolite molecular sieves Ag-ETS-10 have been used for O_2 adsorption and separation [56], and the purity of O_2 can exceed 99.7% after the separation of air. In addition, solid-state adsorbents of crystalline metal–organic frameworks (MOFs) with a high surface area have also been reported for O_2 separation. A redox-active organometallic molecule, ferrocene, was used as a precursor to form O_2-selective Fe^{2+} species within a thermochemically stable, highly porous MOF, MIL-101 (materials from Institute Lavoisier [MIL]) to improve O_2-adsorption and O_2/air-separation selectivity at room temperature [57]. The as-prepared material was exceptionally selective toward oxygen, but showed almost rare absorption of nitrogen, argon, and carbon dioxide, indicating great O_2 separation ability. In addition, a novel MOF-based composite membrane (PAN-γ-CD-MOF-PU membrane) was successfully fabricated by a facile and fast spin-coating method [58]. The formation of a unique γ-CD-MOF layer greatly improved the separation ability of CO_2, and the selectivity of CO_2/O_2 is 154.28. New strategies should be focused on the mitigation of carbonate formation and the influence of CO_2 more efficiently through developments in materials science and fuel cell operation protocols.

14.5 The Improvement of AEMFC Output

In only a decade, the reported AEMFC performance showed significant improvements. This striking performance improvement is mainly owing to the increased progress made in the development of AEMs with higher anion conductivity.

14.5.1 Electrode Optimization

The CL has great influences on the cell performance, which is composed of the catalyst, ionomer, and interface. In comparison with the optimization of operating conditions, the design of CL components and their morphological structure is crucial for optimizing the mass transfer and catalytic performance of the fuel cell [17]. Unlike PEMFCs, the electrode optimization of AEMFCs is more difficult. Nafion® possesses unique membranes and ionomers, while there is no standard commercial membrane and ionomer with high performance for AEMFCs. Generally, there are two effective methods to prepare the AEMFC CL. The commonly used method is to disperse catalyst particles with dissolved anionic polymers to produce the catalyst ink. This method is highly dependent on the removal of solvents to construct the physical structure of the ionomer and combine it with the catalyst. If the solvent is not suitable or the solvent is not completely removed, it will affect the density of the CL or the structure of the ionomer, which is not conducive to ion transport [59]. Only a few groups can obtain a peak power density of more than 1000 mW cm^{-2} through this method. The second method is to integrate the large solid PTFE, ionomer particles, and catalyst to fabricate the electrode. Such solid ionomers are rarely used, and only grafted and quaternized ETFE powders by radiation have been reported [18]. The advantages and disadvantages of these two methods are not fully understood. Moreover, the different methods need to be further explored to optimize the catalyst interface and fuel cell performance.

14.5.2 Catalyst Optimization

The anode and cathode CLs are positioned at the place where the chemical reactions take place. The CL acts as an active surface for the electrochemical reactions as well as a transport site for reactants and products of chemical reactions. AEMFCs enable the utilization of non-precious metal catalysts, which can further decrease the cost. However, the output of AEMFCs with non-precious metal catalysts has been inferior to that of platinum-group metal catalysts. In recent reports, AEMFCs used Pt–Ru/C at the anode and Pt/C at the cathode, indicating the obvious improvement of cell performance [32, 60]. Li and coworkers successfully synthesized non-platinum metal catalyst-carbon supported Ag nanoparticles (Ag/C) by a solution phase-based nanocapsule method [61]. A single H_2/O_2 cell performance with the Ag/C cathode catalyst demonstrated an open circuit potential of 0.98 V and a peak power density of 190 mW cm^{-2} at 80 °C, indicating the resulted catalyst was a promising and an alternative non-precious metal catalyst. Silver nanowires (Ag NWs) were synthesized using the polyol synthesis method, which demonstrated an extraordinarily high electrocatalytic activity toward the ORR in a three-electrode cell [62]. A peak power density of 164 mW cm^{-2} at 60 °C was achieved under the use of supportless Ag NWs as the cathode catalyst in an H_2/O_2 single cell. In addition, a novel and promising non-precious metal catalyst containing cobalt sulfide (Co–S) and graphene was synthesized and used as the cathode catalyst to fabricate the MEA in AEMFC [63]. A peak power density of 31 mW cm^{-2} was achieved at 30 °C under ambient pressure and still needed to be further improved.

14.5.3 Optimization of Operation Conditions

Besides the intrinsic property of AEM, it is worth noting that the performance of the fuel cell could be greatly affected by the operating conditions [6]. For example, the B-$C_{30\%}$-P AEM reached 197 mW cm^{-2} at 80 °C, whereas 152 mW cm^{-2} was obtained at 60 °C under a similar condition [64]. 1% GA membrane with 0.1 MPa backpressure showed the peak power density of 179 mW cm^{-2}, whereas the output of 146 mW cm^{-2} was achieved under no backpressure condition [65]. The application and significance of backpressure in AEMFCs have been widely studied by several researchers [66, 67]. The result of adding backpressure to the cathode side was a clear increase in performance. In addition, the operation of fuel cells under low relative humidity (RH) conditions gives substantial cost and performance benefits [68]. Such conditions are particularly desirable for automotive applications in which the size of the radiator and humidity are critical cost factors for an efficient fuel cell system. Nonetheless, it is not currently feasible to operate AEMFCs at low RH conditions because current materials for MEA cannot provide sufficient water for the ORR [69]. Based on this, further effort should be focused on the optimization of catalyst binder, MEA structure, and operating conditions in subsequent research.

14.6 Conclusions

AEMFC is a promising fuel cell technology, which can enable the future use of low-cost cells and non-precious metal catalysts. Recent work has shown that AEMFCs can achieve comparable initial performances to those of state-of-the-art PEMFCs. However, one of the major challenges is the lack of commercially available ionomers that could meet the general requirements of AEMFCs. Ionomers are expected to play a key factor for the output of AMECs, and more efforts should be focused on the ionomer conductivity, durability, and cost. In addition, while a fuel cell is operated in ambient air, the OH^- formed in the cathode electrode can react with the ambient CO_2 to form HCO_3^- or CO_3^{2-} anions, which significantly affect the cell performance. The presence of carbonates and bicarbonates is an issue that must be addressed to achieve high-performance AEMFCs. Studies on how to eliminate the effects of carbon dioxide in fuel cells are still required. This chapter summarizes the advancements of ionomers and carbonation elimination in AEMFCs, and it is important for researchers in the field to develop new strategies to limit the impact of effects and design ultrahigh-performance AEMFCs operating in ambient air.

References

1 Pan, Z.F., Chen, R., An, L., and Li, Y.S. (2017). Alkaline anion exchange membrane fuel cells for cogeneration of electricity and valuable chemicals. *J Power Sources* 365: 430–445.

2 Liu, L., Wang, C., He, Z. et al. (2021). An overview of amphoteric ion exchange membranes for vanadium redox flow batteries. *J Mater Sci Technol* 69: 212–227.

3 Talukdar, K., Helmly, S., Schulze, M. et al. (2019). Enveloping of catalyst powder by ionomer for dry spray coating in polymer electrolyte membrane fuel cells. *J Power Sources* 424: 82–90.

4 Hayashi, K. and Furuya, N. (2004). Preparation of gas diffusion electrodes by electrophoretic deposition. *J Electrochem Soc* 151: 354–357.

5 Furuya, N. (2003). A new method of making a gas diffusion electrode. *J Solid State Electrochem* 8: 48–50.

6 Yang, W., Xu, P., Li, X. et al. (2021). Mechanically robust semi-interpenetrating polymer network via thiol-ene chemistry with enhanced conductivity for anion exchange membranes. *Int J Hydrog Energy* 46: 10377–10388.

7 Fedkiw, P.S. and Her, W.-H. (1989). An impregnation-reduction method to prepare electrodes on Nafion SPE. *J Electrochem Soc* 136: 899–900.

8 Zhang, K., Gong, S., Zhao, B. et al. (2018). Bent-twisted block copolymer anion exchange membrane with improved conductivity. *J Membr Sci* 550: 59–71.

9 Shah, K., Shin, W.C., and Besser, R.S. (2004). A PDMS micro proton exchange membrane fuel cell by conventional and non-conventional microfabrication techniques. *Sensors Actuators B Chem* 97: 157–167.

10 Park, J.-Y., Lee, J.-H., Kim, J. et al. (2008). Stable operation of air-blowing direct methanol fuel cells with high performance. *J Power Sources* 179: 1–8.

11 Bhosale, A.C., Ghosh, P.C., and Assaud, L. (2020). Preparation methods of membrane electrode assemblies for proton exchange membrane fuel cells and unitized regenerative fuel cells: a review. *Renew Sust Energ Rev* 133: 110286–110305.

12 Chen, Z., Fu, T., Kong, X. et al. (2021). A novel impregnation-reduction method combined with galvanic replacement for fabricating low cost MEA with high performance for PEM fuel cells. *J Electrochem Soc* 168: 34522.

13 Mehmood, A. and Ha, H.Y. (2012). An efficient decal transfer method using a roll-press to fabricate membrane electrode assemblies for direct methanol fuel cells. *Int J Hydrog Energy* 37: 18463–18470.

14 Breitwieser, M., Bayer, T., Büchler, A. et al. (2017). A fully spray-coated fuel cell membrane electrode assembly using Aquivion ionomer with a graphene oxide/cerium oxide interlayer. *J Power Sources* 351: 145–150.

15 Kim, S., Ahn, C.-Y., Karuppannan, M. et al. (2022). Structural modification of electrode for anion exchange membrane fuel cell by controlling ionomer dispersion. *Int Energy Res* 46: 6471–6479.

16 Chen, N. and Lee, Y.M. (2021). Anion exchange polyelectrolytes for membranes and ionomers. *Prog Polym Sci* 113: 101345–101380.

17 Mustain, W.E. (2018). Understanding how high-performance anion exchange membrane fuel cells were achieved: component, interfacial, and cell-level factors. *Curr Opin Electrochem* 12: 233–239.

18 Poynton, S.D., Slade, R.C.T., Omasta, T.J. et al. (2014). Preparation of radiation-grafted powders for use as anion exchange ionomers in alkaline polymer electrolyte fuel cells. *J Mater Chem A* 2: 5124–5130.

19 Zhang, J., Zhu, W., Huang, T. et al. (2021). Recent insights on catalyst layers for anion exchange membrane fuel cells. *Adv Sci* 8: 2100284–2100310.

20 Bunazawa, H. and Yamazaki, Y. (2008). Influence of anion ionomer content and silver cathode catalyst on the performance of alkaline membrane electrode assemblies (MEAs) for direct methanol fuel cells (DMFCs). *J Power Sources* 182: 48–51.

21 Gu, S., Cai, R., Luo, T. et al. (2009). A soluble and highly conductive ionomer for high-performance hydroxide exchange membrane fuel cells. *Angew Chem Int Ed* 48: 6499–6502.

22 Kumar, M., Douglin, J.C., and Dekel, D.R. (2021). Crosslinked quaternary phosphonium-functionalized poly(ether ether ketone) polymer-based anion-exchange membranes. *J Membr Sci* 626: 119167–119180.

23 Gao, X., Yu, H., Jia, J. et al. (2017). High performance anion exchange ionomer for anion exchange membrane fuel cells. *RSC Adv* 7: 19153–19161.

24 Shin, M.-S., Byun, Y.-J., Choi, Y.-W. et al. (2014). On-site crosslinked quaternized poly(vinyl alcohol) as ionomer binder for solid alkaline fuel cells. *Int J Hydrog Energy* 39: 16556–16561.

25 Wang, X.Q., Lin, C.X., Liu, F.H. et al. (2018). Alkali-stable partially fluorinated poly(arylene ether) anion exchange membranes with a claw-type head for fuel cells. *J Mater Chem A* 6: 12455–12465.

26 Zhu, L., Pan, J., Wang, Y. et al. (2016). Multication side chain anion exchange membranes. *Macromolecules* 49: 815–824.

27 Ul Hassan, N., Mandal, M., Huang, G. et al. (2020). Achieving high-performance and 2000 h stability in anion exchange membrane fuel cells by manipulating ionomer properties and electrode optimization. *Adv Energy Mater* 10: 2001986–2001994.

28 Lee, Y.M., Chen, N., Hu, C. et al. (2020). Poly(alkyl-terphenyl piperidinium) ionomers and membranes with outstanding alkaline membrane fuel cell performance of 2.58 W cm^{-2}. *Angew Chem Int Ed* 60: 7710–7718.

29 Cha, M.S., Park, J.E., Kim, S. et al. (2020). Poly(carbazole)-based anion-conducting materials with high performance and durability for energy conversion devices. *Energy Environ Sci* 13: 3633–3645.

30 Penga, H., Lia, Q., Hua, M. et al. (2018). Alkaline polymer electrolyte fuel cells stably working at 80 °C. *J Power Sources* 390: 165–167.

31 Wang, Y., Wang, G., Li, G. et al. (2014). Pt–Ru catalyzed hydrogen oxidation in alkaline media: oxophilic effect or electronic effect? *Energy Environ Sci* 8: 177–181.

32 Wang, L., Bellini, M., Miller, H.A., and Varcoe, J.R. (2018). A high conductivity ultrathin anion-exchange membrane with 500+ h alkali stability for use in alkaline membrane fuel cells that can achieve 2 W cm^{-2} at 80 °C. *J Mater Chem A* 6: 15404–15412.

33 Omasta, T.J., Wang, L., Peng, X. et al. (2018). Importance of balancing membrane and electrode water in anion exchange membrane fuel cells. *J Power Sources* 375: 205–213.

34 Omasta, T.J., Park, A.M., LaMann, J.M. et al. (2018). Beyond catalysis and membranes: visualizing and solving the challenge of electrode water accumulation and flooding in AEMFCs. *Energy Environ Sci* 11: 551–558.

35 Ran, J., Fu, C., Ding, L. et al. (2018). Dual hydrophobic grafted chains facilitating quaternary ammonium aggregations of hydroxide conducting polymers: a theoretical and experimental investigation. *J Mater Chem A* 6: 5714–5723.

36 Jeon, J.Y., Park, S., Han, J. et al. (2019). Synthesis of aromatic anion exchange membranes by Friedel–Crafts bromoalkylation and cross-linking of polystyrene block copolymers. *Macromolecules* 52: 2139–2147.

37 Gao, X., Yu, H., Qin, B. et al. (2019). Enhanced water transport in AEMs based on poly(styrene–ethylene–butylene–styrene) triblock copolymer for high fuel cell performance. *Polym Chem* 10: 1894–1903.

38 Liu, D., Xu, M., Fang, M. et al. (2018). Branched comb-shaped poly(arylene ether sulfone)s containing flexible alkyl imidazolium side chains as anion exchange membranes. *J Mater Chem A* 6: 10879–10890.

39 Gonzalez, J.P., Wang, L., Santiago, E. et al. (2017). High performance radiation-grafted anion conducting polymer electrolytes for energy applications. *231st ECS Meeting.*

40 Jin, Z., Guo, Y., and Qiu, C. (2022). Electro-conversion of carbon dioxide to valuable chemicals in a membrane electrode assembly. *Sustainability* 14: 5579–5603.

41 Ziv, N., Mondal, A.N., Weissbach, T. et al. (2019). Effect of CO_2 on the properties of anion exchange membranes for fuel cell applications. *J Membr Sci* 586: 140–150.

42 Ziv, N., Mustain, W.E., and Dekel, D.R. (2018). The effect of ambient carbon dioxide on anion-exchange membrane fuel cells. *ChemSusChem* 11: 1136–1150.

43 Kimura, T. and Yamazaki, Y. (2011). Effects of CO_2 concentration and electric current on the ionic conductivity of anion exchange membranes for fuel cells. *Electrochemistry* 79: 94–97.

44 Nightingale, E.R. (1959). Phenomenological theory of ion solvation. *J Phys Chem* 63: 1381–1387.

45 Connors, B.W. and Ransom, B.R. (1984). Chloride conductance and extracellular potassium concentration interact to modify the excitability of rat optic nerve fibres. *J Physiol* 355: 619–633.

46 Kiss, A.M., Myles, T.D., Grew, K.N. et al. (2013). Predicting carbonate and bicarbonate ion transport in alkaline anion exchange membranes. *J Electrochem Soc* 160: 994–999.

47 Inaba, M., Matsui, Y., Saito, M. et al. (2011). Effects of carbon dioxide on the performance of anion-exchange membrane fuel cells. *Electrochemistry* 5: 322–325.

48 Li, G., Wang, Y., Pan, J. et al. (2015). Carbonation effects on the performance of alkaline polymer electrolyte fuel cells. *Int J Hydrog Energy* 40: 6655–6660.

49 Fukuta, K., Inoue, H., Watanabe, S., and Yanagi, H. (2009). In-situ observation of CO_2 through the self-purging in alkaline membrane fuel cell (AMFC). *ECS Trans* 19: 23–27.

50 Unlu, M., Zhou, J., and Kohl, P.A. (2009). Anion exchange membrane fuel cells: experimental comparison of hydroxide and carbonate conductive ions. *Electrochem Solid-State Lett* 12: 27–30.

51 Gottesfeld, S., Dekel, D.R., Page, M. et al. (2018). Anion exchange membrane fuel cells: current status and remaining challenges. *J Power Sources* 375: 170–184.

52 Yanagi, H. and Fukuta, K. (2008). Anion exchange membrane and ionomer for alkaline membrane fuel cells (AMFCs). *ECS Trans* 16: 257–262.

53 Zheng, Y., Omasta, T.J., Peng, X. et al. (2019). Quantifying and elucidating the effect of CO_2 on the thermodynamics, kinetics and charge transport of AEMFCs. *Energy Environ Sci* 12: 2806–2819.

54 Adams, L.A., Poynton, S.D., Tamain, C. et al. (2008). A carbon dioxide tolerant aqueous-electrolyte-free anion-exchange membrane alkaline fuel cell. *ChemSusChem* 1: 79–81.

55 Rigdon, W.A., Omasta, T.J., Lewis, C.A., and Mustain, W.E. (2015). Reaction dependent transport of carbonate and bicarbonate through anion exchange membranes in electrolysis and fuel cell operations. *ECS Trans* 69: 1–9.

56 Shi, M., Kim, J., Sawada, J.A. et al. (2013). Production of argon free oxygen by adsorptive air separation on Ag-ETS-10. *AIChE J* 59: 982–987.

57 Zhang, W., Banerjee, D., Liu, J. et al. (2016). Redox-active metal-organic composites for highly selective oxygen separation applications. *Adv Mater* 28: 3572–3577.

58 Fan, S.T., Qiu, Z.J., Xu, R.Y. et al. (2021). Ultrahigh carbon dioxide-selective composite membrane containing a gamma-CD-MOF layer. *ACS Appl Mater Interfaces* 13: 13034–13043.

59 Faid, A.Y. and Sunde, S. (2022). Anion exchange membrane water electrolysis from catalyst design to the membrane electrode assembly. *Energy Technol* 10: 2200506–2200531.

60 Wang, L., Magliocca, E., Cunningham, E.L. et al. (2017). An optimised synthesis of high performance radiation-grafted anion-exchange membranes. *Green Chem* 19: 831–843.

61 Xin, L., Zhang, Z., Wang, Z. et al. (2013). Carbon supported Ag nanoparticles as high performance cathode catalyst for H_2/O_2 anion exchange membrane fuel cell. *Front Chem* 1: 16–21.

62 Zeng, L., Zhao, T.S., and An, L. (2015). A high-performance supportless silver nanowire catalyst for anion exchange membrane fuel cells. *J Mater Chem A* 3: 1410–1416.

63 Arunchander, A., Peera, S.G., Giridhar, V.V., and Sahu, A.K. (2016). Synthesis of cobalt sulfide-graphene as an efficient oxygen reduction catalyst in alkaline medium and its application in anion exchange membrane fuel cells. *J Electrochem Soc* 164: 71–80.

64 Shang, C., Wang, Z., Wang, L., and Wang, J. (2020). Preparation and characterization of a polyvinyl alcohol grafted bis-crown ether anion exchange membrane with high conductivity and strong alkali stability. *Int J Hydrog Energy* 45: 16738–16750.

65 Yang, W., Liu, S., Yan, J. et al. (2021). Metallo-polyelectrolyte-based robust anion exchange membranes via acetalation of a commodity polymer. *Macromolecules* 54: 9145–9154.

66 Mandal, M., Huang, G., Hassan, N.U. et al. (2020). Poly(norbornene) anion conductive membranes: homopolymer, block copolymer and random copolymer properties and performance. *J Mater Chem A* 8: 17568–17578.

67 Wang, Z., Mandal, M., Sankarasubramanian, S. et al. (2020). Influence of water transport across microscale bipolar interfaces on the performance of direct borohydride fuel cells. *ACS Appl Energy Mater* 3: 4449–4456.

68 Leonard, D.P., Maurya, S., Park, E.J. et al. (2020). Asymmetric electrode ionomer for low relative humidity operation of anion exchange membrane fuel cells. *J Mater Chem A* 8: 14135–14144.

69 Hyun, J., Jo, W., Yang, S.H. et al. (2022). Tuning of water distribution in the membrane electrode assembly of anion exchange membrane fuel cells using functionalized carbon additives. *J Power Sources* 543: 231835–231844.

15

Applications of Anion Exchange Membranes Excluding Fuel Cells

Xiuqin Wang and Rob G.H. Lammertink

University of Twente, MESA+ Institute for Nanotechnology, Faculty of Science and Technology, Soft Matter, Fluidics and Interfaces, Enschede, 7522, NB, The Netherlands

15.1 AEMs in Alkaline Water Electrolysis

Hydrogen (H_2) is widely accepted as the core of the world's energy architecture of the future and is considered the cleanest fuel. Alkaline water electrolyzers (AWEs) and proton exchange membrane water electrolyzers (PEMWEs) are commercially available for the hydrogen processing industry. AWEs are a well-established technology, which use an alkaline solution as the liquid electrolyte, a porous diaphragm separator, and platinum group-free electrocatalysts for hydrogen evolution reaction (HER) and oxygen evolution reaction (OER). The history of the production of H_2 through AWE technology spans over 100 years, dating back to when Dmitry Lachinov developed the first industrial AWE device [1]. Commercial porous diaphragm Zirfon was used in traditional AWEs as the separator between electrode compartments to isolate the hydrogen and oxygen gases and to prevent the direct contact of the cathode and anode [2]. However, it suffers from low efficiency of hydrogen production, leakage of the electrolyte, reduced voltage efficiency, as well as other problems [3]. Water electrolysis based on a solid polymer electrolyte membrane has a compact cell design, as shown in Figure 15.1. It enables the production of pressurized H_2 while reducing further compression and storage costs [4]. However, PEMWE operates under an acidic environment and relies on expensive noble metal catalysts and perfluorosulfonic acid (PFSA) membranes (e.g. Nafion), which increases the cost of H_2 production and limits large-scale H_2 production applications.

As an alternative to traditional AWEs or PEMWEs, anion exchange membrane water electrolyzers (AEMWEs) hold several advantages. For example, the AEMWE uses inexpensive non-noble metal catalysts (e.g. Ni and Cu), simple electrolytes (like

Alkaline Anion Exchange Membranes for Fuel Cells: From Tailored Materials to Novel Applications, First Edition. Edited by Jince Thomas, Alex Schechter, Flavio Grynszpan, Bejoy Francis, and Sabu Thomas.

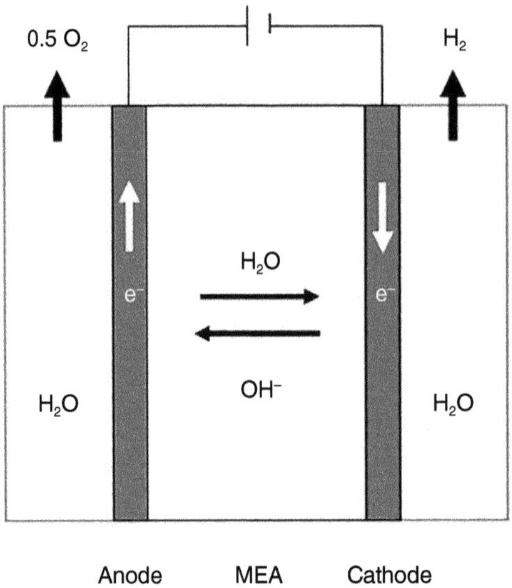

Figure 15.1 Schematic diagram of an AEMWE.

low-concentrated alkaline solution or distilled water to replace the use of highly corrosive electrolytes), cheap membranes, and construction materials and sustains high current densities which are comparable to those of PEMWE [3, 5]. As a crucial component of the AEMWE, the anion exchange membranes (AEMs) play an important role in determining the electrolysis performance and long-term stability of the electrolyzer. An ideal AEM should provide low ohmic and contact resistances, improved stability, low gas crossover, and good mechanical and chemical stabilities [6].

15.1.1 Working Principle

Generally speaking, water splitting technology involves a fuel cell working in reverse mode. The zero-gap AEMWE cell, as illustrated in Figure 15.1, consists of an AEM and two electrodes [3]. Pure water or dilute alkaline electrolyte solution (e.g. K_2CO_3, NaOH, or KOH) is typically provided in both anode and cathode chambers. The AEM is located in the middle of the electrolyzer and serves as a selective medium to transport OH^-, while separating the anode and cathode compartments. Water is converted into H_2 gas and OH^- ions at the cathode by combining electrons. The OH^- ions are transported through the AEM toward the anode and are converted into water and O_2 via the OER, releasing the electrons. During this water splitting process, the OER and HER occur at the anode and cathode, respectively.

The overall electrochemical processes in AEMWE are as follows:

$$\text{Cathode reaction: } 2H_2O + 2e^- \rightarrow H_2 + 2OH^- \; E_0 = -0.828\,\text{V}$$
$$\text{Anode reaction: } 2OH^- \rightarrow 1/2O_2 + H_2O + 2e^- \; E_0 = 0.401\,\text{V}$$
$$\text{Overall reaction: } H_2O \rightarrow H_2 + 1/2O_2 \; E_0 = 1.23\,\text{V}$$

15.1.2 Research Progress of AEMs for AEMWE

Several kinds of AEMs have been reported for the AEMWE. Cationic groups like quaternary ammonium or imidazolium are normally attached to the polymer backbones to fabricate the AEMs. Commercially available AEMs such as A201 (Tokuyama), Fumasep® (Fumatech), Aemion™ (Ionomr), Sustainion® (Dioxide Materials), and Orion TM1 (Orion Polymer) are currently widely used in alkaline water electrolysis [7]. For example, Tokuyama-A201 demonstrated a high conductivity of 32 mS cm^{-1} at 30 °C and could achieve a current density of 399 mA cm^{-2} at 1.8 V when assembled in an AEMWE running with deionized water at 50 °C (Figure 15.2a) [8]. Furthermore, Comotti and coworkers [11] found that electrolysis based on A201 membranes with non-platinum catalysts run continuously for at least 1000 h, which confirms its high chemical and mechanical stability. The relatively low-cost Fumasep® FAA-3-PK-75 membrane, which is composed of quaternary ammonium-functionalized polyketone-based polymer materials, is also a potential candidate for the AEMWE. For example, Lim et al. [12] found that FAA-3-PK-75-based AEMWE showed a current density of 983 mA cm^{-2} at 1.8 V (90 °C) by using standard noble catalysts. But the performance of FAA-3-50-based AEMWE using a non-noble $NiMn_2O_4$ catalyst showed a significant voltage decay after operating at 50 °C for 1000 h, as investigated by Carbone's group [13]. For a better comparison, Bessarabov and coworkers [9] systematically studied the long-term stability performance of three commercial AEMs, as shown in Figure 15.2b, A201 exhibited high stability, while the FAA-3 became mechanically weak due to strong swelling. This work indicated that mechanical stability seems to be a big issue for AEMs applied in water electrolysis. Even though, Sustainion® membranes as depicted in Figure 15.2c have shown improved long-term performance for AEMWE applications [10]. But the fact is they are still far from the durability requirement of the AEMWE. As the target stability requirement for AEMs in the NEWELY project, the membrane area-specific resistance should be $\leq 0.07\,\Omega\,cm^2$ after 2000 h real or simulated operation in an electrolyzer [7].

Functionalized polymeric material AEMs based on polysulfone [14], poly(vinyl benzyl chloride) [15], polystyrene [16], polyaromatics [17, 18], etc. have been recently

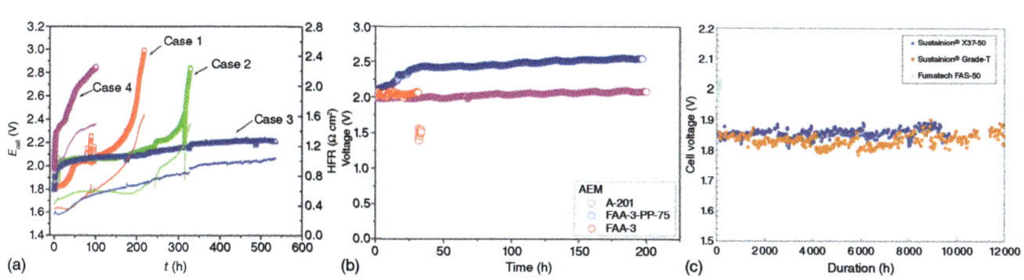

Figure 15.2 Long-term performance of commercialized AEM-based water electrolysis. (a) Tokuyama-A201: tested with noble catalysts in deionized water at 50 °C. Source: Leng et al. [8]/American Chemical Society. (b) A201, FAA-3-PK-75 and FAA-3: tested with Ni-based catalysts in 1% K_2CO_3 at 60 °C. Source: Vincent et al. [9]/American Chemical Society. (c) Sustainion® membranes: tested with Ni-based catalysts in KOH solution at 60 °C. Source: Motealleh et al. [10]/American Chemical Society.

explored for applications in the AEMWE. For example, the group of Zhuang developed an AEM electrolyzer using self-cross-linking quaternized polysulfone (QPSU) and non-precious metal catalysts [19]. This was the first time that the AEMWE used pure water, achieving a current density of 400 mA cm^{-2} at 70 °C under a cell voltage of 1.8–1.85 V. Later on, the effect of cations on AEMWE performance was studied by the group of Ramani [20]. A series of AEMs based on functionalized polysulfone with three different cationic groups (quaternary ammonium and imidazolium groups) were fabricated. The study showed that the trimethylammonium-functionalized polysulfone-based AEMWE had the highest performance (a current density of 400 mA cm^{-2} was obtained at 1.8 V working with ultrapure water at 50 °C). However, the corresponding electrolyzer showed poor long-term stability due to severe AEM chemical degradation. The reason is that the chemical structure of the sulfone group in polysulfone-based AEMs is an electron-withdrawing group which has a negative effect on the membrane stability [21]. Besides, the cationic groups close to the polysulfone backbone further trigger the degradation reaction since trimethylammonium is also an electron-withdrawing group [22]. A better AEMWE performance was achieved by using low-density polyethene-based AEMs, which were prepared by UV-induced grafting of vinylbenzyl chloride and quaternization with 1,4-diazabicyclo(2.2.2)octane [23]. A constant hydrogen production rate of about 30 cc min^{-1} over more than 500 h was obtained with a membrane electrode assembly (MEA) area of 5 cm^2. This electrolysis performance is higher than that of the commercial membrane from Tokuyama. However, the problem of low long-term stability still needs to be further improved for future AEMWE applications in harsh alkaline condition.

There are many studies showing that AEMs containing aryl-ether bonds (like poly(sulfone) and poly(ether ketone)) degrade severely under alkaline conditions, while AEMs with ether-bond-free backbones (like poly(aromatic), poly(olefin)) remain stable [24]. To confirm this, a series of twisted ether-free poly(arylene piperidinium) AEMs used in the AEMWE were developed by the Yan group [25]. These AEMs showed a high conductivity of 37 mS cm^{-1} at 30 °C, and the corresponding electrolysis cell demonstrated a high current density of about 1 A cm^{-2} at 2.5 V when fed with 1 M KOH at 50 °C. The durability test was carried out at a current density of 200 mA cm^{-2}, confirming good electrochemical stability, where the voltage hardly changed while continuously working for 500 h. A further systemic investigation on the performance and durability of these polymers with an ether-bond-free backbone was conducted by Prof. Yu Seung Kim's group [18, 26]. The chemical structures of these ether-bond-free polymers are shown in Figure 15.3a. The study showed that the best electrochemical performance was based on a poly(phenylene) (HTMA-DAPP) membrane with a poly(fluorene) (FLN55) ionomeric binder when the electrolysis was run at a current density of about 1 A cm^{-2} at 2 V fed with 1 wt% K_2CO_3 solution [18]. The durability of HTMA-DAPP AEM in Figure 15.3b shows a stable cell performance with a voltage decay rate of ~1 μV h^{-1}. In this case, there are no signs of cross-cell leakage or electronic-short failures for currents up to 2 A cm^{-2}. Their study also demonstrated that ionomeric binders have a remarkable influence on the AEMWE performance. As depicted in Figure 15.3c, the electrochemical

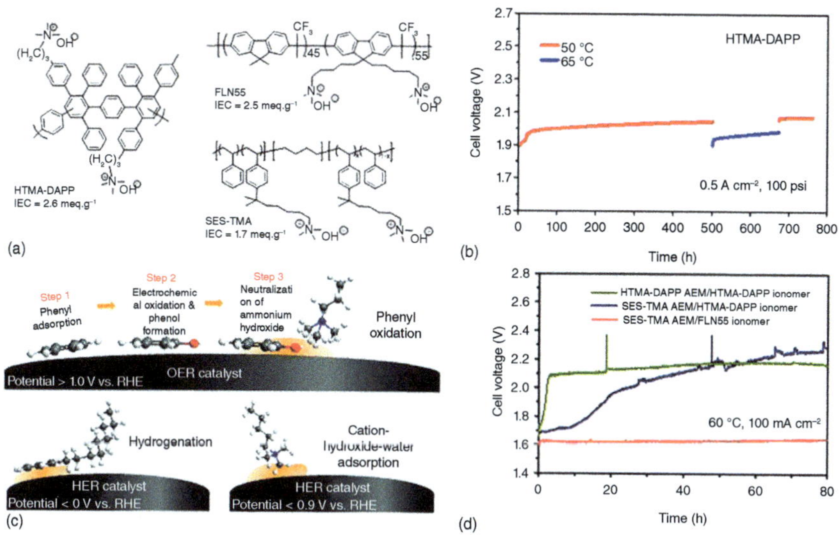

Figure 15.3 (a) Chemical structure of ether-bond-free backbone polymers, (b) durability of the HTMA-DAPP cell, (c) electrochemical oxidation, and (d) the cell durability test with differing ionomer combinations. Source: Li et al. [26]/Royal Society of Chemistry.

oxidation of the adsorbed phenyl group can be observed when the ionomer containing a benzene ring makes contact with the catalyst. The AEMWE employing trimethyl alkylammonium-functionalized poly(styrene-ethylene-*b*-styrene) triblock copolymer AEM with an FLN55 ionomer showed a more stable performance than with the HTMA-DAPP ionomer (Figure 15.3d) [26]. Thus, for future development of the AEMWE, AEMs with ether-bond-free backbone and ionomer design based on polyolefin combined with the investigation of ionomer and catalyst interactions are urgently needed.

15.1.3 Summary and Future Perspectives

Although great progress has been made in developing AEM materials, the corresponding AEMWE processes are still far from the performance of the PEMWE. The ion-solvating membrane based on polybenzimidazole (PBI, as shown in Figure 15.4, neither a porous diaphragm nor an AEM that attains its high conductivity from a homogeneous ternary electrolyte system of polymer/water/KOH) was recently developed for high alkaline stability membranes for the AEMWE [27]. The challenge of balancing the ionic conductivity and mechanical stability of AEMs at present could be improved by the introduction of micro-phase-separated structures, cross-linked structures, and polytetrafluoroethylene (PTFE)-reinforced structures [26, 28]. Besides, membranes with acid–base electrolytic hydrogenation technology may also be an important direction for future development of AEMs [29, 30] with the increasing concern on membrane chemical stability, especially for the AEMs operated at high pH, temperature, and voltage for a long time [20]. The design of AEMs based on ether-bond-free backbones and cationic groups with steric hindrance could greatly

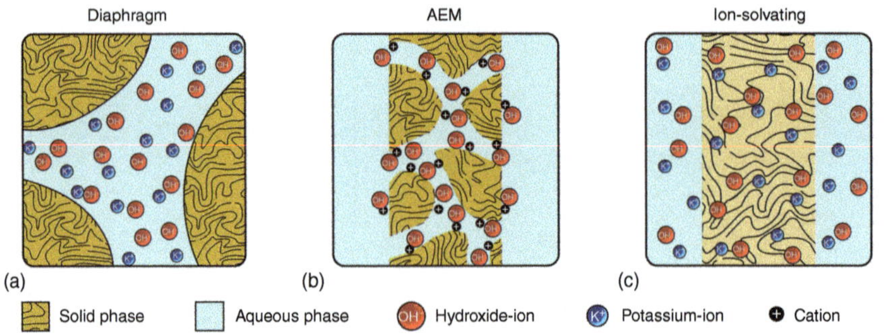

Figure 15.4 The illustration of ion-solvating membrane and comparison. Source: Kraglund et al. [27]/Royal Society of Chemistry.

improve the stability of the membranes. Ionomers with non-rotatable phenyl groups, such as fluorene or carbazole, that have lower phenyl group adsorption energy, are attractive for the further improvement of the overall AEMWE performance. Still, understanding AEM degradation mechanism, using less expensive materials (catalysts and membranes), and focusing on ionomer and catalyst interactions are needed in the future to achieve long-lasting AEMWEs with highly efficient hydrogen production capabilities.

15.2 AEMs in CO_2 Electrolysis

Besides the applications involving hydrogen production, AEMs can also be used as an ion conductor and separator for the electrocatalytic reduction of CO_2. The electrochemical conversion devices are similar to a water electrolysis cell, but with different feeds for the anode and cathode. Figure 15.5 shows the electrolyzer diagram for the electrochemical reduction of CO_2 to CO. As one of the most promising approaches in the broad scheme of carbon capture and utilization, CO_2 can be converted into organic feedstocks such as formic acid (HCOOH), carbon monoxide (CO), methane (CH_4), ethylene (C_2H_4), and ethanol (C_2H_5OH) through CO_2 electrolysis (CO_2E) [31, 32].

The overall electrochemical processes in CO_2E are as follows:

$$\text{Cathode reaction: } CO_2 + H_2O + 2e^- \rightarrow CO + 2OH^-$$
$$\text{Anode reaction: } 2OH^- \rightarrow 1/2O_2 + H_2O + 2e^-$$
$$\text{Overall reaction: } CO_2 \rightarrow CO + 1/2O_2$$

Several kinds of commercial AEMs, including Sustainion® X37~50 Grade T, Sustainion® X37~50 Grade 60, and Fumasep® FAA-3-20, were investigated for electrochemical CO_2 reduction [33]. Among them, Sustainion® X37~50 Grade 60 demonstrated the highest Faradaic efficiency of 72.7% (at 88 mA cm^{-2}) and energy efficiency of 19.9% (at 4.2 V). However, the Sustainion® membrane is prone to the crossover of CO_2 products at high current densities. To some degree, the energy

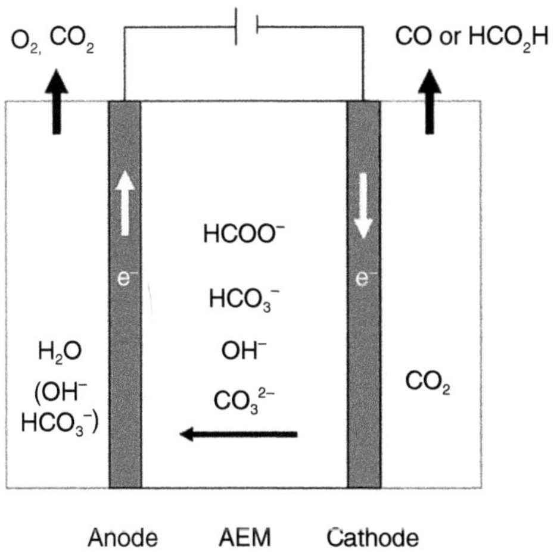

Figure 15.5 Schematic diagram of a CO_2E.

efficiency is still low since the anode needs a high overpotential for water oxidation [34]. Besides, commercialized AEMs often suffer from low OH^- conductivity and poor chemical stability under alkaline condition.

The group of Zhuang largely contributes to the development of AEMs for CO_2E. They developed a CO_2 electrolyzer based on highly conductive and stable quaternized poly(*N*-methyl-piperidine-*co*-*p*-terphenyl) (QAPPT) membranes [35]. The QAPPT membrane exhibits higher conductivity (137 mS cm^{-1} at 80 °C) and is more stable under alkaline conditions than the commercialized Sustainion® membrane. By feeding pure water and CO_2 into the electrolyzer, a high current density of 0.5 A cm^{-2} was achieved at 3 V and 60 °C. The Faradaic efficiency of CO production was higher than 85%. Recently, the group of Varcoe developed a series of radiation-grafted AEMs with high ionic conductivity and durability for CO_2E [36]. The AEMs with the benzyl-*N*-methylpiperidinium head group was assembled in a CO_2 electrolyzer and operated for 200 h at 150 mA cm^{-2} and showed a high CO selectivity of 80% at around 3.1 V. Since there are still no commercially available AEMs for CO_2E and there are only a few AEMs developed for CO_2E, further studies on design of high-performance AEMs are required in the future. The research on novel AEM materials for CO_2E should focus on reducing CO_2 and product crossover, and improving chemical stability and mechanical properties.

15.3 AEMs in Redox Flow Batteries

Redox flow batteries (RFBs) are one of the most promising candidates for grid-scale energy storage due to their wide power range, long lifetime, and flexible discharge duration [37]. The energy capacity of the system is directly proportional to the concentration of the active redox species and electrolyte volume, while the number of cells in the stack and electrode area determine the system power. As the energy is stored in fluid materials, RFBs act more like a regenerative fuel cell than a conventional battery [38]. The technology of RFBs has passed over 100 years since it was initially developed in the United States [39]. Now, there are various flow battery systems in the market. Among them, vanadium redox flow batteries (VRFBs) are the most promising ones that use the same electrolyte in both half-cells, thereby eliminating cross-contamination and electrolyte maintenance problems. Membranes play an important role in VRFBs, wherein the membrane works as a separator to avoid the mixing of the charged vanadium species and prevent battery self-discharge [38]. An ideal membrane should have high proton conductivity, limited water migration, low swelling ratio, low vanadium permeability, good stability, and low cost [40].

The membranes in VRFBs can be divided into ion exchange membranes (IEMs) and non-ionic porous membranes [41]. Normally, a porous membrane has low ionic resistance in concentrated electrolyte solutions and is inexpensive [42]. However, a porous structure could lead to a high crossover of electroactive species and result in a loss of capacity and battery performance. PEMs and AEMs show lower vanadium crossover and better VRFB performance compared to batteries based on porous membranes [43]. Considering the membrane costs, large-scale grid systems of VFRB based on PEM (e.g. Nafion) are limited. AEMs with positively charged functional groups are attractive alternatives to PFSA membranes since they repulse positively charged species from the membrane, owing to the Donnan exclusion mechanism [44, 45]. However, the low vanadium crossover evidently comes at the expense of reduced proton conductivity. Moreover, the degradation of aromatic membranes may be mitigated by using AEMs when the membrane is exposed to highly oxidative and acidic media of the charged electrolyte [46].

15.3.1 Working Principle

Basically, in the VRFB system, vanadium species undergo chemical reactions to various oxidation states via reversible redox reactions and converted into electricity

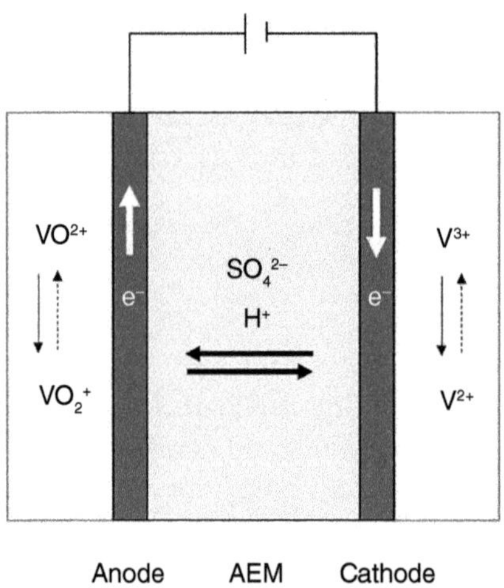

Figure 15.6 Schematic diagram of a VRFB.

directly [47]. The electrolyte solution in VRFBs typically contain vanadium in a range of 1–3 M, dissolved in a 1–2 M H_2SO_4 solution. The vanadium ions involved in the system include VO^{2+}, VO^{2+}, V^{3+}, and V^{2+}. The charging and discharging processes of VRFB using an AEM are illustrated in Figure 15.6. During charging, the positive electrode contains V^{3+} ions, while the negative side hosts V^{2+}. During discharging, reduction occurs at the cathode and oxidation occurs at the anode. H^+ along with SO_4^{2-} and HSO_4^- diffuse across the AEM [46]. The positive charged group in AEMs could repel charged V species via the Donnan effect. The AEM should possess high SO_4^{2-} and HSO_4^- permeability, leading to preferential water transfer through the membrane.

The overall electrochemical processes in the VRFB are seen as follows:

$$\text{Cathode reaction}: V^{3+} + e^- \Leftrightarrow V^{2+}$$
$$\text{Anode reaction}: VO^{2+} + H_2O \Leftrightarrow 2H^+ + VO_2{}^+ + 2e^-$$
$$\text{Overall reaction}: VO^{2+} + H_2O + V^{3+} \Leftrightarrow 2H^+ + VO_2{}^+ + V^{2+}$$

15.3.2 Research Progress of AEMs for VRFBs

The VRFB cell performance using commercial AEMs (including AHA and AFN membranes from ASTOM Co. and APS from Asahi Glass Co.) has been investigated by Choi et al. [48] It was found that the AFN membrane showed a higher current efficiency (96.1%) than Nafion 117 (93.1%) in the VRFB system, indicating its potential as a separator for VRFB. Moreover, Hwang and Ohya [49] introduced a cross-linked structure into the commercial AEM (Selemion™ type II-b) using accelerated electron radiation, and an overall energy efficiency of 82% was achieved. The long-term stability of those commercial AEMs in VRFB was not provided.

Many new types of AEMs with low cost and high performance have been widely explored to increase the performance of VRFB. The group of Yang developed a series of novel AEMs with long side chains and without ether bonds in the backbone for VRFB [50]. The poly(*p*-terphenyl-*N*-methylpiperidine)-quaternary ammonium membrane exhibited much higher ion selectivity than Nafion 115. The corresponding VRFB demonstrated a Coulombic efficiency above 99% at current densities of 80–160 mA cm^{-2}. Yun et al. [51] evaluated the feasibility of using quaternized cardo-polyetherketone (QPEK-C) AEMs for VRBF applications. A low vanadium permeability of $8.2 \pm 0.2 \times 10^{-9}$ cm^2 s^{-1} and ionic conductivity of 5.6 ± 0.5 mS cm^{-1} at 30 °C were obtained. However, a 15% decline in the content of cationic groups for QPEK-C AEM was observed using nuclear magnetic resonance (NMR) after operation for 100 h. In order to improve the stability of the cationic groups, Xing et al. [52] developed imidazolium cation-based polysulfone (PSf-Im) AEMs for VRFB (Figure 15.7). Four kinds of imidazolium cations were compared. It was found that C_2-substituted imidazolium-based AEMs remain intact in both *ex situ* (1.5 M $(VO_2)_2$ SO_4/3 M H_2SO_4 solution for 90 days) and *in situ* (cycling at a current of 120 h mA cm^{-2}) stability tests, while unsubstituted imidazolium-based AEMs degraded severely under the same testing condition. After assembling a PSf-Im AEM into VFRB and continuous operation for 3638 h (4800 cycles), 34% of energy efficiency was lost.

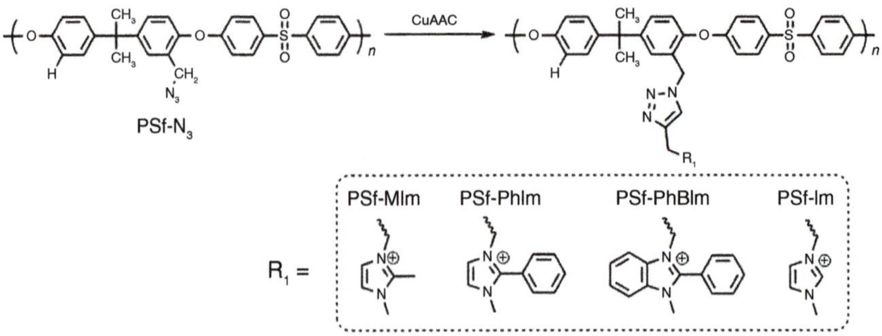

Figure 15.7 Polysulfone-based AEMs with C_2-protected imidazolium cations for the vanadium flow battery. Source: Adapted with permission from Xing et al. [52]. Copyright 2021 Elsevier.

PBI-based AEMs have attracted great attention in the field of VRFBs due to their stable polymer backbone and reduced vanadium permeability. Tang et al. [53] reported a quaternized PBI-based AEM that contained a long flexible side-chain with well-controlled morphology, as shown in Figure 15.8. The prepared AEMs showed a low area resistance of $0.43\,\Omega\,cm^{-2}$ and a vanadium ion permeability of 6.9×10^{-7} $cm^2\,min^{-1}$, which were even lower than Nafion 115 ($0.61\,\Omega\,cm^{-2}$ and 3.2×10^{-6} $cm^2\,min^{-1}$). As a result, the VRFB based on these AEMs have higher energy efficiency (82.7% at $80\,mA\,cm^{-2}$) than those using Nafion 115 (77.4%). The cell performance hardly changed after cycling at $100\,mA\,cm^{-2}$ for 400 times with the

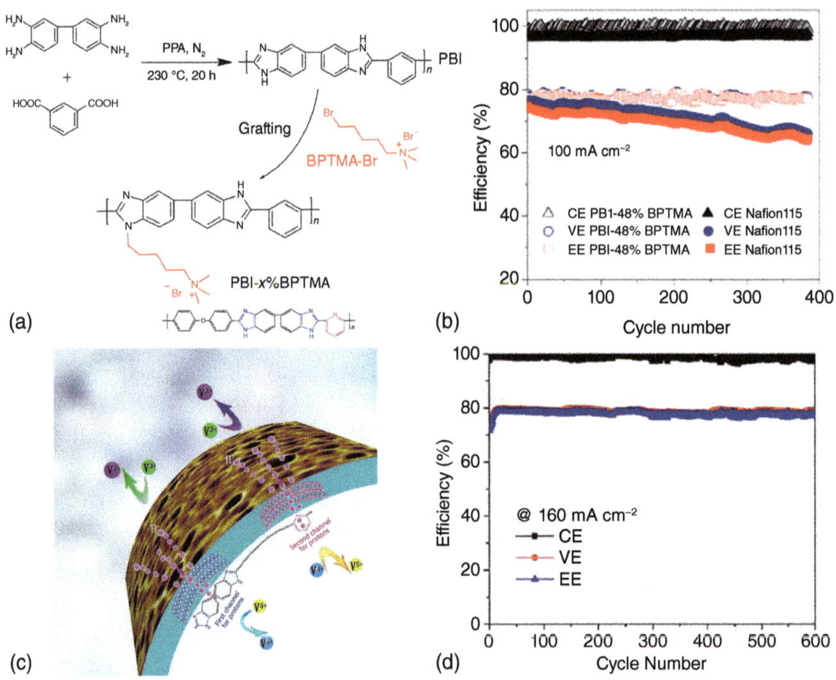

Figure 15.8 (a) The chemical structure and (b) cell performance of graft PBI-based AEMs. Source: Tang et al. [53]/Elsevier. (c) Illustration of the B-PBI membrane with dual proton transport channels and (d) cell performance. Source: Chen et al. [54]/Elsevier.

cut-off voltage controlled between 1.00 and 1.65 V, demonstrating good stability of PBI-based AEM in VRFBs. More efforts to efficiently improve the performance of PBI membranes was conducted by Chen et al. [54]. Dual-proton transport channels were proposed by introducing pyridine groups (channel 1) into the PBI (imidazole rings channel 2) membrane (Figure 15.8), which facilitates the proton transport across the membrane. Donnan repulsion to vanadium ions still provided low vanadium ion permeability. Accordingly, a Coulombic efficiency of 99% and a voltage efficiency of ~80% were obtained, which confirmed its promise for vanadium flow battery.

15.3.3 Summary and Future Perspectives

There is not a single optimal AEM that meets the requirements for all applications of RFBs since the cell configuration and operating conditions vary across different RFB technologies. Although the AEMs could efficiently reduce vanadium permeation in VRFBs, the drawbacks of reduced proton conductivity and poor stability still limit their commercial application. Design and preparation of AEMs with microphase-separated structures seem to be essential for fabricating highly efficient ion conducting pathway and is recommended to solve the issue of low conductivity [55]. The issue of long-term stability is another difficulty that needs to be overcome. Stationary energy storage systems often need to operate over thousands of hours, while the commonly used aromatic AEMs are prone to degradation [42]. The degradation mechanism of AEMs in VRFB suggests that the stability of the AEMs is affected both by the molecular structure of the polymer and the functional groups [56, 57]. The PBI-based AEM material discussed above seems a good candidate due to its stable backbone structure [53]. Besides, cationic groups with large steric hindrance are also suggested to improve the stability of AEMs [52].

Various new types of membranes have been applied to improve the final performance of VRFBs. For example, amphoteric ion-exchange membranes (AIEMs) contain both acidic and alkaline groups to provide intra- and intermolecular acid–base (or hydrogen-bond) interactions, resulting in low vanadium ion permeability and high ionic conductivity [58, 59]. Besides, the introduction of inorganic particles like zero-dimensional materials (nanoparticles), one-dimensional materials (nanotubes), two-dimensional materials (nanosheets), and other materials into the membranes can improve the mechanical properties while suppressing the crossover of vanadium ions and is being explored to achieve high performance [60]. The sulfuric acid-doped PBI porous membrane is another promising candidate for VFRBs because acid treatment could improve its proton conductivity [61]. The exploration of AIEMs, composites, and acid-doped membranes in VRFBs is relatively new and needs further design and targeted synthesis of membrane materials.

15.4 AEMs in Alkali Metal–Air Batteries

Rechargeable alkali metal–air batteries (AMABs) (e.g. Al^-, Zn^-, and Mg–air) that combine the advantages of portable traditional batteries and fuel cells facilitate the wide utilization of flexible and wearable electronic devices. Among them, rechargeable Zn–air batteries (ZABs) are highly promising due to the abundance of zinc, minimal environmental impact, and good safety characteristics [6]. Rechargeable ZABs are regarded as promising candidates for energy storage due to their high specific energy density, safety, and environmental friendliness [62]. Commercial rechargeable ZABs commonly have a significant drawback of zincate ion (e.g. $Zn(OH)_4^{2-}$) crossover due to the use of porous polyolefin membranes (e.g. Celgard® 5550) as the separator, leading to an increase in cell polarization and decreased battery cycling efficiency [63]. The utilization of AEMs in rechargeable ZABs was suggested to solve this problem. There are some advantages of using AEMs: (i) reducing dendrites; (ii) avoiding leaching of catalysts; and (iii) preventing zincate ion permeation ($Zn(OH)_4^{2-}$) [64]. For practical applications, the AEMs should have high ionic conductivity, low zincate permeation, and good chemical/electrochemical stability. However, the use of AEMs in rechargeable ZABs is restricted by their poor long-term stability in strong alkaline conditions: a problem similarly limiting their applications in AEM fuel cells [61]. Normally, a typical AEM exhibits a 10% loss in performance after 1000 h, far from the requirements for practical applications [65].

15.4.1 Working Principle

Figure 15.9 illustrates the structure of the rechargeable ZAB cell. The battery comprises a positive air electrode, an AEM, and a negative zinc electrode assembled in an alkaline electrolyte. During discharge, the zinc metal is oxidized and converted to soluble zincate ions (e.g. $[Zn(OH)_4]^{2-}$). The zincate ions decompose into insoluble zinc oxide when the zincate ion becomes supersaturated in the electrolyte. During charging, the reaction proceeds in the opposite direction.

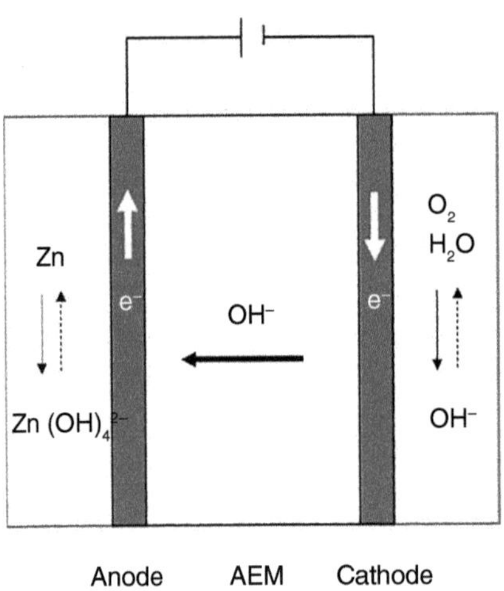

Figure 15.9 Schematic diagram of a rechargeable ZAB.

The overall electrochemical processes in the rechargeable ZAB are seen as follows:

$$\text{Cathode reaction: } O_2 + 2H_2O + 4e^- \Leftrightarrow 4OH^-$$
$$\text{Anode reaction: } 2Zn + 8OH^- \Leftrightarrow 2Zn(OH)_4^{2-} + 4e^-$$
$$Zn(OH)_4^{2-} \Leftrightarrow 2ZnO + 4OH^- + 2H_2O$$
$$\text{Overall reaction: } 2Zn + O_2 \Leftrightarrow 2ZnO$$

15.4.2 Research Progress of AEMs for rechargeable ZABs

There are only a few reports regarding the practical applications of the AEMs in AMABs. Commercialized AEMs, like A201 (Tokuyama) and FAA membrane (Fumatech), have been tested in ZABs. The study showed a rapid decrease in voltage for the A201-based battery during the initial 400 min, and capacity loss was obvious during discharge [66]. This was ascribed to the loss of water, which caused a decrease in ionic conductivity and resulted in high ohmic resistance of the battery [67]. A second reason for the decreased conductivity concerns the cationic groups in the AEMs which may have been partially degraded under highly concentrated alkaline conditions [6].

In order to fully meet the requirements for the application of AEMs in ZABs, novel membrane materials with various synthetic approaches were developed. Recently, Zhang et al. [67] prepared quaternized nanocellulose/GO membranes (QAFCGO) with a conductivity of 33 mS cm^{-1} at room temperature and tested them in the ZABs. At a current density of 60 mA cm^{-2}, the QAFCGO-based ZAB exhibits a high peak power density (44 mW cm^{-2}) and better cycling stability performance than that of ZAB based on the A201 membrane (33 mW cm^{-2}). Low zincate diffusion coefficient (~10^{-8} cm^2 min^{-1}), high specific discharge capacity (~800 mAh g^{-1}), and high specific power (1000 mWh g^{-1}, for 2.5 mA cm^{-2} discharge current density) were obtained by Abbasi et al. [68] where the functionalized poly(phenylene oxide) AEM with various cationic groups was used. Moreover, the AEMs demonstrated good alkaline stability in 7 M KOH solution at 30 °C for at least 150 h. The group of Tsehaye developed the ultraviolet curing method to prepare AEMs (PPO-6CH_2-3x) from hexyl diallyl ammonium-grafted PPO and the diallyl piperidinium [69]. A high conductivity of 10 mS cm^{-1} at 20 °C and a low zincate ion diffusion coefficient of 2.3×10^{-12} m^2 s^{-1} were achieved for the PPO-6CH_2-3x membrane. The membrane was assembled in ZAB and exhibited a maximum power density of 153 mW cm^{-2}, making it a promising membrane material for ZABs.

As mentioned earlier, the conductivity of the AEMs can be improved by developing AEMs with hydrophilic/hydrophobic phase-separated morphologies. Wang et al. [70] prepared poly(vinyl alcohol)/guar hydroxypropyltrimonium chloride (PGG-GP) AEMs. A high conductivity of 123 mS cm^{-1} was obtained at room temperature due to the construction of a microphase-separated structure. Furthermore, the flexibility of PGG-GP AEM made it possible for use in flexible ZABs. A maximum power density of 50.2 mW cm^{-2} at a current of 48 mA cm^{-2} and good cycling stability (9 h at 2 mA cm^{-2}) were achieved. Instead, flexible AEMs are used for the application of flexible ZABs to achieve high OH^- transport capacity, and the membranes with KOH-doped systems are also performed. In AEMWE, these membranes are called ion-solvating membranes which are mostly based on PBI materials, as shown in Figure 15.8 [27]. Here, these membranes are normally classified as gel polymer electrolyte (GPE) [71]. Most GPEs are based on polyvinyl alcohol (PVA) or polyacrylic acid (PAA) materials in which the alkaline solution is stored forming a three-dimensional network structure. They suffer from relatively poor mechanical stability and loss of doped KOH, leading to a large ohmic polarization and resultant degradation of battery operation (more than 6 M KOH electrolyte is needed). Further crosslinking or replacing KOH with tetraethylammonium hydroxide may be needed [62, 71, 72]. An efficient method to block zincate ion crossover seems to be using ion-selective AEMs that could conduct OH^- and restrict permeation of large-size $Zn(OH)_4^{2-}$ [62].

15.4.3 Summary and Future Perspectives

Although the promise of AEMs for rechargeable ZABs has been identified for a long time, it still remains unclear whether AEMs are applicable in ZABs. The standardized testing methodologies are still lacking for evaluating the key properties of AEMs. The main challenge for the application of AEMs in rechargeable ZABs concerns the poor chemical stability under alkaline conditions. Compared to the AEMs that are used in fuel cells, where the AEMs operate at elevated temperatures and low humidity conditions, this problem seems less severe. Similar strategies as those developed for fuel cells can be adopted for alkaline-stable AEMs in rechargeable ZABs, as shown in Figure 15.10 [64]. Considering the impressive progress of high-performance AEM materials in fuel cells, a promising potential can be foreseen toward the development of low-cost and stable AEMs for rechargeable ZABs.

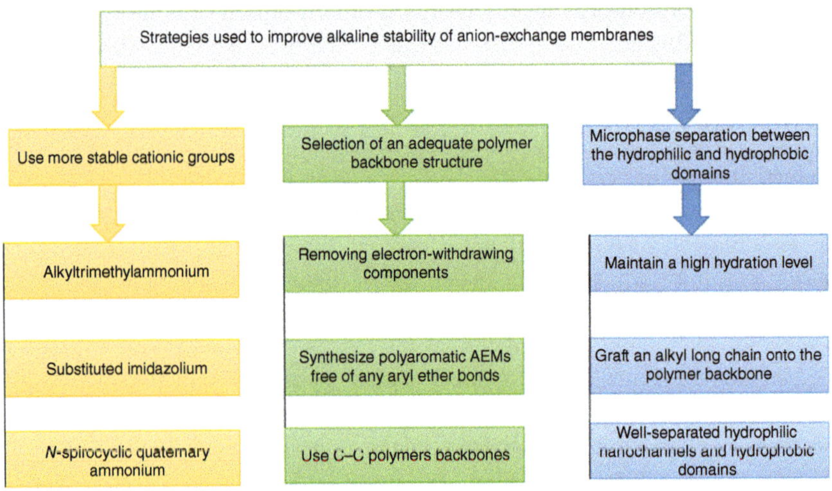

Figure 15.10 Strategies used to prepare alkaline stable AEMs. Source: Tsehaye et al. [64]/ MDPI/CC by 4.0/Public Domain.

15.5 AEMs in Reverse Electrodialysis

Reverse electrodialysis (RED) technology can be used to harvest energy directly from salinity gradients [73, 74]. In 1954, Pattle first developed the RED technology and proved that energy can be harvested through mixing river water with seawater [75]. Then, Weinstein used a simple mathematical model to prove the feasibility of power transfer from RED [76]. Normally, the AEMs are combined with cation exchange membranes (CEMs) in a RED cell. AEMs are used to selectively transport anions, while CEMs transport cations.

15.5.1 Working Principle

A schematic illustration of a RED cell for energy harvesting is shown in Figure 15.11. The working principle of RED relies on the electrochemical equilibrium of the co-ion exclusion theorized by Zoungrana [77]. Basically, IEMs are stacked together in an alternating pattern between a cathode and an anode with fresh water and salt water flowing between the IEMs [78]. An electrochemical potential difference between the various water streams drives the cations through the CEMs and the anions through the AEMs. The transported ions are converted into electrons through a reversible redox reaction at the electrodes.

15.5.2 Research Progress of AEMs for RED

Since the AEMs in RED cells are commonly in the Cl^- form, the membrane resistance is higher than that of the AEMs used for fuel cells which are in the OH^- form. Dlugolecki et al. [79] compared the RED performance using various commercialized

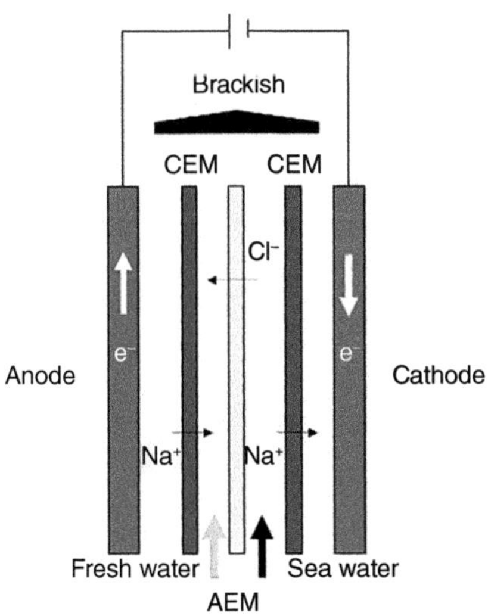

Figure 15.11 Schematic diagram of a RED cell.

AEMs, such as Fumasep®, Neosepta®, Ralex®, and Selemion™ membranes. The best benchmarked commercial AEMs reach a power density of more than $5\,W\,m^{-2}$, indicating a higher power density can be obtained by using AEMs with low membrane resistance and high permselectivity. However, the preferred membrane area resistance is lower than $1–2\,\Omega\,cm^2$, and permselectivity higher than 95% seems difficult to improve.

Güler developed two homogenous AEMs based on polyepichlorohydrin (PECH) with various amounts of chlorine (PECH-H, 37 wt% and PECH-C, 25 wt%) [80, 81]. It was found that there was a statistical correlation between the thickness of the membrane and its area resistance. The PECH-C-based AEM demonstrated a power density of $0.316\,W\,m^{-2}$ in a RED system. A low area resistance ranged from 0.82 to $2.05\,\Omega\,cm^2$, and high permselectivities ranging from 87 to 90% were obtained using 0.017 and 0.507 M NaCl synthetic feed solution. After optimizing the membrane thickness, a power density of $1.27\,W\,m^{-2}$ was achieved by feeding sea and river water into the RED cell. This value is higher than that of cells based on commercial Neosepta® AMX membranes ($1.1\,W\,m^{-2}$) [81]. Most recently, Cho et al. [82] introduced three kinds of cationic groups (tetramethyl ammonium, 1-methyl-imidazolium, and 1-azabicyclo [2,2,2] octane) into poly(arylene ether sulfone) (PAES) to fabricate AEMs and investigated the effect of functional groups on RED performance. The imidazolium-functionalized PAES (IMD-30-PAES) showed higher permselectivity than the other two AEMs, while their conductivities were similar. The IMD-30-PAES demonstrated a maximum power density of $1.2\,W\,m^{-2}$, which is also higher than that of a commercial Neosepta® AMX membrane. The area resistance of $1.65–3.86\,\Omega\,cm^2$ still seems to be somewhat high. Therefore, Lee et al. [83] developed poly(ethylene)-reinforced AEMs based on crosslinked quaternized polystyrene and poly(phenylene oxide) for RED. The fabricated AEMs exhibited a low area resistance of $0.69–1.67\,\Omega\,cm^2$, and the corresponding RED stack provided a maximum power density of $1.82\,W\,m^{-2}$, which is higher than that of the AMV (Selemion™) membrane ($1.50\,W\,m^{-2}$).

15.5.3 Summary and Future Perspectives

Most reports focus on the technological aspects of RED rather than the membranes. Although stack design could optimize the overall performance of RED cells, the effect is limited when higher flow rates or thinner spacers are applied [84]. Therefore, the enhancement of membrane properties would definitely be one of the main challenges for developing high-performance RED cells. Instead of flat AEMs, tailor-made AEMs like ridges, waves, and pillars are widely used in RED [85]. Besides, cheaper pseudo-homogeneous, porous, grafted, and other types of membranes were studied [85]. It is noted that natural water is usually used for large-scale application of RED cells, which introduces issues related to fouling and RED power density [86]. Future designs of AEMs for the RED process should perhaps consider the anti-fouling property of the membranes and monovalent-ion selectivity. Besides, it is worth pointing out that the reports on RED often ignore the economic analysis of this process. The energy density between seawater and fresh water is quite low, and it seems that RED may not be an ideal technology for energy harvesting. Alternative processes like pressure-retarded osmosis provide better large-scale efficiencies [87, 88].

15.6 AEMs in Electrodialysis

Electrodialysis (ED) is a separation technology used in various industries, including the chlor-alkali process, demineralization of saline water, and waste water treatment [89]. In 1890, Maigret et al. first developed ED to demineralize sugar syrup using carbon as the electrode and permanganate paper as the membrane. Then, Juda et al. developed IEMs for ED in 1950 [90]. ED was considered a mature desalination technology for over 50 years [91]. During ED, no phase change occurs and the ED technology has advantages of simple operation, small footprint, and low pretreatment requirements [92]. The specific requirements for the AEMs used in ED process include high permselectivity, especially the selectivity between monovalent and bivalent anions [93].

15.6.1 Working Principle

Figure 15.12 shows a schematic diagram of a typical ED cell. Similar to RED, a set of alternating AEMs and PEMs are placed between the cathode and the anode to fabricate an ED cell. AEMs and PEMs act as barriers for cations and anions, respectively. The cathode and the anode of the ED cell then deliver a direct current (DC) electric field to migrate the charged ions through the IEMs, obtaining enriched and diluted outlet streams [90]. ED can be operated in batch mode or continuous mode, depending on the scale of the process. Continuous operation eliminates the need for batch-wise replacement of concentrate compartments [94].

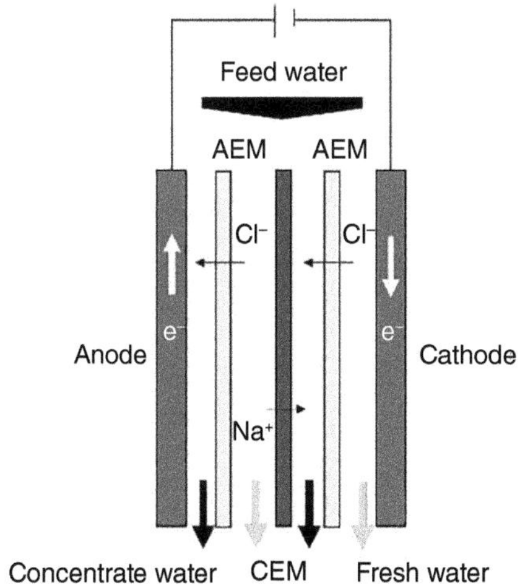

Figure 15.12 Schematic diagram of an ED cell.

15.6.2 Research Progress of AEMs for ED

Commercially available AEMs such as ACS (Astom, Japan) have been applied in ED processes [95]. The separation factor of Cl^-/SO_4^{2-} is low (about 0.4), which needs to be further improved for a more efficient separation process [96]. In order to improve the separation factor for Cl^-/SO_4^{2-}, Goel et al. [97] prepared a series of crosslinked CrPSf-*x* AEMs based on chloromethylated polysulfone and 1-4-diazabicyclo[2.2.2]octane functionalized graphene oxide, as shown in Figure 15.13. The AEM becomes dense after crosslinking and thus may differentiate the ions via the size sieving effect. The ED performance was studied by using a mixed salt solution of 0.05 mol l^{-1} ($NaCl + Na_2SO_4$). A high separation factor (5.7) for Cl^-/SO_4^{2-} was obtained for the prepared CrPSf-3 membrane, a value higher than that of polysulfone-based (1.7) and commercial Fumasep®FAB-PK-130 AEMs (2.01). Besides, the CrPSf-3 membrane with an effective area of 12 cm^2 showed a high Cl^- flux of 2.97 mol m^{-2} h^{-1} at a constant current density of 12.0 mA cm^{-2} and maintained other parameters constant (e.g. applied current density, flow rate, temperature, and feed volume), which has the potential to replace the use of commercial AEMs.

Khan et al. [98] developed new AEMs for ED from brominated poly(2,6-dimethyl-1,4-phenylene oxide) (BPPO) and dimethylethanolamine (DMEA). The IEC of the prepared membranes ranged from 0.6 to 1.38 mmol g^{-1}. Two AEMs named DMEA-10 and DMEA-15 were selected for desalination in ED cells and showed better ED performance than commercial Neosepta® AMX membranes. In Khan's further research, they found that using methyl(diphenyl)phosphine (MDPP) instead of DMEA could increase the IEC of the AEMs (1.09–1.52 mmol g^{-1}) [99]. The ED performance of the prepared MDPP-based membranes was evaluated by using 0.1 M NaCl as the feed solution. The water recovery and desalination of MPDD-43 can reach up to 98% and 98.49%, respectively, which is higher than that of AMX (97% and 97.0%). The group of

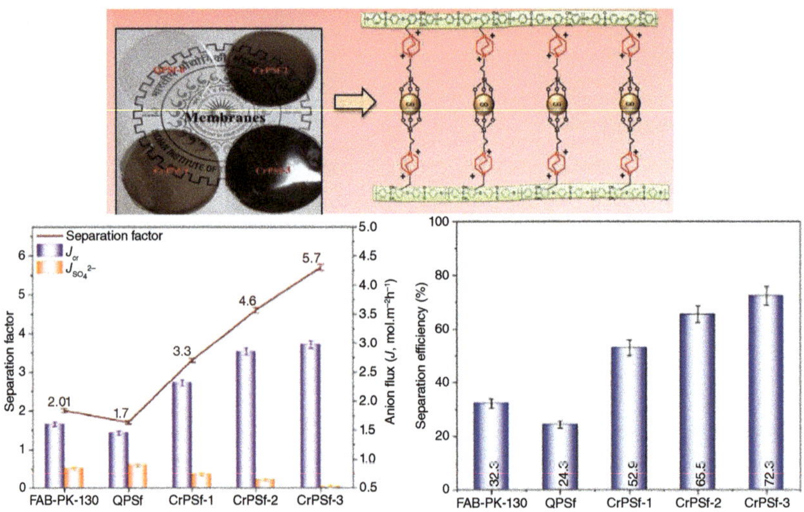

Figure 15.13 The chemical structure and ED performance of CrPSf-x AEMs conducted at fixed current density, 12.0 mA cm^{-2}. Source: Reproduced with permission from Goel et al. [97]. Copyright 2021 Elsevier.

Chattopadhyay used poly(vinylidenefluoride) (PVDF)-blended AEMs to recover carboxylic acid from binary or ternary mixtures through ED. [100] A current efficiency of 63.7% and low energy consumption of 5.13 kWh kg^{-1} could be achieved by using ED.

ED is often used for highly concentrated wastewater treatment, which may lead to fouling issues. Surface modification is an efficient method to protect the AEMs from fouling [101]. Mulyati et al. [102] used poly(sodium 4-styrene sulfonate) (PSS) to modify the surface of the Neosepta® AMX membrane and found that the antifouling potential of the AEMs was significantly improved due to the hydrophilicity and density of charged groups on the modified membrane surface. A sandwich-like AEM composed of upper/bottom bilayers of polydopamine (PDA) and alternating bilayers of PSS/hydroxypropyltrimethyl ammonium chloride chitosan-nano silver particles (HACC-Ag Np) was also fabricated by Hao et al. using the layer-by-layer method [103]. The modified AEMs with 4.5 bilayers of PSS showed a high permselectivity of 5.1 for Cl^-/SO_4^{2-}, much higher than that of a commercial Fujifilm membrane (0.98). The anti-fouling property was strongly improved due to the electrostatic repulsion and hydrophilicity of PDA in the sandwich structure. In addition, antibacterial activities were also achieved by anchored silver nanoparticles, showing the modified AEM is potentially applicable in ED.

15.6.3 Summary and Future Perspectives

ED is effective in separating ions in the process of water treatment and desalination. However, water pollutants such as suspended particles, colloids, bacteria, and large organic molecules hamper the process. Integrating ED with other water treatment technologies is strongly suggested to remove these pollutants and produce highly purified water [104]. Besides, fouling in ED still occurs under certain conditions despite the use of reduced fouling surfaces. Further research on the fouling mechanism on the membrane is crucial to alleviate the fouling issues during long-term operation. The introduction of crosslinking the AEM matrix or coating a thin layer on the surface of the AEM is considered to present an effective way to improve ED performance [105].

15.7 AEMs in Diffusion Dialysis

To date, many methods including crystallization, solvent extraction, ion exchange, and membrane technology are used to recover acid from industrial effluents [106]. Considering these methods, membrane technology is regarded as an efficient and simple approach to recover acid. In 1861, Burlakova proposed the concept of dialysis to separate small molecules from large ones using a semi-permeable membrane [107]. The IEM is employed for dialysis, and the process is defined as diffusion dialysis (DD). The first diffusion dialyzer was invented in 1950, and 30 years later, the DD technology was applied as an industrial membrane process [108]. DD using AEMs has already been widely used in industry for recovering acid because it requires low energy consumption, low operation cost, high product quality, and environmental friendliness [108]. The AEMs used in DD contain positive charge groups that enable the migration of anions and repel cations. To obtain high DD performance, AEMs should present limited water permeability, high proton permeability, and strong rejection of salts [106].

15.7.1 Working Principle

The working principle of DD using AEM for acid recovery is shown in Figure 15.14. The anions such as Cl^- and NO_3^- in the acidic waste solution can transport through the AEM, while the cations are blocked owing to the Donnan criteria of co-ion rejection [109]. Unlike other cations, H^+ can still transport through the AEM due to its small size and co-transport with anions. As a result, the acid can be collected in the receiving side, while larger size metal cations are kept on the feed side.

15.7.2 Research Progress of AEMs for DD

To date, commercial AEMs such as Selemion™ DSV (Asahi Glass), Neosepta® AFX/N (Tokuyama), and DF120 (Shandong Tianwei Membrane Technology Co.) are used for acid recovery [108]. However, the acid permeability and ion selectivity of these AEMs still need further improvement for the applications. Cationic groups are attached to the polymer backbones (polysulfone, poly(2,6-dimenthyl-1,4-phenylene oxide), and PVA) to fabricate AEMs for acid recovery [110–113]. Afsar et al. [114] fabricated a series of AEMs based on bromo-methylated poly(phenylene oxide) and dimethylamino pyridine. The obtained AEMs had proton dialysis coefficients ranging from 0.37 to 20.3 ($10^{-3}\,m\,h^{-1}$) at 25 °C. Besides, the membranes showed a higher separation factor of H^+/Fe^{2+} (73–351) compared to commercial DF-120 membrane (18). Sharma et al. [115] reported a series of crosslinked poly(2,6-dimethyl-1,4-phenylene ether) AEMs for acid recovery by DD. They found that hydrate functionality with a non-overlapping set of lone pairs on oxygen improved the proton mobility via the Grotthuss mechanism inside the membrane. A high HCl recovery efficiency of 92.82% and a separation factor of ~117 can be achieved. The group of Lin

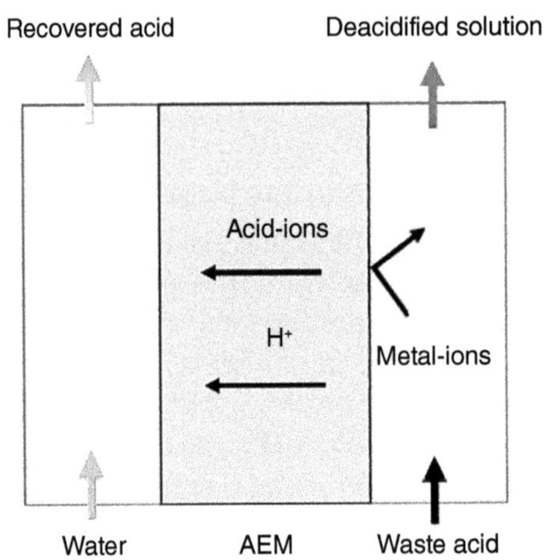

Figure 15.14 Schematic diagram of a DD cell.

developed novel AEMs based on a porous chloromethyl polyethersulfone (CMPES) membrane substrate and 1,4-dimethylpiperazine (DMP) for acid recovery [116]. The optimal AEM (DPES-3h) possessed a proton dialysis coefficient of $43.2\times10^{-1}\,m\,h^{-1}$ and an acid/salt separation factor of 29.6 in simulated $HCl/FeCl_2$ solution at 25 °C, exceeding the commercial DF-120 AEM. These AEMs have the potential for acid recovery from industrial acidic effluents using DD.

15.7.3 Summary and Future Perspectives

Efforts have been made over the years to enhance the properties of the AEMs used in the DD process for acid recovery. The acid permeability, acid selectivity, and membrane stability are crucial factors that could determine the efficiency of acid recovery. The trade-off effect for AEMs regarding acid permeability and selectivity still needs to be solved. Highly stable fillers can be used for the membrane matrix to improve acid recovery and stability. In addition, pore-filled, nanofiber-, and microporous-supported AEMs are also investigated for acid recovery. These could reduce ion transport resistance and thus improve the acid recovery capacity and efficiency [104]. Moreover, the mechanism of ion transportation through the AEMs should be further investigated. In order to improve the potential for practical applications of DD technology, various extensions of the technology including applying pressure, an electric field, or continuous operation are suggested to integrate DD process using AEMs [106].

15.8 AEMs in Microbial Fuel Cells

The conventional methods for wastewater treatment, e.g. flocculation/froth flotation, are highly energy-intensive, complex, and typically environmentally unfriendly [117]. Microbial fuel cells (MFCs) are a kind of energy conversion devices which can convert the chemical energy of organic matter in sludge or wastewater to electric energy using enriched electroactive bacteria [118]. In comparison to the low-temperature (<100 °C) polymer electrolyte membrane fuel cells, which rely on expensive precious metals as electrode catalysts, the use of microbes in the MFC enables lower cost fuel cell devices. The development of MFCs is beneficial to environmental protection, energy harvesting, and water purification. The first MFC was developed by Potter from Durham University–UK in 1911 [119]. After 20 years, Cohen succeed in producing stacked MFCs which showed a high voltage of 35 V. Afterward, the technology of MFCs has grown vigorously [120]. Normally, IEMs work as the separator and ion conductor in MFCs. The state-of-the-art PEMs, such as Nafion, are widely used in MFCs. However, the PEMs have the drawbacks of pH splitting and inefficient proton transport due to the low proton concentration, which limits their practical application for MFCs [121, 122]. As an alternative approach toward overcoming the drawbacks of PEMs, AEMs show a higher rate of proton transfer by using phosphate or carbonate as the proton carrier and pH buffer in MFCs [121]. Therefore, MFCs based on AEMs generally exhibit a higher power density than the MFCs based on PEMs.

15.8.1 Working Principle

A schematic representation of a typical MFC based on AEMs is presented in Figure 15.15. Normally, an AEM-based MFC is composed of an anode, an AEM, and a cathode. In the anode (anaerobic) chamber, the fuel is oxidized by the microorganisms generating protons and electrons. The electrons transfer to the cathode chamber via an external circuit, while protons react rapidly with dibasic anions (e.g. HPO_4^{2-} and CO_3^{2-}) to form monobasic anions (e.g. $H_2PO_4^-$ and HCO_3^-). The monobasic anion diffuses from the anode to the cathode through an AEM, releasing the proton to the cathode, and then transfer back to the anode chamber as dibasic anion via diffusion and electromigration [123].

15.8.2 Research Progress AEMs for MFCs

A few AEMs are designed and prepared for MFC applications. Ultrex® AMI-7001 is a commercially available AEM that has been assembled in a two-chamber, air-cathode cube MFC for the evaluation of its potential MFC application [124]. A maximum power density of 610 mW m^{-2} was reported at a current density of ~0.13 mA cm^2, which is higher than that of Nafion. Besides, the AEMs provide a high Coulombic efficiency of ~70%, in comparison to the values obtained with Nafion, other CEMs, and ultrafiltration membranes. The low internal resistance and the proton charge-transfer facilitated by phosphate were responsible for the high performance of AEM-based MFCs. Except for the commercialized membranes, researchers strived to fabricate novel AEM materials that exhibit properties and performance superior to those of the Ultrex® AMI-7001 or Nafion. As an alternative, QPSU was developed [125]. The QPSU AEM produced an improved power density of 810 mW m^{-2} at 0.753 V compared to the AMI-7001, when tested in a single-chamber MFC at 30 °C. Additionally, AEMs based on poly(ether-imide) [126], poly(phenylene oxide)

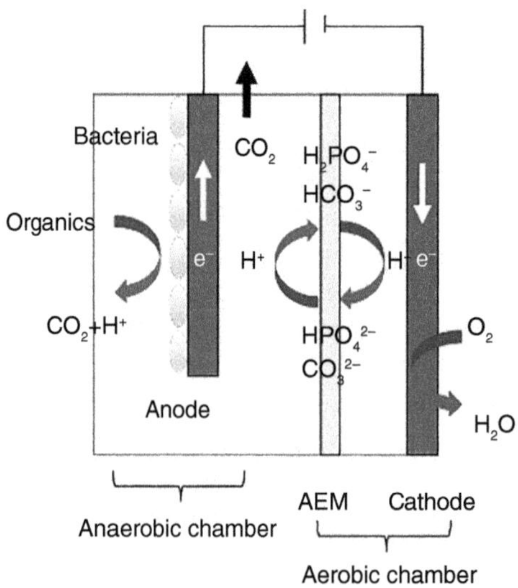

Figure 15.15 Schematic diagram of an MFC.

[122], polystyrene-block-poly(ethylene-ran-butylene)-block-polystyrene) [127], etc. were developed to achieve high-performance MFCs. Recently, the group of Logan reported an AEM-based MFC with a maximum power density of $8.8 \pm 0.5\,W\,m^{-2}$ over a current density range of $35–50\,A\,m^{-2}$, which is the highest value ever achieved using acetate in a 50 mM phosphate buffer medium [128]. Their work also confirmed that the selectivity of OH^- transport through the AEMs without the catholyte is critical to limit biofilm acidification by neutralizing the protons produced at the anode.

15.8.3 Summary and Future Perspectives

Although higher power density can be achieved in an AEM-based MFC with buffer anions, the AEMs still suffer from substrate crossover [121, 124, 129]. This could promote biofilm formation on the cathode surface and reduce the MFC performance. Membranes with high hydrophilicity often show more resistance to biofouling. This method may be used to fabricate novel AEMs to solve the biofouling problem by modification with nanoparticles, activated carbon, clay, and bio-based carbon materials. Moreover, AEMs are liable to deform compared to the PEM-based single-chamber air-cathode MFCs, which could generate a void space between the cathode and the membranes, resulting in increased internal resistance of MFCs [130]. The utilization of highly concentrated phosphate buffers in MFC is uneconomic, which could increase the cost of treating the wastewater. Bicarbonate buffers present an alternative with lower cost and comparable buffering capacity compared to phosphate buffer. In the future, efforts should be made to deal with the drawbacks of biofouling formation, poor dimensional stability, and buffer issues in order to utilize AEMs as a separator for practical applications.

15.9 AEMs in Other Applications

Given the increasing applications of recombinant adenoviruses for gene therapy and vaccination, AEM-based devices may play a crucial role in the separation/purification of biotechnological products [131]. Traditional chromatographic methods have limitations in terms of high-pressure drop and slow processing rates due to pore diffusion and channeling of the feed through the bed. The benefits of AEMs adsorbers are fast, gentle, and effective isolation. Therefore, AEM-based adsorbers or AEM chromatography hold the promise to maintain high efficiencies both at high flowrates and for use with large biomolecules that have small diffusivities. Their use is also accompanied by reduced protein degradation and denaturation [132]. Frequently, ligands containing quaternary amine groups or diethylaminomethyl (DEAE) are used to produce strong AEM adsorbers, which enable high capacities at high ionic strength due to the presence of secondary hydrogen bonding interactions. Sartobind Q1(Sartorius AG), Mustang Q1 (Pall Corporation) are commercial examples of these [133]. The use of AEMs was described for the purification of bacteriophage putative recombination initiation defect 1 (PRD1) from bacterial culture and alpha herpesvirus from the cell culture supernatant of porcine kidney or Madin–Darby bovine kidney cells [134]. Regarding proteins, a tremendous amount of literature exists demonstrating the applicability of AEM adsorbers as an alternative technique for the concentration and purification of adenoviruses and as a promising solution for downstream processing of other viral vectors [135].

15.10 Summary

The applications of AEMs in AEMWE, CO_2E, VRFBs, ZABs, RED, ED, DD, MFCs, and other fields are summarized in this chapter. Significant progress has been made in designing novel AEM materials. For electrochemical devices like AEMWE, CO_2E, VRFBs, ZABs, and RED, AEMs with high conductivity and long-term alkaline stability are crucial. For water treatment applications, such as ED, DD, MFC, and other biological applications, AEMs with anti-fouling properties and high ion selectivity are especially important. The trade-off effect between different membrane properties still needs to be handled. One of the main challenges for the application of AEMs lies in the development of low-cost and high-performance AEMs.

Abbreviations

AEM	anion exchange membrane
AEMWE	anion exchange membrane water electrolyzer
AIEM	amphoteric ion-exchange membrane
AMAB	alkali metal–air battery
AWE	alkaline water electrolyzer
CO_2E	CO_2 electrolysis
CEM	cation exchange membrane
DD	diffusion dialysis
ED	electrodialysis
IEM	ion exchange membrane
MEA	membrane electrode assembly
MFC	microbial fuel cell
PEM	proton exchange membrane
PEMWE	proton exchange membrane water electrolyzer
PFSA	perfluorosulfonic acid
RED	reverse electrodialyzer
RFB	redox flow battery
VRFB	vanadium redox flow battery
ZAB	Zn–air battery

References

1 Yu, Z.Y., Duan, Y., Feng, X.Y. et al. (2021). Clean and affordable hydrogen fuel from alkaline water splitting: past, recent progress, and future prospects. *Adv Mater* 33: 2007100.

2 Kreuter, W. and Hofmannz, H. (1998). Electrolysis: the important energy transformer in a world of sustainable energy. *Int J Hydrog Energy* 23: 661–666.

3 Zakaria, Z. and Kamarudin, S.K. (2021). A review of alkaline solid polymer membrane in the application of a AEM electrolyzer: materials and characterization. *Int J Energy Res* 45: 18337–18354.

4 Ayers, K.E., Anderson, E.B., Capuano, C.B. et al. (2010). Research advances towards low cost, high efficiency. *ECS Trans* 1: 3–15.

5 Abbasi, R., Setzler, B.P., Lin, S. et al. (2019). A roadmap to low-cost hydrogen with hydroxide exchange membrane electrolyzers. *Adv Mater* 31: 1805876.

6 Varcoe, J.R., Atanassov, P., Dekel, D.R. et al. (2014). Anion-exchange membranes in electrochemical energy systems. *Energy Environ Sci* 7: 3135–3191.

7 Henkensmeier, D., Najibah, M., Harms, C. et al. (2021). Overview: state-of-the art commercial membranes for anion exchange membrane water electrolysis. *J Electrochem Energy Convers Storage* 18: 024001–024018.

8 Leng, Y., Chen, G., Mendoza, A.J. et al. (2012). Solid-state water electrolysis with an alkaline membrane. *J Am Chem Soc* 134: 9054–9057.

9 Vincent, I., Kruger, A., and Bessarabov, D. (2017). Development of efficient membrane electrode assembly for low cost hydrogen production by anion exchange membrane electrolysis. *Int J Hydrog Energy* 42: 10752–10761.

10 Motealleh, B., Liu, Z., Masel, R.I. et al. (2021). Next-generation anion exchange membrane water electrolyzers operating for commercially relevant lifetimes. *Int J Hydrog Energy* 46: 3379–3386.

11 Pavel, C.C., Cecconi, F., Emiliani, C. et al. (2014). Highly efficient platinum group metal free based membrane-electrode assembly for anion exchange membrane water electrolysis. *Angew Chem Int Ed* 53: 1378–1381.

12 Lim, A., Kim, H.-J., Henkensmeier, D. et al. (2019). A study on electrode fabrication and operation variables affecting the performance of anion exchange membrane water electrolysis. *J Ind Eng Chem* 76: 410–418.

13 Carbone, A., Zignani, S.C., Gatto, I. et al. (2020). Assessment of the FAA-3-50 polymer electrolyte in combination with a $NiMn_2O_4$ anode catalyst for anion exchange membrane water electrolysis. *Int J Hydrog Energy* 45: 9285–9292.

14 Li, H., Yu, N., Gellrich, F. et al. (2021). Diamine crosslinked anion exchange membranes based on poly(vinyl benzyl methylpyrrolidinium) for alkaline water electrolysis. *J Membr Sci* 633: 119418.

15 Li, H., Kraglund, M.R., Reumert, A.K. et al. (2019). Poly(vinyl benzyl methylpyrrolidinium) hydroxide derived anion exchange membranes for water electrolysis. *J Mater Chem A* 7: 17914–17922.

16 Park, H.J., Lee, S.Y., Lee, T.K. et al. (2020). N_3-butyl imidazolium-based anion exchange membranes blended with poly(vinyl alcohol) for alkaline water electrolysis. *J Membr Sci* 611: 118355.

17 Li, D., Park, E.J., Zhu, W. et al. (2020). Highly quaternized polystyrene ionomers for high performance anion exchange membrane water electrolysers. *Nat Energy* 5: 378–385.

18 Motz, A.R., Li, D., Keane, A. et al. (2021). Performance and durability of anion exchange membrane water electrolyzers using down-selected polymer electrolytes. *J Mater Chem A* 9: 22670–22683.

19 Xiao, L., Zhang, S., Pan, J. et al. (2012). First implementation of alkaline polymer electrolyte water electrolysis working only with pure water. *Energy Environ Sci* 5: 7869–7871.

20 Parrondo, J., Arges, C.G., Niedzwiecki, M. et al. (2014). Degradation of anion exchange membranes used for hydrogen production by ultrapure water electrolysis. *RSC Adv* 4: 9875–9879.

21 Amel, A., Zhu, L., Hickner, M., and Ein-Eli, Y. (2014). Influence of sulfone linkage on the stability of aromatic quaternary ammonium polymers for alkaline fuel cells. *J Electrochem Soc* 161: F615–F621.

22 Arges, C.G. and Ramani, V. (2013). Two-dimensional NMR spectroscopy reveals cation-triggered backbone degradation in polysulfone-based anion exchange membranes. *Proc Natl Acad Sci* 110: 2490–2495.

23 Faraj, M., Boccia, M., Miller, H. et al. (2012). New LDPE based anion-exchange membranes for alkaline solid polymeric electrolyte water electrolysis. *Int J Hydrog Energy* 37: 14992–15002.

24 Choe, Y.K., Fujimoto, C., Lee, K.S. et al. (2014). Alkaline stability of benzyl trimethyl ammonium functionalized polyaromatics: a computational and experimental study. *Chem Mater* 26: 5675–5682.

25 Yan, X., Yang, X., Su, X. et al. (2020). Twisted ether-free polymer based alkaline membrane for high-performance water electrolysis. *J Power Sources* 480: 228805.

26 Li, D., Motz, A.R., Bae, C. et al. (2021). Durability of anion exchange membrane water electolyzers. *Energy Environ Sci* 14: 3393–3419.

27 Kraglund, M.R., Carmo, M., Schiller, G. et al. (2019). Ion-solvating membranes as a new approach towards high rate alkaline electrolyzers. *Energy Environ Sci* 12: 3313–3318.

28 Lin, C.X., Zhuo, Y.Z., Hu, E.N. et al. (2017). Crosslinked side-chain-type anion exchange membranes with enhanced conductivity and dimensional stability. *J Membr Sci* 539: 24–33.

29 Mayerhöfer, B., McLaughlin, D., Böhm, T. et al. (2020). Bipolar membrane electrode assemblies for water electrolysis. *ACS Appl Energy Mater* 3: 9635–9644.

30 Rajesh, A.M., Kumar, M., and Shahi, V.K. (2011). Functionalized biopolymer based bipolar membrane with poly ethylene glycol interfacial layer for improved water splitting. *J Membr Sci* 372: 249–257.

31 Valenti, G., Melchionna, M., Montini, T. et al. (2020). Water-mediated electro-hydrogenation of CO_2 at near-equilibrium potential by carbon nanotubes/cerium dioxide nanohybrids. *ACS Appl Energy Mater* 3: 8509–8518.

32 Qiao, J., Liu, Y., Hong, F., and Zhang, J. (2014). A review of catalysts for the electroreduction of carbon dioxide to produce low-carbon fuels. *Chem Soc Rev* 43: 631–675.

33 Feng, D., Jiang, W., Zhang, C. et al. (2021). A membrane reactor with microchannels for carbon dioxide reduction in extraterrestrial space. *Catalysts* 12: 3.

34 Gabardo, C.M., O'Brien, C.P., Edwards, J.P. et al. (2019). Continuous carbon dioxide electroreduction to concentrated multi-carbon products using a membrane electrode assembly. *Joule* 3: 2777–2791.

35 Yin, Z., Peng, H., Wei, X. et al. (2019). An alkaline polymer electrolyte CO_2 electrolyzer operated with pure water. *Energy Environ Sci* 12: 2455–2462.

36 Giron Rodriguez, C.A., Joensen, B.O., Moss, A.B. et al. (2023). Influence of headgroups in ethylene-tetrafluoroethylene-based radiation-grafted anion exchange membranes for CO_2 electrolysis. *ACS Sustain Chem Eng* 11: 1508–1517.

37 Li, Z. and Lu, Y.C. (2020). Material design of aqueous redox flow batteries: fundamental challenges and mitigation strategies. *Adv Mater* 32: 2002132.

38 Prifti, H., Parasuraman, A., Winardi, S. et al. (2012). Membranes for redox flow battery applications. *Membranes* 2: 275–306.

39 Bartolozzi, M. (1989). A historical bibliography. *J Power Sources* 27: 219.

40 Wang, T., Jeon, J.Y., Han, J. et al. (2020). Poly(terphenylene) anion exchange membranes with high conductivity and low vanadium permeability for vanadium redox flow batteries (VRFBs). *J Membr Sci* 598: 117665.

41 Shi, Y., Eze, C., Xiong, B. et al. (2019). Recent development of membrane for vanadium redox flow battery applications: a review. *Appl Energy* 238: 202–224.

42 Dang, H.-S. and Jannasch, P. (2016). Anion-exchange membranes with polycationic alkyl side chains attached via spacer units. *J Mater Chem A* 4: 17138–17153.

43 Sánchez-Díez, E., Ventosa, E., Guarnieri, M. et al. (2021). Redox flow batteries: status and perspective towards sustainable stationary energy storage. *J Power Sources* 481: 228804.

44 Shanahan, B., Böhm, T., Britton, B. et al. (2019). 30 μm thin hexamethyl-p-terphenyl poly(benzimidazolium) anion exchange membrane for vanadium redox flow batteries. *Electrochem Commun* 102: 37–40.

45 Park, S.G., Kwak, N.S., Hwang, C.W. et al. (2012). Synthesis and characteristics of aminated vinylbenzyl chloride-co-styrene-co-hydroxyethyl acrylate anion-exchange membrane for redox flow battery applications. *J Membr Sci* 423–424: 429–437.

46 Tempelman, C.H.L., Jacobs, J.F., Balzer, R.M., and Degirmenci, V. (2020). Membranes for all vanadium redox flow batteries. *J Energy Storage* 32: 101754.

47 Wang, W., Luo, Q., Li, B. et al. (2013). Recent progress in redox flow battery research and development. *Adv Funct Mater* 23: 970–986.

48 Choi, H.S., Oh, Y.H., Ryu, C.H., and Hwang, G.J. (2014). Characteristics of the all-vanadium redox flow battery using anion exchange membrane. *J Taiwan Inst Chem Eng* 45: 2920–2925.

49 Hwang, G.J. and Ohya, H. (1997). Crosslinking of anion exchange membrane by accelerated electron radiation as a separator for the all-vanadium redox flow battery. *J Membr Sci* 132: 55–61.

50 Che, X., Tang, W., Dong, J. et al. (2021). Anion exchange membranes based on long side-chain quaternary ammonium-functionalized poly(arylene piperidinium)s for vanadium redox flow batteries. *Sci China Mater* 65: 683–694.

51 Yun, S., Parrondo, J., and Ramani, V. (2014). Derivatized cardo-polyetherketone anion exchange membranes for all-vanadium redox flow batteries. *J Mater Chem A* 2: 6605–6615.

52 Xing, Y., Geng, K., Chu, X. et al. (2021). Chemically stable anion exchange membranes based on C_2-protected imidazolium cations for vanadium flow battery. *J Membr Sci* 618: 118696.

53 Tang, W., Yang, Y., Liu, X. et al. (2021). Long side-chain quaternary ammonium group functionalized polybenzimidazole based anion exchange membranes and their applications. *Electrochim Acta* 391: 138919.

54 Chen, D., Qi, H., Sun, T. et al. (2019). Polybenzimidazole membrane with dual proton transport channels for vanadium flow battery applications. *J Membr Sci* 586: 202–210.

55 Shin, D.W., Guiver, M.D., and Lee, Y.M. (2017). Hydrocarbon-based polymer electrolyte membranes: importance of morphology on ion transport and membrane stability. *Chem Rev* 117: 4759–4805.

56 Yuan, Z., Li, X., Zhao, Y., and Zhang, H. (2015). Mechanism of polysulfone-based anion exchange membranes degradation in vanadium flow battery. *ACS Appl Mater Interfaces* 7: 19446–19454.

57 Mohanty, A.D., Tignor, S.E., Krause, J.A. et al. (2016). Systematic alkaline stability study of polymer backbones for anion exchange membrane applications. *Macromolecules* 49: 3361–3372.

58 Chen, D., Chen, X., Ding, L., and Li, X. (2018). Advanced acid-base blend ion exchange membranes with high performance for vanadium flow battery application. *J Membr Sci* 553: 25–31.

59 Sha'rani, S.S., Abouzari-Lotf, E., Nasef, M.M. et al. (2019). Improving the redox flow battery performance of low-cost thin polyelectrolyte membranes by layer-by-layer surface assembly. *J Power Sources* 413: 182–190.

60 Xiong, P., Zhang, L., Chen, Y. et al. (2021). A chemistry and microstructure perspective on ion-conducting membranes for redox flow batteries. *Angew Chem Int Ed* 60: 24770–24798.

61 Geng, K., Tang, H., Ju, Q. et al. (2021). Symmetric sponge-like porous polybenzimidazole membrane for high temperature proton exchange membrane fuel cells. *J Membr Sci* 620: 118981.

62 Tsehaye, M.T., Alloin, F., Iojoiu, C. et al. (2020). Membranes for zinc-air batteries: recent progress, challenges and perspectives. *J Power Sources* 475: 228689.

63 Hwang, H.J., Chi, W.S., Kwon, O. et al. (2016). Selective ion transporting polymerized ionic liquid membrane separator for enhancing cycle stability and durability in secondary zinc-air battery systems. *ACS Appl Mater Interfaces* 8: 26298–26308.

64 Tsehaye, M.T., Alloin, F., and Iojoiu, C. (2019). Prospects for anion-exchange membranes in alkali metal–air batteries. *Energies* 12: 4702.

65 Li, Y. and Dai, H. (2014). Recent advances in zinc–air batteries. *Chem Soc Rev* 43: 5257–5275.

66 Fu, J., Zhang, J., Song, X. et al. (2016). A flexible solid-state electrolyte for wide-scale integration of rechargeable zinc–air batteries. *Energy Environ Sci* 9: 663–670.

67 Zhang, J., Fu, J., Song, X. et al. (2016). Laminated cross-linked nanocellulose/graphene oxide electrolyte for flexible rechargeable zinc–air batteries. *Adv Energy Mater* 6: 1600476.

68 Abbasi, A., Hosseini, S., Somwangthanaroj, A. et al. (2019). Poly(2,6-dimethyl-1,4-phenylene oxide)-based hydroxide exchange separator membranes for zinc–air battery. *Int J Mol Sci* 20: 3678.

69 Tsehaye, M.T., Choi, N.H., Fischer, P. et al. (2022). Anion exchange membranes incorporating multi N-spirocyclic quaternary ammonium cations via ultraviolet-initiated polymerization for zinc slurry–air flow batteries. *ACS Appl Energy Mater* 5: 7069–7080.

70 Wang, M., Xu, N., Fu, J. et al. (2019). High-performance binary cross-linked alkaline anion polymer electrolyte membranes for all-solid-state supercapacitors and flexible rechargeable zinc–air batteries. *J Mater Chem A* 7: 11257–11264.

71 Wei, Y., Shi, Y., Chen, Y. et al. (2021). Development of solid electrolytes in Zn–air and Al–air batteries: from material selection to performance improvement strategies. *J Mater Chem A* 9: 4415–4453.

72 Fang, W., Zhao, J., Zhang, W. et al. (2021). Recent progress and future perspectives of flexible Zn–air batteries. *J Alloys Compd* 869: 158918.

73 Logan, B.E. and Elimelech, M. (2012). Membrane-based processes for sustainable power generation using water. *Nature* 488: 313–319.

74 Ramon, G.Z., Feinberg, B.J., and Hoek, E.M.V. (2011). Membrane-based production of salinity-gradient power. *Energy Environ Sci* 4: 4423.

75 Pattle, R.E. (1954). Production of electric power by mixing fresh and salt water in the hydroelectric pile. *Nature* 174: 660.

76 Weinstein, J.N. (1976). Electric power from differences in salinity: the dialytic battery. *Science* 191: 557.

77 Zoungrana, A. and Çakmakci, M. (2020). From non-renewable energy to renewable by harvesting salinity gradient power by reverse electrodialysis: a review. *Int J Energy Res* 45: 3495–3522.

78 Hong, J.G., Zhang, B., Glabman, S. et al. (2015). Potential ion exchange membranes and system performance in reverse electrodialysis for power generation: a review. *J Membr Sci* 486: 71–88.

79 Dlugolecki, P., Nymeijer, K., Metz, S., and Wessling, M. (2008). Current status of ion exchange membranes for power generation from salinity gradients. *J Membr Sci* 319: 214–222.

80 Güler, E., Elizen, R., Vermaas, D.A. et al. (2013). Performance-determining membrane properties in reverse electrodialysis. *J Membr Sci* 446: 266–276.

81 Güler, E., Zhang, Y., Saakes, M., and Nijmeijer, K. (2012). Tailor-made anion-exchange membranes for salinity gradient power generation using reverse electrodialysis. *ChemSusChem* 5: 2262–2270.

82 Cho, D.H., Lee, K.H., Kim, Y.M. et al. (2017). Effect of cationic groups in poly(arylene ether sulfone) membranes on reverse electrodialysis performance. *Chem Commun* 53: 2323–2326.

83 Lee, Y.J., Cha, M.S., Oh, S.G. et al. (2019). Reinforced anion exchange membrane based on thermal cross-linking method with outstanding cell performance for reverse electrodialysis. *RSC Adv* 9: 27500–27509.

84 Safronova, E.Y., Golubenko, D.V., Shevlyakova, N.V. et al. (2016). New cation-exchange membranes based on cross-linked sulfonated polystyrene and polyethylene for power generation systems. *J Membr Sci* 515: 196–203.

85 Mei, Y. and Tang, C.Y. (2018). Recent developments and future perspectives of reverse electrodialysis technology: a review. *Desalination* 425: 156–174.

86 Rijnaarts, T., Moreno, J., Saakes, M. et al. (2019). Role of anion exchange membrane fouling in reverse electrodialysis using natural feed waters. *Colloid Surf A Physicochem Eng Aspects* 560: 198–204.

87 Patel, S.K., Biesheuvel, P.M., and Elimelech, M. (2021). Energy consumption of brackish water desalination: identifying the sweet spots for electrodialysis and reverse osmosis. *ACS ES&T Eng* 1: 851–864.

88 Yip, N.Y. and Elimelech, M. (2014). Comparison of energy efficiency and power density in pressure retarded osmosis and reverse electrodialysis. *Environ Sci Technol* 48: 11002–11012.

89 Chandra, A., Tadimeti, J.G.D., Bhuvanesh, E. et al. (2018). Switching selectivity of carboxylic acids and associated physico-chemical changes with pH during electrodialysis of ternary mixtures. *Sep Purif Technol* 193: 327–344.

90 Mohammadi, R., Tang, W., and Sillanpää, M. (2021). A systematic review and statistical analysis of nutrient recovery from municipal wastewater by electrodialysis. *Desalination* 498: 114626.

91 Zhang, Y., Van der Bruggen, B., Pinoy, L., and Meesschaert, B. (2009). Separation of nutrient ions and organic compounds from salts in RO concentrates by standard and monovalent selective ion-exchange membranes used in electrodialysis. *J Membr Sci* 332: 104–112.

92 Li, H. and Zou, L. (2011). Ion-exchange membrane capacitive deionization: a new strategy for brackish water desalination. *Desalination* 275: 62–66.

93 Luo, T., Abdu, S., and Wessling, M. (2018). Selectivity of ion exchange membranes: a review. *J Membr Sci* 555: 429–454.

94 Ward, A.J., Arola, K., Thompson Brewster, E. et al. (2018). Nutrient recovery from wastewater through pilot scale electrodialysis. *Water Res* 135: 57–65.

95 Quéméneur, F., Schlumpf, J.P., Firdaous, L. et al. (2002). Modification of ionic composition of natural salt-waters by electrodialysis. *Desalination* 149: 411–416.

96 Zhang, Y., Liu, R., Lang, Q. et al. (2018). Composite anion exchange membrane made by layer-by-layer method for selective ion separation and water migration control. *Sep Purif Technol* 192: 278–286.

97 Goel, P., Bhuvanesh, E., Mandal, P. et al. (2021). Di-quaternized graphene oxide based multi-cationic cross-linked monovalent selective anion exchange membrane for electrodialysis. *Sep Purif Technol* 276: 119361.

98 Khan, M.I., Zheng, C., Mondal, A.N. et al. (2017). Preparation of anion exchange membranes from BPPO and dimethylethanolamine for electrodialysis. *Desalination* 402: 10–18.

99 Khan, M.I., Luque, R., Akhtar, S. et al. (2016). Design of anion exchange membranes and electrodialysis studies for water desalination. *Materials* 9: 365.

100 Mandal, P., Mondal, R., Goel, P. et al. (2022). Selective recovery of carboxylic acid through PVDF blended anion exchange membranes using electrodialysis. *Sep Purif Technol* 292: 121069.

101 Pismenskaya, N., Bdiri, M., Sarapulova, V. et al. (2021). A review on ion-exchange membranes fouling during electrodialysis process in food industry, Part 2: Influence on transport properties and electrochemical characteristics, cleaning and its consequences. *Membranes* 11: 811.

102 Mulyati, S., Takagi, R., Fujii, A. et al. (2012). Improvement of the antifouling potential of an anion exchange membrane by surface modification with a polyelectrolyte for an electrodialysis process. *J Membr Sci* 417–418: 137–143.

103 Hao, L., Liao, J., Jiang, Y. et al. (2018). "Sandwich"-like structure modified anion exchange membrane with enhanced monovalent selectivity and fouling resistant. *J Membr Sci* 556: 98–106.

104 Jiang, S., Sun, H., Wang, H. et al. (2021). A comprehensive review on the synthesis and applications of ion exchange membranes. *Chemosphere* 282: 130817.

105 Ran, J., Wu, L., He, Y. et al. (2017). Ion exchange membranes: new developments and applications. *J Membr Sci* 522: 267–291.

106 Zhang, C., Zhang, W., and Wang, Y. (2020). Diffusion dialysis for acid recovery from acidic waste solutions: anion exchange membranes and technology integration. *Membranes* 10: 169.

107 Burlakova, E.B., Molochkina, E.M., and Nikiforov, G.A. (2008). Applications of dialysis. *Oxid Commun* 4: 758–775.

108 Luo, J., Wu, C., Xu, T., and Wu, Y. (2011). Diffusion dialysis-concept, principle and applications. *J Membr Sci* 366: 1–16.

109 Xu, T. (2005). Ion exchange membranes: state of their development and perspective. *J Membr Sci* 263: 1–29.

110 Lin, X., Kim, S., Zhu, D.M. et al. (2017). Preparation of porous diffusion dialysis membranes by functionalization of polysulfone for acid recovery. *J Membr Sci* 524: 557–564.

111 Ersoz, M., Gugul, I.H., and Sahin, A. (2001). Transport of acids through polyether-sulfone anion-exchange membrane. *J Colloid Interface Sci* 237: 130–135.

112 Khan, M.I., Mondal, A.N., Emmanuel, K. et al. (2016). Preparation of pyrrolidinium-based anion-exchange membranes for acid recovery via diffusion dialysis. *Sep Sci Technol* 51: 1881–1890.

113 Wang, C., Wu, C., Wu, Y. et al. (2013). Polyelectrolyte complex/PVA membranes for diffusion dialysis. *J Hazard Mater* 261: 114–122.

114 Afsar, N.U., Erigene, B., Irfan, M. et al. (2018). High performance anion exchange membrane with proton transport pathways for diffusion dialysis. *Sep Purif Technol* 193: 11–20.

115 Sharma, J., Misra, S.K., and Kulshrestha, V. (2021). Internally cross-linked poly(2,6-dimethyl-1,4-phenylene ether) based anion exchange membrane for recovery of different acids by diffusion dialysis. *Chem Eng J* 414: 128776.

116 You, X., Chen, J., Pan, S. et al. (2022). Piperazine-functionalized porous anion exchange membranes for efficient acid recovery by diffusion dialysis. *J Membr Sci* 654: 120560.

117 Crini, G. and Lichtfouse, E. (2018). Advantages and disadvantages of techniques used for wastewater treatment. *Environ Chem Lett* 17: 145–155.

118 Sharma, V. and Kundu, P.P. (2010). Biocatalysts in microbial fuel cells. *Enzym Microb Technol* 47: 179–188.

119 Potter, M.C. (1911). Electrical effects accompanying the decomposition of organic compounds. *Proc R Soc London Ser B Contain Papers Biol Charac* 84: 260–276.

120 Gajda, I., Greenman, J., and Ieropoulos, I.A. (2018). Recent advancements in real-world microbial fuel cell applications. *Curr Opin Electrochem* 11: 78–83.

121 Li, W.W., Sheng, G.P., Liu, X.W., and Yu, H.Q. (2011). Recent advances in the separators for microbial fuel cells. *Bioresour Technol* 102: 244–252.

122 Mohanty, A.K., Song, Y.E., Jung, B. et al. (2020). Partially crosslinked comb-shaped PPO-based anion exchange membrane grafted with long alkyl chains: synthesis, characterization and microbial fuel cell performance. *Int J Hydrog Energy* 45: 27346–27358.

123 Scott, K. (2016). Membranes and separators for microbial fuel cells. *Microb Electro Fuel Cells* 153–178.

124 Kim, J.R., Cheng, S., Oh, S.E., and Logan, B.E. (2001). Power generation using different cation, anion, and ultrafiltration membranes in microbial fuel cells. *Environ Sci Technol* 41: 1004–1009.

125 Elangovan, M. and Dharmalingam, S. (2017). Application of polysulphone based anion exchange membrane electrolyte for improved electricity generation in microbial fuel cell. *Mater Chem Phys* 199: 528–536.

126 Elangovan, M. and Dharmalingam, S. (2016). Preparation and performance evaluation of poly(ether-imide) based anion exchange polymer membrane electrolyte for microbial fuel cell. *Int J Hydrog Energy* 41: 8595–8605.

127 Koók, L., Žitka, J., Bakonyi, P. et al. (2020). Electrochemical and microbiological insights into the use of 1,4-diazabicyclo[2.2.2]octane-functionalized anion exchange membrane in microbial fuel cell: a benchmarking study with Nafion. *Sep Purif Technol* 237: 116478.

128 Rossi, R. and Logan, B.E. (2021). Using an anion exchange membrane for effective hydroxide ion transport enables high power densities in microbial fuel cells. *Chem Eng J* 422: 130150.

129 Pandit, S., Ghosh, S., Ghangrekar, M.M., and Das, D. (2012). Performance of an anion exchange membrane in association with cathodic parameters in a dual chamber microbial fuel cell. *Int J Hydrog Energy* 37: 9383–9392.

130 Zhang, X., Cheng, S., Huang, X., and Logan, B.E. (2010). Improved performance of single-chamber microbial fuel cells through control of membrane deformation. *Biosens Bioelectron* 25: 1825–1828.

131 Peixoto, C., Ferreira, T.B., Sousa, M.F.Q. et al. (2008). Towards purification of adenoviral vectors based on membrane technology. *Biotechnol Prog* 24: 1290–1296.

132 Charcosset, C. (2006). Membrane processes in biotechnology: an overview. *Biotechnol Adv* 24: 482–492.

133 Weaver, J., Husson, S.M., Murphy, L., and Wickramasinghe, S.R. (2013). Anion exchange membrane adsorbers for flow-through polishing steps: Part II. Virus, host cell protein, DNA clearance, and antibody recovery. *Biotechnol Bioeng* 110: 500–510.

134 Kalbfuss, B., Wolff, M., Geisler, L. et al. (2007). Direct capture of influenza a virus from cell culture supernatant with sartobind anion-exchange membrane adsorbers. *J Membr Sci* 299: 251–260.

135 Saxena, A., Tripathi, B.P., Kumar, M., and Shahi, V.K. (2009). Membrane-based techniques for the separation and purification of proteins: an overview. *Adv Colloid Interf Sci* 145: 1–22.

16

Research Challenges and Future Directions on Anion Exchange Membranes for Fuel Cells

Jince Thomas[1], Parthiban Velayudham[2], Ramesh K. Singh[3], Sabu Thomas[1], Alex Schechter[2,4], and Flavio Grynszpan[2]

[1] *Mahatma Gandhi University, International and Inter University Centre for Nanoscience and Nanotechnology, Kottayam 686560, Kerala, India*
[2] *Ariel University, Department of Chemical Sciences, Ramat HaGolan St 65, Ariel Research Park, Ariel 40700, Israel*
[3] *Vellore Institute of Technology, CO_2 Research and Green Technologies Centre, Vellore 632014, Tamil Nadu, India*
[4] *University of West Bohemia, New Technologies – Research Centre (NTC), Teslova 1198/9, Pilsen 30100, Czech Republic*

16.1 Prelude to Anion Exchange Membranes

Fuel cell technology utilization for energy generation is an environmentally sustainable approach that can contribute toward the attainment of carbon neutrality. Fuel cells are a key technology in the H_2 economy, generating clean energy from H_2 through a series of electrochemical reactions involving an anode, cathode, electrolyte, separated by an electrolyte and connected by an external circuit [1]. Recent technological advancements have demonstrated the effectiveness of fuel cells as a power source for medium-to-large-scale backup power and vehicle applications, such as automobiles and buses. In particular, fuel cell-powered electric vehicles offer a more extended driving range and shorter refueling time than rechargeable battery-powered vehicles. For instance, Toyota has reported that its hybrid fuel cell vehicle can travel up to 830 km using 5 kg of H_2 stored in 700 bar pressure gas tanks [2–4]. From this standpoint, we examine the current progress in the development of fuel cells based on polymer electrolyte membranes. These fuel cells run at relatively low temperatures (60–90 °C), start up quickly, are portable, and have earned significant attention for their potential commercial applications. Low-temperature fuel cells can be categorized into two classes, based on their electrolytes: proton exchange membrane (PEM) and anion exchange membrane (AEM)-based cells. The PEM-based cells have been under constant development for

Alkaline Anion Exchange Membranes for Fuel Cells: From Tailored Materials to Novel Applications, First Edition. Edited by Jince Thomas, Alex Schechter, Flavio Grynszpan, Bejoy Francis, and Sabu Thomas.

decades and are commercially ready with mature technology. However, due to their intrinsic strong acid environment, these cells require critical components such as noble metal electrocatalysts (Pt, Ir, and Ru) for the anode and cathode reactions, graphitic or Ti-based bipolar plates, and perfluorosulfonic acid polyelectrolytes. Precious metal catalysts and PEM are crucial components in PEM-based devices, but their usage contributes significantly to the high cost of fuel cell systems [5]. Moreover, the limited availability of these catalyst elements on Earth exacerbates the issue of their scarcity. Although proton exchange membrane fuel cells (PEMFCs) have achieved commercial success, their cost has restricted their broad propagation. Fuel cells that integrate AEMs present a viable substitute to PEMFCs, providing comparable efficiency at a lower cost. Hydroxide (OH^-) ion redox reactions have gained attention as an alternative to using H^+ in ion transport. Fuel cells that utilize OH^- transporting membranes are known as anion exchange membrane fuel cells (AEMFCs). Although the working principles of AEMFCs differ slightly from those of PEMFCs, their overall redox fundamentals remain the same. In an AEMFC, water and oxygen are introduced at the cathode where O_2 is reduced to OH^-. The OH^- then travels through the AEM to meet the fuel, MeOH, or H_2, which gets oxidized to generate water and electrons, which flow through the system producing the desired electrical current. The water produced at the anode will diffuse back to the cathode to replace the consumed water during the cathodic reaction. The essential elements of AEMFCs comprise AEM, catalyst layers, bipolar plates, and gas/liquid diffusion layers. Among the components of AEMFCs, the AEM and catalyst layer are particularly crucial and have received extensive research attention. In recent decades, a significant number of non-precious metal electrocatalysts have been developed with even better electrocatalytic rate than noble metal benchmarks [6]. However, these electrocatalysts often do not perform well in the AEMFCs due to the undesirable properties of the AEM and the inadequate design of the AEM electrode's three-phase interfacial structure. The AEM is a solid polymer electrolyte that effectively conducts anions and acts as a separator between the cathode and anode. It is composed of positively charged groups situated on the side chains or in the polymer backbone, which permit efficient anion transport. Several commercial AEMs have been reported, such as Aemion from Ionomr, Sustainion from Dioxide Materials, PiperION from Versogen, and FAA-X from Fumatech. However, these AEMs do not have a competitive ionic conductivity compared to state-of-the-art PEMs. Hence, only a few AEMs can provide a peak power density of over $1\,W\,cm^{-2}$ in fuel cells. Moreover, the commercially available AEMs fall short of meeting the device lifespan requirements due to their lack of stability, including alkaline, mechanical, and electrochemical stability.

Although some issues with engineering have caused delays in the implementation of AEMFCs, there has been significant progress in overcoming the challenges they face when being incorporated into devices. In fact, a lot of work has been done to address the obstacles that AEMs experience during the integration process. It is worth noting that some AEMFCs have achieved power densities of $1–3\,W\,cm^{-2}$, which is comparable to those of PEMFCs. Most of the advancements in this field have occurred after 2015, and it is exciting to foresee what the next decade will bring regarding further progress for AEMFCs. During the initial stages of AEM research, the practical implementation of AEMFCs, the most significant challenges were the AEMs' insufficient ionic conductivity and alkaline stability. However, researchers have made considerable progress, and effective strategies have been developed to overcome these challenges. These strategies include constructing microphase separation and regulating micropores to facilitate ion conduction rates and reduce hindrance. As a result, the conductivity has increased from less than $10\,mS\,cm^{-1}$ to exceed $100\,mS\,cm^{-1}$. Furthermore, developing an aryl ether-free backbone and stable *N*-heterocyclic ammonium groups has taken the alkaline stability to a new level. The ex-alkaline stability of the membranes has surpassed 2000 h, demonstrating the remarkable progress made in this field as we approach 2023 [7–16]. The previous chapters of this book thoroughly explored the significance of the polymer backbone in AEM design, along with different ways of integrating cations into the polymeric chain. This involves using chemical reactions such as the Menschutkin reaction, copper-catalyzed azide-alkyne cycloaddition (CuAAC) Click reaction, Suzuki coupling reaction, Ullmann coupling reaction, Leuckart reaction, and Paal–Knorr reaction in the fabrication process.

16.2 Progress in AEM Development

In the realm of AEM development, there are two main types of polymer matrices: ether-containing polyelectrolytes and ether-free polyelectrolytes. In the past, AEMs were primarily made using aryl ether polymer backbones. These were easily prepared using nucleophilic aromatic substitution (S_NAr) reactions between oxygenated nucleophiles and aromatic carbon electrophiles. This method gently creates covalent bonds between aryl moieties and heteroatoms, resulting in polyarylethers with excellent engineering properties like high processability, molecular weight, mechanical strength, and low synthetic cost. By modifying through halogenation, amination, and quaternization, various structures of AEMs can be prepared, including comb-shaped or cross-linked arrangements [17–21]. Some examples of AEMs that contain ether links include those made from poly(arylene ether phosphine oxide) with aliphatic cationic side chains [22], poly(hydroxy-triptycene ether sulfone) with pyrazolium cross-linkers [23], or polyethylene (PE) glycol with imidazolium side chains [24]. These AEMs have decent ionic conductivities of around 50–$100\,mS\,cm^{-1}$ and a peak power density of approximately $1\,W\,cm^{-2}$. Nevertheless, polyarylethers face a significant problem regarding their membrane stability. This is due to the presence of electrophilic groups, such as aryl ethers and aryl sulfones, which are reactive under basic conditions and at high temperatures. As a result, these groups can facilitate the OH^--mediated hydrolysis of ethers and quaternary carbons. This can lead to substituting the carbon near the electronegative oxygen atom with hydroxide, resulting in the degradation of the polymer backbone (as shown in Figure 16.1). This ultimately weakens the chemical stability and mechanical integrity of AEMs [18, 25]. Moreover, acidic phenols and benzyl alcohols resulting from polymer fragmentation can neutralize the OH^- in fuel cells, thereby

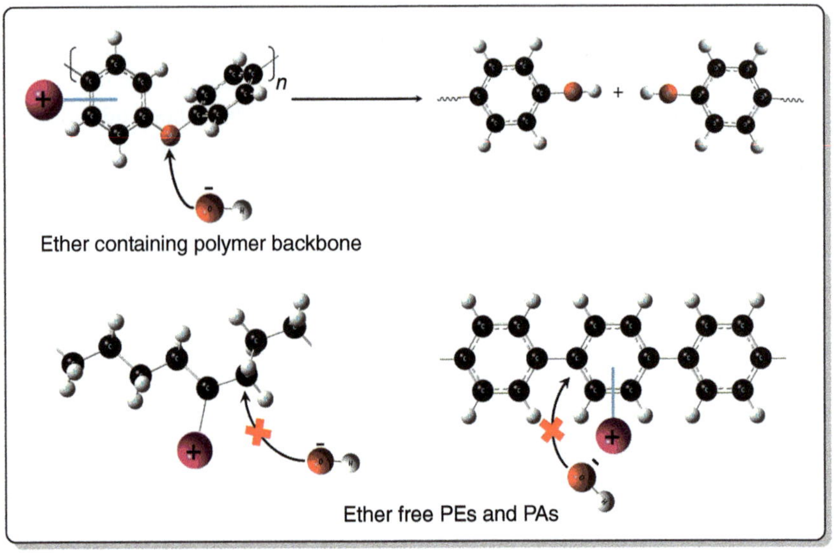

Figure 16.1 The schematic illustration of the degradation mechanism of ether-containing AEMs, mediated by OH^-.

degrading AEM performance. Due to this instability, only four studies have shown the membrane conductivity that retained over 70% of its initial value upon testing in alkaline conditions for up to 750 h [2]. Currently, the primary focus in the field is to create AEMs that are chemically and mechanically durable that do not rely on an ether backbone. These AEMs have proven to be much more durable in tests with alkaline-stable cations when compared to ether polymer chains. Ether-free AEMs have been shown to maintain membrane conductivities of over 93% of their initial values after testing in alkaline conditions for approximately 2100 h, making them the most effective option [26–31]. When examining ether-free polyelectrolytes, the two most significant and intensively researched types are polyarylene-based AEMs (PA-AEMs) and PE-based AEMs (PE-AEMs).

16.2.1 Polyarylene-Based AEMs

Polyarylenes (PAs) are polymers with aromatic backbones or cationic rings like imidazolium or benzimidazolium. Examples of PAs include polyimidazoliums, polybenzimidazoliums, polyfluorenes, and polyphenylene. In the initial stages of PAs for AEMFCs between 2015 and 2017, the peak power density values of ether-free cells were below $0.5\,W\,cm^{-2}$, which was lower than those of ether-containing cells. However, the peak power density values of PAs have been increasing since 2018. As of 2021, they have reached $2.58\,W\,cm^{-2}$ [32] and above[33, 34]. High-performing PAs commonly use cyclic ammonium, tetramethyl ammonium (TMA), *N*-spirocyclic, imidazolium, phosphonium, guanidinium, and piperidinium cations as their main cations. The most popular method for developing the AEM is Friedel–Crafts polycondensation, as it enables the attachment of substituents to aromatic rings. Alternatively, some studies have used metal-mediated coupling reactions and Diels–Alder cycloaddition approaches, which allow for direct C—C bond and aromatic ring formation, respectively. The Holdcroft group conducted one of the first studies on PAs, in which they synthesized polybenzimidazoliums through the condensation of amino aryl and acid aryl compounds. Despite the resulting cells having low conductivity and power density, their research proved the effectiveness of condensation reactions in producing polyelectrolytes for AEMs [35]. In a study conducted by the Jannasch group, Friedel–Crafts polycondensation was used to create stable and highly conductive AEMs. Copolymers were formed by reacting trifluoromethanesulfonic acid with biphenyl, piperidinone, or trifluoroacetone monomers. The piperidine moieties were then quaternized, resulting in the formation of PAs containing *N*-spirocyclic quaternary ammonium. These PAs showed an impressive conductivity of $102\,\mathrm{mS\,cm^{-1}}$ despite the cations being degraded via β-elimination and ring-opening nucleophilic substitution [36]. This approach was further developed by the Yan and Lee groups, who used *p*-terphenyl instead of biphenyl and replaced the *N*-spirocyclic quaternary ammonium with *N, N*-dimethylpiperidinium. Remarkably, the resulting polymer retained a conductivity of over $75\,\mathrm{mS\,cm^{-1}}$ for up to 2000 h [37]. The Friedel–Crafts polycondensation method is a reliable way for polymerizing cationic rings into cationic polymers. Though current research has focused on readily available monomers, developing and synthesizing new aromatic rings with cations could yield exceptional AEMs with distinct and valuable properties. Apart from the Friedel–Crafts polycondensation method, various other reactions such as Diels–Alder polymerization [38], CuAAC Click reaction [39], Suzuki coupling reaction [40], Ullmann coupling reaction [41], Paal–Knorr reaction [42], and Suzuki–Miyaura reaction [7] have been explored for developing effective AEMs.

The Kim's group's research on poly(ammonium phenylenes) synthesized through Diels–Alder polymerization yielded impressive results. They utilized phenyl-substituted cyclopentadienone as the diene and 1,4-bis(ethynyl)benzene as the dienophile to produce a high-molecular weight polymer with $Mn > 50\,000$. After reducing ketone groups and functionalizing with cations, they successfully created polyelectrolytes for AEMs with an initial conductivity of $120\,\mathrm{mS\,cm^{-1}}$. Even after being subjected to 1000 h in 4 M NaOH at 80 °C, over 75% of the initial conductivity was retained [43]. Another notable structural analog of the PA backbone is a porous polymer containing Tröger's base units. With its V-shaped bridged bicyclic aromatic diamine units, this polymer exhibited microporosity and phase segregation, resulting in a high conductivity of $164\,\mathrm{mS\,cm^{-1}}$ [13]. Also, the ether-free nature ensured both alkaline and dimensional stability, making these polymers a promising subject for further research.

16.2.2 Polyethylene-Based AEMs

Polyethylenes (PEs) are polymers that do not contain neither ether nor aromatic moieties. They are formed by polyaddition reactions that create C—C single bonds, which give PEs high number-average molecular weights. PE-based AEMs are durable and resistant to OH^- attack. Their backbone is free from aromatic moieties, which prevents the formation of acidic phenols at the cathode and ensures optimal performance in cells. PEs have unique advantages in terms of mechanical strength due to their high-molecular-weight linear chains that improve their maximum tensile strain and strength. Compared to non-PE-based membranes, PE-based membranes have a lower glass transition temperature and greater resilience, making them more durable. Consequently, ether-free PEs have significantly enhanced the peak power density of cells, i.e. showing a similar increasing trend of peak power density to that of PA-based cells. The step-by-step development of ether-free polymers used in AEMs, including the methods used to create them, is shown in Figure 16.2. Some examples of AEMs based on PE include ammonium PEs created through radiation grafting [44], imidazolium PEs made through ring-opening metathesis polymerization (ROMP) [45], ammonium PEs produced through Ziegler–Natta polymerization [46], and piperidinium PEs formed through ROMP [47]. The earliest examples were pyrrolidinium polymers made by the Jannasch group using radical polymerization of monomers generated via the diallylation of a cyclic amine. The resulting polymers were then quaternized and cyclopolymerized through a radical initiation process mediated by an oxidizing agent [48].

The Hickner group successfully utilized Ziegler–Natta polymerization to produce ammonium-functionalized PE-AEMs, resulting in copolymers that proved highly effective in a cell, achieving a peak power density of $1.01\,W\,cm^{-2}$ at $60\,°C$ [49]. Ziegler–Natta polymerization is well-known for producing cost-effective AEMs with exceptional alkaline and electrochemical stability. An efficient

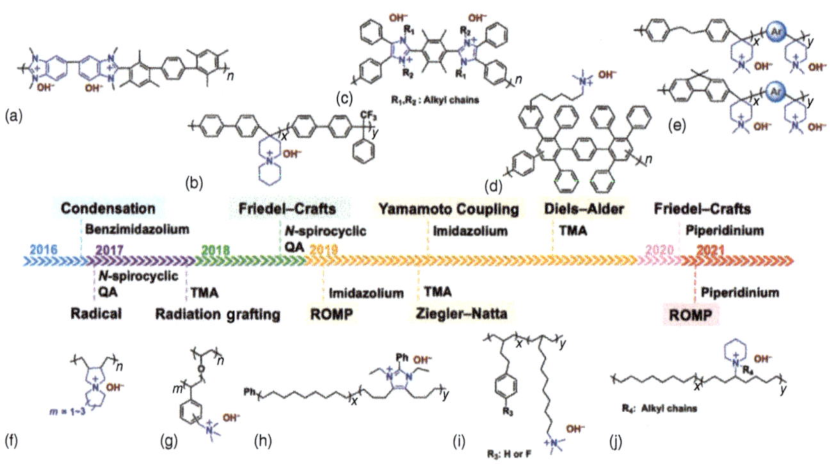

Figure 16.2 The timeline progression of ether-free polymers used in AEMs and the processing methods. (a) Polybenzimidazolium, (b) polyarylpiperidinium, (c) polyimidazolium, (d) polyarylammonium, (e) polydiphenylethanepiperidinium and polyfluorenepiperidinium, (f) polyallylpiperidinium, (g) polyethylene-*co*-tetrafluoroethyleneammonium, (h) polyethyleneimidazolium, (i) polyethyleneammonium, and (j) polyethlenepiperidinium. Source: Reproduce R with permission from Chen et al. [2].

and straightforward method to enhance the features and capabilities of membranes is by using radiation grafting. The Varcoe group conducted an experiment where they activated low-density PE membranes using electron beam bombardment, which allowed them to install ammonium groups. As a result, their PE-AEM had a hydroxide conductivity of 145 mS cm^{-1} and a peak power density of 1.45 W cm^{-2} at 80 °C [50]. The ROMP process involves using cyclooctadienes and norbornenes as typical monomers to create polymer products with an all-carbon backbone free of ether bonds. The Coates group created new materials by combining cyclooctene-fused imidazolium monomers through ROMP to form imidazolium polymers. These polymers proved to be highly stable, with less than 1% loss of cations or 4% loss in conductivity, when tested in PE-AEM cells after 30 days. At 80 °C, they demonstrated high conductivity levels of 134 mS cm^{-1} [45]. The same group also developed a piperidinium-modified polymer using piperidine-functionalized cyclooctadiene. This PE-based polymer exhibited high conductivity levels of up to 100 mS cm^{-1} in PE-AEM cells [47].

Although ether-free PAs and PEs are more stable and conductive than ether-containing polymers, some concerns surround their practical application. The synthesis of ether-free polymers is usually more complex than the S_NAr-based synthesis of aryl ether polymers. For example, multiple synthetic steps are required to synthesize the required well-defined monomers in the case of PAs prepared by polycondensation, such as Yamamoto polymerization and Diels–Alder polymerization. Friedel–Crafts polymerization is simpler regarding monomer synthesis, but it requires using trifluoromethanesulfonic acid, one of the most potent acids. Similarly, preparing PEs for AEMs is also challenging. Synthesizing the necessary ionic-moiety-bearing cyclooctene monomers requires multi-step synthesis in the case of PEs designed by ROMP, and ROMP requires expensive ruthenium-based catalysts. Direct polymerization of ethylene with polar monomers is complex in Ziegler–Natta polymerization due to the low tolerance of early transition metal catalysts that are optimal for ethylene polymerization. Therefore, ionic moieties are typically added to PEs via post-modification using intensive electromagnetic radiation or polymerizing α-olefins with halide functional groups.

The development of AEMs progresses through various backbone structures, including main chains, block copolymers, long side chains, cross-linked and network backbones, organic–inorganic composites, and cationic functional groups.

16.2.3 Main Chain-Based AEMs

The primary chain of the AEM undergoes halogenation via chloromethylation or bromination, after which it is quaternized by means of polymerization. In the early stages of AEMs, commonly used polyphenoxyether, polyaryl ether sulfone, poly(arylene ether ketone), and poly(aryl piperidinium) are used as the main chains due to their affordability, ability to form membranes well, and ease of modification. However, the chemical stability of these back chain-based membranes leads to the breakdown of benzyl trimethylammonium and ether-containing main chains in strongly alkaline environments. Within the main chain structure, electron withdraw groups ,such as C=O and S=O, can speed up the breaking of aryl ether bonds and the degradation of benzyl quaternary ammonium groups. To address these issues, Zhang employed the Leuckart reaction to reduce the C=O group in traditional poly(arylene ether ketone) and created an exceptionally stable conductive polymer with C–NH_2 groups that donate electrons [51]. In addition, the polymeric backbone can be synthesized through direct polymerization of monomers without the presence of ether bonds. A milestone in developing ether-free AEM has been achieved by using methylated polybenzimidazole to prepare a polydialkylated phenylene–benzimidazole backbone [52].

16.2.4 Block Copolymer Based AEMs

It is widely recognized that most AEMs have difficulty transporting anions with low ion exchange capacity (IEC) due to inadequate microstructure control of the amorphous polymer. Researchers have combined two or more polymers to overcome this challenge, forming block-type multi-cationic clusters of anion exchange polyelectrolytes. This approach enhances the separation of hydrophilic and hydrophobic phases. The block copolymer facilitates the synergistic combination of ionic and non-ionic polymers at the molecular level, promoting the formation of ion transport channels and improving OH^- transport efficiency. Zhu et al. [53] created a triblock copolymer (PSm-PDVPPA2n-PSm) using a unique monomer (*N,N*-diethyl-3-(4-vinylphenyl) propan-1-amine) with a trialkyl amine group and polystyrene trithiocarbonate via reversible addition-break chain transfer. The alkyl spacer group links the quaternary ammonium group to the backbone, while the nearby benzene ring shields the ion center, resulting in excellent stability for the AEM. After immersing the membrane in a 10 M NaOH solution at 80 °C for 20 days, it still maintained about 93% of its initial OH^- conductivity, which was quite remarkable. In another work of Zhu et al. [54], a fluorine-containing hydrophobic block was integrated into the QA diblock copolymer styrene to create two diblock copolymers (PVBC263-*b*-P4FSy and PVBC263-*b*-PPFSy). The resultant AEM had excellent ionic conductivity (86.1 $mS\,cm^{-1}$ at room temperature) and IEC (4.27 $mmol\,g^{-1}$ at 90 °C). Furthermore, the presence of rigid structures such as diphenyl, fluorenyl, aromatic pyrene units, and aromatic heterocyclic compound carbazole with a tricyclic structure improved the microphase separation, thereby enhancing the dimensional stability and electrical conductivity of AEM [55]. The π–π stacking of electron-rich aromatic pyrene units is expected to increase the aggregation of adjacent cations and the development of well-connected ion transport channels [29]. Combining two or more oligomers through copolymerization makes it possible to create a block structure in the AEM that leads to a well-balanced separation between hydrophilic and hydrophobic phases. This results in AEM with exceptional high electrical conductivity. However, there is still a concern about the cationic groups being too close to the main chain. Even if the cationic functional groups have steric hindrance, they may not fully protect the ether-containing segment of the polymer backbone from OH^- attacks in an alkaline environment.

16.2.5 Long Side-Chain AEMs

The high rigidity of the polymer backbone can limit the mobility and activity of charged groups, ultimately impacting their performance. Researchers have incorporated alkyl or other flexible chains to overcome this obstacle while creating unique structures with long ion-bearing segments. This approach enhances the separation distance between the cationic groups and the polymer skeleton and leads to the formation of a structure with microphase separation morphology. As a result, the structure can self-assemble into hydrophilic channels, promoting efficient ion transport. By experimenting with different alkyl chain lengths and cationic groups in various positions, researchers can create a range of hydrophilic/hydrophobic pathways. Wang's group [56] has successfully created a highly flexible anion exchange polyelectrolyte using amphophilic cationic side chains with three ether bonds. This was accomplished by forming an interconnected network throughout the hydrophobic backbone through ion clustering. The AEM's remarkable toughness and swelling resistance can be attributed to the strong hydrogen bonding between the OH^- ion and pyridine nitrogen while maintaining high conductivity levels of up to 131.8 mS cm^{-1} (80 °C). Even after being immersed in 2 M KOH for 672 h (60 °C), the ionic conductivity and IEC remained at 78.0% and 84.2%, respectively. Xu's group [57] has suggested utilizing double cationic and fluorinated side chains to create enhanced microphase separation structures, building upon the flexible ether bond-containing side chain. Molecular dynamics simulations have shown that piperidine and ether bonds in the longer side chains can interact strongly with water molecules, promoting OH^- transport. The resulting fluorinated AEM has demonstrated the highest conductivity (168.46 mS cm^{-1} at 80 °C) with a similar IEC, attributed to the formation of ion transfer channels. Researchers are now concentrating on polymers with dense functional groups and hyperbranched structures to create more prominent microphase separation structures. Grafting high-density cationic groups to the polymer's main chain has created numerous transport sites, significantly improving OH^- transport efficiency [58–60]. Long-side chains inevitably induce the formation of hydrophilic and hydrophobic phase separation structures. These structures are crucial for enhancing AEM's conductivity and chemical stability, as the hydrophilic domains are well-connected. However, increasing the hydrophilic region can improve water absorption, but it may also cause excessive membrane expansion and affect mechanical stability.

16.2.6 Cross-linked AEMs

As the IEC and water uptake increase in AEMs, their ionic conductivity usually increases. However, this could lead to decreased mechanical robustness of the membrane. It is imperative to undergo cross-linking to minimize swelling and establish a compact network structure in the AEM. This step is crucial in enhancing its mechanical stability, thus rendering it an indispensable part of the process that should not be disregarded. Cross-linking involves physical and chemical methods to form intermolecular bonds within the membrane. Physical cross-linking introduces ion–ion interactions, van der Waals forces, or hydrogen bonds between polymeric matrices and cations. The synthesis pathways for different cross-linked AEMs are shown in Figure 16.3.

Additionally, chemical cross-linking involves creating covalent bonds between polymer chains using either oligomers or end groups. The cross-linking process can occur in a one-step synthesis during polymerization or through a post-polymerization cross-linking step. Covalent cross-linking is a promising approach to balance the AEM's ionic conductivity and mechanical stability. Various cross-linking techniques have been reported, such as mercaptan-ene chemistry, Menshutkin reaction, ROMP, olefin metathesis, and thermal cross-linking. The technique of chemical cross-linking is commonly used in current AEM research. In Yang's study [61], the addition of 1-vinyl imidazole as a cross-linking agent and cationic group resulted in remarkable improvements in the mechanical strength and alkali-resistant stability of the AEM. Kim's group [62] developed *x*-PBB-SEBS membranes by cross-linking by combining SEBS with poly(biphenylbenzyl *n*-methylpiperidine). The *x*-PBB-SEBS membranes have exceptional properties due to the excellent phase separation behavior of SEBS and PBB. These properties include an outstanding elongation at a break of 70.3% and a high Young's modulus of 486 MPa. The water retention of the *x*-PBB-SEBS membranes increases with an increase in the cross-linking degree. This enhances the interaction between water and ionic conducting groups, resulting in high ionic conductivity values of up to 146.25 mS cm^{-1} (80 °C). Cross-linking the two polymers without an aryl ether structure has also given the AEM excellent

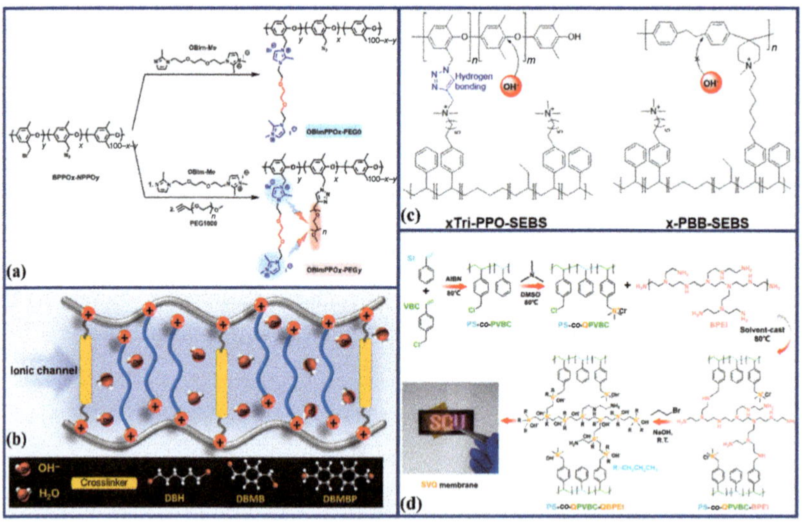

Figure 16.3 The cross-linked AEM synthesis. (a) Grafted a flexible side chain with an alkoxy group onto the polyphenylene oxide backbone using polyethylene glycol; (b) influence of flexible (4,4′-dihydroxydiphenyl ether) and rigid (1,6-dibromohexane, 1,4-bis (bromomethyl) benzene) cross-linking agents on the development of ion transport channels in the AEM; (c) *x*-PBB-SEBS membranes by cross-linking SEBS with poly(biphenylbenzyl *n*-methylpiperidine); (d) semi-IPN AEM (PS-*co*-PVBC).
Source: Reproduce with permission from Chen et al. [16]. Copyright 2023 Elsevier.

alkali stability. The conductivity remains above 99%, and the IEC remains almost unchanged after a 720 h treatment with 2 M KOH solution at 80 °C. In recent years, the semi-interpenetrating polymer network (semi-IPN) has emerged as a successful approach to balancing the AEM's ionic conductivity and mechanical stability. This is achieved by cross-linking one polymer while leaving the other non-cross-linked [63]. The work of Zhao [64] utilized styrene and 4-vinylchlorobenzene (VBC) as monomers, both highly stable, and branched polyethylenimine (BPEI) as a cross-linking agent. BPEI offers the benefits of both semi-IPN and cross-linking methods, which allows it to serve as a supplementary substrate for creating interconnected ion transport channels over long distances. This eliminates the need for additional cross-linking agent monomers. Using natural polymerization cross-linking, a uniform semi-IPN AEM (PS-*co*-PVBC) was successfully synthesized, resulting in a cross-linking network that effectively protects cations from OH^- attacks. This AEM is finely tuned and has exceptional hydroxide conductivity (106 mS cm^{-1}, 80 °C), maintaining a residual conductivity of over 81% even after 720 h of immersion in 1 M NaOH (60 °C). These findings highlight the benefits of combining a pure carbon skeleton polymer with a semi-IPN network structure, improving the AEM's long-term applicability in the AEMFC field. Cross-linking can improve the stability of the membrane, but it must be controlled carefully to avoid compromising the IEC and alkali stability. Cross-linking limits water uptake and may slow down OH^- migration.

16.2.7 Organic–Inorganic Composite AEMs

Developing organic–inorganic hybrid materials presents a promising avenue for enhancing AEM performance. This involves generating ion nanoaggregates within the membrane that exhibit exceptional ionic conductivity. There are two types of hybrid membranes, each determined by the forces that exist between the two phases. The first method involves weaker interaction forces, such as hydrogen bonds, van der Waals forces, and electrostatic forces. In contrast, the second approach entails synthesizing the two phases with a chemical bond. The numerous active sites on the surface and inner pores of inorganic nanomaterials allow for the grafting of cationic groups, effectively increasing the ion conduction capacity of the AEM. Shang et al. [65] developed the AEM by combining organic and inorganic components to create a highly conductive material. They used a silicon-based material to form the inorganic phase and created a high-density mesoporous molecular sieve (SBA-NH_2) precursor. To increase the number of cations in the void, they added a crown ether to the pore channel of the molecular sieve. By cross-linking the crown ether-functionalized SBA-NH_2 to the main chain of PVA through the sol–gel method, they created the AEM. Incorporating a mesoporous structure resulted in a significant number of pores, which facilitated the formation of highly ordered ion channels and improved the transport efficiency of OH^-. The resulting AEM exhibited superior ionic conductivity (107 mS cm^{-1} at 80 °C), and the ionic conductivity value decreased by only 2% after 168 h of immersion in 6 M KOH (80 °C). Chen et al. [66] have developed a cutting-edge multidimensional AEM that incorporates a one-dimensional polyphenylene oxide polymer chain into a two-dimensional layered covalent organic framework (COF). The membrane's surface is coated with sulfonated graphene oxide using ionic and π–π interactions between benzene rings, creating a sandwich structure that effectively prevents excessive expansion, nucleophilic attacks, and degradation of the quaternary ammonium group. Remarkably, after an alkali treatment (2 M NaOH, 80 °C at 840 h), the membrane's ionic conductivity remained at 92% of its original value. Molecular dynamics simulation results suggest that the well-organized hydrogen bond network within the COF channel facilitates directional transfer of OH^- along the low barrier channel, resulting in excellent OH^- conductivity (151 mS cm^{-1} at 80 °C). Overall, this AEM's distinctive sandwich structure exhibits excellent ion conductivity, making it a highly promising strategy for the preparation of composite membranes in the future. Organic–inorganic composite membranes exhibit superior chemical and mechanical stability compared to AEMs from conventional organic polymers and enhance the conductivity and stability through rational architectural design.

16.2.8 AEMs Based on Cationic Functional Groups

In order to achieve long-term stable operation in an alkaline environment and optimize the electrochemical performance of AEMs, the presence of high-performing cationic functional groups is crucial. At present, this is achieved by fine-tuning the design of the molecular structure to minimize the number of electron-absorbing groups, enhance steric hindrance, and encourage the hydration of OH^- ions. Such a design considers the degradation mechanism of these groups, ultimately strengthening the alkaline stability of the cationic functional groups. Throughout the preceding chapters of this book, a thorough description has been conducted on various AEMs featuring cationic head groups such as quaternary ammonium, guanidinium, imidazolium, triazole, phosphonium, and metal cations. Although AEMs containing quaternary phosphonium and guanidinium exhibit exceptional stability, their development is constrained by low ionic conductivity and inadequate compatibility, which could benefit from further investigation in the future.

16.2.9 Challenges Developing Long-Lasting AEMs

Regarding high-performance polyelectrolytes for use in AEMs, there are three critical factors to consider: chemical stability, mechanical stability, and ionic conductivity. The approach to achieving high-performance AEMs is demonstrated in Figure 16.4, which outlines the applicable standards and methods. The polyelectrolytes must have ether-free backbones and alkaline-stable cations to ensure optimal chemical stability. Adequate mechanical stability requires sufficient mechanical properties and low water uptake. Lastly, a high IEC and an ion percolating network are necessary to achieve excellent ionic conductivity. Achieving a balance between these three properties can be challenging since the most conductive polymers may not be the most durable AEMs. Nonetheless, some promising approaches have been identified in recent years that could lead to developing the next generation of AEMs.

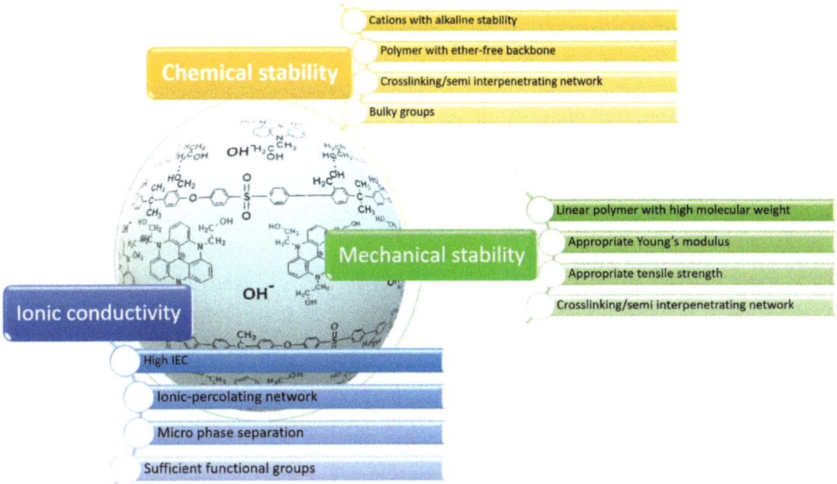

Figure 16.4 The criteria for high-performance AEMs are chemical stability, mechanical stability, and ionic conductivity.

16.2.10 Chemical Stability

A key obstacle that significantly limits the durability of cells is the chemical stability of AEMs, which encompasses both alkaline and oxidative stability. The headgroup influences chemical stability. Significantly, notable progress has been accomplished in advancing alkali-stable membrane materials in recent years, as discussed in the previous sections.

16.2.10.1 Alkaline Stability

The assessment of membrane stability is crucial, but there is currently no widely accepted protocol for evaluating it. The testing conditions for alkaline stability can vary between studies, which makes it difficult to compare different membranes due to differences in solution concentration (ranging from 1 to 10 M aqueous NaOH or KOH) and temperature range (from room temperature to 100 °C). The customary method for evaluating *ex situ* membrane stability involves monitoring changes in the IEC or conductivity during immersion in aqueous alkali over a period of time. Functional groups play a significant role in the stability of alkaline membranes. Various techniques are utilized to enhance the stability of the head group, including inserting a spacer between the head group and backbone, using a more prominent head group, and incorporating electron-donating groups. Recent literature suggests that AEMs can maintain stable fuel cell functioning for more than 2000 h under alkaline conditions. However, relying solely on changes in conductivity is not an accurate way to assess alkaline stability. A comprehensive analysis of the IEC, conductivity, mechanical properties, and water absorption behavior of alkali-aged membranes is necessary to fully comprehend their stability properties. Furthermore, it would be advantageous to develop new *ex situ* alkaline stability test protocols that consider operating environments that are closer to real working conditions, as well as create more alkali-stable membranes.

16.2.10.2 Oxidative Stability

It is important to note that, apart from considering alkali stability, it is crucial to pay attention to oxidation stability. Numerous literature sources have documented that the deterioration of membranes occurs when subjected to the reactions with radicals, predominantly superoxide and hydroxyl radicals. Ramani and coworkers [67], Kruczała and coworkers [68], and Dekel and coworkers [69] utilized advanced techniques during AEMFC operation to detect the formation and existence of radicals in alkaline membranes. *In situ*, fluorescence techniques were used by Ramani and coworkers to identify superoxide radicals, while Kruczała and coworkers used an electron paramagnetic resonance spectrometer. Ramani and coworker [70] also discovered that the degradation of the tested membrane was more pronounced near the oxygen evolution electrode during AEM water electrolyzer operation. These findings reveal that oxygen and OH^- ions can spontaneously generate superoxide and hydroxyl radicals within the AEMs, resulting in membrane backbone degradation and thinning. These observations underscore the growing concern of oxidation degradation. To improve the resilience of membrane structures against oxidation, it is possible to incorporate radical scavengers such as ceria and sulfide groups. Membranes that possessed sulfide groups exhibited exceptional oxidative durability and retained their properties for more than 50 h in Fenton's solution [71]. The inclusion of phenolic groups in the monomer solution augmented the AEM oxidative stability [72]. Nevertheless, it is essential to remember that these groups cannot be restored and only provide limited defense as sacrificial sites. Although Fenton's reagent is a widely used method for assessing oxidation stability, it may not be the most suitable approach for evaluating the stability of AEMs in an alkaline environment. This is because Fenton's reagent only generates hydroxyl and hydroperoxyl radicals under acidic conditions, while more hydroxyl and superoxide radicals are generated in an alkaline environment. Future testing protocols should be developed to demonstrate the chemical stability of AEMs accurately.

16.2.11 Ionic Conductivity

Upon comparing OH^- and H^+, it has been observed that OH^- possesses a larger ionic radius, resulting in a slower movement through a membrane than H^+. This leads to lower levels of conductivity for OH^-. To improve the conductivity levels, there are two possible options: augmenting the number of conduction points among nearest neighbors or optimizing the conduction pathways. A practical method to achieve this is by elevating the IEC, which can significantly enhance the ionic conductivity of membranes. In the preceding chapters of this book, significant challenges related to the ionic conductivity of AEMs have been explored. However, researchers have made notable strides in overcoming these hurdles through the use of hybrid polymer designs, cutting-edge additives, sophisticated computational models, and semi-interpenetrating membrane structures. Current research demonstrates exceptional progress in polymer architecture, functionalized additives, ion transport models, ion-selective channels, and cross-linking techniques. As a result, the ionic conductivity of AEMs has seen a remarkable improvement, surpassing 150 mS cm^{-1}. The polymer components of AEMs exhibit both hydrophobic and hydrophilic properties, resulting in a well-defined microphase separation structure that enables ion transport. Cations, which are hydrophilic and tend to be solvated, benefit from the presence of water molecules that enhance their hydration. The membrane's homogeneous hydrogen bond network enables the continuous transfer of OH^-, while different phase separation structures contribute to anion diffusion through water channels. Understanding these mechanisms is crucial in determining the relationship between polymer structure and charge transport efficiency. It can guide the design of systems that enhance the ionic conductivity of functional AEMs.

16.2.12 Mechanical and Dimensional Stability

Aside from chemical degradation, physical degradation presents a considerable challenge in practical applications. Achieving successful membrane electrode fabrication, handling, assembly, and durability of polymer electrolyte membranes necessitates prioritizing mechanical integrity. Membranes are generally vulnerable to physical degradation, leading to membrane creep and microcrack fractures as frequent problems. Over time, the AEM in the device can undergo deformation as a result of heat and stress caused by compressive force. This can result in a thinner membrane and, with continued use, can lead to permanent deformation, which ultimately causes cell failure. Another common cause of membrane failure is microcracking due to fluctuations in temperature and humidity. In order to avoid these issues, it is crucial to utilize sturdy membranes that are capable of withstanding mechanical and swelling stress. Furthermore, the membrane should possess some degree of elasticity (elongation) to prevent cracking. Incorporating high molecular weight polymers can significantly enhance the mechanical robustness of AEMs. Polymers with high molecular weights have longer chains than their low molecular weight counterparts. These longer chains increase entanglement, making the polymer network more robust and deformation-resistant. This can improve the resistance of AEMs to mechanical stress and prevent the occurrence of cracks or tears. Additionally, these polymers display less swelling and creep behavior when exposed to solvents or high temperatures. By adding these polymers to AEMs, the membrane's mechanical stability can be maintained over time as it reduces the chances of dimensional changes caused by swelling caused by water uptake.

Recent scientific breakthroughs have led to significant advancements in hydroxide conductivity. Researchers have achieved remarkable conductivity levels, ranging from 150 to 200 mS cm^{-1}, by carefully manipulating molecular topology. Additionally, they have explored various strategies to combat degradation in alkaline environments, including the use of alkyl spacers, the elimination of electron-withdrawing and benzylic positions, the design of block structures and cross-linking, the creation of microphase separations, and the avoidance of ether bonds. The development of the ether-bond-free aryl backbone, paired with stable *N*-heterocyclic ammonium groups, is particularly noteworthy for its ability to maintain high conductivity and alkali resistance stability. The membrane community's research efforts have led to the creation of multiple commercial AEMs and accelerated the development of AEMFC technology. Recently, alkaline membranes have witnessed notable progress thanks to significant advancements in molecular engineering, topological regulation, and innovative synthetic techniques. These breakthroughs have resulted in membranes that exhibit exceptional anion conductivity of over 150 mS cm^{-1}, are capable of withstanding alkaline stability for more than 2000 h, and possess robust mechanical properties, including tensile stress of over 40 MPa and elongation at break of more than 20%. With such developments, the potential for further progress in alkaline membrane-based technologies is immense.

Furthermore, it is imperative to engage in additional research and development efforts to maximize the compatibility of AEMs with various fuel cell components, including membrane electrode assembly (MEA), catalysts, electrodes, and bipolar plates; these efforts will prove invaluable in surmounting current obstacles and unlocking the full potential of AEM fuel cells.

16.3 Durability of Anion Exchange Membrane Fuel Cells

In addition to ionic conductivity, alkaline stability, and mechanical properties, achieving optimal performance from AEMFC devices requires considering the compatibility of AEMs with various fuel cell components such as the MEA, catalysts, electrodes, and bipolar plates. The MEA consists of three distinct layers: the AEM, a catalyst, and a gas diffusion layer (GDL). The catalyst layer is the primary location where the electrochemical reaction takes place. It contains catalysts and an ionic polymer binder, also known as an ionomer. Numerous factors can affect the performance of the device; among them, three crucial elements significantly impact its efficiency: water distribution, the carbonation effect, and the membrane–electrode interface.

16.3.1 Water Management

Effective water management is of utmost importance in the AEMFC, as it significantly influences both the ionic conductivity and electrochemical reaction activity, hence impacting the overall power performance and durability. It is widely recognized that in AEMFCs, water is depleted at the cathode through an oxygen reduction reaction (ORR) and generated at the anode through a hydrogen oxidation reaction (HOR). Additionally, the transportation of OH^- from the cathode to the anode, along with eight water molecules, occurs via electro-osmotic drag at a significantly higher rate than that of the PEMFC [73, 74]. Therefore, effective water management through membrane design, stream humidity, and temperature control is paramount in the AEMFC. At high current densities, the cathode and anode can suffer from drying out and flooding, respectively, which can limit the achievable power and current densities [75]. The anode experiences hindered catalyst sites and gas transport during severe flooding, while the inadequate water supply at the cathode can cause ionomer dry-out and degradation [76]. A recent study emphasized the balance voltage, at which water generation and phase-change-induced flow are balanced, preventing cell flooding [77]. To design an MEA for an AEMFC, it is essential to ensure sufficient water transfer from the anode to the cathode to compensate for water consumption during operation. Therefore, it is necessary to engineer the nanotechnology concept to the membrane with appropriate water diffusion properties to maintain proper water balance in the AEMFC.

Omasta et al. [78, 79] conducted a study using operando neutron imaging investigate the distribution of water dynamics at constant current density by varying the anode and cathode dew points. Their findings were strong experimental proof of delicate water balance for long-lasting AEMFCs and suggest that the anode flooding take place with increase in dewpoints and cathode has no effect with dew point. To tackle the problem of anode flooding and achieve long-term durability, a new GDL was designed with improved hydrophobicity [78, 79]. The new GDL proved effective as the system's stability was improved over 1000 h of continuous operation at 600 mA cm^{-2}. The hydrophobicity design of catalyst layers was reported by Hyun et al. [80] to mitigate cathode dryness and suppress anode water flooding. The incorporation of hydrophilic carbon into the cathode and hydrophobic carbon into the anode resulted in a significant increase by a factor of 1.7 enhancement of the peak current density .

16.3.2 Carbonation Effect

Within AEMFC applications, hydroxide-swelled AEMs and ionomers are capable of absorbing CO_2 from the feed stream. Within the cathode, the ORR facilitates the production of hydroxide anions, which then interact with CO_2, resulting in the formation of a combination of bicarbonate (HCO_3^-) and carbonate (CO_3^{2-}). When the carbonate mixture reaches the anode, it balances the external circuit charge. However, it also causes the accumulation of carbonation at the anode, leading to the carbonate diffusion effect. As a result, carbonated anions create a carbonation gradient and pH gradient across the membrane, which can affect electrocatalysts. One positive impact of this process is that the lowered pH reduces polymer degradation caused by nucleophilic attack. Despite this, the negative influence of carbonation is more harmful to the cell output and its long-term endurance. The major drawback of carbonation is the resulting lower ion mobility compared to hydroxide ions. This is because the carbonate anion is larger in size (2.7 Å) than hydroxide ions (1.7 Å) [81].

During the early stages of AEMFC research, it was believed that the ability of AEMs to transport carbonates freely would have little effect on the operation of cells, except for the reduced mobility of those anions compared to hydroxide, which would cause an increase in Ohmic losses in the cell. However, recent experimental and theoretical work has allowed for a more complete understanding of the impact of CO_2 on AEMFC performance. It has been discovered that several mechanisms lead to a reduction in cell voltage upon exposure to CO_2. The first mechanism is connected to carbonate mobility, as mentioned before. The second mechanism that affects AEMFC operation is related to the inability of bicarbonate anions to directly oxidize H_2 at the anode at usual AEMFC anode potentials. This results in the carbonates produced at the cathode not being consumed at the anode and, therefore, not being immediately released to the anode gas flow field as CO_2. Instead, there is a time lag between CO_2 exposure and CO_2 release, during which the carbonates accumulate at the anode. This accumulation leads to a decrease in the pH of that electrode. As the pH drops, more carbonates accumulate, eventually resulting in the release of CO_2. This drop in the anode pH causes a Nernstian increase in the anode potential, ultimately reducing the overall cell voltage. The third mechanism is related to the fact that (bi)carbonates cannot react directly with H_2. The anode has a specific IEC, so when carbonates accumulate, it creates a concentration gradient in the anode. As a result, there will be regions in the anode where the OH^- concentration is low. In addition, since OH^- is no longer the only charge carrier, the anode reaction must obtain reacting OH^- anions through both migration and diffusion. This forces the anode current density to be concentrated near the anode/AEM interface, resulting in a higher effective local current density of the anode. This also causes higher reaction overpotentials.

16.3.3 Membrane–Electrode Interface

The membrane–electrode interface refers to the region where the AEM interacts with the catalyst layers on both the anode and cathode sides of the fuel cell. This interface is crucial for efficiently transporting hydroxide ions from the cathode to the anode and vice versa. The catalyst layers on either side of the membrane assist in the electrochemical reactions involving hydrogen and oxygen, producing water and electricity. This subject has been covered by an increasing number of articles describing strategies to build effective interfaces between the membrane and the electrodes. Ion conductivity, electron transfer, water diffusion, and gas mass transfer all rely on proper interfacial compatibility. Research has shown that the amount of the ionomer present is crucial in adjusting the electrode's microstructure. Typically, 10–20 wt% content is considered optimal, but slight variations can occur depending on the ionomer structures used. Ethanol, propanol, and isopropanol are the most commonly used solvents for slurry preparation, with a dielectric constant (ε) ranging from 18.3 to 24.3. The polarizability of a solvent molecule determines its ε value, which, in turn, affects the state of the slurry – whether it is a solution, colloid, or precipitate. An AEMFC cathode catalyst layer has been developed with a modified catalyst ink using methanol, propanol, isopropanol, and tetrahydrofuran. Among these, the tetrahydrofuran-based catalyst layer resulted in a power density of $0.18\,W\,cm^{-2}$. This is due to the effect of tetrahydrofuran on the microstructure of the catalyst layer, which resulted in the emergence of a nanopore gradient ranging from 10 to 100 nm [82]. This gradient improved the catalyst layer's water transport ability and moisture adsorption. Park et al. reported that the permeability of the ionomer in wet conditions affects the power density of the fuel cell, specifically with regard to hydrogen and oxygen gas. They synthesized an ionomer by modifying poly(phenylene oxide) with six-membered dimethyl piperidinium (PPO-DMP) to address this issue. The PPO-DMP ionomer exhibited improved flexibility compared to the low-performing fumion ionomers, resulting in a peak power density of 0.34 W cm^{-2} [83]. Additionally, in AEMFC applications, covalently locked interfaces between the membrane and the ionomer-containing catalyst layer are reported to reduce catalyst layer peeling off. The presence of a covalent bond at the interface between the membrane and catalyst layer prevents interfacial delamination of the catalyst layer. When comparing the preparation of physically reinforced MEA to the advanced triple-phase boundary structure, it can be observed that the latter exhibits a greater propensity for enhanced durability. A compact interface plays a significant role in reducing the overall cell resistance, encompassing both AEM resistance and contact resistance. Wang et al. [84] developed a method of enhancing gas permeability in catalyst layers using a rigid anthracene-modified imidazolium side chain structure. This method created microscopic gas channels, which led to a 130% increase in gas permeability. Additionally, the researchers used ultraviolet light-induced cross-linking structures to prevent self-aggregation of the Pt/C particles. This resulted in a peak power density of 0.49 W cm^{-2} due to sufficient oxygen penetrating the ionomer onto the catalytic reaction interface.

16.4 Future Directions

The field of AEM technology is rapidly growing, and there is no doubt that continued investment in research and development will lead to significant advancements in the coming years. The impact of AEMs on the world is expected to be transformative and revolutionizing various industries. The focus of AEMFCs in the coming years will revolve around the enhancement of alkaline-resistant AEMs, ionomers, catalysts, and MEA fabrication.

16.4.1 Expansion of AEM Development

One of the primary design strategies employed in the practical implementation of advanced AEMs involves the preparation of a polymer backbone devoid of aryl ether moieties. Mitigating aryl ether cleavage reactions in quaternized polymers containing aryl ethers has proven to be a challenging endeavor, with limited success thus far. Polyolefinic and aryl ether-free polyaromatics are backbone polymers that do not contain aryl ethers. Polyolefinic AEMs, such as PE, polynorbornene, ethylene tetrafluoroethylene, and polystyrene-block copolymers, have certain advantages over polyaromatic-based polymers. They are better suited for applications that require water resistance and the ability to form films. However, highly quaternized polyolefinic AEMs have inferior high-temperature properties (>80 °C) compared to polyaromatic-based polymers. This makes them less suitable for use in high-temperature operations of AEMFCs. Developing polyolefinic AEMs with improved stability through crystallinity, cross-linking, and hydrophobic cationic functional groups could enhance AEMFC performance and durability. Reinforced AEMs can also improve dimensional stability. Polyaromatic polymers, such as polyphenylene, polyfluorene, and poly(alkyl phenylene), free from aryl ether, have various benefits at high temperatures (≥80 °C). Unfortunately, they often become brittle due to the absence of flexible ether linkage in their backbone and low molecular weight. Several strategies can be implemented to overcome this challenge, such as achieving high molecular weight, minimizing chain branches, and incorporating kinked structures and reinforcement.

New developments in AEMs have led to the creation of aryl ether-free polyolefinic and polyaromatics, but there are concerns about their practical application. Polymers without ether links are more complex to synthesize than aryl ether polymers using S*N*Ar. For PAs, multiple steps are needed for well-defined monomers via Yamamoto or Diels–Alder polymerization. Friedel–Crafts reaction is more straightforward but requires potent triflic acid. Making PEs for AEMs is also tricky as ionic-moiety-bearing cyclooctene monomers need multi-step synthesis. Ziegler–Natta polymerization with polar monomers is complex due to the low tolerance of early transition metal catalysts.

Another design strategy for creating effective AEMs is to select appropriate cationic groups. The most commonly used cationic group is trimethyl ammonium, and adding alkyl spacers between the polymer backbone and side chain has proven to increase the stability of these groups. However, other cationic functional groups like piperidinium and triazatriangulenium carbocation have also been successfully implemented and should be further researched. Further research should be conducted to prepare polymers with stable cationic functional groups, including those in ionic liquid compounds. It is also essential to explore other possibilities of carbocation on AEMs and enhance the stability of known cationic groups, which are areas that require continued research. Recently, new carbocationic species have emerged in the search for AEMs with improved alkaline-stability, in addition to the existing cationic head groups. According to a recent study conducted by the Schechter and Grynszpan groups [85], the use of triazatriangulenium carbocation rises as a highly promising alternative for the development of AEMs. The research suggests that this type of extended π carbonations exhibits exceptional stability, making them a preferred choice for efficiently transporting anions in AEMs. The research conducted by Schechter and coworkers [85] sheds light on the process of creating triazatriangulenium carbocation, specifically regarding the production of 4,8,12-tri-ethano-4,8,12-triazatriangulenium tetrafluoroborate ($TATA^+$ OH)-based AEMs. This development is a significant step toward achieving AEMs with superior alkaline stability. The complete pathways for the fabrication of their proposed AEM are shown in Figure 16.5. As shown in Figure 16.5, the $TATA^+$ OH cation was prepared using the tris(2,6-dimethoxyphenyl)-methylium tetrafluoroborate precursor. The precursor for the synthesis of $TATA^+$ OH is synthesized following a reported procedure [86]. Additionally, a commercially available polysulfone was modified through the standard chloromethylation procedure [87]. Finally, the working membrane was assembled through the solution-casting method. The controllable microporous structure (Figure 16.6a) in the membranes provides the ease of transport of

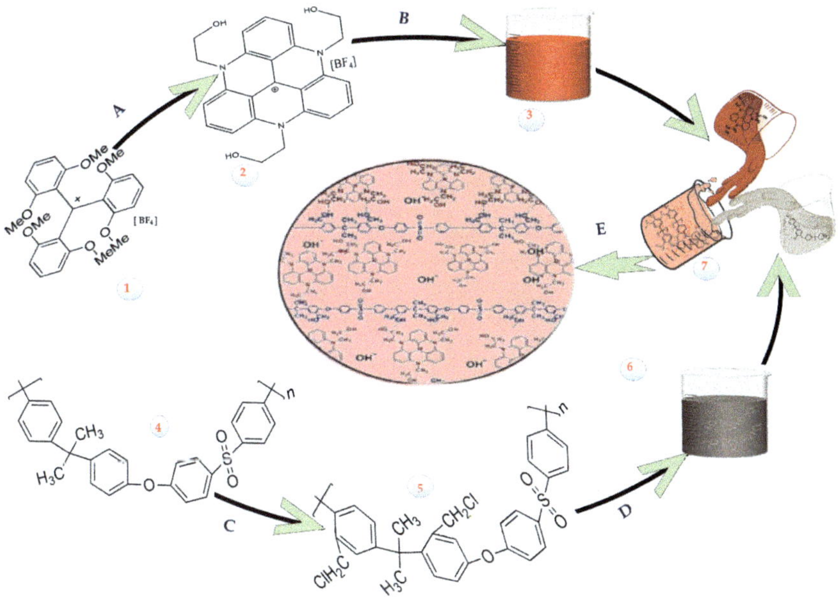

Figure 16.5 The fabrication process of TATA$^+$-OH-based AEM. (1) Tris(2,6-dimethoxyphenyl)-methylium tetrafluoroborate (starting material for TATA$^+$-OH synthesis); (2) 4,8,12-tri-ethano-4,8,12-triazatriangulenium (TATA$^+$-OH); (3) dispersion of TATA$^+$-OH in NMP; (4) polysulfone; (5) chloromethylated polysulfone (CMPSF); (6) dispersion of CMPSF in NMP; (7) mixing and sonication of CMPSF and TATA$^+$-OH solutions; reaction conditions: (A) ethanolamine/N_2, 190 °C; (B) stirred TATA$^+$-OH; (C) paraformaldehyde, $(CH_3)_3SiCl$, $SnCl_4/N_2$, 72 h; (D) stirred CMPSF; (E) solvent evaporation and drying of a casted membrane. Source: Reproduce with permission from Thomas et al. [85].

hydroxide ions on it. The efficiency of hydroxide ion transport of these fabricated membranes was calculated from the Arrhenius equation through Arrhenius plots of the conductivity analysis, which has provided in Figure 16.6b. In Figure 16.6b, the low conductivity activation energy (E_a) for 20 wt.% TATA$^+$ OH incorporated in the membrane, i.e. 6.56 kJ mol^{-1}. Lower values of E_a indicate the better transport of hydroxide ions. Additionally, Figure 16.6c indicates exceptional alkaline stability of the membranes. Furthermore, according to Figure 16.6c, the membranes showed remarkable alkaline stability, with a mere 12% total hydroxide conductivity loss over 3 M KOH solution at 60 °C after 840 h.

Following a thorough investigations and analysis, it is suggested that triazatriangulenium-based cations could be a high alternative to traditionally used quaternary ammonium cations, guanidinium cations, imidazole cations, etc., for the fabrication of stable AEMs. Further research on triazatriangulenium-based cations is essential to determine their practicality, cost-effectiveness, and scalability for use in electrochemical devices and applications. In the forthcoming years, more triazatriangulenium-based AEMs will emerge to address stability concerns. Moreover, additional research is necessary on the fabrication of membrane electrode assemblies using triazatriangulenium cation-based AEMs.

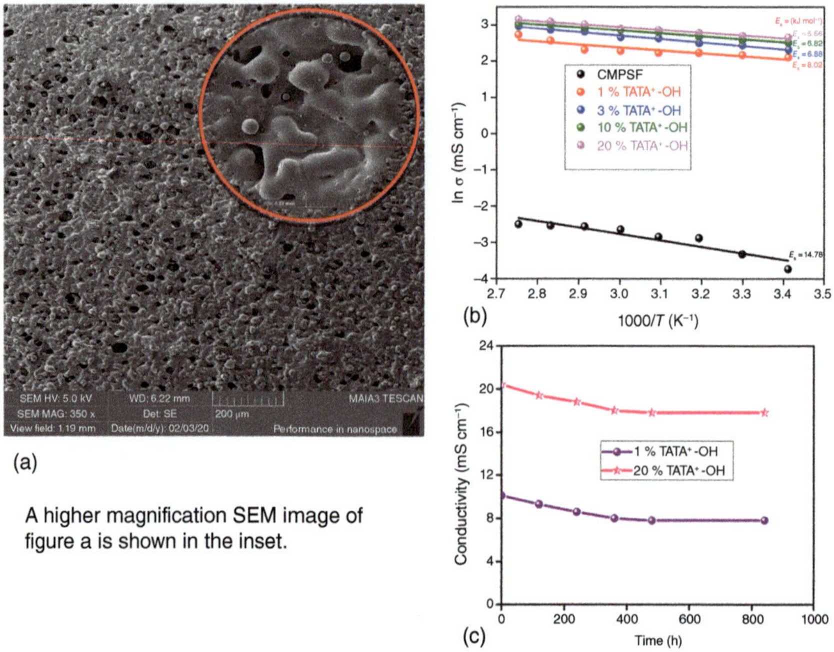

Figure 16.6 The scanning electron image of the microporous network structure of the membranes (a); Arrhenius plots of the membranes' conductivity (b); hydroxide conductivity in terms of alkaline stability of membranes as a function of soaking time in 3 M KOH at 60 °C (c). Source: Reproduce with permission from Thomas et al. [85]. Copyright 2021 Elsevier.

During the design phase of new AEMs, it is important to ensure that the membrane can operate effectively under low relative humidity (RH) conditions. However, the current performance of AEMFCs under low RH conditions is limited by the electrodes. Achieving reliable low RH operation is crucial for automotive fuel cell applications. One common approach to enhance the AEMs' performance is to improve their phase-separated morphology. This can be attained by synthesizing multiblock copolymers or by introducing hydrophobic polymer backbones. It is worth noting that the hydroxide conductivity of AEMs is influenced by the concentration of cationic functional groups as well as the mesoscale structure of the polymer system. Specifically, the ionic conductivity and stability of the AEMs may be affected by the order–disorder transition in the nano-phase-separated domains. When designing AEMs, it is crucial to take into account their interaction with catalyst layers. The full understanding of the interface between AEMs and electrode faces remains somewhat elusive. For AEMs to perform optimally over an extended period, a robust interface is necessary to facilitate the smooth transportation of water and hydroxide ions through the interface between AEM and catalyst layers.

16.4.1.1 Ionomer Development

Designing ionomers separately for the anode and cathode catalyst layers is imperative, as they exhibit distinct catalyst–ionomer interactions arising from varied electrode potentials. In the case of anode ionomeric binders, preventing cation–hydroxide water co-adsorption on the surface of catalysts at HOR potentials is crucial for the performance and durability of AEMFCs. The commonly used methylammonium functional group is known to be highly adsorbed, which can have a negative impact on the system. To prevent this, developing less adsorbing cationic groups is a plausible approach. However, whether alkaline stable cyclic cationic groups are a better alternative is still unclear. Studies suggest that more bulky cations might have a lower adsorption energy. Another approach is to reduce the interfacial contact between anode ionomers and anode catalysts using particulate ionomers. This can decrease ionomer IECs and lead to a better performance-durability balance in AEMFCs. Another factor that affects AEMFC performance is phenyl adsorption, which can be mitigated by using ionomers that are free of phenyl groups. However, creating phenyl-free ionomers can be a technical challenge. Instead, ionomers with a non-adsorbing phenyl group can be used as an effective alternative. The main concern with ionomeric binders for cathodes is their susceptibility to electrochemical oxidation at high electrode potentials. The phenyl group in ionomers is particularly vulnerable to this process, as it is converted to acidic phenol. One potential solution is to prepare phenyl-free ionomers; however, it is uncertain whether ionomers with non-adsorbing phenyl groups are adequate as they too can be oxidized at high electrode potential. It is possible that substituted phenyl groups or a favorable stacking configuration could help reduce the oxidation rate.

16.4.1.2 Catalyst Development

The development of catalysts for high-performance and durable AEMFCs can be grouped into platinum group metal (PGM)-free anode and cathode catalysts, as well as low-PGM anode and cathode catalysts. To create PGM-free anodes, Ni-based materials have been suggested [88]. However, the main challenges associated with Ni-based catalysts include their intrinsically low hydrogen oxidation activity; their tendency to surface oxidation (passivation), which leads to a gradual loss of catalytic activity over time; and their specific interaction with anode ionomers. The development of PGM-free anode catalysts should be prioritized in the next few years. These catalysts should take into account the inherent activity of hydrogen oxidation and mass-transfer obstacles. Various types of PGM-free ORR catalysts have been developed. These include Metal-Nitrogen-Carbon (M-N-C) [89], metal oxides [90], and silver-based catalysts [91] are considered the leading compounds for alkaline oxygen reduction catalysts. Of these, M-N-C catalysts have demonstrated similar performance to that of Pt/C [92]. Nevertheless, their efficacy and endurance in fuel cells have yet to be thoroughly assessed. One way to improve the mass transfer limitations of thick layers of M-N-C catalysts is to create hybrid materials with dense oxide or silver catalysts. Research in this area is needed. It is important to evaluate the stability of non-carbon-based metal oxide or silver-based catalysts in MEAs and improve their catalytic activity. While studies have been conducted to assess the durability of PGM catalysts in ORR electrocatalysts, further research is needed to understand and improve the durability of PGM-free or low-PGM catalysts for developing long-lasting AEMFC systems.

16.4.1.3 Membrane Electrode Assembly Developments

The fabrication methodology of the MEA plays a pivotal role in determining the performance and durability of AEMFCs. However, the lack of information on the optimized MEA structures causes hindrance in the production of highly performing MEAs. Two methods for fabricating MEAs have proven to be highly effective and durable. The first method, known as the Los Alamos standard method, utilizes a homogeneous ionomeric binder. The second method, developed by USC/Surrey, employs quaternized ionomers with limited solubility. The management of water within the MEA remains a challenge for AEMFCs. Excessive water can cause flooding, which impedes reactant transport, while insufficient water can lead to membrane drying and performance loss. Furthermore, mitigating carbonation of the AEM and carbon dioxide crossover is critical for maintaining high cell efficiency and longevity. Addressing these challenges is vital for adopting AEMFCs widely in various applications, such as portable electronics, transportation, and stationary power generation. Another challenge is the development of efficient and stable catalysts for the ORR at the cathode. While platinum-based catalysts have been the standard, their high cost and limited availability make them less practical for widespread AEMFC adoption. Researchers are actively exploring alternative catalyst materials that maintain high activity and stability in alkaline conditions. Additionally, optimizing the electrode structure and ionomer content within the MEA to enhance mass transport and minimize the effects of water management is an ongoing challenge. Addressing these challenges is crucial for the continued advancement and commercialization of AEMFCs as a sustainable energy solution.

16.5 Concluding Remarks

Over the past two decades, significant progress has been made by researchers in the areas of structural framework, cation configuration, and membrane morphology, all aimed at achieving a balance between the ionic conductivity and stability of the AEM. The implementation of systematic molecular design and meticulous regulation of ion conduction channels has led to promising results in sustainable energy applications. This chapter delves into the research challenges and future directions in the field of AEMs for fuel cells, underlining the importance of interdisciplinary research and collaboration between material scientists, chemists, physicists, and engineers.

Several strategies have been discussed to enhance the AEM's performance, including the categorization of AEMs, such as main chain, block, long side chain, cross-linked, and organic–inorganic composite AEM. Incorporating flexible, hydrophobic, or densely charged side chains reinforces micro-phase separation and accelerates anion transport. Cross-linking between ionic conducting groups minimizes swelling and enhances membrane stability. At the same time, introducing inorganic composite fillers optimizes the micro/nanostructure of the membrane to accelerate OH^- conduction and improve stability.

The future of AEMs in fuel cells is bright, as novel materials, innovative synthesis techniques, and advanced characterization methods continue to be developed. The development of efficient and cost-effective AEM-based fuel cells has the potential to revolutionize clean energy generation, making it more accessible and environmentally friendly. As we embark on this exciting journey of discovery and innovation, we must remain committed to sustainability and environmental responsibility.

This book offers insightful information on the fast evolving and complex area of AEMs for electrochemical energy applications and serves as a guide for researchers. By addressing the outlined challenges and exploring the suggested future directions, we can contribute to realizing efficient and environmentally friendly energy conversion technologies, bringing us closer to a more sustainable and prosperous future.

Acknowledgments

The authors would like to acknowledge the support of the Planning & Budgeting Committee/ISRAEL Council for Higher Education (CHE) and Israel National Research Center for Electrochemical Propulsion (INREP). Flavio Grynszpan is the incumbent of the Cosman endowment for organic chemistry research.

References

1 Mehta, V. and Cooper, J.S. (2003). Review and analysis of PEM fuel cell design and manufacturing. *J Power Sources* 114: 32.

2 Chen, H., Tao, R., Bang, K.T. et al. (2022). Anion exchange membranes for fuel cells: state-of-the-art and perspectives. *Adv Energy Mater* 12: 2200934.

3 Kalghatgi, G., Agarwal, A.K., Leach, F., and Senecal, K. (ed.) (2022). Engines and fuels for future transport. In: *Engines and Fuels for Future Transport; Energy, Environment, and Sustainability*. Singapore: Springer.

4 Kalghatgi, G., Levinsky, H., and Colket, M. (2018). Future transportation fuels. *Prog Energy Combust Sci* 69: 103.

5 Li, J., Liu, C., Ge, J. et al. (2023). Challenges and strategies of anion exchange membranes in hydrogen-electricity energy conversion devices. *Chem Eur J* 29: e202203173.

6 Luo, Z., Zhang, H., Yang, Y. et al. (2020). Reactant friendly hydrogen evolution interface based on di-anionic MoS_2 surface. *Nat Commun* 11: 1.

7 Huang, J., Yu, Z., Tang, J. et al. (2022). A review on anion exchange membranes for fuel cells: anion-exchange polyelectrolytes and synthesis strategies. *Int J Hydrog Energy* 47: 27800.

8 Cai, Z.H., Gao, X.L., Gao, W.T. et al. (2022). Effect of hydrophobic side chain length on poly(carbazolyl terphenyl piperidinium) anion exchange membranes. *ACS Appl Energy Mater* 5: 10165.

9 Wang, Q., Huang, L., Wang, Z. et al. (2022). High conductive anion exchange membranes from all-carbon twisted intrinsic microporous polymers. *Macromolecules* 55: 10713.

10 Zelovich, T., Dekel, D.R., and Tuckerman, M.E. (2023). Functional groups in anion exchange membranes: insights from ab initio molecular dynamics. *J Membr Sci* 678: 121638.

11 Allushi, A., Bakvand, P.M., and Jannasch, P. (2023). Polyfluorenes bearing *N,N*-dimethylpiperidinium cations on short spacers for durable anion exchange membranes. *Macromolecules* 56: 1165.

12 Arges, C.G. and Zhang, L. (2018). Anion exchange membranes' evolution toward high hydroxide ion conductivity and alkaline resiliency. *ACS Appl Energy Mater* 1: 2991.

13 Yang, Z., Guo, R., Malpass-Evans, R. et al. (2016). Highly conductive anion-exchange membranes from microporous Tröger's base polymers. *Angew Chem* 128: 11671.

14 He, G., Li, Z., Zhao, J. et al. (2015). Nanostructured ion-exchange membranes for fuel cells: recent advances and perspectives. *Adv Mater* 27: 5280.

15 Song, W., Zhang, X., Yang, C. et al. (2023). Alkaline membranes toward electrochemical energy devices: recent development and future perspectives. *ACS Cent Sci* 9 (8): 1538–1557.

16 Chen, X., Zhan, Y., Tang, J. et al. (2023). Advances in high performance anion exchange membranes: molecular design, preparation methods, and ion transport dynamics. *J Environ Chem Eng* 11: 110749.

17 Fang, J. and Shen, P.K. (2006). Quaternized poly(phthalazinon ether sulfone ketone) membrane for anion exchange membrane fuel cells. *J Membr Sci* 285: 317.

18 Arges, C.G. and Ramani, V. (2013). Two-dimensional NMR spectroscopy reveals cation-triggered backbone degradation in polysulfone-based anion exchange membranes. *Proc Natl Acad Sci U S A* 110: 2490.

19 Mohanty, A.D., Tignor, S.E., Krause, J.A. et al. (2016). Systematic alkaline stability study of polymer backbones for anion exchange membrane applications. *Macromolecules* 49: 3361.

20 Varcoe, J.R. and Slade, R.C.T. (2005). Prospects for alkaline anion-exchange membranes in low temperature fuel cells. *Fuel Cells* 5: 187.

21 Merle, G., Wessling, M., and Nijmeijer, K. (2011). Anion exchange membranes for alkaline fuel cells: a review. *J Membr Sci* 377: 1.

22 Ran, J., Ding, L., Chu, C. et al. (2018). Highly conductive and stabilized side-chain-type anion exchange membranes: ideal alternatives for alkaline fuel cell applications. *J Mater Chem A Mater* 6: 17101.

23 Kim, Y., Wang, Y., France-Lanord, A. et al. (2019). Ionic highways from covalent assembly in highly conducting and stable anion exchange membrane fuel cells. *J Am Chem Soc* 141: 18152.

24 Yu, W., Zhang, J., Liang, X. et al. (2021). Anion exchange membranes with fast ion transport channels driven by cation-dipole interactions for alkaline fuel cells. *J Membr Sci* 634: 119404.

25 Cheng, J., He, G., and Zhang, F. (2015). A mini-review on anion exchange membranes for fuel cell applications: stability issue and addressing strategies. *Int J Hydrog Energy* 40: 7348.

26 Zhang, F., Zhang, Y., Sun, L. et al. (2023). A π-conjugated anion-exchange membrane with an ordered ion-conducting channel via the McMurray coupling reaction. *Angew Chem Int Ed* 62: e202215017.

27 Ge, J., Chen, T., Wan, X. et al. (2023). Highly ion-conductive and alkaline stable anion exchange membranes containing bulky aryl units and symmetric bis-piperidinium branches. *ACS Appl Polym Mater* 5 (6): 4554–4564.

28 Liu, Q., Zhang, S., Tian, L. et al. (2023). "Windmill" shaped branched anion-conducting poly(aryl piperidine) with extra molecular interaction sites as new anion exchange membranes. *J Power Sources* 564: 232822.

29 Wei, C., Yu, W., Zhang, Y. et al. (2023). Alkaline anion exchange membrane containing pyrene-based π-π stacking interactions. *J Power Sources* 553: 232247.

30 Chen, N., Hu, C., Wang, H.H. et al. (2021). Chemically & physically stable crosslinked poly(aryl-*co*-aryl piperidinium)s for anion exchange membrane fuel cells. *J Membr Sci* 638: 119685.

31 Zhu, H., Li, Y., Chen, N. et al. (2019). Controllable physical-crosslinking poly(arylene 6-azaspiro[5.5] undecanium) for long-lifetime anion exchange membrane applications. *J Membr Sci* 590: 117307.

32 Chen, N., Hu, C., Wang, H.H. et al. (2021). Poly(alkyl-terphenyl piperidinium) ionomers and membranes with an outstanding alkaline-membrane fuel-cell performance of 2.58 $W cm^{-2}$. *Angew Chem Int Ed* 60: 7710.

33 Chen, H., Bang, K.T., Tian, Y. et al. (2023). Poly(ethylene piperidinium)s for anion exchange membranes. *Angew Chem Int Ed* 62: e202307690.

34 Wang, T., Zhao, Y., Ma, R. et al. (2023). Highly conductive and dimensionally stable anion exchange membranes enabled by rigid poly(arylene alkylene) backbones. *Synth Met* 296: 117385.

35 Wright, A.G., Fan, J., Britton, B. et al. (2016). Hexamethyl-*p*-terphenyl poly(benzimidazolium): a universal hydroxide-conducting polymer for energy conversion devices. *Energy Environ Sci* 9: 2130.

36 Pham, T.H., Olsson, J.S., and Jannasch, P. (2018). Poly(arylene alkylene)s with pendant *N*-spirocyclic quaternary ammonium cations for anion exchange membranes. *J Mater Chem A Mater* 6: 16537.

37 Wang, J., Zhao, Y., Setzler, B.P. et al. (2019). Poly(aryl piperidinium) membranes and ionomers for hydroxide exchange membrane fuel cells. *Nat Energy* 4: 392.

38 Hibbs, M.R., Fujimoto, C.H., and Cornelius, C.J. (2009). Synthesis and characterization of poly(phenylene)-based anion exchange membranes for alkaline fuel cells. *Macromolecules* 42: 8316.

39 Yang, W., Chen, J., Yan, J. et al. (2022). Advance of click chemistry in anion exchange membranes for energy application. *J Polym Sci* 60: 627.

40 Mohanty, A.D., Lee, Y.B., Zhu, L. et al. (2014). Anion exchange fuel cell membranes prepared from C-H borylation and Suzuki coupling reactions. *Macromolecules* 47, 1973.

41 Ahmed Mahmoud, A.M. and Miyatake, K. (2022). Highly conductive and alkaline stable partially fluorinated anion exchange membranes for alkaline fuel cells: effect of ammonium head groups. *J Membr Sci* 643: 120072.
42 Zhou, Y.C., Bao, R.Y., Liu, Z. et al. (2021). Electrospun modified polyketone-based anion exchange membranes with high ionic conductivity and robust mechanical properties. *ACS Appl Energy Mater* 4: 5187.
43 Park, E.J., Maurya, S., Hibbs, M.R. et al. (2019). Alkaline stability of quaternized Diels-Alder polyphenylenes. *Macromolecules* 52: 5419.
44 Wang, L., Bellini, M., Miller, H.A., and Varcoe, J.R. (2018). A high conductivity ultrathin anion-exchange membrane with 500+ h alkali stability for use in alkaline membrane fuel cells that can achieve 2 W cm^{-2} at 80 °C. *J Mater Chem A Mater* 6: 15404.
45 You, W., Padgett, E., MacMillan, S.N. et al. (2019). Highly conductive and chemically stable alkaline anion exchange membranes via romp of trans-cyclooctene derivatives. *Proc Natl Acad Sci U S A* 116: 9729.
46 Zhang, M., Shan, C., Liu, L. et al. (2016). Facilitating anion transport in polyolefin-based anion exchange membranes via bulky side chains. *ACS Appl Mater Interfaces* 8: 23321.
47 You, W., Ganley, J.M., Ernst, B.G. et al. (2021). Expeditious synthesis of aromatic-free piperidinium-functionalized polyethylene as alkaline anion exchange membranes. *Chem Sci* 12: 3898.
48 Olsson, J.S., Pham, T.H., and Jannasch, P. (2017). Poly(*N,N*-diallylazacycloalkane)s for anion-exchange membranes functionalized with *N*-spirocyclic quaternary ammonium cations. *Macromolecules* 50: 2784.
49 Zhu, L., Peng, X., Shang, S.L. et al. (2019). High performance anion exchange membrane fuel cells enabled by fluoropoly(olefin) membranes. *Adv Funct Mater* 29: 1902059.
50 Wang, L., Peng, X., Mustain, W.E., and Varcoe, J.R. (2019). Radiation-grafted anion-exchange membranes: the switch from low- to high-density polyethylene leads to remarkably enhanced fuel cell performance. *Energy Environ Sci* 12: 1575.
51 Zhang, F., Li, T., Chen, W. et al. (2021). Electron-donating C-NH_2 link backbone for highly alkaline and mechanical stable anion exchange membranes. *ACS Appl Mater Interfaces* 13: 10490.
52 Thomas, O.D., Soo, K.J.W.Y., Peckham, T.J. et al. (2011). Anion conducting poly(dialkyl benzimidazolium) salts. *Polym Chem* 2: 1641.
53 Zhu, M., Zhang, M., Chen, Q. et al. (2017). Synthesis of midblock-quaternized triblock copolystyrenes as highly conductive and alkaline-stable anion-exchange membranes. *Polym Chem* 8: 2074.
54 Zhu, M., Zhang, X., Wang, Y. et al. (2018). Novel anion exchange membranes based on quaternized diblock copolystyrene containing a fluorinated hydrophobic block. *J Membr Sci* 554: 264.
55 Xu, F., Chen, Y., Cao, X. et al. (2022). Comb-shaped polyfluorene with variable alkyl chain length for application as anion exchange membranes. *J Power Sources* 545: 231880.

56 Wang, X., Li, J., Chen, W. et al. (2021). Polybenzimidazole ultrathin anion exchange membrane with comb-shape amphiphilic microphase networks for a high-performance fuel cell. *ACS Appl Mater Interfaces* 13: 49840.

57 Xu, G., Pan, J., Zou, X. et al. (2023). High-performance poly(biphenyl piperidinium) type anion exchange membranes with interconnected ion transfer channels: cooperativity of dual cations and fluorinated side chains. *Adv Funct Mater* 2302364.

58 Wu, X., Chen, N., Hu, C. et al. (2023). Fluorinated poly(aryl piperidinium) membranes for anion exchange membrane fuel cells. *Adv Mater* 35: 2210432.

59 Wu, X., Chen, N., Klok, H.-A. et al. (2022). Branched poly(aryl piperidinium) membranes for anion-exchange membrane fuel cells. *Angew Chem* 134: e202114892.

60 Li, L., Jiang, T., Wang, S. et al. (2022). Branched anion-conducting poly(arylene alkylene)s for alkaline membrane fuel cells. *ACS Appl Energy Mater* 5: 2462.

61 Yang, K., Xu, J., Shui, T. et al. (2020). Cross-linked poly (aryl ether ketone) anion exchange membrane with high ion conductivity by two different functional imidazole side chain. *React Funct Polym* 151: 104551.

62 Min, K., Lee, Y., Choi, Y. et al. (2022). High-performance anion exchange membranes achieved by crosslinking two aryl ether-free polymers: poly(bibenzyl *N*-methyl piperidine) and SEBS. *J Membr Sci* 664: 121071.

63 Zeng, L., Zhao, T.S., Wei, L. et al. (2016). Polyvinylpyrrolidone-based semi-interpenetrating polymer networks as highly selective and chemically stable membranes for all vanadium redox flow batteries. *J Power Sources* 327: 374.

64 Zhao, Z., Zhang, M., Du, W. et al. (2022). Strong and flexible high-performance anion exchange membranes with long-distance interconnected ion transport channels for alkaline fuel cells. *ACS Appl Mater Interfaces* 14: 38132.

65 Shang, C., Liu, Y., Zeng, X. et al. (2022). Preparation and characterization of organic-inorganic hybrid anion exchange membrane based on crown ether functionalized mesoporous SBA-NH_2. *Int J Hydrog Energy* 47: 14141.

66 Chen, J., Guan, M., Li, K., and Tang, S. (2021). High-performance COF-based composite anion exchange membrane sandwiched by GO layers for alkaline H_2/O_2 fuel cell application. *J Ind Eng Chem* 104: 136.

67 Parrondo, J., Wang, Z., Jung, M.S.J., and Ramani, V. (2016). Reactive oxygen species accelerate degradation of anion exchange membranes based on polyphenylene oxide in alkaline environments. *Phys Chem Chem Phys* 18: 19705.

68 Wierzbicki, S., Douglin, J.C., Kostuch, A. et al. (2020). Are radicals formed during anion-exchange membrane fuel cell operation? *J Phys Chem Lett* 11: 7630.

69 Wierzbicki, S., Douglin, J.C., Singh, R.K. et al. (2023). Operando EPR study of radical formation in anion-exchange membrane fuel cells. *ACS Catal* 13 (4): 2744–2750. https://doi.org/10.1021/ACSCATAL.2C05843.

70 Parrondo, J. and Ramani, V. (2014). Stability of poly(2,6-dimethyl 1,4-phenylene) oxide-based anion exchange membrane separator and solubilized electrode binder in solid-state alkaline water electrolyzers. *J Electrochem Soc* 161: F1015.

71 Bu, F., Zhang, Y., Hong, L. et al. (2018). 1,2,4-Triazole functionalized poly(arylene ether ketone) for high temperature proton exchange membrane with enhanced oxidative stability. *J Membr Sci* 545: 167.

72 Ye, N., Xu, Y., Zhang, D. et al. (2018). Inhibition mechanism of the radical inhibitors to alkaline degradation of anion exchange membranes. *Polym Degrad Stab* 153: 298.

73 Lu, J., Barnett, A., and Molinero, V. (2019). Effect of polymer architecture on the nanophase segregation, ionic conductivity, and electro-osmotic drag of anion exchange membranes. *J Phys Chem C* 123: 8717.

74 Cao, H., Pan, J., Zhu, H. et al. (2021). Interaction regulation between ionomer binder and catalyst: active triple-phase boundary and high performance catalyst layer for anion exchange membrane fuel cells. *Adv Sci* 8: 2101744.

75 Otsuji, K., Shirase, Y., Asakawa, T. et al. (2022). Effect of water management in membrane and cathode catalyst layers on suppressing the performance hysteresis phenomenon in anion-exchange membrane fuel cells. *J Power Sources* 522: 230997.

76 Willdorf-Cohen, S., Mondal, A.N., Dekel, D.R., and Diesendruck, C.E. (2018). Chemical stability of poly(phenylene oxide)-based ionomers in an anion exchange-membrane fuel cell environment. *J Mater Chem A Mater* 6: 22234.

77 Shi, L., Setzler, B.P., and Yan, Y. (2022). Understanding the ebalance for water management in hydroxide exchange membrane fuel cells. *J Power Sources* 536: 231514.

78 Peng, X., Kulkarni, D., Huang, Y. et al. (2020). Using operando techniques to understand and design high performance and stable alkaline membrane fuel cells. *Nat Commun* 11 (1): 1–10.

79 Omasta, T.J., Park, A.M., Lamanna, J.M. et al. (2018). Beyond catalysis and membranes: visualizing and solving the challenge of electrode water accumulation and flooding in AEMFCs. *Energy Environ Sci* 11: 551–558. https://doi.org/10.1039/C8EE00122G.

80 Hyun, J., Jo, W., Yang, S.H. et al. (2022). Tuning of water distribution in the membrane electrode assembly of anion exchange membrane fuel cells using functionalized carbon additives. *J Power Sources* 543: 231835.

81 Vega, J.A., Chartier, C., and Mustain, W.E. (2010). Effect of hydroxide and carbonate alkaline media on anion exchange membranes. *J Power Sources* 195: 7176.

82 Zhang, J., Pei, Y., Zhu, W. et al. (2021). Ionomer dispersion solvent influence on the microstructure of Co–N–C catalyst layers for anion exchange membrane fuel cell. *J Power Sources* 484: 229259.

83 Park, H.J., Chu, X., Kim, S.P. et al. (2020). Effect of *N*-cyclic cationic groups in poly(phenylene oxide)-based catalyst ionomer membranes for anion exchange membrane fuel cells. *J Membr Sci* 608: 118183.

84 Wang, B., Pan, J., Zou, X. et al. (2022). UV-crosslinkable anthracene-based ionomer derived gas "expressway" for anion exchange membrane fuel cells. *J Mater Chem A Mater* 10: 13355.

85 Thomas, J., Francis, B., Thomas, S. et al. (2021). Dependable polysulfone based anion exchange membranes incorporating triazatriangulenium cations. *Solid State Ionics* 370: 115731.

86 Martin, J.C. and Smith, R.G. (1964). Factors influencing the basicities of triarylcarbinols. the synthesis of sesquixanthydrol. *J Am Chem Soc* 86: 2252.

87 Toiserkani, H., Yilmaz, G., Yagci, Y., and Torun, L. (2010). Functionalization of polysulfones by click chemistry. *Macromol Chem Phys* 211: 2389.

88 Biemolt, J., Douglin, J.C., Singh, R.K. et al. (2021). An anion-exchange membrane fuel cell containing only abundant and affordable materials. *Energy Technol* 9 (4): 2000909. https://doi.org/10.1002/ENTE.202000909.

89 Sivakumar, P., Subramanian, P., Maiyalagan, T. et al. (2017). Electrochemical oxygen reduction activity of metal embedded nitrogen doped carbon nanostructures derived from pyrolysis of nitrogen-rich guanidinium salt. *J Electrochem Soc* 164 (7): F781–F789. https://doi.org/10.1149/2.0651707JES/XML.

90 Sebastian, M.M., Velayudham, P., Schechter, A., and Kalarikkal, N. (2022). Spinel nickel ferrite nanoparticles supported on a 1T/2H mixed-phase MoS_2 heterostructured composite as a bifunctional electrocatalyst for oxygen evolution and oxygen reduction reactions. *Energy and Fuels* 36 (14): 7782–7794. https://doi.org/10.1021/ACS.ENERGYFUELS.2C01191.

91 Konar, R., Das, S., Teblum, E. et al. (2021). Facile and scalable ambient pressure chemical vapor deposition-assisted synthesis of layered silver selenide (β-Ag_2Se) on ag foil as a possible oxygen reduction catalyst in alkaline medium. *Electrochim Acta* 370: 137709. https://doi.org/10.1016/J.ELECTACTA.2020.137709.

92 Subramanian, P., Mohan, R., and Schechter, A. (2017). Unraveling the oxygen-reduction sites in graphitic-carbon Co–N–C-type electrocatalysts prepared by single-precursor pyrolysis. *ChemCatChem* 9 (11): 1969–1978. https://doi.org/10.1002/CCTC.201700324.

Index

Alkaline Anion Exchange Membranes for Fuel Cells: From Tailored Materials to Novel Applications, First Edition. Edited by Jince Thomas, Alex Schechter, Flavio Grynszpan, Bejoy Francis, and Sabu Thomas.

q

r

s